Operator Theory: Advances and Applications
Vol. 132

Sonja Kovalevsky (1850–1891)

Introduction

Софья Васильевна Ковалевская (**Sonja Kovalevsky**) was born in Moscow in 1850 and died in Stockholm in 1891. Between these years, in the then changing and turbulent circumstances for Europe, lies the all too brief life of this remarkable woman. This life was lived out within the great European centers of power and learning in Russia, France, Germany, Switzerland, England and Sweden. To this day, now 150 years after her birth, her influence for and contribution to mathematics, science, literature, women's rights and democratic government are recorded and reviewed, not only in Europe but now in countries far removed in time and distance from the lands of her birth and being.

This volume, dedicated to her memory and to her achievements, records the Proceedings of the Marcus Wallenberg Symposium held, in memory of Sonja Kovalevsky, at Stockholm University from 18 to 22 June 2000. The symposium was held at the Department of Mathematics with its excellent library and lecture halls providing favourable working conditions.

Within these pages are contained a *curriculum vitae* for Sonja Kovalevsky, a list of all her scientific publications, together with a copy of the moving and elegant obituary notice written by her friend and protector Gösta Mittag-Leffler. These papers are followed by a leading article entitled *Sonja Kovalevsky: Her life and professorship in Stockholm*, written especially for this volume by Jan-Erik Björk in preparation for his major address to the Symposium.

The scientific papers contributed to the Symposium include short articles on two of her mathematical contributions: the first based on her doctoral thesis (1874); the second based on the outstanding and significant publication which led to her of the award of the Bordin Prize (1888).

The main body of the scientific work of the Symposium is to be seen in the 23 contributions in pure and applied mathematics, and in mathematical physics resulting from the lectures delivered within the program of the Symposium.

Finally, there are recorded the lists of the Symposium participants together with titles of their lectures.

The Organizing Committee of the Marcus Wallenberg Symposium (P. Kurasov and N. Elander) takes this opportunity to thank the many organizations for financial and material support without whose aid it would not have been possible to mount the Symposium:

- The Marcus Wallenberg Foundation

together with

- The Royal Swedish Academy of Sciences
- The Swedish Natural Science Research Council

- The Wenner-Gren Foundation
- Stockholm University
- The Department of Mathematics, Stockholm University

The Editorial Board for these Proceedings (S. Albeverio, N. Elander, W.N. Everitt and P. Kurasov) is grateful to all the authors who submitted their work for publication, and to all the referees called upon to assess these publications in support of the exacting standards applied by the Board for acceptance of manuscripts as contributions to this volume.

All the participants will recall the presence of the bust of Sonja Kovalevsky which graced the main lecture theater throughout the meeting, and now remembered in the leading photograph of this volume.

Towards the end of the Symposium a short but moving pilgrimage was made to the grave of Sonja Kovalevsky, in the North Cemetery of Stockholm; this ceremony took place during white nights when Stockholm was blooming and is recorded in the photograph of the grave placed within this volume.

Finally the Editors thank the publishers Birkhäuser Verlag, Basel, for their help and expert support in the closing stages of the preparation of the contents of this volume.

Lund, December 2001

The Editors

Part I

Sonja Kovalevsky

Operator Theory:
Advances and Applications, Vol. 132, 3–4

Curriculum Vitae

1850, **3(15) January** – Sonja Krukovskaya (Софья Васильевна Круковская) is born, from 1856 Korvin-Krukovskaya (Корвин-Круковская).

1853 –**1858** – living with the parents in Kaluga.

1858 – moves to Palibino (Палибино), now in Pskov region.

1868, Spring – meets Vladimir Kovalevsky.

1868, **15(27) September** – marriage with Vladimir Kovalevsky (Владимир Онуфриевич Ковалевский), moves to St. Petersburg for studies.

1869, April – departure for Heidelberg.

1870, October – moves to Berlin, begins her studies with K. Weierstraß.

1874, July – receives the Philosophy Doctor degree from the university of Göttingen, *summa cum laude.*

1878, **5(17) October** – her daughter, Софья Владимировна Ковалевская, is born.

1879, **28 December (9 January, 1880)** – lecture of S. Kovalevsky at the VI-th Meeting of Russian Naturalists and Medical Doctors in St. Petersburg "On the reduction of Abel integrals of the third rank to elliptic ones."

1881, **17 (29) March** – elected member of Moscow Mathematical Society.

1882, **24 July** – elected member of the Mathematical Society in Paris.

1883, **15 (27) April** – death of Vladimir Kovalevsky.

1883, **22 August (4 September)** – lecture of S. Kovalevsky at the VII-th Meeting of Russian Naturalists and Medical Doctors in St. Petersburg "On the refraction of light in crystals."

1883, **18 November** – arrival in Stockholm.

1884, **11 February** – the first lecture at Stockholm university.

1884 – appointment as associate professor at Stockholm university.

1888, **24 December** – The Paris Academy of Sciences awards Bordin Prize to S. Kovalevsky for her paper "On the rotation of a solid body around fixed point."

After П.Я. Кочина, Софья Васильевна Ковалевская, Moscow, Nauka, 1981 (in Russian).

1889, 7(19) November – elected as a correspondent member of the St. Petersburg Academy of Sciences at the meeting of the physical-mathematical section of the Academy, confirmed by the General Assembly of the Academy on **2(14) December**.

1889 – The Stockholm Academy of Sciences awards a prize to S. Kovalevsky for the second memoire on the rotation of solid body.

1889 – appointed as a professor at Stockholm university.

1891, 10 February – S. Kovalevsky died.

Operator Theory:
Advances and Applications, Vol. 132, 5–5

Scientific Publications

(1) Zur Theorie der partiellen Differentialgleichungen. (*Inaugural-Dissertation*, 1874; *Journal für die reine und angewandte Mathematik*, **80** (1875), 1–32.)

(2) Über die Reduction einer bestimmten Klasse Abel'scher Integrale 3ten Ranges auf elliptische Integrale. (*Acta mathematica*, **4** (1884), 393–414.)

(3) Om ljusets fortplanting uti ett kristalliniskt medium. (*Öfversigt af Kongl. Vetenskaps-Akademiens Förhandlingar*, **41** (1884), 119–121.)

(4) Sur la propagation de la lumière dans un milieu cristallisé. (*Comptes rendus des séances de l'académie des sciences*, **98** (1884), 356–357.)

(5) Über die Brechung des Lichtes in cristallinischen Mitteln. (*Acta mathematica*, **6** (1883), 249–304.)

(6) Zusätze und Bemerkungen zu Laplace's Untersuchung über die Gestalt der Saturnsringe. (*Astronomische Nachrichten*, **111** (1885), 37–48.)

(7) Sur le problème de la rotation d'un corps solide autour d'un point fixe. (*Acta mathematica*, **12** (1889), 177–232.)

(8) Sur une propriété du système d'équations différentielles qui définit la rotation d'un corps solide autour d'un point fixe. (*Acta mathematica*, **14** (1890), 81–93.)

(9) Mémoire sur un cas particulier du problème de la rotation d'un corps pesant autour d'un point fixe, où l'intégration s'effectue à l'aide de fonctions ultraelliptiques du temps. (*Mémoires présentés par divers savants à l'académie des sciences de l'institut national de France*, **31** (1890), 1–62.)

(10) Sur un théorème de M. Bruns. (*Acta mathematica*, **15** (1891), 45–52.)

Operator Theory:
Advances and Applications, Vol. 132, 7–10

Sophie Kovalevsky

Biographical Note

by G. Mittag-Leffler

Sophie Corvin-Kroukovsky was born in Moscow, 15 January 1850.[1] In September 1868 she married Woldemar Kovalevsky, later a distinguished palaeontologist, and, in the following Spring, moved with him to Heidelberg. There she assiduously followed courses for three full semesters, studying mathematics and physics under the supervision of Kirchhoff, Königsberger, du Bois-Reymond and Helmholtz. From the age of 15 she had been imbued with a passion for the study of mathematics; indeed, upon her arrival at Heidelberg she had long been familiar with the fundamentals of geometry and infinitesimal calculus. In October 1870 she moved to Berlin, where she remained — except for short absences motivated by journeys to France — for four years in the study of mathematics under the special direction of Weierstraß.

During the Summer of 1874 Sophie Kovalevsky was granted, by the faculty of philosophy in Göttingen, the degree of Doctor of Philosophy, *in absentia* and without the customary oral examination. For her Thesis she had presented her memoir, *Zur Theorie der partiellen Differentialgleichungen*, published in the Crelle Journal, volume 80. But further, she had already presented to the faculty two other works; the first: *Über die Reduction einer bestimmten Klasse Abel'scher Integrale* 3ten *Ranges auf elliptische Integrale*, published later in this journal (*Acta Mathematica*), volume 4; the other : *Zusätze und Bemerkungen zu Laplace's Untersuchung über die Gestalt der Saturnsringe*, published in the *Astronomische Nachrichten*, volume 3. After her advancement she returned to Russia. On the 17 October 1878 she gave birth to a daughter who has survived her. In March 1883 Woldemar Kovalevsky died in Moscow in tragic circumstances. She found herself at that time in Paris: upon learning of the death of her husband, and the circumstances surrounding this, she fell ill and was prostrate for a month, between life and death.

Towards the end of June, recovering finally from her illness, she rejoined in Berlin her loyal friend and professor, Weierstraß. She resumed her mathematical work with enthusiasm and finished an investigation which she published under the title : *Über die Brechung des Lichtes in cristallinischen Mitteln*, this journal

This article appeared in the Swedish periodical Acta Mathematica 16 (1893), 385–392. This English translation, from the original article in French, has been made by Professor Lawrence Markus, School of Mathematics, University of Minnesota in Minneapolis, USA.

[1] This date (new calendar) was furnished by her birth certificate, a copy of which, delivered by the ecclesiastical court at Moscow on 17 February 1851, has been discovered among her papers.

(*Acta Mathematica*), volume 6. In the present volume of this publication, M. Vito Volterra has again taken up this same question: he has shown that the functions given by Sophie Kovalevsky as general integrals of the Lamé differential equations, do not satisfy these equations, and he has given explanations for this fact.

Some years before the death of her husband, Sophie Kovalevsky had expressed the desire to devote herself to a career in education as a professor in a university. Having known of her wishes and sharing the high opinion which M. Weierstraß had of the exceptional talent of his student, I had undertaken during the Autumn of 1880 the project of appointing Sophie Kovalevsky as my *docent* (professor agrégé) at the University of Helsingfors, where I occupied the Chair of Mathematics. This project foundered; but during the Spring of 1881 I was called to the newly established university at Stockholm. I immediately opened negotiations with the university authorities, with the goal of designating Mme Kovalevsky as my professor agrégé, if she consented to this.

For herself, the principal difficulties she faced, before she could realize her ambitions, arose from the disappearance and death of her husband. In a letter of 5 August 1883, M. Weierstraß informed me that she was disposed to offer a mathematical course at Stockholm, but that, in order to begin, she wanted the course to have no aspect of publicity. In December 1883 Sophie Kovalevsky arrived in Stockholm, and during the Spring semester of 1884, before a limited but attentive audience, she expounded in German the theory of partial differential equations. Graced with the success of this course, as well as the impression produced on the circle of the intellectuals of Stockholm by her sympathetic personality and the geniality of the lecturer, it became possible for me to obtain additional sufficient funds to name Sophie Kovalevsky as professor of higher analysis at the University of Stockholm for a period of five years. In spite of the short time that she had lived in Sweden, she already possessed a sufficient grasp of our language to be able to teach in Swedish from her start as a professor at the university.

Before the expiration of the five year period, Sophie Kovalevsky received from the Institute of France the prize Bordin for her work: *Sur le problème de la rotation d'un corps solide autour d'un point fixe* (this journal (*Acta Mathematica*), volume 12, and *Mémoirs présentés par divers savants à l'Académie des sciences de l'Institut national de France*, volume 31).

These circumstances facilitated my efforts to gather the funds necessary for the definitive establishment of the chair of higher analysis at the University of Stockholm. This having been accomplished by the Spring of 1889, our university was able to assure the continuing services of Sophie Kovalevsky with the title of professor *for life.*

That was not to be for a long duration.

Sophie Kovalevsky had spent the vacation of the Winter 1890–91 in the Midi, at the Mediterranean coast of France. During her return journey she had caught a cold, and on the sixth of February 1891, having delivered in the morning the opening lecture of the year, she was forced to retire to bed as she could no longer remain standing. She died the tenth of February, in the morning, from a severe pleurisy which was actually a form of influenza, and which, from its beginning, defied all the skills of the doctors.

It is superfluous to reiterate to the readers of this review the mathematical works of Sophie Kovalevsky. One will find below a complete list of her scientific works as well as the courses that she presented at the University of Stockholm.[2]

The portrait placed at the front of this volume was created in Stockholm, based on a photograph dated in the year 1887, the epoque in which Sophie Kovalevsky arrived at the apogee of her career as mathematician, the professor and scientist.

As a mathematician, Sophie Kovalevsky belonged to the school of Weierstraß. She was filled with enthusiasm and belief in the ideas of her master, this venerable old gentleman who has survived the death of his beloved student. She wished, through her own mathematical works and new discoveries, to demonstrate the range and the extent of the doctrines of Weierstraß. As a professor, she forced herself, with a truly contagious zeal, to expose the fundamental concepts of this doctrine to which she attributed the greatest importance, even for the resolution of the central problems of life. Constantly, and with an obvious joy, she communicated the extraordinary riches of her knowledge and the profound insights of her spirit of divination to those of her students who would only show the ability and desire to draw on this source. As a person, she was extremely simple. She combined, with an instruction extending into the divers branches of the humane sciences, a sure intelligence, warm and sympathetic, from that personal soul that lies within each of us: also many men and women, none the less remarkable, under the influence of that interest that she inspired, even from the first meeting, confessed their most intimate feelings and thoughts, their hopes and their doubts as researchers, the hidden weaknesses of new beliefs, the motivations on which would be founded future expectations; as for the rest, they have confided many times their dreams of happiness and the pains caused by the deceptions of the heart. These qualities which she brought forth into the career of professor made the understanding on which base rested her relation with her students.

More than other sciences, mathematics demands from those who are called to develop it, through new conquests in the domain of scholarship, a powerful imagination. Clarity of thought, by itself, has never made these discoveries. The best work of mathematicians is an art, a high art, perfect, daring as the most secret dreams of the imagination, clear and pure as abstract thought. The genius of mathematics and the genius of art touch, and this could even explain why these two kinds of genius arise, if rarely, within the same human. Sophie Kovalevsky had hesitated, since her youth, between mathematics and literature. Earlier, she had published some literary sketches and collaborated under anonymity on several such works. During the period of exhaustion which followed the publication of her investigation on the problem of rotating bodies she had desired to produce a literary work which would be original and of enduring value, and she published

[2]In this volume the List of scientific publications preceeds the present article.

at Christmas 1889 both in Swedish and Danish, the book Russian Life, the Sisters Rajevsky. A slightly different edition had been published earlier in Russian. It is a description of the paternal home of her own youth. The literary critics of Russia and the Scandinavian countries were unanimous in declaring that Sophie Kovalevsky was equal in style and thought with the best writers of Russian literature. This success and the joy of discovering the path to open her heart — after having been able to address only a tiny number on her works of mathematics — determined Sophie to devote herself more seriously to literature, and she did this with the burning zeal that she applied to all her activities. She began various works, but only one could be finished and it would be published, with the aid of a few brief notes which she gave some days before her death. It is a novel which appears in different languages, a psychological study on contemporary Russia, which connoisseurs have declared to be totally remarkable.

This death has not only destroyed a literary future which seemed full of extraordinary promise, it has interrupted various works of mathematics, most notably the conclusion of her investigations into the problems of rotating bodies.

Sophie Kovalevsky will hold a position of eminence in the history of mathematics, and her posthumous works that will soon be published will preserve her name in the history of literature. But it is perhaps neither as a mathematician nor as a literary figure, excluding all else, that one should appreciate and judge this woman of such spirit and originality; as a *personality*, she was still more remarkable than one could believe from her work alone. All those who have known and approached her, to whatever circle, and to whatever part of the world they belong, they will remain constantly under the lively and forceful impression that was produced by her personality.

October 1892

Operator Theory:
Advances and Applications, Vol. 132, 11–53

Sonja Kovalevsky: Her Life and Professorship at Stockholm

J.-E. Björk

Софья Васильевна Ковалевская (1850–1891)

1. Introduction

Among the extensive list of publications devoted to Sonja Kovalevsky, one should foremost mention the biography written by Pelageya Kochina [9]. It is based upon material from archives in Moscow and the Institute of Mittag-Leffler, and includes memories from her daughter Sofya Kovalevskaya (Софья Владимировна Ковалевская). Among other titles in the list of references one may mention the popularized biography directed to school pupils written by V. Vorontzova. Among memories from persons who met Sonja we mention Elizaveta Litvinova and Julia Lermontova. The memorial article [12] by Gösta Mittag-Leffler published in Acta Mathematica 1892 describes her life and scientific achievements.[1] The article [13] is mainly devoted to the mathematics of Weierstraß but contains also sections devoted to Sonja Kovalevsky. Her correspondence with Weierstraß is documented in [3]. The book [4] contains details about Sonja's connections with French mathematicians. The book [8] together with Kochina's biography covers most of Sonja's life.

This article contains material which is not covered by previous publications from Sonja Kovalevsky's time in Stockholm where she lived during the last seven years of her life. Her nomination as professor at Stockholm University was a historical landmark. This present article contains translated excerpts from official letters written by Mittag-Leffler to the Board of Stockholm University prior to her nomination as associate professor in June 1884, and some excerpts from Mittag-Leffler's diary which describe events prior to her nomination as full professor in 1889. More extensive material from these diaries have been translated into English by Lars Hörmander in [6].

The mathematical work by Sonja is treated in [2, 14] while this article is mainly devoted to her interest in politics and literature. My ambition has not been to present a biography in the strict sense but to offer a fairly detailed account of special episodes from her life. Sections 3–7 deal mainly with her mathematical life in Stockholm including excerpts from correspondence with Gösta Mittag-Leffler. Sections 16 and 17 deal with Sonja's relation to politics where the main focus is on her relation to the German socialist Georg Vollmar and Hjalmar Branting.

The last section contains excerpts from the memorial article by Ellen Key written after Sonja's decease. Ellen was a close friend of Sonja and her text offers a concise and yet passionate presentation of Sonja's life and character.

Here follows a brief account of the more specialized sections which can be read independently of each other.

Professorship in Stockholm. It was a veritable historical event when Sonja was offered a position as professor in mathematics — at a time when women in Sweden were not supposed to gain enough for their own living. Her nomination on June 28, 1884 was limited to a period of five years; on June 6, 1889 she was appointed for a

[1] English translation of this article is reproduced in this volume.

lifetime position. There was much opposition during the spring of 1889 before the final decision was made. The debate was intense and at the very end it became extreme, although Sonja's scientific merits were not questioned. Letters of recommendation from Bjerknes, Hermite and Beltrami had already been approved by the Council of Stockholm University. The two conceivable competitors — *Lars Edvard Phragmén* (1863-1937) and *Ivar Bendixson* (1861–1935) — had declared that they considered Sonja to be superior and they did not apply when the position was opened in March 1889. Later both became professors in Stockholm; Phragmén was appointed to Sonja's chair in 1892, and Bendixson became professor at the Royal Institute of Technology in 1900 — five years later he moved to the University of Stockholm where he was professor until 1927.

Instead the strident opposition against Sonja relied upon ideology and politics. Those who opposed her tried, for example, to scandalize her friendship with Hjalmar Branting. An exciting excerpt from Gösta Mittag-Leffler's diaries from May 11, 1889 illustrates what was going on.

Retzius har varit hos Hammarskjöld och helt och hållet vunnit denne genom sina utfall mot socialisterna. Han hade sagt att han motarbetade fru Kovalevsky, därför att hon umgicks intimt med Hjalmar Branting, gift bort fröken Kjellberg med Georg Vollmar samt utövade en ytterst agiterande socialistisk verksamhet här i Stockholm. Och allt detta har gått i Hammarskjöld, samt övertygat honom om att Retzius ej handlat av personliga motiv då han motsatte sig fru Kovalevskys anställning.

[Professor Retzius has paid a visit to the president of the Council, Mr. Hammarskjöld, and has succeeded to turn him into his line after an agitation against socialists. Retzius said that he opposed Mrs. Kovalevsky because of her intimate relation with Hjalmar Branting, that she has arranged the marriage between Julia Kjellberg and Georg Vollmar, and that she exercises an unremitting search for socialism here in Stockholm. Hammarskjöld has swallowed this which therefore has convinced him that Retzius had no personal motive when he opposed Mrs. Kovalevsky's nomination.]

A few weeks later the debate turned to common sense even though Branting's reputation was under severe pressure at this time. He was charged with having published an article which was considered an insult against the church. As the responsible editor of the magazine *Tiden* he was sentenced to prison for three months during the same year. However, Gösta Mittag-Leffler had much support; among those were Ellen Key's elder brother Axel Key who was professor of medicine and a member of the Board of Stockholm University, and the famous explorer Nordenskiöld who was the chairman of the Academy of Sciences. Eventually the opposition against Sonja's nomination receded and there were no official reservations when her appointment was announced in June, 1889.

Gösta's letter to the Council of Stockholm University. In May 1884 Gösta Mittag-Leffler sent a letter to the Board of Stockholm University containing a proposal that Sonja Kovalevsky be appointed to the position of associate professor for a period of five years. In due course this proposal was accepted, notwithstanding the quite exceptional circumstances for a female to be appointed to a teaching position at a university. Translation of this letter is given in this section, together with some details of the terms and conditions of the associate professorship.

Lecture on the Dirichlet problem. Sonja Kovalevsky gave her first lecture at Stockholm University on 11 February, 1884. The subject of this landmark lecture was the so-called Dirichlet problem; at that time one of the outstanding problems in mathematical analysis.

Scientific work. Sonja's two great achievements were the article on partial differential equations from her dissertation in 1874 and the work on rigid bodies which gave her the Bordin Prize. The publication about the equations of Lamé — published in Acta Mathematica 1885 — turned out to contain serious errors. Even though this article has some merits, which rely on Sonja's presentation of Weierstraß method for solving partial differential equations by the method of integration, it was discovered by *Vito Volterra* that Sonja's proposed solution did not solve the problems which had been raised by Lamé around 1850. Volterra's article was published in Acta Mathematica in 1892. The mathematical difficulties to treat the radiation problem are subtle and it was not until work by J. Grünwald, I. Fredholm, N. Zeilon and G. Herglotz that the original problems of Lamé were settled. The article [5] discusses the mathematics of double refraction starting from historical comments up to contemporary results.

Correspondence with Gösta Mittag-Leffler. There exists an extensive collection of letters and other written material between Gösta Mittag-Leffler and Sonja, preserved at the Mittag-Leffler Institute at Djursholm, near Stockholm. This archive contains mostly correspondence of an official character; however there are letters of a more personal form.

Elizaveta Litvinova. She studied mathematics in Zürich with Hermann Schwarz as teacher. Her life and career is described in this section. Sonja and Elizaveta were both confronted with the same difficulties to get positions in Russia after their return from Germany and Switzerland. But after ten years of struggle Elizaveta achieved a historical landmark when she became the first woman in Russia who got permission to teach in the higher school system. Litvinova has published many articles concerning the teaching of mathematics at all levels of the school system and she is considered to be one of the historically leading educationalists in Russia.

My Russian childhood. Sonja met the famous author Fyodor Dostoyevski in 1865 when he visited her sister Anyuta in St. Petersburg. Sonja never forgot the conversations with him from her youth. One chapter in her autobiography *My Russian childhood* is devoted to Dostoyevski. She mastered English, French and German and read literature from early childhood while her own career as an author of literature started in her later years. Several of her own novels were published posthumously — in Swedish by Ellen Key and in Russian by Maksim Kovalevsky.

A detailed review of Sonja's literary work can be found in Kochina's biography [9].

Meeting with Mary Ann Evans. This great British author wrote under male pseudonym *George Eliot.* This section contains excerpts from two articles published by Sonja in Stockholms Dagblad in 1885. The first describes her debate with the philosopher Herbert Spencer during a party at Mary's home in London 1871 while Sonja was a student at Berlin. The second shows her talent to unite passion with an intellectual approach. Even for readers who are not aquainted with Mary's famous novels such as *The Mill on the Floss*, it is fascinating to pursue Sonja's intense words when she reflects upon Mary's authorship.

August Strindberg. Sonja adored novels and dramas by this great Swedish author. She introduced Strindberg to the Russian public long before his books were translated. However, Strindberg held awkward view-points concerning the role of women in society. Six months after Sonja's appointment as an associate professor in mathematics he wrote an article in a Swedish journal published at Christmas 1884, where he complained about her nomination. Some translated excerpts from Strindberg's text are given below; it was very unpleasant for Sonja to be offended in this way; but her reply was subtle. A few months later she published the article about Mary Ann Evans where her debate with Herbert Spencer is included. In those days it was known to everybody in Stockholm that Spencer was Strindberg's spiritual father. Sonja's article aroused admiration and demolished Strindberg; however, Strindberg could not restrain from mentioning Sonja Kovalevsky in her role as a mathematician. One of his books, which was published sixteen years after her decease, contains a passage which in my opinion is an ugly attack upon her scientific achievement; it will be described at the end of this section on Strindberg.

Vladimir Kovalevsky. Sonja and Vladimir were married in 1868. Their daughter was born in 1878. For several years they worked together for the newspaper *Novoe Vremya* where Sonja published articles on new scientific discoveries — such as those of Pasteur, Siemens and Graham Bell. The assassination of Tsar Alexander II in March 1881 created a turmoil in Russia. In the summer of the same year Sonja went to Europe where she lived with her sister Anyuta in Paris for almost two years. Vladimir isolated himself and had a nervous breakdown which led to his suicide in March 1883. This section is devoted to Vladimir's life which seems to be appropriate since biographical notes about Sonja Kovalevsky tend to question the strength of her marriage with Vladimir; sometimes this negligence goes even further. In the recent article [1] published by the European Mathematical Society Newsletter appears the following unfounded sentence: *Meanwhile, Vladimir's death caused relief in mathematical circles.* I have not found a single vestige in reliable literature which would support this phrase.

Julia Lermontova. She lived with Sonja during the years of studies in Germany. Julia obtained a doctor's degree in chemistry at Göttingen in 1874 — the same year that Sonja's thesis was presented. Later she became godmother of Sonja's daughter. Excerpts from her memories of Sonja and her husband Vladimir occur in this section.

Anne Charlotte Leffler's biography. Sonja's first steps towards literary production of her own were encouraged by the sister of Gösta Mittag-Leffler, Anne Charlotte. When Sonja arrived in Stockholm in 1883, Anne Charlotte Leffler was a recognized author in Sweden who had published several successful novels and dramas. It was also Anne Charlotte who introduced Sonja to the public in Sweden in an extensive newspaper article published one month after her nomination as associate professor in June 1884. Later Anne Charlotte and Sonja wrote some dramas together. The novel *Sonja Kovalevsky: What I have experienced with her and what she has told about herself* was written by Anne Charlotte after Sonja's decease. This section contains excerpts from a review written by Ellen Key, which contains critical remarks concerned with the description of Sonja's character in Leffler's novel.

Maksim Kovalevsky. He was not a relative of Sonja's husband. Maksim met Sonja for the first time in Stockholm in 1888 while he was invited to deliver lectures on the history of politics. His lectures were sponsored by a foundation where Sonja and Anne Charlotte Leffler were members of the committee. Like Sonja, Maksim left Russia after the assassination of the Tsar in 1881. He became a close friend of Sonja during the last years of her life. This section includes a letter written to Maksim by Sonja's daughter only a few weeks before her mother's decease.

Politics and Society. The content of this section is mainly Sonja's friendship with the German socialist *Georg Vollmar*. They met for the first time in Paris in 1882. Vollmar — considered as one of the founders of revisionistic socialism opposed to marxism — also played a significant role in Sweden's political history. His visit to Stockholm in 1885 inspired *Hjalmar Branting* when he established the politics of the Social Democratic Party in Sweden. Thirty years later Branting became the first elected Prime Minister ever representing a socialist party. In 1921 he received the Nobel Peace Prize together with Christian Lange from Norway for his international contributions after World War I. Branting was ten years younger than Sonja; he studied mathematics at Uppsala University in his early student years; he became one of Sonja's friends during her time in Stockholm. More will be said about Vollmar and Branting within this section — including excerpts from some letters Sonja wrote to Vollmar.

Politics in Sweden. Sonja played an important and significant role in supporting the development of the ideals of democracy in Sweden. She was well connected with the work of two great political idealists of that time; Hjalmar Branting and Georg Vollmar.

Impressions from Sweden. This section contains excerpts from an article written by Sonja in 1890. Her article is truly prophetic in view of the development in Sweden during several decades after its publication when a new society was formed in Sweden by the socialdemocratic and the liberal parties.

Ellen Key's commemorative words on Sonja. This article ends with Ellen's commemorative words on Sonja. Ellen was a teacher in the school which Sonja's daughter attended for seven years. Later she became internationally famous for her book *The Century of Children* which was published in 1900 and translated into many languages. Ellen often visited Sonja's apartment at Sturegatan 56 where she lived with her daughter, and was present during the last days of Sonja's life. Her brother Axel Key, who was a professor of medicine, was among other doctors who tried to rescue Sonja, but in those days without antibiotics it was not possible to bring about an end to the growing inflammation in her lungs. The last moments of Sonja's life are described in Ellen's obituary.

I would like to use this opportunity to thank Lennart Börjeson and Rooney Magnusson for valuable information and help.

2. Biographical survey

Sofya Vasil'evna was born on 15 January, 1850 in Moscow. Her parents were Vasily Vasil'evich and Elizaveta Fyodorovna Corvin-Kroukovsky. Her sister Anyuta was six years older while her brother Fyodor was five years younger; her father was a general in the artillery of the Russian imperial army. When Sonja was still a child the family moved to an estate at Palibino situated about 600 kilometers south of St. Petersburg and close to the border of Lithuania. The estate has been restored and serves nowadays as the Kovalevsky Museum, it offers an opportunity to feel the atmosphere where Sonja grew up. Her life is historically interesting — she met many famous persons of her time. Her achievements in mathematics have served as an inspiration for generations of female students. In addition to such evident facts, the words below expressed by Yuri Manin — who received the Rolf Schock prize from the Royal Swedish Academy of Sciences in 1998 — explain the essential point.

It was her unique sensibility to catch essential characteristics — be it in company with other people or when confronted with external events in politics, science or literature, which explains why Sonja Kovalevsky stays present in our contemporary time.

Sonja herself summed up the essential goal of her own life:

Je pense que ma destinée est de servir la vérité dans la science, mais aussi de travailler pour la justice en ouvrant de nouveaux chemins pour les femmes.

[I think my destiny is to serve truth in science, and also to work for justice in opening new paths for women.]

Outside the closed circles among mathematicians, Sonja often had to defend her mathematical activity. Confronted with persons who regarded mathematics as a dry and dull subject, and who told her that they could not understand why she devoted so much of her life to such a subject, Sonja used to respond with the following poetic words:

Den äkta matematiken är den minst torra av alla vetenskaper. Den öppnar för den skapande fantasin och spekulativa kraften hela världssystemet. Den sida av ämnet som kan förefalla torr utgör blott de grenar på vilka man klättrar upp och ned i världsträdet.

[The genuine mathematics is the least dry of all sciences. To the creative imagination and speculative power it opens the whole system of the world. That aspect of the subject which might seem dry consists only of those branches along which one climbs up or down in the world's tree.]

Women and mathematics. Much has been written about Sonja's role as a woman in the world of mathematics. More or less speculative accounts have been published since the middle of the 1890's. Ann Koblitz has studied the literature which has tried to create a myth around her life — especially publications in the United States with titles such as *Sonja Kovalevsky's sorrow.* In her article *Changing views of Sofia Kovalevsky* Ann Koblitz writes:

The portrayal of Sonja Kovalevsky's supposed tragic life and her unhappiness in her professional success fit the prevailing sociopolitical and psychological mood of the time. For at least fifteen years after her death, Kovalevsky was one of the warnings held up for the "New Women" as the feminists were often called.

Koblitz points out that similar accounts continue to be published. She is critical of Don Kennedy's book *The little sparrow* when she says:

The reader gets the idea that Sonja killed herself or pined to death from unrequited love.

Sonja's historical role has been emphasized in the article by Koblitz:

Mathematicians of both sexes will have to keep an eye on what is happening in lower levels of the school system, and on what is happening in the academic community as a whole. It does not help much for men to be supportive toward their female colleagues, if prospective generations of women are being socialized into their notion that they are inherently unfit to do mathematics.

I think it is the responsibility of all mathematicians and scientists, as well as that of historians of science like myself, to keep the models of successful women scientists like Kovalevsky in the public eye. Only in this way can the future of women in mathematics remain bright.

Sonja was interested in many disciplines. Foremost she was devoted to mathematics and to literature, but she also wrote as journalist and followed discussions about society and politics. The remaining part of this section will mainly focus on her interest in areas outside mathematics.

Other sciences. Sonja was interested in new achievements in technology and had a broad education which included medicine and chemistry. While she lived in St. Petersburg and Moscow between 1875 and 1881 she wrote many articles — foremost in *Novoye Vremya*. Together with her husband Vladimir she translated contemporary scientific publications; she recognized events that were truly important. Only a few months after Graham Bell's invention of the telephone she reported on his discovery in *Novoye Vremya*; in the same article she also described the most recent typewriter by Remington. In another article she wrote about new discoveries concerning electric light by Siemens whom she knew personally from her time in Berlin. She met Pasteur in Paris and wrote many articles about his important achievements; her explanations of fermentation were so clear that Pasteur's methods were applied to improve the production of beer in Russia. While Sonja and Vladimir lived in Moscow she worked with a company to make electric street-lighting. Her article on sources of energy — especially concerning solar energy — gives the reader of today a reminder of the past. The idea that energy resources like coal and oil are limited is for example not of recent origin.

After her arrival at Stockholm in November 1883, Sonja wrote many articles in Swedish newspapers. Some of them became widely recognized and were translated to other languages — especially German. Among these articles one is devoted to the Scandinavian educational system which was unique in those days — namely schools for adults which started in Denmark around 1860. Sonja's article (see e.g. [10]) — based upon her visit to Tärna Folkhögskola in Västmanland in 1886, and discussions with teachers and pupils — was published in several countries all around Europe.

3. Professorship at Stockholm

Introduction. Sonja's appointment as associate professor in June 1884 was an exclusive event. At this time there were only four professors at Stockholm University — except for Gösta Mittag-Leffler in mathematics, there were professorships in chemistry, zoology and history of arts. This last chair was held by Viktor Rydberg who was a great poet and a learned historian. Astronomy was the responsibility of the Academy of Sciences. In Stockholm there was also the Royal Institute of Technology; during some terms Sonja held a temporary professorship there teaching courses in mechanics. The student level of ability was relatively high; Sonja's pupils had in most cases passed undergraduate examinations at the University of Uppsala. The outstanding Scandinavian mathematician in those days was Sophus Lie, but he was mainly active in Germany. In Sweden the level among the leading mathematicians was fairly modest. Albert Viktor Bäcklund who was professor in mechanics at Lund was the strongest Swedish mathematician around 1885; it was only later that mathematics in Sweden started to blossom around 1900. Among those who attended Sonja's lectures in the late eighties were Ivar Fredholm (1866–1927) who passed his doctor's degree in 1898. Ivar Bendixson (1861–1935) had already started his career when Sonja arrived; he was docent at Stockholm University and was next to Gösta Mittag-Leffler her closest colleague. There was also Lars Edvard Phragmén (1863–1937) who had published articles in Acta Mathematica already in 1884. He succeeded to Sonja's professorship and is famous for his results concerning the growth of complex analytic functions, especially the Phragmén-Lindelöf principle which he established in joint work with Ernst Lindelöf around 1905.

Gösta Mittag-Leffler's merit was foremost that he created a very good climate for research. His lectures were excellent and he had a broad knowledge of mathematics and an extraordinary gift for understanding what was truly essential in contemporary research. When Acta Mathematica started in 1882 it meant that Sweden obtained very good contacts with leading mathematicians in Europe. For Sonja and Gösta it was the mathematics of Weierstraß which gave them the main inspiration.

Sonja and Gösta. They met each other for the first time at Paris in 1876. A few years later — while Gösta was professor at the department of mathematics in Helsinki, he attended a lecture given by Sonja at a congress in St. Petersburg in January 1880. Sonja's lecture on Abelian functions impressed him greatly and he wrote a letter to Carl Johan Malmsten, who had been the leading Swedish mathematician while he was professor at Uppsala from 1842 until 1866; he remained active for many years thereafter and died in 1886.

Mrs. Kovalevsky lectures with an extraordinary skill and precision. I fully understand why Karl Weierstraß regards her as his most talented and brilliant student.

In those days it was difficult for women even to attend lectures at universities. The idea to appoint a female teacher at a university in Russia was unimaginable; however the situation in Stockholm was different. When Gösta was appointed as professor of mathematics in 1881, Stockholm University was not a university in the formal sense; salaries were paid through private donations and from the City of Stockholm. Soon after his own appointment, Gösta began to plan for an invitation in order that Sonja could deliver lectures in Stockholm. In a letter from June 1881 Sonja gave her response:

I can assure you that if a position as Privatdocent will be offered to me, I would accept it with my full heart, I would also like to say that I would feel less embarrassed if my duties are limited to teach mathematics to female students in order to prepare them for higher studies at a university.

Abelian functions. Before her arrival in Stockholm, Sonja did not receive much support in Russia for advocating the new ideas and methods which Weierstraß brought into mathematical analysis. A critical issue in those days was whether functions — especially in the complex domain — should be founded on geometry or on a more formal treatment using series expansions. Analysis in this modern sense had started with Niels Henrik Abel's pioneering work *Mémoire sur une propriété générale d'une classe très étendue de fonctions transcendantes*, dating from 1826. Thirty years later when Karl Weierstraß began his career with work related to Abel-Jacobi functions he said:

Niels Henrik Abel died early and was therefore not able to deduce the consequences of his great discoveries. His predecessor Jacobi did not fully realize that the true significance of his own work had to be traced back to Abel's results. To consolidate and to develop the achievements which already have been obtained — to describe in full generality Abel's generalized elliptic functions and to investigate their properties — is an important area for mathematics even today.

This is how Weierstraß described what he considered to be the central issue for mathematics when his own career started. Gösta Mittag-Leffler presented his dissertation about analytic functions at Uppsala in 1872 — a fairly modest work concerning the argument principle in complex function theory. He went on to Paris in 1873 to study further; but there Hermite told him:

You have made a mistake; you should attend the lectures by Weierstraß in Berlin — he is superior to all of us.

Gösta therefore visited Berlin in 1874 and 1875.

In those days there were strong opinions about what was "good or bad" in mathematics. A cleavage existed between a more intuitive conception of complex analytic functions and Weierstraß' solid constructions, based upon power series, uniform continuity and convergence and his precise meaning of analytic continuation. It was not until the very end of the nineteenth century that Weierstraß' precise notions in analysis were accepted by a majority of mathematicians. Sonja — who always adored the mathematics created by Weierstraß, found herself quite isolated after her return to Russia in 1875 — especially when she had moved to Moscow. In a letter to Gösta — dated in November 1881 — she writes:

Recently I had a very intense debate with several of the professors at Moscow University who claimed that the theory of Abelian functions is not sufficiently developped and therefore cannot be applied to serious problems — and that this theory was so dry that it was out of the question to give a course on this subject.

When Sonja came to Stockholm she established her position in respect of Abelian functions; during five consecutive terms in Stockholm she lectured on these functions.

Arrival in Stockholm. In June 1883 Gösta succeeded in persuading the Board of Stockholm University to allow Sonja to give lectures in the Spring term 1884 — but without salary. Sonja came to Stockholm in November 1883. At that time she had recovered from the extreme emotions caused by the tragic events which had led her husband Vladimir to commit suicide in March of the same year. She had spent the summer in Odessa at the home of Vladimir's elder brother Alexander Kovalevsky together with her daughter Sofya who was five years old at that time. Julia Lermontova — who was godmother to Sofya also visited Odessa regularly; thus it was no problem for Sonja to travel alone to Stockholm since her daughter was taken good care of in Russia.

The Queen of Science. After the arrival in Stockholm, Sonja wrote a letter to Weierstraß in December 1883 describing her first time in Stockholm.

Du mußt wissen, daß Stockholm die komischste kleine Stadt in der Welt ist, wo alles über jeden sogleich bekannt ist und jeder kleinste Vorfall sofort die Proportionen eines Weltereignisses annimmt. Um Dir eine Idee zu geben was nun eigentlich Stockholm ist, werde ich Dir anführen in welcher Weise eine Zeitung am nächsten Tage meine Ankunft bekündigte:

Es lautet ungefähr folgendermaßen: Es ist nicht der Besuch eines nichts bedeutenden Fürsten oder sonstiger hoher Persönlichkeit den wir heute unseren Lesern anzukündigen haben. Nein, es handelt sich um etwas ganz unvergleichbar anderes. Die Fürstin der Wissenschaft, Frau Sophie von Kovalevsky hat unsere Stadt mit ihrem Besuch beehrt und beabsichtigt, Vorlesungen an unserer Universität zu halten.

At the end of the letter Sonja writes about her studies in the Swedish language.

Hauptsächlich beschäftige ich mich nun mit der schwedischen Sprache und habe in diesen zwei Wochen bedeutende Fortschitte gemacht, obgleich die schwedische Sprache recht schwierig ist, viel mehr so, als ich es glaubte. Aber ich kann schon verstehen, was ich lese und ein wenig auch, was um mich herum gesprochen wird. Selbst sprechen kann ich natürlich noch nicht.

Remark. Already in September 1884 Sonja started to give lectures in Swedish. To master mathematical vocabulary in Swedish she received help from Bendixson, who together with Gösta, was Sonja's closest colleague in Stockholm.

The first lecture. This historical moment took place on February 11, 1884; Sonja's lecture was delivered in a building in the center of Stockholm. Gösta has described this event in his diary:

Sonja Kovalevsky gave her lecture in German. The subject was partial differential equations. The audience was large, and everyone understood the historical significance of this unique lecture. There were not only students, teachers, professors from the Academy and officials of the Stockholm University — but also many visitors who were curious to watch a person spoken of as the "Queen of Science" in the newspapers. Sonja was nervous in the beginning and had difficulties to speak fluently — but after a short while she improved. When her lecture had finished she received widespread applause from the audience. It was obvious from the very start that she would be an excellent lecturer.

Sonja gave twelve lectures during the academic term in the Spring of 1884. She also participated in seminars - held at Gösta's home. In May 1884 Gösta wrote a letter to the Council of Stockholm University — requesting that Sonja should be offered a position as associate professor for a period of five years. Her salary would be 4 000 swedish crowns — to be compared with 6 000:- which was the usual salary for a professor at the universities in Lund and Uppsala at this time. But Sonja would be less involved with examination procedures as compared to professors at official state universities. Gösta's request was supported by several other leading scientists in Sweden — foremost by Carl Johan Malmsten and by the professor of mechanics at the Royal Institute of Technology — Hjalmar Holmgren. The next section contains excerpts from Gösta's letter to the Board of the university wherein he requests that Sonja should be appointed.

4. Mittag Leffler's letter to the Council of Stockholm University

Gösta Mittag-Leffler sent a letter to the Council of Stockholm University on the 2nd of May 1884. Here follows an excerpt which describes Sonja's mathematical activity.

The Council of Stockholm University had previously allowed Mrs. Sonja Kovalevsky — born Corwin-Krukovski — to deliver lectures on the Theory of Partial Differential Equations. These lectures have been held without cost during the Spring term for students and other visitors.

The lectures have been held twice a week and have been regularly attended by sixteen students, who also follow my lectures. In spite of the abstract and difficult nature of the subject chosen by Mrs. Kovalevsky and even though she has brought her presentation as far as possible up to the present level of mathematical science, I can confirm that her lectures have been truly understood by most of her students. She has succeeded to rouse their interest for the subject she has treated, as well as for mathematics in general.

Mrs. Kovalevsky has also participated in the mathematical seminars which have taken place in my home every other week. She has given talks herself and supervised talks by the pupils and given valuable contributions during the subsequent discussions. The work at these mathematical seminars has led to the completion of several investigations of decisive importance for the development of mathematical science. Much of this progress has relied on the contributions from Mrs. Kovalevsky's knowledge in different branches of higher mathematics which she has given to the mathematical community at our university, and by her sound judgement and sharp-wittedness when she has helped her pupils in their work.

Under these circumstances it seems desirable for future needs of our University to obtain an assurance of continuing service from Mrs. Kovalevsky. Especially since it would be impossible for me alone to carry out the expanding scientific education.

As I have previously pointed out to the Board, the nature of mathematics — with its abstract and difficult access of knowledge — implies that those students who really endeavor true and enduring knowledge in higher mathematics are in great need — more than in most or perhaps in any other science — of the guidance which oral lectures can communicate.

In his letter Gösta then describes how Sonja's position would be financed; about fifty percent of her salary was covered by private donations. He finishes the request by reporting that Sonja has committed herself to teach in Swedish in the next term and that she would accept the position as associate professor in Higher Analysis for a period of five years. Gösta finishes his letter by the words:

My proposal to the Board is entirely motivated by my wish to increase as much as possible the quality of mathematics at the University — and in our country as a whole.

5. A lecture on the Dirichlet problem

Introduction. In 1821 the British physicist Green announced his famous integral formulas which suggested solutions to several boundary value problems in mathematical physics. Gauss and later on Dirichlet supplied mathematical proofs to Green's formulas which had been motivated upon physical grounds. Dirichlet announced the principle asserting that when $\Omega \subset \mathbf{R}^3$ is a bounded open set with a smooth boundary, then every $u \in C^0(\partial\Omega)$ has a unique harmonic extension to the interior. It was not until Weierstraß made a critical study of earlier results in the calculus of variations, that Dirichlet's proof began to be questioned. During the years 1870 until the beginning of 1900 work was done to consolidate Dirichlet's claim. It was Carl Neumann who found the first rigorous proof in the case when Ω is convex. The method was to reduce the proof to an integral equation. Eventually Ivar Fredholm found the complete proof for bounded domains with C^1-boundary in $\mathbf{R}^3$. His work on integral equations was not only motivated by the Dirichlet problems, but the problem to construct fundamental solutions to an extensive class of both elliptic and hyperbolic equations in mathematical physics. The doctoral thesis of Fredholm was completed at Uppsala in 1898. He was one of the students who had attended Sonja's lectures regularly in the late eighties; Dirichlet's problem occurs among the topics in her lectures. The collection of notes from her lectures contains several hundred pages devoted to different subjects during her years of teaching. Here follows an excerpt from a lecture on Dirichlet's result.

Der Satz von Dirichlet lautet:

Satz *Ist die Function v eindeutig, endlich und stetig für jeden Punkt auf der Oberfläche S eines gewissen Raumes T gegeben, so läßt sie sich immer und nur in einer Weise für das Innere so bestimmen, daß sie auch da eindeutig, endlich und stetig variabel ist, und den partiellen Differentialgleichungen $\Delta v = 0$ Genüge leistet.*

Later in the lecture notes Sonja points out that the proof consists in studying the integral

$$S(v) = \iiint \left[\left(\frac{\partial u}{\partial x}\right)^2 + \left(\frac{\partial u}{\partial y}\right)^2 + \left(\frac{\partial u}{\partial z}\right)^2 \right] d\omega$$

and gives the straightforward demonstration that if a function u minimizes this integral with prescribed boundary values, then $\Delta(u) = 0$ in the interior domain. Afterwards she presents the critical remarks by Weierstraß concerning the proof based upon calculus of variation together with specific examples. More precisely she discusses the weak point in Dirichlet's proof, namely that one cannot claim the existence of a minimizing function in the variational problem. In the notes she writes:

Solche Fälle kommen oft in der Variationsrechnung vor. Das hat Weierstraß als erster bemerkt, und Schwarz hat es dann in Crelles Journal genauer untersucht... Nun wollen wir einen anderen Fall betrachten, wo dasselbe stattfindet und wo doch kein wirkliches Minimum erreicht wird.

Kovalevsky's way of presenting contemporary problems was appreciated by her audience in Stockholm. In a similar fashion other topics were presented during the eleven terms she gave lectures in Stockholm.

Her own research was focused on the use of analytic series to represent functions and solutions to differential equations with analytic coefficients.

6. Scientific work

Sonja Kovalevsky's most important articles are

1. Zur Theorie der partiellen Differentialgleichungen.
   (*Inaugural-Dissertation*, 1874; *Journal für die reine und angewandte Mathematik*, **80** (1875), 1–32.)
2. Über die Brechung des Lichtes in cristallinischen Mitteln.
   (*Acta Mathematica*, **6** (1883), 249–304.)
3. Sur le problème de la rotation d'un corps solide autour d'un point fixe.
   (*Acta Mathematica*, **12** (1889), 177–232.)

The first article was written for her doctor's thesis, submitted to Göttingen in 1874 before its publication. Prior to Sonja's work, Cauchy had studied this subject. It is an impressive work which started a new era in the theory of partial differential equations.[2]

The second paper has a more involved history. Sonja started to work on this paper in 1881 after a visit to Weierstraß. It is concerned with the Cauchy problem where light is supposed to possess two velocities in each direction. Around 1850 the French physicist G. Lamé had found a system of equations which suited the problem, where he described the oscillations by the equations

$$\frac{\partial^2 u}{\partial t^2} + \partial \times a\partial \times u = 0 \quad \text{and} \quad \operatorname{div}(u) = 0.$$

Here t is the time variable, $x = (x_1, x_2, x_3)$ and $\partial = (\partial_1, \partial_2, \partial_3)$ denotes differentiation in the x-variables, while a is a diagonal matrix with elements $a_1 > a_2 > a_3$. In suitable x-coordinates the equations can be written in a form adapted to electromagnetic theory. When light is uniformly distributed Lamés equation is related to

[2]These results are described in detail in a separate article by H. Shapiro published in this volume.

the wave equation with three space variables. Euler discovered that the spherical waves

$$v(t,x) = f(t - |x|)/|x|$$

solve the wave equation

$$v_{tt} - \Delta v = 4\pi f(t)\delta(x)$$

where f is an arbitrary function and $\delta(x)$ the Dirac measure. Of course Euler did not express it in this form; but his original result can easily be translated into this form. In a crystal with non-uniform distribution the wave-surface generated by a point source is given by an equation $t = \lambda(x)$ where $\lambda(x)$ is a homogeneous function of degree one. Lamé tried to extend Euler's solution in the special case $\lambda(x) = |x|$ to obtain similar formulas for the spherical waves in his model. The calculations presented in his book *Leçons sur la théorie de l'élasticité des corps solides* (Paris 1853) appeared to give correct solutions.

However, the problem is subtle and the original formulas of Lamé cannot give the desired solutions. Sonja's article starts with a discussion of Lame's work; at the end of the introduction, on page 253 in her paper, she points out that a complete solution must take into account certain singularities.

Außerdem gibt es noch einen Punkt im Raum in welchem Lamés Formeln nicht mehr im Stande sind, die Erscheinungen zu beschreiben. In den Coordinatenanfangspunkten, d.h. im Schwingungsmittelpunkte selbst, wird jede der Größen X,Y,Z unendlich groß. Hier müßten also die Schwingungen unendlich groß sein und zwar nach allen Richtungen hin.

Having faced the problem, Sonja continues to describe the strategy of proof using the method of integration introduced by Weierstraß to solve partial differential equations.

Ich werde hier den Inhalt einer noch nicht publicierten Abhandlung mitteilen, welche mir von Weierstraß im Jahre 1881 zur Verfügung gestellt, aber viel früher von ihm verfaßt wurde.

Sonja conceived that solutions would be singular but finally she did not succeed to find anything really new beyond Lamé. Roughly speaking this can be explained by saying that her methods using series expansions of tentative solutions outside the singularities, did not give sufficient information about these singular solutions.

Vito Volterra. Volterras work [15] was sent to Acta Mathematica in April 1892 and published the same year. In the introduction he first points out that Poincaré's recent studies of Huyghens' principle in optics eventually lead to similar equations as in Sonja's paper — and therefore are insufficient to treat the equations of Lamé. Then he continues to discuss Kovalevsky's article and after some geometric considerations Volterra concluded that her functions eventually are the same as those found by Lamé:

Ce sont les mêmes fonctions qui apparaisænt dans le mémoire de Mme Kovalevski. Lorsque on s'aperçoit qu'elles sont polydromes, on voit ainsi que la méthode découverte par Weierstraß pour intégrer des équations linéaires aux dérivées partielles ne peut pas être appliquée pour intégrer les équations de Lamé.

Then Volterra obtained certain results using methods by Kirchhoff, Neumann and Weber. The work by Volterra meant that Sonja's contributions in the second paper are quite limited [5]. What remains is her exposition of the method initiated by Weierstraß which did not implement any new discovery.

Lars Gårding. A historical account of double refraction is presented by Lars Gårding in [5]; it contains an accurate history combined with an eminent mathematical presentation. As we know today the use of the Fourier transform is essential in order to analyse the propagation of waves in crystals with double refraction. The first analysis using the Fourier transform of contemporary standard in optics was performed by Nils Zeilon; his work was published by Uppsala University in a series of papers around 1920. Zeilon became professor at Lund University in 1926 where his colleague was Marcel Riesz.

The Bordin Prize. Sonja's third paper was a brilliant achievement; she found a solution to a problem which had stayed open for almost a century. One refers nowadays to the Kovalevsky gyroscope. Her case has been analysed numerically and there exist nowadays films which show how a rigid body rotates in her case.

The starting point of her article is a rigid body K which can rotate around a fixed point and where gravity is the sole external force. Euler established the equations of motion of K. When the center of gravity is not the fixed point or placed on an axis of symmetry through the fixed point, it is in general not possible to find enough invariants which would imply that one can solve the equations of motion by quadrature.

Kovalevsky found an exceptional case where the equations of motion are integrable; her method was to assume that the solutions to the equations of motion are expressed as Laurent series of the time variable. After some computations, she succeeded to find restrictions on the body in order that such a solution exists and thereafter she used this restriction to find an explicit case where the equations are integrable. The value of her discovery was underlined later on when it was shown that her case is the only which yields integrability except the classical cases treated by Euler and Lagrange. Details of these results are presented in the separate article [2].

7. Correspondence with Gösta Mittag-Leffler

Introduction. There exists an extensive collection of letters and other written notes between Gösta Mittag-Leffler and Sonja, preserved at the Mittag-Leffler Institute. This archive includes mostly correspondence of an official character when they worked together at the University, and with Acta Mathematica. However there are also letters of a more personal character. In 1916 Gösta Mittag-Leffler and his wife Signe donated their house at Djursholm to the Royal Swedish Academy of Science; nowadays it is a world-famous research institute. A description of the Institute occurs in Notices of AMS [7].

At the theatre. Social life in Stockholm was quite intense among all those who were close to the University. The world was small and Sonja met almost everyone who was employed at Stockholm University. The letter below illustrates the open atmosphere outside work. It is an excerpt from a letter Sonja wrote to Gösta in 1887.

Yesterday night I could not keep quiet. Your libidinous sister Anne Charlotte easily persuaded me to go with her to the theatre. But there I had the unexpected joy to see you. Do not try to deny that ! But instead of taking a seat among the audience, you entered the stage to play the role of the "Professor in mathematics" in the comedy by Paul Heyse. The program said the role was played by Mr. Hilberg. However on the stage the performance of those stiff movements, the squinting eyes and that fur you always carry and which is too long for your size — gave such an obvious picture of yourself. During the pause Knut Wicksell even asked me if I did not find that professor Leffler imitated himself too much. Today I devote all my time to work; if I am able to finish my mathematical calculations I will come to dinner at your place; otherwise we meet each other tomorrow.

Regards from Sonja

There were times when Gösta and Sonja did not agree. But both forgave easily and their quarrels used to disappear quickly. Sonja received the Bordin Prize on Christmas Eve in Paris 1888. After the overwhelming ceremonies she had a breakdown. Gösta was in Stockholm preparing for a great welcome; but Sonja did not go to Stockholm. She stayed in France and sent a medical certificate asking for her release from all duties in Stockholm during the Spring term. The subsequent correspondence between Gösta and Sonja describes their feelings and reactions during the first months in 1889. In situations where Gösta "lost control" of his careful planning he could be quite harsh.

Early January 1889 Gösta wrote to Sonja after he had heard about her conditions after the Prize ceremony. He told her that naturally she was allowed to take a sabbatical term and continued to inform her about all preparations which had been made to welcome her back.

Plans were made in Stockholm to welcome you with an official salute. But now this is all over... You did not receive congratulations from the University since we had planned everything for your arrival here in Stockholm. Yesterday I met King Oscar; he spoke about you and expressed his respect and admiration for everything you have achieved.

Sonja responded after a few days:

Dear Gösta!

I just received your letter. How grateful I am for your friendship! I think it is the most precious gift I ever have owned in my life. I am so ashamed that I cannot give you anything in return and prove how much I value this friendship. I have lost my mind ! I receive a flood of congratulations; but the irony is that I have never been so depressed in my life as at this very moment... The only thing I can do is to hide my depression from my surroundings,

Your devoted Sonja

Gösta wrote an immediate reply where he tried to comfort her:

Dear Sonja!

I am so sorry for your situation. If I could only help you there would be nobody who would be as willing to do something as myself.

At this time Gösta had already started to prepare the application for her life time position. He had also given Sonja several tasks concerned with a prize ceremony to be held in Stockholm. But time went on and he did not hear from Sonja. His tone changes in the next letter.

My dear Sonja!

First I must scold you for not having told me what has happened in Paris and for not having properly answered my questions. Did I not ask you to visit the Swedish embassy and Count Lewenhaupt? Did I not tell you how important this affair was? And you never went to the embassy. Therefore you have not been invited to participate in the 60-year anniversary of the King. Neither in Paris nor elsewhere can you continue to play that social role you strive for. I give two reasons:

Firstly, however spiritual you may be, you have no ability whatsoever to keep what you have already gained. Secondly, you lack peace of mind when confronted with social and other external events. When matters are serious these qualities are essential.

Now I will stop moralizing. But please check — as soon as the memoir of your work which won the Bordin Prize has been printed — that you receive a binding of highest quality at Hermann in Paris. Write a letter to the King — you may send me a sketch before; tell him that you had been looking forward to visit him — but that you were unable to come because you had been exhausted after completing so much work. Once the printed work and the letter are prepared — and I should approve everything beforehand — you should visit Lewenhaupt and ask him kindly to forward it to King Oscar.

At the end of his lengthy letter, Gösta adds some further exhortations almost as if Sonja was a child:

Be careful ! And for goodness' sake, do not soil the book as it often happens for you and beware that the corners of the cover are not folded

Sonja had left Paris and stayed in Nice when Gösta wrote this letter. She was busy with her literary manuscripts and wrote to him that she had recovered and felt much better but that she was going to stay in France during the whole term. She also wrote to tell him that she waited impatiently for more letters from Gösta and from her daughter who was taken care of by Therèse Gyldén and Gösta's wife Signe. At last Sonja replied to all Gösta's reminders and began to arrange all the tasks he had asked her to carry out.

Dear Gösta!

I have received your last letter. I acknowledge that all your complaints were justified. My sole excuse is that I lost control of myself in Paris. Tomorrow I will immediately write to Hermite and do everything you have said about the affairs with Count Lewenhaupt as soon as I have a copy of my memoir. You are perfectly right; I am not of use to anything; I cannot handle anything properly; I am unbearable and unsafe; I overestimated my ability to keep my nerves in order. Now they have taken "revanche" and therefore I still feel in a nervous state.

A month later Gösta, his wife Signe and Sonja's daughter Sofya arrived in Paris during the Easter vacation. At this moment everything which has passed between Gösta and Sonja was forgotten. When Gösta returned to Stockholm he became active in preparing for Sonja's life-time position. Excerpts from diaries during this period describe this activitiy and can be read in Hörmander's article [6].

8. Elizaveta Litvinova

Elizaveta Litvinova (1845–1919) studied mathematics at St. Petersburg in the late sixties. Her teacher was A.N. Strannoulski who had taught Sonja before she went to Heidelberg. When Elizaveta had reached a similar level as Sonja he encouraged her to go abroad to study mathematics at university level; but her husband forbade her to go. Elizaveta's studies with Strannoulski had been through private lessons which her husband never became aware of. After some time they divorced and Elizaveta went to Zürich where she attended lectures by Hermann Schwarz. Switzerland — and in particular Zürich — was the center for Russian nihilists who lived there in exile. A decree was then issued in June 1873 by the Russian authorities stipulating that all Russian women who studied abroad should return within a year. The punishment for delayed return entailed that all qualifications from abroad would be invalid in Russia and that official positions would not be available; many students were obliged to halt their studies. Litvinova belonged to those who stayed; she received a higher diploma for her article *Lösung einer Aufbildungsaufgabe* in 1876, which was published by the Academy of Sciences in St. Petersburg in 1879. Hermann Schwarz and Sonja were also friends; Elizaveta and Sonja met each other — first in Zürich and later on in St. Petersburg where both applied to teach mathematics at the recently started High School for Women.

Sonja and Elizaveta received a lower diploma which entitled them to teach in lower classes only. Sonja had reached a considerably higher academic level than Elizaveta. She decided to continue research in order to improve her academic merits further whilst Elizaveta accepted to teach in lower classes. She received a poor salary — payed for every hour of teaching, without holiday pay or pension since she was still regarded as "unqualified".

Elizaveta worked for nine years as a teacher in lower classes. After many petitions from colleagues and university professors in mathematics, who were aware of her qualifications, the department of education finally permitted her to teach in the senior High School system in Russia. But even then she did not receive a full salary since she was still considered to be "formally unqualified". To increase her salary she began to write biographies and articles in popular science. She was active in the Women's movement and participated among four Russian delegates at the International Congress for Women, in Brussels in 1897.

Elizaveta could not pursue research in mathematics under her strained conditions; instead she was engaged in developing new pedagogical methods. During her 35 years as an active teacher of mathematics she wrote seventy articles devoted to philosophical and practical problems in the realm of mathematical education in schools.

Around 1910 she was regarded as one of the leading educationalists in Russia. One of Elizaveta's pupils in those days was Nadezhda Krupskaya who later on became the wife of Lenin; in the years after 1920 she emphasized the methods of education which Elizaveta had put forward. Elizaveta's pedagogy has had a great impact on mathematical education in Russian schools from 1920 until recent years.

The fundamental idea of her pedagogy was that mathematics should not be reduced to learning rules and formulas. Imagination should be stimulated by means of verbally formulated problems; different attempts of proof should occur at every level. The central role of mathematics should be to teach school pupils to *think* in a logical way. She gave many examples to illustrate how to obtain generalizations and how to build up general principles from individual cases.

Details of her last years are unclear. Presumably Elizaveta lived with her sister in the countryside during the years after the October revolution in 1917; probably she died in 1919.

Elizaveta wrote a biography of Sonja Kovalevsky entitled *Her life and her scientific work* published in St. Petersburg in 1894; in 1899 she published an article about her memories of Sonja. The two magazines Zhenskoe Obrazovanie (Education for Women) followed by Obrazovanie contain the majority of her pedagogical articles.

9. My Russian Childhood

Sonja wrote the original text in Russian; it was gradually translated into Swedish during the fall 1889. The Swedish title was *Systrarna Rajevski*. The book was a great success when it was first published at Christmas, 1889. Excerpts from Sonjas's original Russian manuscript were published the year after in the magazine Vestnik Evropu. The whole book was published in Russia in 1893. The English translation appeared in 1895 and within a few years there were also editions in Polish, German, French and even in Japanese.

A review of Sonja's book. Carl David af Wirsén — who was one of the leading members of the Swedish Academy of Letters and considered a demanding critic — praised Sonja's book in a review published in *Vårt Land* in January 1890. Wirsén emphasized the colorful description of emotions and thoughts of the novel's *Tanja* — which for example describes Sonja's passion for simple poems in her youngest years — and later her encounter with the works of great poets like Pushkin — and how she gradually became interested in science and mathematics. The chapter devoted to Dostoyevski and his meetings with Sonja and her syster Anyuta arose much interest in those days. Sonja's novel gave also a vivid portrait of Dostoyevski. Wirsén finished his review by the words:

Such good descriptions as in this novel are rare to find. Nothing is artificial — nothing is obscure — everything is spontaneous — everything has a natural freshness. We sincerely hope that a continuation of these notes from the Russian life will not fail to appear

[En så god skildring som denna får man ej ofta läsa. Här finner man ingen förkonstling, intet bortkrånglat, allt är osökt, allt har en naturlig friskhet. Vi hoppas att en fortsättning av dessa anteckningar ur ryska livet ej må utebliva...]

The Polish uprising. Palibino was not very far away from the places where fighting took place between Poland and Russia around 1860. Sonja's memories of this time were published in the Swedish magazine *Nordisk Tidskrift för Kultur och Vetenskap* (*Nordic Journal for Culture and Science*), in December 1890.

Poetry. Sonja wrote some poems. After her death they were kept by Julia Lermontova who published one of Sonja's poem's in 1892. The whole collection was eventually published by Sonja's daughter in 1951; it contains about ten poems and is reproduced in [10].

The Nihilist. This novel describes events which took place in Russia during 1860. It was published shortly after her death in Swedish edited by Ellen Key. A Russian version edited by Maksim Kovalevsky was published in Geneva 1892. The novel remained forbidden for a long period in Russia and was not published there until 1928.

For a more complete account of Sonja's literary work we refer to Kochina's biography [9].

10. Meeting with Mary Ann Evans

Introduction. Mary Ann Evans (1819–1880) wrote her novels under the male pseudonym George Eliot. The novels described in Sonja's article are *The Mill on the Floss* — published 1860 — *Daniel Deronda* (1876) and *Middlemarch* (1872).

Sonja and Vladimir visited England for two months during Christmas in 1870. The greatest event for Sonja during this journey occurred when she met Mary Ann Evans; Sonja had started to read her novels early. English had been the first foreign language taught to her at Palibino. Sonja and Vladimir met the director of the British Museum — Mr. Rallston; Sonja enjoyed discussing Mary's novels with him; eventually he suggested to her to write to the famous author — the response came quickly. It turned out that Mary already had heard about Sonja's talents from the British mathematician Hill who had also followed lectures in physics and mathematics by Königsberger at Heidelberg.

Sonja published two consecutive articles in the dayly newspaper *Stockholms Dagblad* in April 1885. The first describes her meeting with Mary in 1871; the second her conversation with Mary in the fall 1880. When Sonja wrote these articles she had mastered the Swedish language but received also help from Ellen Key to polish the prose; the Swedish text is therefore of a high literary standard. Sonja's articles were much appreciated and admired.

The first article The excerpt describes when Sonja met Herbert Spencer during a tea-party at Mary's home.

I felt truly happy when I received the letter from Mary Ann. The great author had thought about me for more than a year! When I first came to her house at John Wood's Road, I was confused and embarrassed — her appearance was so different from my own imagination of her... She was very thin and had a nose which was enormous for a female face — dressed in a black skirt in thin transparent cotton emphasizing her skinny neck. Could it really be true that this old woman wrote all those novels I had read with such a passion from my early childhood? But already her first spoken word brought me back to my previous dreams of her — I have never listened to words so rich and full of sympathy. I do not remember the subject of our first talk — I only know that I liked her and felt as if I spoke to an old friend.

It is not possible to describe in words the magic Mary transmitted to those around her; but I know that many others have had the same experience as I did myself. Turgenev told me once when we talked about Mary: I know she looks ugly but I do not see it when I speak to her. After an hour our first meeting was over. Mary invited me to come the following Sunday and told me that I was going to meet some people who certainly would be of interest for me.

Discussion with Spencer Sonja first described her arrival next Sunday.

About twenty persons came to Mary's house next Sunday. One of her guests was Herbert Spencer who had supported Mary when she began to publish her novels. But I had never met him and had no idea that it was of him Mary was speaking, when she presented me to an old chap among the guests, addressing the following words:

"Here my dear friend, you will meet somebody who is a negation of all your philosophical principles, for Mrs. Sophie Kovalevsky is a mathematician."

Then Mary turned to me: "I must tell you that my old friend here denies even the possibility of an existing female mathematician. He admits that a few exceptional women may posses a talent which supersedes the average level of the male sex. But he claims that such women always will be attracted to literature and arts — they will never be enrolled in a world of abstractions. Now you should try to fill in new thoughts in his brain..."

The old chap sat beside me looking at me with some curiosity. The guests started to debate about the usual questions concerning women's skills and rights — and what would happen if more women should enter scientific studies... The old chap started to argue against everything I tried to insert into these discussions — adding sarcastic comments. Later I understood that his intention was to provoke a more intensive debate. I started to defend myself, but it was not a difficult task. I had passed my childhood a long time ago - during the last five years I had been fighting hard to obtain the rights to study my beloved subject — mathematics. In those days I was so convinced that truth was entirely on my side that all my shyness disappeared when the issue was to defend women's right to perform studies in science.

The dispute continued. Mary encouraged me to speak — all the other guests stopped talking — listening only to the conversation between me and the old chap. Perhaps I should have hesitated when this took place; but I continued to defend myself against all comments, opinions and sometimes sarcastic objections from the old chap.

After three quarters of an hour Mary turned to me and said with a smile in her face: "Mrs. Kovalevsky has fought well and bravely for women's rights. If Mr. Spencer does not consider himself defeated I would say that he is extremely stubborn..."

I was amazed — suddenly I understood that I had been debating with the prominent British philosopher!

The second article. Mary and Sonja did not meet again until Sonja visited London in the fall 1880. The second article first describes the author's new life after her marriage with Mr. Cross. Sonja also tells how she recognizes Mary in a different way several years after they first met; now she feels as if they were almost of the same age. Sonja had also experienced the fragility of life — her delivery when Sofya was born had been difficult, after which Sonja's heart and lungs had become weak. The following passage contains Sonja's comments upon Mary's novels and ends with the answer she received from Mary.

We talked a lot about old and contemporary literature. Then I turned to her novels. I wanted to tell her that there was one crucial point in her novels I could not fully understand — why the heroes and heroines in her novels die just at a moment when the psychological conflict has become more tense and complex than ever — at a stage when the reader inpatiently waits to see how life develops and is eager to read about the consequences of good or bad actions from earlier pages of the novel.

After a short break I continued: "At such a crucial moment in your novels death appears and puts everything on an equal basis..."

Mary kept silent. Then I began to speak about her novel The Mill on the Floss: "The reader understands that the heroine Maggie during a moment of ecstasy can sacrifice her own love to save her brother. In a moment — when a person is crushed by unforeseen bliss it seems possible — almost easy — to deny oneself. At such a moment suffering seems so remote and is painted with colours so different from reality, that a sacrifice can be made.

But how will Maggie continue when the tense emotional feelings she had at a certain moment are replaced by the unavoidable reality later on. A boring almost deadly monotony — a lonely life without love during an endless sequence of days and years will burden her. Will she not see her sacrifice take shape, when she realizes that she succeeded in killing her lover's devotedness and when torment of jealousy becomes a reality for her. Will she not surprise herself by passionately demanding the return of happiness? And if she actually persists in her self-sacrifice, what will the consequences be..."

After another short break I continued: "The reader wants to see Maggie after the turmoil, and get to know whether privation makes a mortal more noble and more worthy of reverence. Or perhaps it is a fact that one cannot give up passion without loosing the best quality of the heart. Eventually there only remains a fantasy, just as callous for its own suffering as for other's pain and joy..."

That is what the reader wants to know! I almost cried, while Mary still kept silent. I continued my interpretations

"The novel is not developed further — instead comes the flood and a huge black wave bringing Maggie together with her brother — putting an end to all their struggles — all their sacrifices and all their expectations".

At this moment I felt almost exhausted. But after a while I continued to discuss some other novels. "In Middlemarch Mr. Casanbon dies before Dorothea has lost anything of her youth and her eagerness to fulfill the infertile and futile deed that she has been chained to by ill-considered devotion. In Daniel Deronda Gwendoline's troublesome husband gets drowned during a trip in a gondola, exactly at a moment when their married life has become almost impossible — at a moment when the reader is eager to know how the poor Gwendoline can retreat from the terrifying position her vanity has placed her in. Always death enters at the right moment, puts everything in order and cuts off the entangled knot."

At last I stopped my long talk and awaited, with some dread, for what Mary was going to say. Then she began to talk.

"Perhaps you are right; but did you not encounter the same in real life. I think personally that death is far more logical than we usually make a conception of. At a moment when the situation has become too appalling — when no possible outcome can be seen — when the most contradictory duties stand against each other — then death comes - opening a new way and blessing everything which seemed irreconcilable."

In the final part of her article Sonja writes about Mary's last years — especially her marriage with Mr. Cross. In the very last sentence of her article Sonja cites Mary's last words when they took farewell:

It is belief in death which gives me courage to live.

11. August Strindberg

August Strindberg (1849–1912). In June 1884 when Sonja was appointed associate professor, Strindberg was sent to trial for his novel *Getting married.* Many people defended him; among those were Anne Charlotte and Gösta Mittag-Leffler. Later Strindberg went to the continent. It was therefore a surprise to many people that he wrote about Sonja and her appointment after his arrival in Paris. Here follows an excerpt from his article which was published in *Dagens Krönika* (*The Dayly Chronicle*) just before Christmas.

I cannot deny that I consider the Countess who started a firm for delivery of goods far more worthy than the Lady who became a professor of mathematics. The Countess serves as a good example for the future, while the Lady of mathematics may only create unnecessary ambitions in the minds of young girls... To invite a Russian Lady to Stockholm was only an expression of old-fashioned gallantry — and did not respond to the need of mathematics for the citizens in Stockholm. At this moment the world has far more need of able mothers than professors in mathematics — such abnormalities can be produced at any desired amount if one allows persons with special talent of mathematics to be narrowly educated into mathematical monsters...

Sonja received a copy of this article from Anne Charlotte while she was in Berlin preparing her lectures for the next semester in Stockholm. It was unpleasant for her to be scandalized in this manner — especially since her position at Stockholm had been criticized by teachers and professors at Uppsala University. Her response to Strindberg's article was subtle; instead of entering into a meaningless debate, she wrote about her meeting with Herbert Spencer which took place 14 years before! It was known to everybody in Stockholm that Strindberg regarded Spencer as a very great philosopher — almost as his spiritual godfather. Sonja's brilliant article demolished Strindberg's statement; this action impressed many readers.

Gradually Strindberg understood that Sonja could not be regarded as a "mere student of Weierstraß". In a newspaper article published in 1885 — where Strindberg describes Sweden as a company directed by foreigners — he starts to speak about the King of French origin, the Queen of German origin, inserting the insulting phrase: *Does she really speak the Swedish language?* Then he goes on with an extensive list of famous persons: "The greatest historian, Geijer from Austria — the greatest poet, Snoilsky from Poland — the greatest scientist Nordenskiöld from Finland. Strindberg's exclusive list ends with *The most learned woman* — Kovalevsky from Russia.

Sonja liked Strindberg's novels and dramas; she introduced him to Russian readers — long before his books were translated. In the *Nya Idun* (*New Idun*) Society she declared that one should ignore his angry attitude and pay attention to his creative power as an author. Strindberg's two masterpieces — Miss Julie and The Father were published in 1887–88. Sonja adored them. At that time Strindberg's peculiar views on women had been scrutinized and blamed in a most efficient way in a widespread article written by Hjalmar Branting. Strindberg's views on women were therefore not taken seriously in the late eighties; Sonja herself could make jokes about this to her girl friends. In a note to Anne Charlotte she once wrote:

Nej, vet du, vore bra att en gång ha varit fader. Då vet man vad de stackars männen får lida från de elaka kvinnorna. Fick jag bara träffa Strindberg och hålla honom i handen.

You know, it would be good to be a father once. Then one should be able to realize how the poor men suffer from naughty women. If I only could meet Strindberg and keep his hand in mine.

The blue books. It was not until 1907 that Strindberg once more mentions Sonja in his series of Blue books where he speculates over different scientific subjects, among those astronomy — a subject which fascinated Strindberg. He also devoted considerable time and effort to learning mathematics — especially during his years in Paris. In view of his high standard in language and in areas like arts, photography and to some extent even chemistry, it is remarkable that Strindberg never seemed to have understood fundamental rules of mathematics — such as the distinction between axioms, definitions and deductive results. When he mentions Sonja in 1907 he ends up with a misleading commentary. Here follows the section where Sonja's name appears.

It is a fact that the Moon only shows one side against Earth... but to say that this depends upon its rotation is a lie! The motion of the Moon has so far defied attacks by mathematicians, because its 19 years cycle is connected to the unsolvable three-body problem. In 1890 all it was claimed be solved by Mrs. Kovalevsky, but that was a mere lie...

[Att han — månen — bara vänder en sida till jorden, det är ett faktum, men att orsaken skulle bero av rotationen, det är lögn ! Ty under ett månadslopp måste den vända andra sidan till åt någon punkt på jorden. Månbanan har hittills trotsat alla matematiska attentat, ty dess 19-åriga (!) rörelse står i sammanhang med det olösliga tre-kropparproblemet. Detta problem uppgavs 1890 vara löst av fru Kowalevsky, men det var bara lögn...]

Remark. Ignoring Strindberg's initial claim that Lagrange's explanation of the Moon's monthly rotation around an axis in his famous work from 1765 should be invalid, it is the last sentence which illuminates how rumors can be misinterpreted and used to make false accusations. There is an explanation why Strindberg refers to the Three Body Problem; it was the subject of an international competition initiated by the Swedish King Oscar II. The prize-winner was Henri Poincaré. Just during the publication in Acta Mathematica — after Poincaré's work already had been nominated for the Golden Medal — Lars Edvard Phragmén discovered some serious errors in the manuscript; this caused great upset. The necessary corrections in a new printing were carried out jointly by Phragmén and Poincaré.

Strindberg was therefore well aware of the fact that the Three Body Problem was essential for mathematicians. Sonja had received the Bordin Prize in Paris for her work on rotating rigid bodies. If one wants to be kind to Strindberg, one could say that he confused these two events — one prize in Paris and the other in Stockholm — but with prize winners from switched places; but in retrospect Strindberg's text was ugly. He published his text in a book and not as a mere article in a magazine, at a time when Sonja no longer could defend herself. Strindberg's text gave less informed readers the lasting idea that Sonja's scientific merits were exaggerated.

12. Vladimir Kovalevsky

Introduction. Details about Vladimir's scientific career can be found in [9], Kochina's biography contains excerpts from several letters between Sonja and Vladimir.

Vladimir's life. He was born 1842 in the vicinity of Palibino. Vladimir had a gift for languages and began to translate German and English texts into Russian when he was sixteen years old and studied law for a couple of years in St. Petersburg, around 1860. His friend Boborykin described him as a student;

The most conspicuous trait was his capacity to catch ideas from any scientific discipline, his dialectic sharpness, and his desire to participate in all kinds of political movements.

Vladimir finished his studies when he was nineteen years old. Then he went abroad for some time where he visited London. He went home to participate at the front on the Polish side during the uprising in 1863. When it was defeated by Russian troops he returned to St. Petersburg where he started a small publishing company. In 1866 he published a translation of Herzog's political book *Who is to Blame*; the second edition was confiscated and burnt by the censorship. It was an economic blow for him and he began to work as a correspondent for his living. In the same year 1866 he visited Italy where he met Garibaldi, and wrote about the Italian war in Russian newspapers. On his return he was engaged to Mariya Shelgunov who had been active in political movements since the early sixties and had been arrested several times, but after a short time their engagement was broken. Later on Mariya was sentenced to exile and had to leave Russia.

Vladimir decided to start a more stable life. At the end of 1867 he met Sonja; she was then 17 years old and her father — general Korvin-Krukovski did not like the idea of marriage; but Sonja was determined to marry him. Vladimir together with his elder brother, Alexander Kovalevsky — who was a recognized scientist — was persuaded to enter studies in science.

Vladimir had kept his publishing company and started to edit books like the Principles of Geology and books in biology and zoology. He managed to get into contact with Darwin and received an unpublished manuscript for which Vladimir was supposed to read proofs. While reading he translated Darwin's work to Russia; the result was that *Variation of animals and plants under domestication* was published in Russia shortly before the original came out in England. Darwin himself did not oppose this since he wanted more than anything else that his ideas were spread as much as possible. The second volume was translated by Sonja and Vladimir together.

Gradually Sonja's father softened and agreed to their marriage which took place in Palibino in September 1868; next spring they went to Heidelberg. In the fall 1869 they made an extensive journey to England where they stayed at Darwin's home for some time.

After the return to Heidelberg, Vladimir decided to study paleontology and the next years consist of intensive studies. Sonja lived with Julia Lermontova in Berlin while Vladimir obtained his doctor's degree at Jena University in 1872. His scientific publications between 1870–75 were remarkable and aroused much attention. He did pioneering work, applying Darwin's theory on a zoological basis.

In the fall 1874 Sonja and Vladimir returned to Russia — where they spent some time at Palibino and then settled in St. Petersburg. Sonja's father died next year; Sonja's inheritance gave them economic independence for several years. However the estate at Palibino was lost when Sonja's younger brother Fedya was ruined in gambling. Vladimir faced similar difficulties as Sonja — he could not find a position at the university in St. Petersburg since his foreign degree was not recognized. Instead he started up his publishing company once more. Both he and Sonja worked extensively with the magazine *Novoye Vremya*. Sonja became well known in St. Petersburg in those days — not only for her expository articles on scientific achievements — but also as a critic in theatre and literature. She became acquainted with many prominent Russian authors — Turgenev was a close friend of her, Dostoyevski, Tolstoy and Chekhov were among the visitors to the Kovalevsky's home. Sonja and Vladimir overestimated their fortune and engaged themselves in building projects where they lost much money. Sofya was born in 1878; during the first year after the birth of Sofya, Sonja was not occupied with anything else other than her newborn child.

In the fall 1880 Vladimir finally obtained an offer from Moscow University where he became a docent in geology; during the same period Sonja takes up mathematics again. She had been encouraged by Chebyshev and had taken part at some scientific meetings in St. Petersburg. In October 1880 she visited Weierstraß to discuss new problems.

During the early winter 1881 Sonja realizes that their economic situation is very weak. Vladimir had engaged himself in a company prospecting oil — but instead of gaining a salary he had been cheated and was left owing the company considerable sums. This tense situation was aggravated when Alexander II was assassinated in March 1881. During the same time Vladimir was troubled by another blow — former friends published an article in Zürich where he was accused of having served the secret police during travels abroad and claimed that he had committed plagiarism in respect of Darwin and his scientific work.

Vladimir had a nervous breakdown and was threatened by the oil company and feared a trial. Sonja left Moscow and travelled to Germany with their daughter during the Summer. Letters from this period between Sonja and Vladimir show that they still hoped for a happy future. But a few months later Vladimir urged Sonja to go to Paris and stay there with her sister Anyuta. From this time onwards he isolated himself; not even his elder brother Alexander was able to help him. On one occasion he sold jewelry which Sonja had kept from her family home and misused obligations which in reality belonged to her; he confessed these acts in a letter to Alexander expressing his feelings of guilt. By the end of 1882 there was no more sight of Vladimir. At this time Sonja lived at her sister's home in Paris, while her daughter was in Odessa with Alexander's family. On April 27 1883 Vladimir committed suicide by inhaling chloroform.

Vladimir's last letter. On his desk he left a letter addressed to his brother Alexander where he asked him

Tell Sonja that she has been present in my thoughts until the end, how I have been wrong to her, how I have destroyed her life that might have been bright and happy without me. My last appeal is that Anyuta takes care of her and the little girl Sofja. Anyuta is the sole person who can do this and plead with her to do so...

13. Memories by Julia Lermontova

Introduction. Julia Lermontova and Sonja lived together for more than four years in Germany. Here follow some excerpts from Julia's memories; Kochina's biography contains a more complete account.

Julia describes the first time after the arrival in Heidelberg as a happy period for everyone. She tells about excursions which Vladimir arranged for Sonja and herself in the weekends; during the week they studied intensely. Then she describes how this idyllic atmosphere was broken when Anyuta and her girlfriend arrived. Anyuta inisted that Vladimir should move; she needed Sonja's flat as a permanent address because she was planning to go to Paris behind the back of the parents. Julia continues:

Sonja visited Vladimir as often as possible outside our place. It was troublesome for them to stay with us; especially since Anyuta and her girlfriend could be nasty to Vladimir — they always tried to point out that his marriage with Sonja was "fictious" — and hence no intimacy was allowed. This caused quarrels and gradually destroyed the good atmosphere Sonja, Vladimir and myself had before.

When Anyuta left for Paris, Sonja and Julia moved to Berlin where they lived together for three years. Vladimir studied at other places — mainly at Jena University. When Sonja had private lessons at the home of Weierstraß, Vladimir travelled to Berlin to escort Sonja after every lesson — this was the request of etiquette in those days.

Sonja and Vladimir In her memories Julia discusses the relation between Sonja and Vladimir. Here follows an excerpt about this:

Sonja always had an almost irresistible need for tenderness. She desired that every conversation should open the heart of the person she spoke to; she always wanted somebody close to her. Someone who could share everything with her. At the same time she made it almost impossible for anyone to live close to her; she was too restless in order to be satisfied by a calm life and yet she dreamed of an existence in harmony and care. Fundamentally she was perhaps too selfish to fully correspond to the demands of those who lived near herself. Vladimir on the other hand was extremely restless; all the time he was possessed by new plans and ideas. God knows if this couple under any circumstances could have lived happily together.

14. Anne Charlotte Leffler's biography

Introduction. When Anne Charlotte Leffler wrote her novel [11] she lived in Italy where she had married few years earlier. The manuscript was finished in the Summer 1892. Shortly after Anne Charlotte shared the same tragic fate as Sonja when she died after accute illness in October 1892. Her novel was published posthumously. When Ellen Key reviewed it in 1893, it was perhaps the most difficult task she was ever confronted with as a literary critic. She had been a close friend both of Anne Charlotte and of Sonja. To begin with Ellen Key brings out credits — emphasizing that Anne Charlotte has described so many details from Sonja's life which otherwise would have been lost. Critical parts in the review are concerned with Leffler's description of Sonja in her role as a scientist.

Ellen Key writes that the reader gets an incorrect impression of Sonja's character in Leffler's description of her, because Leffler disregards the existence of the *Mathematician in the Woman*, while her subjective description is limited to the *Woman in the Mathematician.* Here follow some crucial passages from the review.

Initial remarks. *Leffler has shown the same open mind when she wrote about Sonja Kovalevsky as if she had written a biography over herself. But she did not realize that a truly complex person, with her constantly changing state of mind, cannot be properly presented as compared to people having a one-track mind and more formed characters. By fixing in print passing fancies, self-contradictions or changes of temper when the description of a truly complex personality is at stake, one will involuntarily crystalize them, and thus give solid contours to something which in reality possesses the same floating light and pleasant forms as clouds in heaven.*

The mathematican in the woman. Here is another crucial section from Ellen Key's review.

Sonja's devotion to science gave nobility and firmness to her personality; it was so to speak her spiritual backbone. Sonja's relation to science has been strongly subordinated in the biography by Leffler's decision to limit her account of Sonja to a subjective description. The result is that Sonja's entire personality in the biography appears much less powerful and united than it was in reality. Sonja's scientifically trained, transparently clear and consistent way to think — which has so strongly affected her poetry, her outlook on life and her emotions — are thereby almost lost in the biography. The greater part of her genius has disappeared. Leffler wanted — with all respect — to show the Woman inside the mathematician. But she has not shown the Mathematician inside the woman...

15. Maksim Kovalevsky

Maksim Kovalevsky (1851–1916) never spoke openly about Sonja after her death and kept all letters from her in privacy. According to Pelageya's biography it is likely that he ordered them to be burned before he died.

Sonja and Maksim met in Stockholm in February 1888 when Maksim was invited by the University to deliver lectures in political science. His lectures were successful with hundreds of people in the audience. Maksim was also interested in geography and admired explorers. In those days Nordenskiöld was the most famous person in Sweden after his successful trip with the ship Vega which had clarified the north-east passage all the way from the Atlantic ocean to the Pacific ocean. Maksim also met Sven Hedin who had already made his first journeys to Asia. Hedin was not yet 25 years old; but he had already started his extraordinary career as explorer. A few years later Sven played an important role prior to Sonja's nomination in June 1889, when he helped Gösta Mittag-Leffler in tough debates with the city council of Stockholm, to finance Sonja's salary.

When Maksim came to Stockholm he had left Russia. He had been a professor in political science and law at Moscow University but was forced to leave after the turmoil in March 1881. He moved to Nice in southern France and travelled extensively. He had an international reputation for his profound knowledge of the history of politics and gave for example lectures at Oxford and Chicago.

Sonja's invitation. Sonja was very enthusiastic when Maksim arrived in Stockholm in 1888. Soon after his arrival she invited him to dinner at her home and wrote him a short letter:

Dear Maksim
It is a pity that the word Välkommen [welcome] does not exist in the Russian language. I would very much like to salute you in that manner. I am so happy that you have arrived and I hope that you visit me as soon as possible at my home at

Sturegatan 56. I will be there until three o'clock in the afternoon. In the evening I have invited some friends and then I hope you can come too.

[Многоуважаемый Максим Максимович! Жаль, что у нас нет на русском языке слова *välkommen*, которое мне так хочется сказать Вам. Я очень рада Вашему приезду и надеюсь, что Вы посетите меня немедленно. До 3-х часов я буду дома. Вечером у меня сегодня именно соберутся несколько человек знакомых, и надеюсь, что Вы придете тоже.

Искренне Вас уважающая

Софья Ковалевская.]

The other guests at the dinner were Signe and Gösta Mittag-Leffler, Ellen Key, Anna and Hjalmar Branting. Maksim and Sonja liked each other from the first moment they met. They spent much time together in the spring 1889 when Sonja stayed in France. In 1890 they spent their summer vacation together in Germany and Switzerland. Maksim and Sonja's daughter were also on good terms; all three of them were planning a visit to the Kaukasus in the summer 1891.

Sofya's letter to Maksim. Here follows an excerpt from Sonja's daughter letter to Maksim — written in January 1891 only two weeks before her decease:

Dear Maksim !

Thanks for your letter and Christmas greetings. I am sad to hear that you have been ill and hope you will soon recover; it will be so fun to travel with you and my mother to Russia next Summer. I am taking riding lessons, but only once a week. Next Summer my godmother Julia has promised to give me a horse and then I can ride as much as I please when we arrive.

Next Sunday our class at school performs a stage play; I will take a role. It will be fun, but I feel a bit nervous. Mister Bäcklund returned from St. Petersburg and brought presents from my uncle Alexander. Especially I enjoyed the book To Russian Children by Dostoyevski. The Gyldén family send regards to you. Goodbye dear Maksim Maksimovitch!

Yours Fufa

16. Politics and Society

Before Sonja came to Sweden she had experienced many social and political events in Russia, France and Germany.

Fyodor Mikhaylovich Dostoyevski (1823–1882). Sonja's chapter about this great author in her book My Russian Childhood describes her emotions while Fyodor Mikhaylovich told about the events when he was arrested and kept in prison for a long period in 1850. Especially his memories prior to his proclaimed execution; it was only at the very last moment that the message of mercy was announced while Fyodor Mikhaylovich and his comrades stood in front of a firing squad. It was never the intention of the Tsar to carry out the death penalty for this group of young intellectual nihilists, but to provide an example. After these moments of horror, Fyodor Mikhaylovich and his twenty comrades were sent to prison camps in Siberia; Dostoyevski stayed there 8 years before he was allowed to return to St. Petersburg. When he told about these events to Anyuta and Sonja in 1865 they had not yet been described in his novels; it was only later that he wrote about this experience in his literary works. Sonja was only fifteen years old when she first met Dostoyevski; his stories from real life revealing how tyranny could be used against free thinking had a lasting impact on her.

Maksim Kovalevsky's memory. Maksim Kovalevsky has perhaps come closest in describing Sonja's views on justice, morality, science and religion. In his memorial article about Sonja he writes:

Issues about Monarchy or Republic did not interest her at all. She did not recognize any other aristocracy than knowledge, talent and intelligence. Her early acquaintance with Dostoyevski developed her love for literature; but she was not impressed by his philosophy in the spirit of Rousseau and could not understand his negative attitude against materialism. Among scientists Sonja foremost adored Charles Darwin; she felt sympathy to people representing science in the spirit of Darwin; she despised everyone who denied goodness in people who were not religious

Sonja was ten years old when extensive fighting took place along the border between Poland and Russia quite close to Palibino. In those days also the new era in Russia started when serfdom was abolished. Sonja's elder sister Anyuta went to Paris in 1869 and married one of the leading communards; Anyuta and her husband had radical political views and were convinced marxists. Vladimir and Sonja visited Paris for a short period in April and May 1870 while the city was surrounded. Sonja worked as a nurse during the bombing of the city. They left just before the battle started between the Paris Commune and government forces.

Back to Russia. When Sonja and Vladimir returned to Russia in 1874 the political climate was fairly liberal. But frequent attempts by nihilists upon the life of the Russian Tsar sustained a tense atmosphere. Among those who were affected were hundreds of Russian women who had studied abroad like Sonja. Most of them met with difficulites when they returned home since their academic merits were not recognized; Sonja also encountered this. Her request to become a teacher in mathematics at a High School was rejected; it did not help that prominent mathematicians like Chebyshev declared that Sonja was qualified.

The period after March 1881. When Alexander II was assassinated in March 1881 the social and political climate changed drastically. Friends to Vladimir and Sonja were imprisoned; in one case she was involved — at least indirectly.

Natasha Armfelt. After Sonja's arrival in Heidelberg she urged her cousin Natasha Armfelt to go abroad and study like herself. Natasha spent some time at Sonja's house in Heidelberg, but continued to Zürich where she became associated with the Russian nihilists who lived there in exile. When Natasha returned to Russia she was put in prison for some time. After the events in 1881 she was deported to Siberia where she died a few years later. Tragic destinies like Natasha's were quite frequent in those days.

Sonja's time in Paris. Sonja and her daughter went to Germany in May 1881 where she spent the summer with Karl Weierstraß and his sisters. Afterwards they moved to Anyuta's home in Paris where Sonja stayed for more than a year. At this time the debate about society and politics was intense, all over Europe. Equal rights for men and women and improved conditions for workers in factories and the countryside were the central issues. Marxism dominated in those days over a more revisionistic view on socialism — there was a wide gap between the revolutionary socialism and liberals who encouraged changes in a more gradual way. Sonja attended many discussions and met people with different political views during her stay in Paris from the end of 1881 until spring 1883.

Georg Vollmar. At a dinner in Anyuta's home, Sonja met Georg Vollmar in March 1882. His political views and his commitment to changes in society impressed Sonja; they became good friends from the first moment and their friendship lasted. They met each other many times, both in Berlin and Stockholm. The next section contains more about the relation between Sonja and Georg.

17. Politics in Sweden

Sonja became a member of the women's association *Nya Idun* (*New Idun*) where Anna Branting was one of the leading members; Anna and her husband Hjalmar were a few years younger than Sonja; both were excited about her arrival in Stockholm. Of course Sonja met many other people through *Nya Idun* and in other places, but since Hjalmar Branting has played a very important role in Sweden's political history, the subsequent section about Sonja's social life in Stockholm will be concentrated upon material relating to Branting.

Hjalmar Branting (1860–1925). He had passion for astronomy; already as a teenager he was a member of the astronomical society in Stockholm. He finished High School when he was seventeen years old and his university studies started with mathematics where he passed an examination at Uppsala in 1880. During the same period he was research assistant at the Stockholm Observatory headed by Hugo Gyldén. After this he turned to social science and politics and became a journalist at the magazine Tiden which was a leading publication based upon a socialistic ideology.

The Gävle speech. Hjalmar was one of the founders of the Social Democratic Party in Sweden in 1889. In a historically decisive speech on socialism held in 1886 at the town Gävle — situated 100 kilometers north of Uppsala — he insisted with equally strong emphasis upon a peaceful revolution based on democracy, where rights of vote was the main issue, together with the formation of workers' unions and rules to limit the amount of working time for employed persons. His speeches were inspired by earlier conversations with Georg Vollmar who had presented similar ideas the year before during his visit to Stockholm.

Branting became a member of the Swedish Parliament in 1896. During the first years he was the only member from the Social Democrats; in 1907 he became president of the party. During the World War 1914–1918 he participated in a coalition government; in 1920 he became Prime Minister of Sweden's first Social Democratic government. It was a great triumph for Hjalmar when rules for work was legislated in 1920 — and soon afterwards women were given equal right to vote in general elections.

When bolshevism started to spread Branting took a very firm stand in favour of the ideal of democracy and managed to prevent communists joining the Social Democratic party in Sweden after the Russian revolution had started. In 1921 he was awarded the Nobel Peace Prize together with Lange from Norway.

The Branting monument is situated at Norra Bantorget in central Stockholm; it was created by one of Sweden's greatest artists ever — the sculptor Carl Eldh (1873–1954) who worked on this masterpiece for more than a decade before it was dedicated in 1952.

Julia and Georg Vollmar. In the spring 1884 while Sonja worked in Stockholm, Anne Charlotte Leffler and Julia Kjellberg made an extensive journey in Europe; during this trip Julia met Georg and they married in 1885. Julia was the most radical member of the women's movement in Stockholm; she was a close friend of Karl Marx's daughter whom she met during visits to London.

It was therefore a great event when Georg and Julia visited Stockholm in 1885; Georg had been invited by Hjalmar Branting. Several political meetings took place in Stockholm with Georg as the main speaker; this caused an uproar. Vollmar was described as an extremely dangerous socialist in the conservative press, his speeches impressed the audience very much; Hjalmar Branting assisted — translating every word from German into Swedish. For Sweden's political development Vollmar's acquaintance with Hjalmar Branting played an important role. Hjalmar has described the fundamental importance of Vollmar's political views:

Georg Vollmar was the first one who gave socialism a true face in making it plausible that social progress can be obtained in a democratic spirit which obeys parlamentary rules.

Vollmar later became a member of the German Parliament. For many years he was engaged in improving conditions for farm working. But his political contributions were demolished by the outbreak of World War I. However his name is well known in the history of politics where he is considered as one of the founders of revisionism against marxism.

Sonja's letters to Vollmar. A few months after they first met in Paris, Sonja wrote to Georg in May 1882:

The indignation caused by the injustice around us is so overwhelming that any other interest pales in comparison to the battlefield of economics which is spread out in front of us. The temptation to step forward to the frontier of those who are fighting will be very great...

In the fall 1882 she wrote another letter to Vollmar where she tells about her reflections concerning society and politics.

The problems raised by theoretical socialism together with thoughts about what methods are to be used in the political fight, have been so present in my mind that I can hardly proceed with my own mathematical work which now seems so remote from reality of life. There are periods when I think that my endeavours as a mathematician can only be of interest for a narrow group of people, while we should do our best together for the majority.

During Sonja's travels to Europe in the late eighties she visited Julia and Georg in Berlin. Their marriage was shadowed by a tragedy when they lost their newborn child in 1887 while Georg was temporarily imprisoned because of his political activities. Apart from this Julia and Georg were a happy couple

Sonja's last visit. Sonja stayed with Julia and Georg in late January 1891. Nobody recognized that her health was in such bad shape while she was there. The message that Sonja had died on February 10 in Stockholm came as a shock to everyone who knew her. After Sonja's funeral Georg spoke about her last visit to Berlin.

She came to us from the Sun in southern Europe where she had spent some time during Christmas vacation. She was joyful — seemed happy and was looking forward to all her projects during the forthcoming year — especially the travel to the Kaukasus which was planned for the next Summer, together with her daughter and her friend Maksim Kovalevsky. She spoke with such an enthusiasm about the future. We took farewell and promised to send greetings soon — expecting that we would meet again in Paris, Berlin or Scandinavia. Nobody could realize that she was going to die so soon.

18. Impressions from Sweden

Here follow excerpts from an article which was written in 1890 and published in several languages — including Russian where it appeared the same year (see [10]).

Sweden has never suffered from foreign oppression — there has never been widespread slavery or tyranny such as under the reign of Ivan the Terrible. The persecution of religion has not been as merciless and cruel as in the rest of Europe. Such a relatively free and calm history in the past has developed a temper of logical sense among the Swedes who do not admit a large cleavage between word and action and are not content with empty phrases.

Workers at factories are better off compared to farm workers without their own land in agriculture. However industry has not yet developed very much and therefore industries and their workers form an element of society whose significance cannot be compared with the situation in England, Germany and France. In the countryside there do not exist landlords with large land holdings. In the cities of trade there are wealthy merchants. The rich wholesaler is the characteristic of a successful person in Sweden today; but there is no question of comparison with the colossal fortunes among those as in England and Unites States of America.

The competition — be it for higher official positions or enough to eat — has not reached the acute character compared to the greater countries. A normally gifted person with an ordinary education can find a position which makes it possible to provide for a family. On the other hand, external living conditions are quite simple. Here in Sweden there does not exist the display of luxury or temptations which is developed in capitals like Paris, Berlin and London — leading the minds of people to one single thought, to become rich at any cost. In Sweden there is a plentitude of people who make their living by work and yet have enough spare time to enjoy life.

A rich inner life, a developed fantasy, mixed with a calm and restrictive — one may even say cool — mind, is a major part of the Swedish, and to an even larger extent the Norwegian, temperament. In Ibsen's play Vildanden one of the fictive persons expresses this by saying:

"The need to once and for all create an ideal and then to worship it during the whole life is our national disease."

The Swede is born conservative; each new proposal is met by preconceived distrust. Every thought upon change leads — almost instinctively — to resistance and hostility. It is therefore evident that it is more difficult for Swedes to change their opinions - to be convinced that something is fundamentally wrong with their outlook on life — compared to Russians who cannot remember anything else than an endless chain of contradictions, sudden changes and who never have experienced anything secure, stable and lasting.

But let me repeat — once the Swedes have been convinced about a necessary change — they will not make a halt half way, but consider it as their moral duty to express the change by opinions and reforms.

As far as my personal experience of Sweden reaches I dare say that radical changes of social or economic conditions can be carried out — and this can even happen during a rather short period of time — without rage and struggle. All that is needed is to persuade sufficiently many people about the necessity and the eligibility for every new reform.

Remark When Sonja came to Sweden the extreme poverty around 1860 which included famine in many areas of the country was no longer as widespread. But the emigration to United States was intensive — about 40 000 Swedes emigrated each year in the 1880s due to poverty in agriculture and the decline of mining. Between 1869 and 1919 almost one million emigrated — counting inhabitants of both Norway and Sweden which made up the highest percentage of emigrants in Europe — except for Ireland whose population decreased from eight to four million between 1850 and 1910.

Sonja arrived in Stockholm at a time when Sweden's kingdom for the first time in history was led by a truly liberal king - Oscar II. Sonja met him personally at several occasions. She was aware of her priviliged position in Sweden. Her housemaid Augusta earned a net salary — excluding lodging and meals — which was about 3 percent of Sonja's annual income. Married women had no control of their private economy in those days, and the right to vote was quite restrictive — even among the male population. The slogan "Det är skam det är fläck på Sveriges baner att medborgarrätt heter pengar" [*It is a disgrace and a blemish on the banner of Sweden that civil rights are measured by money*], was introduced by Verner von Heidenstam who received the Nobel Prize for Literature in 1916; this slogan remained relevant in Sweden until 1919. But in a historical perspective Sonja's article gives an accurate description of the situation in Sweden in those days; her comments on the industrial development are perfectly correct. The example below may illustrate this.

LM Ericsson. The world famous telephone company Ericsson has been founded by one single person — Lars Magnus Ericsson (1846–1926). He was born in a poor family in Värmland and lost his parents when he was only ten years old old. He had talents for engineering and started to work for the Swedish telegraph company when he was fourteen years old. Having proved his skills during a period of ten years of work he received a grant from the state which enabled him to obtain an education in modern engineering and physics in Germany. He spent two years abroad but soon after his return to Sweden he heard about Graham Bell's construction of the telephone. Inspired by this invention he started a small firm in central Stockholm with two employed assistants. During fifteen years of creative work, Lars Magnus succeeded to make several complementary inventions — among other things earlier versions of the telephone microphone was improved. This led to a world patent which was the point of departure for L.M. Ericsson's future expansion.

Remark. Lars Magnus should not be confused with another famous Swedish inventor — John Ericsson (1803–1889). They had a similar background and were born in the same area of Sweden; but they were not from the same family. John was an exceptionally gifted engineer; he was only twelve years old when he was employed during the construction of Göta Kanal — the traffic link for ships from the eastern to the western coast of Sweden. He is foremost famous for his development of the propeller, mainly after his arrival in the United States around 1840.

19. Ellen Key's commemorative words on Sonja

Introduction. The subsequent text is an excerpt from Ellen Key's article published in the Swedish journal *Dagny* in 1892. It is difficult to translate her dignified and yet passionate text; Ellen Key is one of Sweden's greatest stylists ever. I can only hope that the subsequent translation has not downgraded her original text much.

Her native country Russia, its development and its destiny was always present in her mind. Her triumphs in life were for a short time dear to her; but there were only two tributes she always mentioned with tears in her eyes:

The call to be a corresponding member of the Academy of Science in St. Petersburg, and when the Norwegian author Jonas Lie proposed a toast expressing his warm feelings to the little girl Tanya Rajevski from her novel My Russian Childhood.

Otherwise she spoke casually about her distinctions — carrying them just as unconcerned as a Queen carries her jewelry. It was a matter of importance to her that no outer limit should be set for the development of female responsibility. She regarded special issues about women as isolated phenomena — regarding them merely as an important part in the grand question of humanity: The greatest possible happiness to the greatest possible number and this including women.

The sole inconceivable matter for Sonja was narrow-mindedness. Her own spirit did not suffer from the common short-sightedness which confuses essentials with unimportant matters. It was precisely her capacity to catch essentials which was the foundation of her psychological acuteness. She noticed the weak points but understood to put them in connection with their merits — describing the entirety as conformity with Nature; in this way she could always judge with tolerance. It never occurred to her to demand from her fellow beings special qualities — it was just as inconceivable for her as to imagine a triangle with four sides. Her clear-sighted tolerance (forbearance) shaped her into an eminent teacher who was able to draw the very best from her pupils. A young girl who was one of her students during her lectures in mathematics wrote after Sonja's decease: I felt as if I was completely seen through by Mrs. Kovalevsky, as if I was made of glass, and yet I felt perfectly confident in front of her mild and understanding look.

Sonja was acquainted with several great authors of her epoch — Turgenev, Tolstoy, Dostoyevski. She had met Mary Ann Evans who in her eyes stood out as the greatest female writer ever. And these intimate relations with literary personalities were brought together with extensive acquaintances in the world of politics and science. Sonja knew practially everything of significant value from her native country, Germany, France and England. Even in recent Scandinavian literature she was more well-read than most Swedes.

But how could she manage all this and find enough time? Unfortunately by overwork; there were periods when she did not sleep more than five hours and she seldom payed much attention to her health.

Sonja was engaged in more projects than ever before at the time of her decease. She was eager to enter into a new great problem of mathematics after the Bordin Prize. She had started to write many literary sketches. The first chapter of her novel Vae Victis was published as late as December 1890. She had notes from the World Exposition in Paris which she had visited in 1889. She wanted to continue writing about her childhood and youth and to describe her experiences with her husband Vladimir during their visit to Paris in the Spring 1871. She was revising a novel from her youth which Dostoyevski once had praised. She was engaged in her sister Anyuta's manuscripts and prepared to publish some of her novels and dramas.

As late as in November Sonja started writing a new drama whose psycological mastery and richness of fantasy perhaps superseded everything she had written before. The manuscript was entitled When death no longer exists.

It was in this high season of seething creativity that death entered the scene. Sonja had been familiar with the thought of death for many years; she was aware of her weak heart, but if there were any period in her life when she would have liked to avoid death it was at this very moment when she was in harmony and full of energy.

She had spent her Christmas holidays in southern France where she had met Russian friends. During the trip back she visited Berlin and caught cold during the travel through Denmark in stormy and cold weather. On Wednesday February 4 she arrived in Stockholm. She did not want to show that she was ill and kept her state of health secret to those around her. On Friday she gave the first lecture of the term. In the evening she was invited to dinner but retired early and it was not until Saturday morning that she went to bed. Even if her disease — pleurisy — was quite serious nobody realized the full extent of its gravity. All of us who surrounded her were following a struggle between life and death during the forthcoming days and nights. Death was caused by suffocation — presumably caused by the violent infection which had produced purulence in the lungs. With a stronger heart the outcome might have been delayed, but the autopsy showed that rescue was not possible.

Sonja was conscious most of the time. She expressed an almost indescribable and patient concern for her surrounding. The near approach of death was not anticipated by the doctors or her close friends during Sunday and Monday but on Tuesday morning on February 10 the heart was paralysed. The very last hours she was unconscious. Death entered as a calm sleep into the vast unknown. Her palish face showed serenity and peace.

At the end of her article Ellen Key summarized the essential and unique character of Sonja.

For Sonja it was neither science nor belles-lettres nor honor which was in the center of existence; the heart was the source of life for her.

For the friends of Sonja it was not her grandeur which made her precious. What made her unprecedented in their eyes was that she possessed that rare standard Goethe once said that every human being should seek to possess:

Grosse Gedanken und ein gutes Herz

The grave of Sonja Kovalevsky.

References

[1] J. Barrow-Green, Sonya Kovalevskaya, *European Math. Soc. Newsletter*, N 35 (2000), 9–11.

[2] J.-E. Björk, Rigid Bodies and the Bordin Prize, this volume, pages 53–58.

[3] R. Bölling (edt.), *Briefwechsel zwischen Karl Weierstraß und Sofja Kowalewskaja*, Akademie Verlag, Berlin, 1993.

[4] J. Détraz, *Sonja Kovalevskaja, 1850–1891: l'aventure d'une mathématicienne*, Belin, Paris, 1993.

[5] L. Gårding, History of the mathematics of double refraction, *Arch. Hist. Exact Sci.*, **40** (1989), 355–385.

[6] L. Hörmander, The first woman professor and her male colleague, *Miscellanea mathematica*, 195–211, Springer, Berlin, 1991.

[7] A. Jackson, The dream of a Swedish mathematician: The Mittag-Leffler Institute, *Notices Amer. Math. Soc.*, **46** (1999), 1050–1058.

[8] A. Koblitz, *A convergence of lives*, second edition, New Brunswick, Rutgers Univ., 1993.

[9] P. Kochina, *Love and mathematics: Sofya Kovalevskaya*, Translated from the Russian by Michael Burov, Mir, Moscow, 1985.

[10] S.V. Kovalevskaya, *Memoirs, novels* (in Russian), Nauka, Moscow, 1974 (P. Kochina edt.).

[11] A.C. Leffler, Sofia Kovalevskaja, Severnui Vestnik, St. Petersburg, 1893.

[12] G. Mittag-Leffler, Sophie Kovalevsky, *Acta Math.*, **16** (1892), 385–392 (see this volume).

[13] G. Mittag-Leffler, Weierstraß et Sonja Kowalewski, *Acta Math.*, **39** (1923), 133–198.

[14] H. Shapiro, The limitations of the Cauchy-Kovalevsky theorem, this volume, pages 59–62.

[15] V. Volterra, Sur les vibrations lumineuses dans les milieux biréfringents, *Acta Math.*, **16** (1892–1893), 153–215.

Dept. of Mathematics
Stockholm Univ.
106 91 Stockholm
Sweden
e-mail: `jeb@matematik.su.se`

Operator Theory:
Advances and Applications, Vol. 132, 55–60

Rigid Bodies and the Bordin Prize

J.-E. Björk

Introduction

In this article we discuss Sonja Kovalevsky's work [7]. First we recall some historical facts concerning motion of rigid bodies. In Section 1 we describe her main result and contributions by other authors which consolidated her discovery when it was proved later that the case found in [7] is unique, i.e. except for previous cases treated by Euler and Lagrange it is only the so-called Kovalevsky gyroscope where the equations of motion can be solved by quadrature.

Let us now discuss dynamics of rigid bodies from a historical perspective prior to Kovalevsky's contribution. In the years around 1880 the subject concerned with the dynamics of rigid bodies was a "hot subject". New technology using machines working with high speed posed problems in engineering.

One may mention that Clerk Maxwell was fascinated by the peculiar effects caused by forces of momentum in rotating bodies. This inspired him when he worked out equations in electrodynamics. In fact, one of Maxwell's theoretical contributions was to link electrodynamics to rigid bodies. He showed how to reduce problems concerned with induction between several electric currents to dynamical equations of rigid bodies. The interested reader may consult the text-book series [3] by August Föppl for illuminating examples and discussions which relate analytic mechanics to technical mechanics and applications to engineering.

Rigid bodies and their history. The presentation below is inspired by Chapter 20 in Volume 2 from [1] by Paul Appell where the reader may find more details. The first study of a the motion of a rigid body rotating around a fixed point was undertaken by d'Alembert in 1749. He realized that one needs a system of six second order differential equations to explain the motion. A precise mathematical treatment was presented by Euler in a publication at the Academy of Berlin in 1759. Euler introduced the ingenious method to work with two coordinate systems at the same time; one fixed to the body and the other in the space where the body moves. He found the equations of motion when no external forces affect the rotating body. Euler's investigations were later on improved and extended by Lagrange, Laplace and Poisson. In particular Lagrange solved the equations of motion when gravity is an external force acting on a rigid body which has two axes of equal momentum around the fixed point, while the center of gravity is placed on the third principal axis through the fixed point. Poisson established general formulas in the spirit of

Euler to express the equations of motion when in the general case, i.e. when a body rotates — or even can move with compound velocity in $\mathbf{R}^3$, while external forces are arbitrary. His famous text-books [11] (Volume 2, Chapter 3) — the first edition appeared in 1811 — laid the foundations for the mathematical theory related to the dynamics of rigid bodies. Later work — especially by Hamilton and Jacobi gave useful links between differential systems in mechanics and geometry. The theory of elliptic functions was developed for several reasons — one major inspiration was precisely to find integrable cases in dynamical systems. The case of a body rotating around a fixed point was one of the central issues. One should mention that studies when the body moves in the space with more than three degrees of freedom were also considered. Here the situation is more complicated since the systems are in general not holonomic.

1. Integrable systems

When Kovalevsky started her work it was an open problem if there could exist any example beyond those of Lagrange and Euler when the six unknown functions which appear in the Euler-Poisson system admit a third algebraic identity. Two algebraic equations hold trivially, i.e. energy is preserved and the vertical component of momentum with respect to the fixed point is constant.

Kovalevsky's strategy in [7] was to consider a body with a symmetry around its fixed point and where the center of gravity is in the plane of symmetry, i.e. perpendicular to the exceptional principal axis. Her example was exposed by Paul Appell in his text-book [1] (part II) from 1896. In the introduction to the chapter *Mouvement d'un solide autour d'un point fixe* Appell writes:

Enfin Madame Kovalevski dans un Mémoire couronné par l'Académie des Sciences, a découvert un nouveau cas d'intégrabilité des équations du mouvement d'un corps solide pesant autour d'un point fixe.

Remark. Concerning rigid bodies which move without a fixed point there exist several situations when the motion can be solved by quadrature. A famous example dealing with a wheel constrained to roll without sliding on an inclined plane is due to Appell. See [2] for a recent account of this example. The impressive work *Theorie des Kreisel* by Klein and Sommerfeld was published in 1910 and contains an extensive list of examples where equations of motion can be solved by quadrature.

2. The Bordin Prize

Sonja Kovalevsky received the Bordin Prize in December 1888. The work was published 1889 in Acta Mathematica [7]. The main result asserts that if K is a rigid body rotating around a fixed point p and gravity is the sole external force, then the equation of motion is integrable if the principal moments of intertia

A, B, C with respect to p satisfy

$$A = B = 2 \cdot C$$

and the center of gravity $\mathbf{o}$ is in the symmetric plane, i.e. if p is placed at the origin then the vector $\mathbf{o}$ is perpendicular to the principal axis which gives the exceptional momentum C. Notice that the distance between p and $\mathbf{o}$ is not restricted.

In this case Kovalevsky found a new algebraic equation of degree four in the six unknown functions which appear in the Euler-Poisson system which makes it possible to solve this system of differential equations by quadrature. This will be discussed in more detail in Section 3. But first we give some comments about work performed after [7] was published. One issue that remained was to integrate the solution found by quadrature in the body coordinates in order to express the solution in the fixed space which is needed to describe the mechanical motion. This was established by Fritz Kötter in [9] and is described under the heading "Work by Kötter" below.

Uniqueness. The fact that the case studied by Kovalevsky gives the only example where the Euler-Poisson system can be solved by quadrature — except for the cases by Euler and Lagrange — was proved in detail by Edouard Husson in [4]. Between 1890 until Husson's thesis was presented at Toulouse in 1905, the non-integrability in the case $A = B$ while $C \neq A/2$ was unclear. Sonja Kovalevsky herself continued to work on the problem of a symmetric rotating body after the Bordin Prize. She found certain equations which seemed to indicate that integrability might exist when $C = A/2n$ for every integer $n \geq 2$. Her article [8] about these investigations was published posthumously in February 1891. Then Roger Liouville submitted a work to the Bordin Prize in 1894 and published later in Acta Mathematica [10] where he announced that integrability holds when $C = A/2n$ for every $n \geq 1$. This incorrect result is mentioned — but without indication of proof — in the first edition from 1896 of the famous text-book *Traité de mécanique rationelle* by Paul Appell. On the other hand, this text-book contains also a presentation of Kovalevsky's example including details concerning the algebraic equation found in her work. It was gradually understood that the singularities which occur when one tries to expand solutions by series *cannot* give algebraic integrability when $C \neq A/2$. In addition Liapunov proved that $C = A/2$ is necessary for integrability for arbitrary initial data. His results are well presented in [6]. Here we shall focus on work by Kötter and Husson.

The article [5] by Husson contains a very detailed proof of this result, i.e. that if the principal moments are all different, then the equations of motion are not integrable. Therefore Husson's work in [4, 5] led to the conclusion the sole integrable cases occur in the situations by Euler, Lagrange and Kovalevsky. In the edition of Paul Appell's text-book from 1911 this correct result is stated as follows:

M.Ed. Husson a demontré que, en dehors des trois cas que nous venons d'étudier (cas d'Euler et de Poinsot, cas de Lagrange et de Poisson, cas de Mme Kowalevski), il est impossible d'obtenir une troisième intégrale algébrique, distincte de celles des *forces vives et des moments, pour le mouvement d'un solide pesant suspendu par un point.*

Restricted initial conditions. One must not confuse Kovalevsky's case with integrable cases which occur when the initial conditions are specialized. Then there exist various cases where the equations of motion are integrable; see Chapter 4 in [2] for a discussion about integrable cases with various restricted initial conditions.

Work by Kötter. In the case studied by Kovalevsky she used the new algebraic equation to express solutions of the Euler-Poisson system by means of hyper-elliptic functions. After this there remained the need to express the solutions in the fixed space in order to describe how the body moves mechanically. This means that one introduces the Euler angles ϕ, ψ, θ and performs integration of the solutions in the moving body-coordinates. The solutions in the fixed space are expressed by theta-functions. This was achieved by Franz Kötter in his article [9] entitled *Sur le cas traité par Mme Kowalevski de rotation d'un corps solide pesant autour d'un point fixe.* In the introduction Kötter refers to Sonja's posthumous article and recalls how she found an algebraic equation relating the six unknown functions in the Euler-Poisson system. Then Kötter begins to discuss the motion of the center of gravity in [8, pp. 209–210]:

Au moyen de ces quatre intégrales il est possible d'exprimer les six grandeurs en question au moyen de fonctions hyperelliptiques de deux arguments. Il résulte encore des équations différentielles du problème que ces arguments sont des fonctions linéaires du temps. Quant aux six cosinus qui manquent encore, Mme Kowalevski déclare qu'on peut les représenter aussi au moyen des fonctions théta, mais elle renonce à faire le calcul à cause des difficultés à exprimer d'une manière convenable.

Remark. The six cosine-functions refer to angles which describe the position of K in the fixed space $\mathbf{R}^3$. On page 210 in [9] Kötter explains a strategy to obtain these cosine-functions by quadrature. The computations are involved but lead to very precise results. At the end of the introduction he writes:

Dans ce qui suit je développerai les résultats auxquels je suis arrivé de cette manière, et même déduirai les quantités obtenues par Mme Kowalevski pour les avoir sous la forme la plus commode pour les développements suivants.

Remark. The equations which govern Kovalevsky's gyroscope have been solved numerically and there are instructive films which show the motion.

3. Sur le probleme de la rotation d'un corps solide autour d'un point fixe

Introduction. Above is the title of [7] which ia an explanation of the work delivered by Kovalevsky to l'Académie des Sciences in December 1888. We shall briefly discuss some key points from her work. The article first recalls the famous Euler-Poisson system.

Le problème de rotation d'un corps solide pesant autour d'un point fixe peut se ramener, comme on sait, à l'intégration du système différentiel suivant:

$$A \cdot \frac{dp}{dt} = (B - C) \cdot q \cdot r + M \cdot g(y_0 \cdot \gamma'' - z_0 \cdot \gamma')$$

$$B \cdot \frac{dq}{dt} = (C - A) \cdot r \cdot p + M \cdot g(z_0 \cdot \gamma - x_0 \cdot \gamma'')$$

$$C \cdot \frac{dr}{dt} = (A - B) \cdot p \cdot q + M \cdot g(x_0 \cdot \gamma' - y_0 \cdot \gamma'')$$

$$\frac{d\gamma}{dt} = r\gamma' - q\gamma'' \quad \text{et} \quad \frac{d\gamma'}{dt} = p\gamma'' - r\gamma \quad \text{et} \quad \frac{d\gamma''}{dt} = q\gamma - p\gamma'$$

Les constantes A, B, C sont les axes principaux de l'ellipsoide d'inertie du corps considéré, relativement au point fixe. M est la masse du corps; g l'intensité de la force de gravité; (x_0, y_0, z_0) les coordonnées du centre de gravité du corps considéré de coordonnées, dont l'origine est au point fixe et dont la direction coïncide avec celle des axes principaux de l'ellipsoide d'inertie.

Remark. The notations above are standard in the literature. More precisely, $p(t)$, $q(t)$, $r(t)$) denote the components of angular velocity in the body coordinates and $\gamma(t), \gamma'(t), \gamma''(t)$ are the cosine functions expressing the position of the gravity vector $\mathbf{e}_3$ in the body coordinates during the rotation. The reader may consult the excellent text-book by Edward Routh [12] for a detailed account about the equations of motion of a rigid body. Routh's famous text-book was published in 1882 and consolidates what was known prior to Kovalevsky's work. Of course more recent literature, especially the text-book by Paul Appell [1] also derive the basic equations related to the dynamics of rigid bodies.

The method used by Kovalevsky in [7] is to ask for solutions where the functions $p, q, r, \gamma, \gamma', \gamma''$ are single valued meromorphic functions of the time variable t. Under this hypothesis Kovalevsky performed certain computations which prove that the poles of p, q, r are at most 1 and the γ-functions admit poles of order ≤ 2. Then she analysed conditions of tentative power series solutions in order that a new algebraic equation exists between the six unknown functions for every choice of initial conditions. The calculations on pages 179–182 in [7] deal with a general case, i.e. here the rigid body need not be symmetric. Then the symmetric case is analysed which means that two of the principal axes with respect to the fixed point p are equal. Kovalevsky puts the center of gravity $\mathbf{o}$ in the plane of symmetry.

Then Kovalevsky begins to analyse this case and finds a system of linear equations which must be satisfied in order that a new algebraic integral exists. This system of linear equations between the coefficients in the tentative series expansions appears on page 183. At the end of this page she concludes:

En effectuant les calculs je me suis assurée que ses conditions ne sont pas remplies dans le cas général; mais que, outre les deux cas déjà connus par Euler et Lagrange, elles sont remplies dans un nouveau cas, où les constantes satisfont aux équations suivantes:

$$A = B = 2 \cdot C \quad \text{et} \quad z_0 = 0.$$

C'est ce cas-là que je me propose d'étudier dans les paragraphes suivants.

The rest of the article is devoted to this case. It would lead us too far to go into details concerning the rather involved calculations which express the solutions. It suffices to recall that Kovalevsky found solutions by quadrature expressing the functions $p, q, r, \gamma, \gamma', \gamma''$ by hyper-elliptic functions of two variables. The interested reader may also consult Kötter's work [9], in particular the equations on page 263.

References

[1] P. Appell, *Traité de Mécanique Rationnelle*, Gautier-Villars, Paris, part I 1893, part II 1896.

[2] V.I. Arnold (edt.), *Dynamical Systems* III, Springer, Berlin, 1988 (translation from the Russian original from 1985).

[3] A. Föppl, *Vorlesungen über Technische Mechanik*, (in 6 volumes), B.G. Teubner Verlag, Leipzig, 1907.

[4] E. Husson, Recherche des intégrales algébriques, *Ann. Fac. Sc. Toulouse*, **8** (1906), 73–152.

[5] E. Husson, Sur un théorème de M.Poincaré, relativement au mouvement d'un solide pesant, *Acta Mathematica*, **31** (1908), 71–88.

[6] P. Kochina, *Love and mathematics: Sofya Kovalevskaya*, Translated from the Russian by Michael Burov, Mir, Moscow, 1985.

[7] S. Kovalevsky, Sur le problème de la rotation d'un corps solide autour d'un point fixe, *Acta Mathematica*, **12** (1889), 177–232.

[8] S. Kovalevsky, Sur une propriété du système d'équations différentielles qui définit la rotation d'un corps solide autour d'un point fixe, *Acta Mathematica*, **14** (1890), 81–93.

[9] F. Kötter, Sur le cas traité par Mme Kowalevski de rotation d'un corps solide pesant autour d'un point fixe, *Acta Mathematica*, **17** (1893), 209–264.

[10] R. Liouville, Sur les équations de la dynamique, *Acta Mathematica*, **19** (1895), 251–284.

[11] S.D. Poisson, *Traité de Mécanique*, Mme veuve Courcier, Paris, 1811.

[12] E.J. Routh, *The advanced part of a treatise on the Dynamics of a system of rigid bodies*, 4-th edition, Macmillan and Co., London, 1884.

Dept. of Mathematics
Stockholm Univ.
106 91 Stockholm
Sweden.
e-mail: `jeb@matematik.su.se`

Operator Theory:
Advances and Applications, Vol. 132, 61–64

The Limitations of the Cauchy-Kovalevsky Theorem

Harold S. Shapiro

The so-called Cauchy-Kovalevsky theorem (C-K theorem) is the fundamental local existence and uniqueness theorem for partial differential equations in the holomorphic category. In one formulation, one considers a system of m quasilinear partial differential equations of first order for unknown functions $u_1, \ldots, u_m$ of n complex variables $z := (z_1, \ldots, z_n)$, which moreover are required to coincide on a complex hyperplane (for example, $\{z_n = 0\}$) in a neighborhood of some point of $\mathbf{C}^n$, with m specified (germs of) holomorphic functions on that hyperplane. The theorem asserts that, if the system is in "normal form" there is one and only one solution for the holomorphic germs $\{u_j\}$.

What this normal form looks like, is given in most textbooks of partial differential equations (see, for instance, [5], p. 74; there the presentation is in the language of real analyticity rather than complex holomorphy, but that is scarcely more than a difference in terminology). Once the system of partial differential equations is given, the question of being in "normal form" is dependent on the hyperplane carrying the initial data being appropriate, in modern terminology "non-characteristic". We will not here insist on details, and for our purposes here it will not be necessary to state the full C-K theorem, since the main point to be made will be illustrated with a simple example.

The *uniqueness* part of the C-K theorem is very simple: from the given system of differential equations and initial data one can compute recursively all partial derivatives of the u_j, evaluated at the "initial point" near which the data are specified, so there is at most one solution, even in the wider category of formal power series. Moreover, this does not even require the normal form assumption, but is true for a much more vast family of initial value problems. The crux of the C-K theorem is, that these readily obtained formal power series solutions converge on some neighborhood of the initial point. The usual method of proof, and that which also yields results of greatest generality, is by the technique of majorizing series, due to Cauchy.

The question of priorities in the C-K theorem seems quite complicated, and I will not try to do justice to it here, but let me record a few facts. The method of majorizing series (called by Cauchy "calcul des limites") was introduced by him in 1839, in order to give a rigorous proof of what we today would call the implicit function theorem for holomorphic functions; shortly thereafter, he used this technique to prove rigorously (local) existence of solutions to initial value problems for holomorphic ordinary differential equations and systems thereof. Continuing in this direction, he proved in a memoir of 1842 an analogous result for partial differential equations, to all intents and purposes the full C-K theorem. (For references, see [2], p. 204, and [3], p. 318; the latter also presents English translations of portions of the memoirs of Cauchy and of Kovalevsky concerning the C-K theorem.)

What seems amazing in retrospect is that Cauchy's seminal 1842 paper remained virtually unknown, even to leading French mathematicians, until 1875 (according to the account in [7], p.240 ff.), at which time Kovalevsky had independently found the theorem, and apparently also Darboux.

In particular, Karl Weierstraß seems to have been unaware of Cauchy's memoir of 1842, and had also discovered some partial cases of the results in that paper. Indeed his first work in this direction is also from around 1842, although he did not publish anything of this until 1894 (see [9], p. 57). And, in 1872 he suggested to his esteemed student Kovalevsky, as a theme for her thesis work, the study of partial differential equations in the holomorphic category. He was aware that convergence proofs for the formal power series solutions fail when the equation (or system) fails to be in "normal form" (what exactly this means, we shall return to shortly) and seems to have believed that this was a mere failure of technique, that with more attention to detail it would be shown that the formal power series solutions always have a polydisk of convergence. See, on this point, his letter of December 15, 1874 to P. du Bois-Reymond (published in vol. 39 of *Acta Mathematica*).

In her Habilitationsschrift [8], published 1875, Kovalevsky gave the famous example showing this is not so. We should add that she also gave by far the most elegant, unified and simple account of the C-K theorem up to that time, in the opinion of several leading mathematicians of the day (including, besides Weierstrass, Poincaré and Hermite). But, whereas this was an impressive performance one could argue that it belongs in the categeory of technical improvements of ideas more or less known earlier to Cauchy. What was her unique, and undisputedly original contribution was the example to show that when the system is not in normal form, the formal power series solution may converge nowhere. Although from our present day vantage point this may appear quite simple, it was not expected, indeed Weierstrass on several occasions expressed his surprise, as we already indicated.

Kovalevsky's example involves the single second order partial differential equation

$$\frac{\partial u}{\partial w} = \frac{\partial^2 u}{\partial z^2} \tag{1}$$

for the holomorphic function $u(z, w)$ of two complex variables (here denoted z, w). Let us place the action on a neighborhood of $(0,0)$ in $\mathbf{C}^2$, and consider various initial value problems for the equation (1).

We could try to solve (1) such that, on the hyperplane $\{z = 0\}$, u and its partial derivatives of first order coincide (all this is local, near $(0,0)$) with the corresponding data belonging to some given holomorphic function $f(z, w)$. This turns out to be a "correct" problem: to be sure, we have here a second order equation but if we rewrite it, introducing additional variables in the standard manner (cf. [5]), as a system of first order equations it is in normal form. Thus, it is covered by the C-K theorem, and we get a unique solution.

If, on the other hand, we try to solve (1) by placing initial data on $\{w = 0\}$, that is, requiring that $u(z, 0)$ equals some given holomorphic germ $g(z)$ near $z = 0$, the equation is not in normal form. It would be if in the right-hand member of (1) we had any linear combination of u and its first z-derivative, the coefficients being holomorphic functions of z, w (or even, of the three variables z, w, u, in the quasi-linear version) but the presence of the second derivative is what violates the normal form required for the applicability of the C-K theorem.

Kovalevsky considers specifically (with other notations) the equation (1) with $u(z, 0) = 1/(1 - z)$ and observes that the unique (formal) solution is

$$\sum_{n=0}^{\infty} \frac{(2n)!}{n!} \frac{w^n}{(1-z)^{(2n+1)}}, \tag{2}$$

and this series does not converge on any neighborhood of $(0, 0)$.

This was an important observation. She remarks that, not even the choice of an *entire* function $f(z)$ as the data $u(z, 0)$ will in general give rise to a convergent power series solution u, since simple calculations show that $u(z, 0) + u(-z, 0)$ must in any case have Taylor coefficients majorized by those of $\exp[Az^2]$ for some positive A. With a little more work, she might have shown (but did not enter into this) that any solution to (1) holomorphic in the bidisk $\{|z| < a, |w| < b\}$ is holomorphically extendible to the "tube domain" consisting of all $(z, w) \in \mathbf{C}^2$ with $|w| < b$ (see the discussion in [6]and [1], where also a remarkable application to lacunary Taylor series, to which Kovalevsky's result inspired young Ivar Fredholm, is presented). Automatic enlargement of the domain of holomorphy of a function on the basis of its satisfying a suitable partial differential equation (e.g. results of Zerner, and Bony-Schapira, cf. [4]) is an important theme, and a latter-day descendant of Kovalevsky's discovery.

In modern versions of the C-K theorem, e.g. in the important case of a single equation, of arbitrary order, it is customary to consider initial data carried, not by a coordinate hyperplane but by an arbitrary complex manifold of complex codimension 1 — then the old fashioned "normal form" requirement disappears and is replaced by that of non-characteristicity of this manifold at the initial point, with respect to the given operator. In this language, Kovalevsky's discovery taught us to be on the lookout for characteristic initial value problems, where formal series solutions may converge nowhere.

There is still a lot more to be said about the C-K theorem, e.g. its use by Holmgren in proving his uniqueness theorem for the Cauchy problem in $\mathbf{R}^n$ (see [5]), Leray's attempt to obtain a global version of C-K (for which there does not yet seem to exist a good introductory account), Hans Lewy's example showing that holomorphy is essential in C-K and C^∞ is not enough (see [5]) but, let those be for another occasion.

References

[1] J.M. Anderson, D. Khavinson and H.S. Shapiro, Analytic continuation of Dirichlet series, *Revista Mat. Iberoamer.* **11**(1955), 453–476.

[2] B. Belhoste, *Augustin-Louis Cauchy: A Biography*, Springer-Verlag, 1991.

[3] G. Birkhoff, *A Source Book in Classical Analysis*, Harvard Univ. Press, 1973.

[4] L. Hörmander, *The Analysis of Linear Partial Differential Operators*, I, Springer-Verlag, 1983.

[5] F. John, *Partial Differential Equations*, Fourth ed., Springer-Verlag, 1982.

[6] D. Khavinson and H.S. Shapiro, The heat equation and analytic continuation: Ivar Fredholm's first paper, *Expositiones Math.*, **12** (1994), 79–95.

[7] A.H. Koblitz, *A Convergence of Lives*, Rutgers Univ. Press, New Brunswuick, N. J., second printing rev., 1988.

[8] S. von Kowalevsky, Zur Theorie der partiellen Differentialgleichungen, *J. für die reine und angew. Math.*, **80**(1875), 1–32.

[9] W. Tuschmann and P. Hawig, *Sofia Kowalewskaja*, Birkhäuser, 1993.

Dept. of Mathematics
KTH
S-10044 Stockholm
Sweden
e-mail: `shapiro@math.kth.se`

Part II

Research Papers

Operator Theory:
Advances and Applications, Vol. 132, 67–76

On Integrability of Many-body Problems with Point Interactions

S. Albeverio, S.M. Fei, and P. Kurasov

Abstract. A study of the integrability of one-dimensional quantum mechanical many-body systems with general point interactions and boundary conditions describing the interactions which can be independent or dependent on the spin states of the particles is presented. The corresponding Bethe ansatz solutions, bound states and scattering matrices are explicitly given. Hamilton operators corresponding to special spin dependent boundary conditions are discussed.

Exactly solvable models of a single quantum particle moving in a local singular potential concentrated at one or a discrete number of points have been extensively discussed in the literature, see e.g. [1, 2, 3] and references therein. In the one-dimensional case, the local singular potential (contact interactions) at, say, the origin ($x = 0$) can be characterized by the boundary conditions imposed on the wave function φ at $x = 0$. There are two classes of such boundary conditions: separated and nonseparated boundary conditions, corresponding to the cases where the perturbed operator is equal to the orthogonal sum of two self-adjoint operators in $L_2(-\infty, 0]$ and $L_2[0,\infty)$ and when this representation is impossible, respectively. The many-body problems with pairwise interactions given by such boundary conditions are generally not exactly solvable [4, 5, 6]. In the present paper we give a systematic description for integrable models of many-body systems with pairwise interactions given by such singular potentials for the case where the boundary conditions are independent as well as for the case where they are dependent on the spin states of the particles.

We first consider the case of spin independent boundary conditions. The family of point interactions for the one-dimensional Schrödinger operator $-\frac{d^2}{dx^2}$ can be described by unitary 2×2 matrices via von Neumann formulas for self-adjoint extensions of symmetric operators, since the second derivative operator restricted to the domain $C_0^\infty(\mathbf{R}\setminus\{0\})$ has deficiency indices $(2,2)$. The nonseparated boundary conditions describing the self-adjoint extensions have the following form

$$\begin{pmatrix} \varphi \\ \varphi' \end{pmatrix}_{0+} = e^{i\theta} \begin{pmatrix} a & b \\ c & d \end{pmatrix} \begin{pmatrix} \varphi \\ \varphi' \end{pmatrix}_{0-}, \tag{1}$$

where

$$ad - bc = 1,\ \ \theta, a, b, c, d \in \mathbb{R}. \tag{2}$$

$\varphi(x)$ is the scalar wave function of two spinless particles with relative coordinate x. (1) also describes two particles with spin s but without any spin coupling between the particles when they meet (i.e. for $x = 0$), in this case φ represents any of the components of the wave function. The values $\theta = b = 0$, $a = d = 1$ in (1) correspond to the case of a positive (resp. negative) δ-function potential for $c > 0$ (resp. $c < 0$). For general a, b, c and d, the properties of the corresponding Hamiltonian systems have been studied in detail, see e.g. [7, 8, 9].

The separated boundary conditions are described by

$$\varphi'(0_+) = q^+\varphi(0_+) \ , \ \varphi'(0_-) = q^-\varphi(0_-), \tag{3}$$

where $q^\pm \in \mathbb{R} \cup \{\infty\}$. $q^+ = \infty$ or $q^- = \infty$ correspond to Dirichlet boundary conditions and $q^+ = 0$ or $q^- = 0$ correspond to Neumann boundary conditions.

To study the integrability of one-dimensional systems of N-identical particles with general contact interactions described by the boundary conditions (1) or (3) that are imposed on the relative coordinates of the particles, we first consider the case of two particles ($N = 2$) with coordinates x_1, x_2 and momenta k_1, k_2 respectively. Each particle has n-'spin' states designated by s_1 and s_2, $1 \leq s_i \leq n$. For $x_1 \neq x_2$, these two particles are free. The wave functions φ are symmetric (resp. antisymmetric) with respect to the interchange $(x_1, s_1) \leftrightarrow (x_2, s_2)$ for bosons (resp. fermions). In the region $x_1 < x_2$, from the Bethe ansatz the wave function is of the form,

$$\varphi = u_{12}e^{i(k_1x_1+k_2x_2)} + u_{21}e^{i(k_2x_1+k_1x_2)}, \tag{4}$$

where u_{12} and u_{21} are $n^2 \times 1$ column matrices. In the region $x_1 > x_2$, the wave function has the form

$$\varphi = (P^{12}u_{12})e^{i(k_1x_2+k_2x_1)} + (P^{12}u_{21})e^{i(k_2x_2+k_1x_1)}, \tag{5}$$

where according to the symmetry or antisymmetry conditions, $P^{12} = p^{12}$ for bosons and $P^{12} = -p^{12}$ for fermions, p^{12} being the operator on the $n^2 \times 1$ column that interchanges $s_1 \leftrightarrow s_2$.

Let $k_{12} = (k_1 - k_2)/2$. In the center of mass coordinate $X = (x_1 + x_2)/2$ and the relative coordinate $x = x_2 - x_1$, we get, by substituting (4) and (5) into the boundary conditions at $x = 0$,

$$u_{21} = Y_{21}^{12}u_{12} \ , \tag{6}$$

$$Y_{21}^{12} = \frac{2ie^{i\theta}k_{12}P^{12} + ik_{12}(a-d) + (k_{12})^2b + c}{ik_{12}(a+d) + (k_{12})^2b - c} \tag{7}$$

for boundary condition (1) and

$$Y_{21}^{12} = \frac{ik_{12} + q}{ik_{12} - q} \tag{8}$$

for boundary condition (3), where $q \equiv q_+ = -q_- \in \mathbb{R} \cup \{\infty\}$.

For $N \geq 3$ and $x_1 < x_2 < \cdots < x_N$, the wave function is given by

$$\begin{aligned} \psi &= u_{12\cdots N} e^{i(k_1 x_1 + k_2 x_2 + \cdots + k_N x_N)} + u_{21\cdots N} e^{i(k_2 x_1 + k_1 x_2 + \cdots + k_N x_N)} \\ &\quad + (N! - 2) \textit{ other terms}. \end{aligned} \tag{9}$$

The columns u have $n^N \times 1$ dimensions. The wave functions in the other regions are determined from (9) by the requirement of symmetry (for bosons) or antisymmetry (for fermions). Along any plane $x_i = x_{i+1}$, $i \in 1, 2, \ldots, N-1$, from similar considerations as above we have

$$\alpha_{\alpha_1 \alpha_2 \cdots \alpha_i \alpha_{i+1} \cdots \alpha_N} = Y^{ii+1}_{\alpha_{i+1}\alpha_i} u_{\alpha_1 \alpha_2 \cdots \alpha_{i+1} \alpha_i \cdots \alpha_N}, \tag{10}$$

where

$$Y^{ii+1}_{\alpha_{i+1}\alpha_i} = \frac{2ie^{i\theta} k_{\alpha_i \alpha_{i+1}} P^{ii+1} + ik_{\alpha_i\alpha_{i+1}}(a-d) + (k_{\alpha_i\alpha_{i+1}})^2 b + c}{ik_{\alpha_i\alpha_{i+1}}(a+d) + (k_{\alpha_i\alpha_{i+1}})^2 b - c} \tag{11}$$

for nonseparated boundary condition and

$$Y^{ii+1}_{\alpha_{i+1}\alpha_i} = \frac{ik_{\alpha_i\alpha_{i+1}} + q}{ik_{\alpha_i\alpha_{i+1}} - q} \tag{12}$$

for separated boundary condition. Here $k_{\alpha_i\alpha_{i+1}} = (k_{\alpha_i} - k_{\alpha_{i+1}})/2$ play the role of spectral parameters. $P^{ii+1} = p^{ii+1}$ for bosons and $P^{ii+1} = -p^{ii+1}$ for fermions, with p^{ii+1} the operator on the $n^N \times 1$ column that interchanges $s_i \leftrightarrow s_{i+1}$.

For consistency Y must satisfy the Yang-Baxter equation with spectral parameter [10, 13, 14, 15, 16], i.e.,

$$Y^{m,m+1}_{ij} Y^{m+1,m+2}_{kj} Y^{m,m+1}_{ki} = Y^{m+1,m+2}_{ki} Y^{m,m+1}_{kj} Y^{m+1,m+2}_{ij},$$

or

$$Y^{mr}_{ij} Y^{rs}_{kj} Y^{mr}_{ki} = Y^{rs}_{ki} Y^{mr}_{kj} Y^{rs}_{ij} \tag{13}$$

if m, r, s are all unequal, resp.

$$Y^{mr}_{ij} Y^{mr}_{ji} = 1, \quad Y^{mr}_{ij} Y^{sq}_{kl} = Y^{sq}_{kl} Y^{mr}_{ij} \tag{14}$$

if m, r, s, q are all unequal.

The operators Y given by (11) satisfy the relation (14) for all θ, a, b, c, d. However the relations (13) are satisfied only when $\theta = 0$, $a = d$ and $b = 0$, that is, according to the constraint (2), $\theta = 0$, $a = d = \pm 1$, $b = 0$, c arbitrary. The case $a = d = 1$, $\theta = b = 0$ corresponds to the usual δ-function interactions, which has been investigated in [10, 11, 12]. The case $a = d = -1$, $\theta = b = 0$ is related to singular interactions of another type (for $a = d = -1$ and $\theta = b = c = 0$ see [7, 8]), which is in fact unitarily equivalent to the δ-interaction, under a non-smooth "kink type" gauge transformation $U = \prod_{i>j} \mathrm{sgn}(x_i - x_j)$. Associated with the separated boundary condition, the operators Y given by (12) satisfy both the relations (13) and (14) for arbitrary q. Therefore with respect to N-particle (either boson or fermion) problems, there are two non-equivelant integrable one parameter families with contact interactions described respectively by one of the following conditions on the wave function along the plane $x_i = x_j$ for any pair of particles with coordinates x_i and x_j,

$$\varphi(0_+) = +\varphi(0_-),\ \varphi'(0_+) = c\varphi(0_-) + \varphi'(0_-)\ ,\ c \in \mathbb{R}\ ; \tag{15}$$

$$\varphi'(0_+) = q\varphi(0_+),\ \varphi'(0_-) = -q\varphi(0_-)\ ,\ q \in \mathbb{R} \cup \{\infty\}\ . \tag{16}$$

The wave functions are given by (9) with the u's determined by (10) and initial conditions. The operators Y in (10) are given respectively by

$$Y^{ii+1}_{\alpha_{i+1}\alpha_i} = \frac{i(k_{\alpha_i} - k_{\alpha_{i+1}})P^{ii+1} + c}{i(k_{\alpha_i} - k_{\alpha_{i+1}}) - c}\ ; \tag{17}$$

and

$$Y^{ii+1}_{\alpha_{i+1}\alpha_i} = \frac{i(k_{\alpha_i} - k_{\alpha_{i+1}}) + 2q}{i(k_{\alpha_i} - k_{\alpha_{i+1}}) - 2q}\ . \tag{18}$$

When $q < 0$, there exist $2^{N(N-1)/2}$ bound states for the case (18) of separated boundary conditions, with wavefunction

$$\psi_{N,\underline{\epsilon}} = u_{\underline{\epsilon}} \prod_{k>l} (\theta(x_k - x_l) + \epsilon_{kl}\theta(x_l - x_k)) e^{q\sum_{i>j}|x_i - x_j|} \tag{19}$$

and eigenvalue $E = -q^2N(N^2-1)/3$, where $u_{\underline{\epsilon}}$ is the spin wave function and $\underline{\epsilon} \equiv \{\epsilon_{kl}\ :\ k > l\}$; $\epsilon_{kl} = \pm$, labels the $2^{N(N-1)/2}$-fold degeneracy.

We consider now the case of spin dependent boundary conditions. For a particle with spin s, the wave function has $n = 2s+1$ components. Therefore two particles with contact interactions have a general boundary condition described in the center of mass coordinate system by:

$$\begin{pmatrix} \psi \\ \psi' \end{pmatrix}_{0+} = \begin{pmatrix} A & B \\ C & D \end{pmatrix} \begin{pmatrix} \psi \\ \psi' \end{pmatrix}_{0-}, \tag{20}$$

where ψ and ψ' are n^2-dimensional column vectors, A, B, C and D are $n^2 \times n^2$ matrices. The boundary condition (20) can include not only the usual contact interaction between the particles, but also a spin coupling of the two particles if the matrices A, B, C, D are not diagonal.

The matrices A, B, C, and D are subject to restrictions due to the required symmetry condition of the Schrödinger operator. For any $u, v \in C^\infty(\mathbb{R} \setminus \{0\})$, $< -\frac{d^2}{dx^2}u, v >_{L_2(\mathbb{R},\mathbb{C}^n)} - < u, -\frac{d^2}{dx^2}v >_{L_2(\mathbb{R},\mathbb{C}^n)} = 0$, which, together with (20) imply

$$A^\dagger D - C^\dagger B = 1,\ B^\dagger D = D^\dagger B,\ A^\dagger C = C^\dagger A, \tag{21}$$

where $\dagger$ stands for the conjugate and transpose. Obviously (1) is the special case of (20) for $s = 0$.

In the following we study quantum systems with contact interactions described by the boundary condition (20), in particular, N-body systems with δ-interactions. We first consider two spin-s particles with δ-interactions. The Hamiltonian is then of the form

$$H = (-\frac{\partial^2}{\partial x_1^2} - \frac{\partial^2}{\partial x_2^2})\mathbf{I}_2 + 2h\delta(x_1 - x_2), \tag{22}$$

where $\mathbf{I}_2$ is the $n^2 \times n^2$ identity matrix, h is an $n^2 \times n^2$ Hermitian matrix. If the matrix h is proportional to the unit matrix $\mathbf{I}_2$, then H is reduced to the usual two-particle Hamiltonian with contact interactions but no spin coupling.

Let e_α, $\alpha = 1, \ldots, n$, be the basis (column) vector with the α-th component as 1 and the rest components 0. The wave function of the system (22) is of the form

$$\psi = \sum_{\alpha,\beta=1}^{n} \phi_{\alpha\beta}(x_1, x_2) e_\alpha \otimes e_\beta. \tag{23}$$

In the center of mass coordinate system, the operator (22) has the form

$$H = -\left(\frac{1}{2}\frac{\partial^2}{\partial X^2} + 2\frac{\partial^2}{\partial x^2}\right)\mathbf{I}_2 + 2h\delta(x). \tag{24}$$

The functions $\phi = \phi(x, X)$ from the domain of this operator satisfy the following boundary condition at $x = 0$,

$$\phi'_{\alpha\beta}(0^+, X) - \phi'_{\alpha\beta}(0^-, X) = \sum_{\alpha,\beta=1}^{n} h_{\gamma\lambda,\alpha\beta}\phi_{\gamma\lambda}(0, X),\ \phi_{\alpha\beta}(0^+, X) = \phi_{\alpha\beta}(0^-, X), \tag{25}$$

$\alpha, \beta = 1, \ldots, n$, where the indices of the matrix h are arranged as $11, 12, \ldots, 1n$; $21, 22, \ldots, 2n$; $\ldots$; $n1, n2, \ldots, nn$. (25) is a special case of (20) for $A = D = \mathbf{I}_2$, $B = 0$ and $C = h$. h acts on the basis vector of particles 1 and 2 by $he_\alpha \otimes e_\beta = \sum_{\gamma,\lambda=1}^{n} h_{\alpha\beta,\gamma\lambda} e_\gamma \otimes e_\lambda$.

The wave functions are still of the forms (4) (resp. 5) in the region $x_1 < x_2$ (resp. $x_1 > x_2$). Substituting them into the boundary conditions (25), we get

$$\left\{\begin{array}{l} u_{12} + u_{21} = P^{12}(u_{12} + u_{21}), \\ ik_{12}(u_{21} - u_{12}) = hP^{12}(u_{12} + u_{21}) + ik_{12}P^{12}(u_{12} - u_{21}). \end{array}\right. \tag{26}$$

Eliminating the term $P^{12}u_{12}$ from (26) we obtain the same relation as (6), $u_{21} = Y_{21}^{12}u_{12}$. Nevertheless the Y operator is given by

$$Y_{21}^{12} = [2ik_{12} - h]^{-1}[2ik_{12}P^{12} + h]. \tag{27}$$

For a system of N identical particles with δ-interactions, the Hamiltonian is given by

$$H = -\sum_{i=1}^{N} \frac{\partial^2}{\partial x_i^2}\mathbf{I}_N + \sum_{i<j}^{N} h_{ij}\delta(x_i - x_j), \tag{28}$$

where $\mathbf{I}_N$ is the $n^N \times n^N$ identity matrix, h_{ij} is an operator acting on the i-th and j-th bases as h and the rest as identity, e.g., $h_{12} = h \otimes \mathbf{1}_3 \otimes \cdots \mathbf{1}_N$, with $\mathbf{1}_i$ the $n \times n$ identity matrix acting on the i-th basis. The wave function in a given region, say $x_1 < x_2 < \cdots < x_N$, is of the form (9), with

$$u_{\alpha_1\alpha_2\cdots\alpha_j\alpha_{j+1}\cdots\alpha_N} = Y^{jj+1}_{\alpha_{j+1}\alpha_j} u_{\alpha_1\alpha_2\cdots\alpha_{j+1}\alpha_j\cdots\alpha_N} \tag{29}$$

and

$$Y^{jj+1}_{\alpha_{j+1}\alpha_j} = [2ik_{\alpha_j\alpha_{j+1}} - h_{jj+1}]^{-1}[2ik_{\alpha_j\alpha_{j+1}} P^{jj+1} + h_{jj+1}]. \tag{30}$$

From the Yang-Baxter equations it is straightforward to show that the operator Y given by (30) satisfies all the Yang-Baxter relations if

$$[h_{ij}, P^{ij}] = 0. \tag{31}$$

Therefore if the Hamiltonian operators for the spin coupling commute with the spin permutation operator, the N-body quantum system (28) can be exactly solved. The wave function is then given by (9) and (29) with the energy $E = \sum_{i=1}^{N} k_i^2$.

For the case of spin-$\frac{1}{2}$, a Hermitian matrix satisfying (31) is generally of the form

$$h^{\frac{1}{2}} = \begin{pmatrix} a & e_1 & e_1 & c \\ e_1^* & f & g & e_2 \\ e_1^* & g & f & e_2 \\ c^* & e_2^* & e_2^* & b \end{pmatrix}, \tag{32}$$

where $a, b, c, f, e_1, e_2 \in \mathbb{C}$, $g \in \mathbb{R}$. We recall that for a complex vector space V, a matrix R taking values in $End_c(V \otimes V)$ is called a solution of the Yang-Baxter equation without spectral parameters, if it satisfies $R_{12}R_{13}R_{23} = R_{23}R_{13}R_{12}$, where R_{ij} denotes the matrix on the complex vector space $V \otimes V \otimes V$, acting as R on the i-th and the j-th components and as identity on the other components. When V is a two-dimensional complex space, the solutions of the Yang-Baxter equation include the ones such as R_q which gives rise to the quantum algebra $SU_q(2)$ and the integrable Heisenberg spin-$\frac{1}{2}$ chain models such as the XXZ model (R corresponds to the spin coupling operator between the nearest neighbor spins in Heisenberg spin chain models) [13, 14, 15, 16]. Nevertheless in general $h^{\frac{1}{2}}$ does not satisfy the Yang-Baxter equation without spectral parameters: $h^{\frac{1}{2}}_{12}h^{\frac{1}{2}}_{13}h^{\frac{1}{2}}_{23} \neq h^{\frac{1}{2}}_{23}h^{\frac{1}{2}}_{13}h^{\frac{1}{2}}_{12}$. But (32) includes the Yang-Baxter solutions, such as R_q, that gives integrable spin chain models (for an extensive investigation of the Yang-Baxter solutions see [17, 18]). Therefore for an N-body system to be integrable, the spin coupling in the contact interaction (28) is allowed to be more general than the spin coupling in a Heisenberg spin chain model with nearest neighbors interactions.

For $N = 2$, from (26) the bound states have the form,

$$\psi^2_\alpha = u_\alpha e^{\frac{c+a\Lambda_\alpha}{2}|x_2-x_1|}, \quad \alpha = 1, \ldots, n^2, \tag{33}$$

where u_α is the common α-th eigenvector of h and P^{12}, with eigenvalue Λ_α, s.t. $hu_\alpha = \Lambda_\alpha u_\alpha$ and $c + a\Lambda_\alpha < 0$, $P^{12}u_\alpha = u_\alpha$. The eigenvalue of the Hamitonian H

corresponding to the bound state (33) is $-(c+a\Lambda_\alpha)^2/2$. We remark that, whereas for the case of the boundary condition (1), for a δ interaction one has a unique bound state, here we have n^2 bound states. By generalization we get the bound state for the N-particle system,

$$\psi_\alpha^N = v_\alpha e^{-\frac{c+a\Lambda_\alpha}{2}\sum_{i>j}|x_i-x_j|},\ \alpha = 1,\ldots,n^2, \tag{34}$$

where v_α is the wave function of the spin part satisfying $P^{ij}v_\alpha = v_\alpha$ and $h_{ij}v_\alpha = \Lambda_\alpha v_\alpha$, for any $i \neq j$.

It is worth mentioning that ψ_α^N is of the form (9) in each of the above regions. For instance comparing ψ_α^N with (9) in the region $x_1 < x_2 \cdots < x_N$ we get

$$k_1 = -i\frac{c+a\Lambda_\alpha}{2}(N-1),\ k_2 = k_1 + ic,\ k_3 = k_2 + ic, \ldots, k_N = -k_1, \tag{35}$$

for $\alpha = 1,\ldots,n^2$. The energy of the bound state ψ_α^N is

$$E_\alpha = -\frac{(c+a\Lambda_\alpha)^2}{12}N(N^2-1). \tag{36}$$

Now we pass to the scattering matrix. For real $k_1 < k_2 < \cdots k_N$, in each coordinate region such as $x_1 < x_2 < \cdots x_N$, the following term in (9) describes an outgoing wave $\psi_{out} = u_{12\cdots N}e^{i(k_1x_1+\cdots+k_Nx_N)}$. An incoming wave with the same exponential as ψ_{out} is given by $\psi_{in} = [P^{1N}P^{2(N-1)}\cdots]u_{N(N-1)\cdots 1}e^{i(k_Nx_N+\cdots+k_1x_1)}$ in the region $x_N < x_{N-1} < \cdots < x_1$. From (29) the scattering matrix S defined by $\psi_{out} = S\psi_{in}$ is given by $S = [X_{21}X_{31}\cdots X_{N1}][X_{32}X_{42}\cdots X_{N2}]\cdots[X_{N(N-1)}]$, where $X_{ij} - Y_{ij}^{ij}P^{ij}$.

The scattering matrix S is unitary and symmetric due to the time reversal invariance of the interactions. $< s_1's_2'\cdots s_N'|S|s_1s_2\cdots s_N >$ stands for the S matrix element of the process from the state $(k_1s_1, k_2s_2, \ldots, k_Ns_N)$ to the state $(k_1s_1', k_2s_2', \ldots, k_Ns_N')$.

The scattering of clusters (bound states) can be discussed in a similar way as in [12]. For instance for the scattering of a bound state of two particles ($x_1 < x_2$) on a bound state of three particles ($x_3 < x_4 < x_5$), the scattering matrix is $S = [X_{32}X_{42}X_{52}][X_{31}X_{41}X_{51}]$.

The integrability of many particles systems with contact spin coupling interactions governed by separated boundary conditions can also be studied. Instead of (20) we need to deal with the case

$$\phi'(0_+) = G^+\phi(0_+),\ \phi'(0_-) = G^-\phi(0_-), \tag{37}$$

where $G^\pm$ are Hermitian matrices. For $G^+ = G^- \equiv G$, $G^\dagger = G$, there is a Bethe Ansatz solution to (9) with $Y_{\alpha_{i+1}\alpha_i}^{ii+1}$ in (29) given by

$$Y_{\alpha_{i+1}\alpha_i}^{ii+1} = \frac{ik_{\alpha_i\alpha_{i+1}} + G}{ik_{\alpha_i\alpha_{i+1}} - G}. \tag{38}$$

Let Γ be the set of n^2 eigenvalues of G. For any $\lambda_\alpha \in \Gamma$ such that $\lambda_\alpha < 0$, there are $2^{N(N-1)/2}$ bound states for the N-particle system,

$$\psi^N_{\alpha\underline{\epsilon}} = v_{\alpha\underline{\epsilon}} \prod_{k>l} (\theta(x_k - x_l) + \epsilon_{kl}\theta(x_l - x_k))e^{\lambda_\alpha \sum_{i>j} |x_i - x_j|}, \tag{39}$$

where $v_{\alpha\underline{\epsilon}}$ is the spin wave function and $\underline{\epsilon} \equiv \{\epsilon_{kl} \ : \ k > l\}$; $\epsilon_{kl} = \pm$, labels the $2^{N(N-1)/2}$-fold degeneracy. The spin wave function v here satisfies $P^{ij}v_{\alpha\underline{\epsilon}} = \epsilon_{ij}v_{\alpha\underline{\epsilon}}$ for any $i \neq j$, that is, $p^{ij}v_{\alpha\underline{\epsilon}} = \epsilon_{ij}v_{\alpha\underline{\epsilon}}$ for bosons and $p^{ij}v_{\alpha\underline{\epsilon}} = -\epsilon_{ij}v_{\alpha\underline{\epsilon}}$ for fermions.

Again $\psi^N_{\alpha\underline{\epsilon}}$ is of the form (9) in each of the regions $x_{i_1} < x_{i_2} < \cdots < x_{i_N}$. For instance comparing $\psi^N_{\alpha\underline{\epsilon}}$ with (9) in the region $x_1 < x_2 \cdots < x_N$ we get $k_1 = i\lambda_\alpha(N-1)$, $k_2 = k_1 - 2i\lambda_\alpha$, $k_3 = k_2 - 2i\lambda_\alpha$,…,$k_N = -k_1$. The energy of the bound state $\psi^N_{\alpha\underline{\epsilon}}$ is $E_\alpha = -\lambda^2_\alpha N(N^2-1)/3$.

We have investigated the integrable models of N-body systems with contact spin coupling interactions. Without taking into account the spin coupling, the boundary condition (1) is characterized by four parameters (separated boundary conditions are a special limiting case of these). Obviously the general boundary condition (20) we considered in this article has much more parameters. A complete classification of the dynamic operators associated with different parameter regions remains to be done. As we have seen, the case $A = D = \mathbf{I}_2$, $B = 0$, $C = h$ corresponds to a Hamiltonian with δ-interactions of the form (22) (for $N = 2$). It can be further shown that (for $N = 2$) the following boundary condition

$$\begin{pmatrix} \psi \\ \psi' \end{pmatrix}_{0+} = \begin{pmatrix} \mathbf{I} & B \\ 0 & \mathbf{I} \end{pmatrix} \begin{pmatrix} \psi \\ \psi' \end{pmatrix}_{0-} \tag{40}$$

corresponds to a Hamiltonian H of the form: $H = -D_x^2(1 + B\delta) - BD_x\delta'$, where B is an $n^2 \times n^2$ Hermitian matrix, D_x is defined by $(D_x f)(\varphi) = -f(\frac{d}{dx}\varphi)$, for $f \in C_0^\infty(\mathbb{R}/\{0\})$ and φ a test function with a possible discontinuity at the origin. The boundary condition

$$\begin{pmatrix} \psi \\ \psi' \end{pmatrix}_{0+} = \begin{pmatrix} \frac{2+iB}{2-iB} & 0 \\ 0 & \frac{2-iB}{2+iB} \end{pmatrix} \begin{pmatrix} \psi \\ \psi' \end{pmatrix}_{0-} \tag{41}$$

describes the Hamiltonian $H = -D_x^2 + iB(2D_x\delta - \delta')$.

Acknowledgements. S.M. Fei would like to thank DFG, Max-Planck-Institute for Mathematics in the Sciences, Leipzig for financial supports and warm hospitalities.

References

[1] S. Albeverio, F. Gesztesy, R. Høegh-Krohn and H. Holden, *Solvable Models in Quantum Mechanics*, New York: Springer, 1988.

[2] S. Albeverio and R. Kurasov, *Singular perturbations of differential operators and solvable Schrödinger type operators*, Cambridge Univ. Press, 1999.

[3] M. Gaudin, *La fonction d'onde de Bethe*, Masson, 1983.

[4] S. Albeverio, L. Dąbrowski and S.M. Fei, *One-dimensional Many-Body Problems with Point Interactions*, to appear in Int. J. Mod. Phys. B.

[5] S. Albeverio, S.M. Fei and P. Kurasov, *N-Body Problems with "Spin"-Related Contact Interactions in One-dimensional*, SFB 265-preprint 1999, to appear in Rep. Math. Phys..

[6] S. Albeverio, S.M. Fei and P. Kurasov, *Gauge Fields, Point Interactions and Few-Body Problems in One Dimension*, preprint 1999.

[7] P. Kurasov, Distribution theory with discontinuous test functions and differential operators with generalized coefficients, *J. Math. Analys. Appl.*, **201** (1996), 297–333.

[8] P.R. Chernoff and R.J. Hughes, A new class of point interactions in one dimension, *J. Funct. Anal.*, **111** (1993), 97–117.

[9] S. Albeverio, Z. Brzeźniak and L. Dąbrowski, Time-dependent propagator with point interaction, *J. Phys.* **A 27** (1994), 4933–4943.

[10] C.N. Yang, *Some exact results for the many-body problem in one dimension with repulsive delta-function interaction*, *Phys. Rev. Lett.*, **19** (1967), 1312–1315.

[11] C.N. Yang, *S matrix for the one-dimensional N-body problem with repulsive δ-function interaction*, *Phys. Rev.*, **168**(1968)1920–1923.

[12] C.H. Gu and C.N. Yang, *A one-dimensional N Fermion problem with factorized S matrix*, Commun. Math. Phys. **122** (1989), 105–116.

[13] V. Chari and A. Pressley, *A Guide to Quantum Groups*, Cambridge University Press, 1994.

[14] C. Kassel, *Quantum Groups*, Springer-Verlag, New York, 1995.

[15] Z.Q. Ma, *Yang-Baxter Equation and Quantum Enveloping Algebras*, World Scientific, 1993.

[16] S. Majid, *Foundations of Quantum Group Theory*, Cambridge University Press, 1995.

[17] S.M. Fei, H.Y. Guo and H. Shi, *Multiparameter Solutions of the Yang-Baxter Equation*, *J. Phys.* **A 25**(1992), 2711–2720.

[18] J. Hietarinta, *All solutions to the constant quantum Yang-Baxter equation in two dimensions*, *Phys. Lett.* A **165**(1992), 245–251.

S. Albeverio
Institut für Angewandte Mathematik
Universität Bonn
D-53115 Bonn
SFB 256; SFB 237;
BiBoS; CERFIM (Locarno);
Acc. Arch., USI (Mendrisio)
e-mail: albeverio@uni-bonn.de

S.M. Fei
Institut für Angewandte Mathematik
Universität Bonn
D-53115 Bonn

Max-Planck-Institute for Mathematics in the Sciences
Inselstr. 22–26
04103 Leipzig
e-mail: fei@uni-bonn.de
e-mail: smfei@mis.mpg.de

P. Kurasov
Dept. of Math.
Stockholm University
10691 Stockholm
Sweden
Current address:
Dept. of Math.
Lund Institute of Technology
Box 118
221 00 Lund
Sweden
e-mail: pak@matematik.su.se

Operator Theory:
Advances and Applications, Vol. 132, 77–86

Generalized Point Models in Boundary Contact Value Problems of Hydroelasticity

Ivan V. Andronov

Abstract. The generalized point model of a short joint of two fluid loaded semi-infinite thin elastic plates is constructed in the form of the zero-range potential.

1. Introduction

The boundary value problems with the generalized impedance boundary conditions (GIBC) require so-called contact conditions to be formulated in all points where the coefficients of GIBC have jumps or the smoothness of the boundary is violated. Such boundary contact value problems appeared first in hydroelasticity [1, 2, 3] when the vibrations of thin elastic plates or shells are described by approximate theories that take into account only some of the possible types of waves. The Kirchhoff theory, for example, assumes that all the particles in the cross-section of a plate move simultaneously and only in the orthogonal to the plate direction. Such assumption brings to GIBC of the 4th order. The point models of cracks [3], stiffeners [1, 2] and other inhomogeneities in thin elastic plates or shells are formulated as second order boundary conditions. These conditions describe the mechanical contact of the parts of elastic construction. However these models do not take into account the real size of the obstacle and therefore have a restricted field of applications.

The use of zero-range potentials [4, 5] extends the class of the point models and allows the models that include corrections for the geometrical size of the inhomogeneity to be formulated.

A peculiarity of the zero-range potentials construction in the boundary contact value problems is in the multichannel character of the scattering problem (see e.g. [6]). One channel corresponds to spatial acoustic waves in fluid and another corresponds to flexure waves in the infinite plate. There could also be presented additional channels associated with infinite stiffeners, cracks or edges of the plate. The multichannel character of the problem induces the matrix structure of the initial operator which is inherited then by the zero-range potentials. The "boundary" conditions in the potential center are parameterized by Hermitian matrices. It is essential that if the "boundary" conditions are written for the coefficients of the local asymptotics of the field, the matrix appears independent of the spectral parameter (frequency) and global characteristics of the problem, it is defined only by the inhomogeneity.

In the theory of zero-range potentials there is no universal tool for the choice of particular parameters adequately to the real inhomogeneity. The choice of parameters is made usually with the use of asymptotics matching procedure or by comparing the far field amplitudes. The noted matrix structure of the main operator gives the possibility to choose the blocks of the parameterization matrix separately. Physically the diagonal blocks of this matrix describe the influence of the inhomogeneity on acoustic pressure and on flexure vibrations separately. Thus, instead of examining the boundary value problem for a fluid loaded elastic construction one can consider two more simple problems: one is for the case of absolutely rigid construction in fluid, the other is for the elastic construction in vacuum. The nondiagonal blocks of the parameterization matrices describe additional interaction of acoustic and flexure channels that appears due to the presence of the inhomogeneity. In a number of cases such interaction is either not presented, or is of smaller order and is neglected by the zero-range potential model.

The constructed models can be then used in the nonstationary problems of scattering, two-dimensional models also can be extended to three-dimensional problems with symmetries, etc.

The generalized point model of a narrow crack is constructed in [7], and the model of a short joint of two semi-infinite plates is presented in this paper. Analytic and numerical analysis of the scattered fields allows the applicability of classical point models to be examined.

2. Formulation of the problem

Consider the mechanical structure composed of two semi-infinite plates $\Pi^{\pm}$ joint along a short segment as shown on Figure 1. Let this structure be in contact to fluid at $\{z > 0\}$. The plates are assumed thin and only flexure displacements are taken into account and are described by the Kirchhoff model. The edges of the plates Π^{+} and Π^{-} are assumed free. On the joint the displacements are continuous. The acoustic pressure $U(x, y, z)$ and flexure displacements $\xi(x, y)$ satisfy the following boundary value problem

$$\begin{gathered} c^2 \Delta U + \omega^2 U = 0, \quad z > 0, \\ D\left(\Delta^2 - m\omega^2\right)\xi + \rho U|_{z=0} = 0, \quad y \neq 0, \\ \xi = \frac{\partial U(x,y,0)}{\partial z}, \quad y \neq 0, \end{gathered} \tag{1}$$

$$\xi_{yyy} + (2-\sigma)\xi_{xxy} = 0, \quad \xi_{yy} + \sigma\xi_{xx} = 0, \quad y = \pm 0, |x| > a, \tag{2}$$

$$\xi(x,+0) = \xi(x,-0), \quad \frac{\partial^j \xi(x,+0)}{\partial y^j} = \frac{\partial^j \xi(x,-0)}{\partial y^j}, \quad |x| < a, \ j = 1,2,3. \tag{3}$$

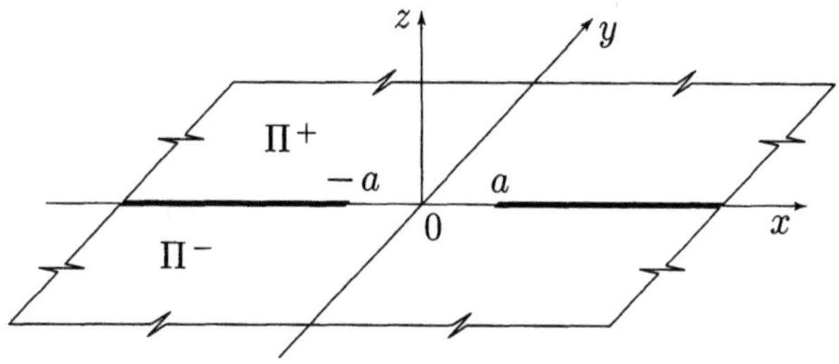

FIGURE 1. Geometry of the problem

Here ω is frequency, c is sound velocity in fluid, D is the bending stiffness of the plate, m is the mass of the plate cross-section, ρ is density of fluid.

The main goal is to construct the point model of such joint, that is spread the conditions (2) to $|x| < a$ and replace (3) by some conditions formulated in a point.

3. Zero-range potentials in isolated plates

Note that the classical point model of the joint is formulated as the continuity condition

$$\xi(0, +0) = \xi(0, -0).$$

It gives the principal order term in the asymptotics of the far field amplitude. This term does not depend on the width of the joint. As it will be shown below (see (9)) the next order terms are only logarithmically smaller. Thus the classical point model describe only very narrow joints.

The generalized point model is constructed first for the isolated mechanical construction. The space is

$$L = L_2(\Pi^+) \oplus L_2(\Pi^-).$$

The initial operator is the biharmonic operator

$$\mathbf{A} = \Delta^2$$

defined on functions from

$$D(\mathbf{A}) = H_f^4(\Pi^+) \oplus H_f^4(\Pi^-).$$

Here subscript $_f$ denotes the free edge boundary conditions

$$\xi_{yyy} + (2 - \sigma)\xi_{xxy} = 0, \quad \xi_{yy} + \sigma\xi_{xx} = 0, \qquad y = 0.$$

The spectral problem for the operator $\mathbf{A}$ describes vibrations of the separated semi-infinite plates Π^+ and Π^-. In order to introduce interaction between flexure displacements in Π^+ and flexure displacements in Π^- the zero-range potential is introduced.

Following the standard scheme [4, 5] one restricts the operator $\mathbf{A}$ to functions that vanish near the origin. Let the operator $\mathbf{A}_0$ be introduced as the restriction of $\mathbf{A}$ to the domain

$$D(\mathbf{A}_0) = \{\xi \in D(\mathbf{A}),\ \xi(0,\pm 0) = 0,\ \xi_x(0,\pm 0) = 0,\ \xi_y(0,\pm 0) = 0\}.$$

It is possible to consider further restrictions of $\mathbf{A}$ in L with vanishing second order derivatives.

The operator $\mathbf{A}_0$ is symmetric. Its deficiency indices can be found equal to 6. The adjoint operator $\mathbf{A}_0^*$ is constructed as follows. Following the Neumann formula functions from the domain of the adjoint operator are decomposed as the sum of deficiency elements and functions from $D(\mathbf{A}_0)$

$$\xi = \sum_{p=\pm}\sum_{j=0}^{2} \left(\left(c_j^p + a_j^p\right) \phi_j^p(i) + \left(c_j^p - a_j^p\right) \phi_j^p(-i) \right) + \xi_0. \tag{4}$$

Here $c_j^\pm$ and $a_j^\pm$ are arbitrary complex coefficients, six functions $\phi_j^\pm(\lambda)$ are the deficiency elements corresponding to spectral parameter λ and function ξ_0 belongs to the domain $D(\mathbf{A}_0)$.

It is preferable to characterize the operator with the help of the asymptotics of functions from $D(\mathbf{A}_0^*)$ in a vicinity of the origin

$$\begin{aligned}
\xi \ \sim\ & c_0^\pm \frac{r^2}{4\pi}\left(\ln(r) + \frac{1+\sigma}{1-\sigma}\left(\ln(r)\cos(2\varphi) - \left(\varphi \mp \frac{\pi}{2}\right)\sin(2\varphi) - \cos(2\varphi)\right)\right) \\
+\ & c_1^\pm \frac{1}{2\pi}\left(\frac{2}{1-\sigma} r\ln(r)\cos\varphi - \frac{1+\sigma}{1-\sigma} r\left(\varphi \mp \frac{\pi}{2}\right)\sin\varphi\right) \\
+\ & c_2^\pm \frac{1}{2\pi}\left(\frac{2}{1-\sigma} r\ln(r)\sin\varphi + \frac{1+\sigma}{1-\sigma} r\left(\varphi \mp \frac{\pi}{2}\right)\cos\varphi\right) \\
+\ & a_0^\pm + a_1^\pm r\cos\varphi + a_2^\pm r\sin\varphi + \cdots, \qquad r \to 0.
\end{aligned} \tag{5}$$

Here two asymptotic representations are combined. When polar angle $\varphi \in (0,\pi)$ the coefficients have superscript $+$ and when $\varphi \in (-\pi, 0)$ the coefficients have superscript $-$.

The operator $\mathbf{A}_0^*$ acts on functions (4) according to the following formula

$$\mathbf{A}_0^*\xi = i\sum_{p=\pm}\sum_{j=0}^{2} \left(\left(c_j^p + a_j^p\right) \phi_j^p(i) - \left(c_j^p - a_j^p\right) \phi_j^p(-i) \right) + \Delta^2\xi_0.$$

The adjoint operator $\mathbf{A}_0^*$ is not symmetric. Its boundary form is given by the following expression (overline denotes complex conjugation)

$$(A^*\xi,\eta) - (\xi, A^*\eta) = \sum_{p=\pm}\sum_{j=0}^{2} \left(c_j^p(\xi)\overline{a_j^p(\eta)} - a_j^p(\xi)\overline{c_j^p(\eta)} \right). \tag{6}$$

Note that if ξ or η in (6) belongs to $D(\mathbf{A}_0)$ then all the coefficients $c_j^\pm$ and $a_j^\pm$ of the corresponding function are equal to zero and the boundary form vanishes. That confirms that the operator $\mathbf{A}_0^*$ is adjoint to $\mathbf{A}_0$.

The self-adjoint operators can be constructed as restrictions of $\mathbf{A}_0^*$ to such functions for which the boundary form vanishes. We shall consider only such restrictions that conserve the flux through the joint, that is

$$c_j^- = -c_j^+, \qquad j = 0, 1, 2.$$

Under this assumption the self-adjoint restrictions of the operator $\mathbf{A}_0^*$ are parameterized by 3×3 Hermitian matrices $\mathbf{S}$ in the form

$$\mathbf{S}\begin{pmatrix} c_0 \\ c_1 \\ c_2 \end{pmatrix} = \begin{pmatrix} a_0^+ - a_0^- \\ a_1^+ - a_1^- \\ a_2^+ - a_2^- \end{pmatrix}. \tag{7}$$

That is the self-adjoint operator $\mathbf{A}_S$ is defined on functions from $D(\mathbf{A}_0^*)$ that have the asymptotics (5) with the coefficients that satisfy the system (7) with a given matrix $\mathbf{S}$.

4. Zero-range potentials in fluid loaded plates

Consider now the operator of the problem for fluid loaded elastic plates. It can be constructed similarly to [6], [7]. First one introduces the main space as the orthogonal sum

$$\mathcal{L} = L_2(\mathbb{R}^3_+) \oplus L.$$

The first component is the space of acoustic pressures $U(x, y, z)$, the second component is the space of flexure displacements $\xi(x, y)$ and coincides with the space used in the previous section. The operator that corresponds to the disjoint plates can be written in matrix form

$$\mathbf{B} = \begin{pmatrix} -c^2\Delta & 0 \\ \dfrac{\rho}{m}\mathbf{J} & \dfrac{D}{m}\mathbf{A} \end{pmatrix}.$$

Here constants c, ρ, m and D are the same as in (1) and $\mathbf{J}$ is the operator, that takes the trace of pressure on the plate

$$U(x, y, z) \xrightarrow{\mathbf{J}} U(x, y, 0).$$

The zero-range potentials for the operator $\mathbf{B}$ can be constructed according to the same scheme of [4], [5] as it was done in the previous section for the operator $\mathbf{A}$. Among these zero-range potentials there is a subset of zero-range potentials that are presented only in the ξ component

$$\mathbf{B}_S = \begin{pmatrix} -c^2\Delta & 0 \\ \dfrac{\rho}{m}\mathbf{J} & \dfrac{D}{m}\mathbf{A}_S \end{pmatrix}.$$

We shall consider only such extensions.

5. Generalized point model

5.1. Idea

The model of the joint can be constructed as the zero-range potential of the operator $\mathbf{B}_S$ with appropriately chosen matrix $\mathbf{S}$. For the choice of the matrix $\mathbf{S}$ one can use the approach of [6], [7]. Namely the matrix $\mathbf{S}$ can be chosen when the problem for isolated construction is considered.

There is no proof that the matrix $\mathbf{S}$ parameterizing the zero-range potential in isolated construction and the matrix $\mathbf{S}$ in the zero-range potential for fluid loaded construction coincide, but the following circumstances allow such a hypothesis to be suggested:

1. The matrix $\mathbf{S}$ that corresponds to the diffraction by a joint of isolated plates is defined only by the length of joint. It does not depend on the global parameters of the problem.
2. Physically matrix $\mathbf{S}$ characterizes the influence of obstacle directly on flexure displacements and this influence does not depend on the presence of fluid.
3. The hypothesis was checked for a number of models, see e.g. [7], [10].

First the problem of flexure waves diffraction in isolated construction is considered in classical formulation and the asymptotics of the far field amplitude by small width of the joint is derived. Then the problem of diffraction by a zero-range potential is analysed and the matrix $\mathbf{S}$ is chosen so that the far field amplitude of the field scattered by the zero-range potential asymptotically coincides with the far field amplitude of the field scattered by the joint. This allows the matrix $\mathbf{S}$ to be determined uniquely.

5.2. Diffraction by a short joint of isolated plates

Consider the diffraction of flexure wave on a joint of two semi-infinite plates Π^+ and Π^-. The flexure vibrations of the plates are described by the Kirchhoff model

$$\left(\Delta^2 - k_0^4\right)\xi(x,y) = 0, \qquad y \neq 0. \tag{8}$$

Here Δ is the two-dimensional Laplace operator and $k_0 = m\omega^2/D$ is the wave number of flexure waves. The edges of the plates are assumed free, that is the boundary conditions (2) are satisfied. On the joint the continuity of the function ξ and all its derivatives by y is assumed as in (3).

The incident plane flexure wave incident at angle φ_0 $(0 < \varphi_0 < \pi)$

$$\xi^i(x,y) = \Theta(y)\exp\left(ik_0(x\cos\varphi_0 - y\sin\varphi_0)\right)$$

generates the field reflected by the free edge of the plate Π^+

$$\xi^r(x,y) = \Theta(y)R(\varphi_0)\exp\left(ik_0(x\cos\varphi_0 + y\sin\varphi_0)\right)$$
$$+\,\Theta(y)T(\varphi_0)\exp\left(ik_0 x\cos\varphi_0 - k_0 y\sqrt{1+\cos^2\varphi_0}\right)$$

and some scattered field $\xi^s(x,y)$. Here $\Theta(y)$ is Heaviside function. The reflection coefficient of plane and inhomogeneous waves are given by the formulae

$$R(\varphi_0) = \frac{i\sin\varphi_0 A_+^2(\varphi_0) - \sqrt{1+\cos^2\varphi_0}A_-^2(\varphi_0)}{L(\varphi_0)},$$

$$T(\varphi_0) = \frac{-2i\sin\varphi_0 A_+(\varphi_0)A_-(\varphi_0)}{L(\varphi_0)},$$

where

$$L(\varphi_0) = i\sin\varphi_0 A_+^2(\varphi_0) + \sqrt{1+\cos^2\varphi_0}A_-^2(\varphi_0),$$

$$A_\pm(\varphi_0) = (1-\sigma)\cos^2\varphi_0 \pm 1.$$

The boundary value problem for the scattered field differs from (8), (2) by the boundary conditions on the joint that become inhomogeneous

$$[\xi^s](x) = -\xi^i(x,+0) - \xi^r(x,+0), \; [\xi^s_y](x) = -\xi^i_y(x,+0) - \xi^r_y(x,+0), \quad |x|<a.$$

The other two conditions can be written as

$$\left[\xi^s_{yyy} + (2-\sigma)\xi^s_{xxy}\right] = 0, \; \left[\xi^s_{yy} + \sigma\xi^s_{xx}\right] = 0, \quad |x|<a.$$

Here by the symbol $[f](x)$ we denote the jump of the function f on the line $y=0$, i.e. $[f](x) = f(x,+0) - f(x,-0)$.

With the help of the Green function for the semi-infinite plate with free edge constructed in [8], the boundary value problem for the scattered field $\xi^s(x,y)$ can be reduced to integral equations on the joint. Asymptotic analysis of these equations with respect to small length of the joint ($k_0 a \ll 1$) allows the far field amplitude $\Psi(\varphi)$ to be found [9]. The three terms of the asymptotics are presented below

$$\begin{aligned}\Psi(\varphi) &= i\frac{\sin\varphi}{L(\varphi)}\frac{\sin\varphi_0}{L(\varphi_0)}\Bigg\{\operatorname{sign}(y)A_+(\varphi)A_+(\varphi_0)\Bigg[\frac{1}{I_0}\\ &- \frac{\chi\cos\varphi\cos\varphi_0}{\ln(k_0a/4)+\gamma+(1-\sigma^2)I_2 - I_1 + i\pi/2}\Bigg]\\ &+ \frac{\chi A_-(\varphi)A_-(\varphi_0)\sqrt{1+\cos^2\varphi}\sqrt{1+\cos^2\varphi_0}}{\ln(k_0a/4)+\gamma-2(1-\sigma)I_2+(2-2\sigma-\sigma^2)I_1+i\pi/2}\Bigg\}. \end{aligned} \tag{9}$$

Here γ is the Euler constant, $\chi = (3+\sigma)(1-\sigma)$ and the following integrals are introduced

$$I_0 = \int\frac{dl}{\ell(l)}, \quad I_1 = \int\frac{dl}{\ell(l)\sqrt{l^4-1}}, \quad I_2 = \int\frac{l^2-\sqrt{l^4-1}}{\ell(l)\sqrt{l^4-1}}\,l^2\,dl, \tag{10}$$

where

$$\ell(l) = \left((1-\sigma)l^2+1\right)^2\sqrt{l^2-1} - \left((1-\sigma)l^2-1\right)^2\sqrt{l^2+1}.$$

5.3. Matrix S

Now one considers the problem of scattering for the operator $\mathbf{A}_S$ and compares the far field amplitude in this problem with the asymptotics (9).

The field scattered by the zero-range potential is searched in the form of decomposition

$$\begin{aligned}\xi^s &= c_0\Big(G(x,y;0,+0) - G(x,y;0,-0)\Big)\\ &+ c_1\Big(G_x(x,y;0,+0) - G_x(x,y;0,-0)\Big)\\ &+ c_2\Big(G_y(x,y;0,+0) - G_y(x,y;0,-0)\Big).\end{aligned}$$

Here $G(x,y;x_0,y_0)$ up to arbitrary multiplier coincides with the Green function from [8]. Comparison of the asymptotics of ξ^s at infinity with the expression (9) allows the desired values of c_0, c_1 and c_2 to be found

$$c_0 = -\frac{4\pi i}{3+\sigma}\frac{A_+(\varphi_0)\sin\varphi_0}{L(\varphi_0)}\frac{k_0^2}{I_0},$$

$$c_1 = \frac{4\pi k_0(1-\sigma)A_+(\varphi_0)\sin\varphi_0\cos\varphi_0}{\ln(k_0a/4)+\gamma-I_1+(1-\sigma^2)I_2+i\pi/2},$$

$$c_2 = \frac{4\pi i k_0(1-\sigma)A_-(\varphi_0)\sqrt{1+\cos^2\varphi_0}\sin\varphi_0}{\ln(k_0a/4)+\gamma+(2-2\sigma-\sigma^2)I_1-2(1-\sigma)I_2+i\pi/2}.$$

Analysis of the asymptotics of the Green function $G(x,y;0,\pm 0)$ allows the coefficients a_j in the asymptotics (5) to be found

$$a_0^+ = a_0^- = -2i\frac{A_+(\varphi_0)\sin\varphi_0}{L(\varphi_0)},$$

$$a_1^+ - a_1^- = -\frac{4k_0\,A_+(\varphi_0)\sin\varphi_0\cos\varphi_0}{\ln(k_0a/4)+\gamma-I_1+(1-\sigma^2)I_2+i\pi/2}\left(\ln(a/2)+\frac{3}{2}\right),$$

$$a_2^+ - a_2^- = \frac{4ik_0\,A_-(\varphi_0)\sqrt{1+\cos^2\varphi_0}\sin\varphi_0}{\ln(k_0a/4)+\gamma+(2-2\sigma-\sigma^2)I_1-2(1-\sigma)I_2+i\pi/2}\ln(a/2).$$

These values of c_j and a_j should satisfy the system (7). From these equations one can determine the matrix **S**. It should be noted that matrix **S** should not depend on the direction of incidence and on the spectral parameter k_0. Such solution is unique

$$\mathbf{S} = \operatorname{diag}\left(0, \frac{2\ln(a/2)+3}{4\pi(1-\sigma)}, \frac{\ln(a/2)}{2\pi(1-\sigma)}\right). \tag{11}$$

Thus the generalized model of the joint in isolated mechanical construction is formulated as the zero-range potential of the operator $\mathbf{A}_S$ with the matrix **S** defined in (11). According to the main hypothesis this model is transformed to fluid loaded construction as the operator $\mathbf{B}_S$.

6. Conclusion

The constructed generalized model reproduces the principal order term and logarithmically smaller corrections in the far field asymptotics in the problem (1)–(3). It is possible to construct the zero-range potential that would reproduce the next order terms (which are $(ka)^2$ times smaller than the principal one) as well. For these the initial operator $\mathbf{A}$ should be restricted to functions with second order derivatives equal to zero in the origin. The deficiency indices of the adjoint operator in that case are equal to 12. To determine the corresponding matrix of parameters it is necessary to derive the asymptotics of the far field amplitude in the case of isolated plate up to terms of order $O\left((k_0a)^2\right)$. Comparing the asymptotics with the amplitude of the field scattered by the zero-range potential and asking matrix $\mathbf{S}$ to be independent of the incidence angle and spectral parameter k_0 allows all the parameters to be determined uniquely.

When constructing the operator $\mathbf{B}_S$ no perturbations were allowed in the fluid component. In the general case it is possible to consider zero-range potentials with components both in internal and external channels of scattering. As discussed in the introduction the parameters of the zero-range potential in external channel can be found when the problem for absolutely rigid plate is examined. Evidently, for absolutely rigid plate neither cracks, nor the joint are noticed by the field and correspondingly the external component of the zero-range potential is not presented in the generalized model.

Finally one can notice that the operators $\mathbf{B}$ and $\mathbf{B}_S$ depend on frequency which is usually considered as the spectral parameter. Following the scheme of [11] it is possible to exclude this dependence.

References

[1] R.H. Lyon *JASA*, 1962, **43**, pp. 1265–1268.

[2] I.P. Konovaljuk and V.N. Krasil'nikov *Scattering and Radiation of Waves*, 1965, **4**, pp. 149–165 (Leningrad State Univ., Leningrad, in Russian).

[3] D.P. Kouzov *Prikladn. Mat. i Mehan. USSR*, 1963, **27**, pp. 1037–1043.

[4] B.S. Pavlov *Uspehi Mat. Nauk USSR*, 1987, **42**, pp. 99–132.

[5] S. Albeverio and P. Kurasov "Singular perturbations of differential operators and solvable Schrödinger type operators", *Cambridge Univ. Press*, 2000, London Math. Soc. Lecture Notes 221.

[6] I.V. Andronov *J. Math. Physics*, 1993, **34**, pp. 2226–2241.

[7] I.V. Andronov *Applicable Analysis*, 1998, **68**, pp. 3–29.

[8] R. Gunda, S.M. Vijayakar, R. Singh, J.E. Farstad *J. Acoust. Soc. Amer.*, 1998, **103**(2), pp. 888–899.

[9] I.V. Andronov *Prikladn. Mat. i Mehan.*, to appear.

[10] I.V. Andronov *J. Vych. Mat. i Mat. Fiz.*, 1992, **31**(2), pp. 285–295; TRANSLATED IN *Comput. Maths Math. Phys.*, 1992, **32**(2), pp. 236–244.

[11] A.V. Badanin, A.A. Pokrovskii *New Zealand J. of Math.*, 1995, **24**, pp. 65–79.

Department of Computational Physics
Faculty of Physics
St.-Petersburg University
Ulianovskaya 1–1
198904 Petrodvorets
Russia
e-mail: iva@aa2628.spb.edu

Operator Theory:
Advances and Applications, Vol. 132, 87–110

Some Spectral Properties of the Heun Differential Equation

P.B. Bailey, W.N. Everitt, D.B. Hinton, and A. Zettl

Dedicated to the achievements and memory of Sonja Kovalevski

Abstract. This paper is concerned with the analytical study of a special boundary value problem for the Heun linear ordinary differential equation, derived from a physical problem first studied by the methods of applied mathematics. This physical problem arises from merging singularities in the plastic deformation of crystalline materials under stress, first studied by Lay and Slavyanov.

The methods used transform the problem into a Sturm-Liouville boundary value problem with singular endpoints. It is shown that this problem can be represented by a self-adjoint differential operator in a Hilbert function space. The properties of this operator prove that the original problem has a simple, discrete spectrum, and that the associated eigenfunctions are complete in this space

1. Introduction

This paper is inspired by the work of Lay and Slavyanov [5], and the lecture given by Sergei Slavyanov at the Kovalevski Symposium held at the Department of Mathematics, University of Stockholm in June 2000.

The standard notations apply: $\mathbb{R}$ and $\mathbb{C}$ denote the real and complex fields; (a, b) and $[a, b]$ denote open and closed intervals of $\mathbb{R}$; the $'$ notation denotes differentiation on the real line $\mathbb{R}$ and d/dz differentiation on the complex plane $\mathbb{C}$. Other notations are introduced below as required.

The starting point for this paper is the special form of the general Heun linear differential equation, given by Lay and Slavyanov; in the notation of [5, Section 1] this special form is

$$L_z y(z) = \lambda y(z) \tag{1}$$

where the independent variable $z \in \mathbb{C}$ (the complex plane), the parameter $\lambda \in \mathbb{C}$ and plays the role of a spectral parameter, and L_z is defined by

$$L_z := z(z-1)(z+s)D^2 + [c(z-1)(z+s) + dz(z+s) + ez(z-1)]D + abz \tag{2}$$

with D denoting the differential expression d/dz.

The six parameters a, b, c, d, e, s in (2) are all real numbers; they have to satisfy the linear relation relationship (the Fuchs identity), see [5, Section 1],

$$a + b + 1 - c - d - e = 0 \tag{3}$$

and the additional conditions

$$c \geq 1 \quad d \geq 1 \quad e \geq 1 \quad a \geq b \quad s > 0. \tag{4}$$

In view of the need in this paper that we have for the application of the Frobenius method of series solution to the differential equation (1), we recall here that the general form of the Heun differential equation in the complex plane $\mathbb{C}$ is, see the authoritative handbook on the Heun differential equation [8, Part A, Chapter 1, Section 1.1],

$$\frac{d^2w(z)}{dz^2} + \left(\frac{\gamma}{z} + \frac{\delta}{z-1} + \frac{\varepsilon}{z-a}\right)\frac{dw(z)}{dz} + \frac{\alpha\beta z - q}{z(z-1)(z-a)}w(z) = 0 \text{ for } z \in \mathbb{C}. \tag{5}$$

Here the point $a \in \mathbb{C}$, with $a \neq 0, 1$ and, in general, the parameters $\alpha, \beta, \gamma, \delta, \varepsilon$ and q all belong to the complex field $\mathbb{C}$. In this form the points $0, 1, a$ are all regular singularities of the equation (5), and provided that the Fuchs identity

$$\alpha + \beta + 1 = \gamma + \delta + \varepsilon \tag{6}$$

is satisfied, see [8, Part A, Chapter 1, (1.1.2)] and (3) above, then the point at ∞ is also a regular singularity of the equation.

The special form (1) of the Heun equation is obtained from the general form (5) by replacing $a \in \mathbb{C}$ with the real parameter $-s < 0$, replacing $\alpha, \beta, \gamma, \delta, \varepsilon$ with the real parameters a, b, c, d, e respectively, and replacing q with the complex spectral parameter λ.

The purpose of the work in [5] is to consider the local analytic behaviour of solutions of the equation (1) near to the singularities 0, 1, and the two point connection, boundary value problem on the interval $(0, 1)$ with the physical boundary conditions, see [5, (1.6)],

$$|y(0)| < \infty \qquad |y(1)| < \infty. \tag{7}$$

It is stated in [5, Section 1] that this boundary value problem has eigenvalues, indexed as $\{\lambda_n(s)\}$, thus indicating their dependence upon the positive parameter s which plays a special role in the applications to physical problems considered later in the paper [5]. Additionally, the qualitative behaviour of the associated eigenfunctions $\{y_n(z)\}$ is also studied.

The initial study of the Heun boundary value problem, considered in this paper, was undertaken with the quantitative and qualitative SLEIGN2 Sturm-Liouville code; see [1]. This code predicted the existence of a discrete spectrum for the Heun problem when the limit-point classification holds at one or both of the endpoints 0 and 1; in analytical terms, but not in physical terms as in [5], this form of the spectrum was far from clear in the initial stages. This Heun boundary value problem, which shows how effectively the code works with problems derived from applied mathematics, has now been added to the list of worked Sturm-Liouville examples available as an integral part of the SLEIGN2 package.

Applying the methods of classical and functional analysis, and calling on established results of Sturm-Liouville theory, we prove in this paper the following theorem:

Theorem 1. *The Lay/Slavyanov boundary value problem consisting of the differential equation* (1) *on the interval* $[0,1]$, *under the parameter conditions* (3) *and* (4), *and the physical boundary conditions* (7) *has a simple, discrete, positive spectrum*

$$\{0 < \lambda_0(s) < \cdots < \lambda_n(s) < \lambda_{n+1}(s) \cdots : n = 0,1,2,\ldots\}.$$

The associated eigenfunctions $\{y_n(z) : z \in [0,1] \text{ and } n = 0,1,2,\ldots\}$ *are orthogonal and complete in the Hilbert space* $L^2([0,1];w)$ *where the positive weight* w *is given by*

$$w(z) = z^{c-1}(1-z)^{d-1}(z+s)^{e-1} \text{ for all } z \in [0,1].$$

These eigenfunctions $\{y_n(z) : z \in [0,1]\}$ *have the property that for* $n = 0,1,2,\ldots$, *the particular eigenfunction* $y_n(z)$ *is real-valued and has exactly* n *zeros in the open interval* $(0,1)$.

Proof. See the results in Corollary 10 and Theorem 11 in Section 11 below. □

In this paper we are concerned with studying the differential equation (1) in the classical Sturm-Liouville form, see Section 4 below, and replacing the physical boundary conditions (7) with structured boundary conditions defined within the associated Hilbert function space, as in the classical text of Naimark [6]. Some attention to these analytical problems for general cases of the Heun differential equation is to be found in the text [8, Part A, Chapter 5, Section 5.2], but not to the problems of classifying the endpoints of the associated Sturm-Liouville differential equation, determining the structured boundary conditions equivalent to the physical boundary conditions given by (7), and then giving an analytic proof of Theorem 1. All these analytical results are obtained in this paper and lead to interesting conjectures that the authors hope to study in subsequent work.

Such results lead to proofs that the eigenvalues and eigenfunctions assumed to exist in the applied studies on the Heun equation, given in such papers as [5] and in the papers and books that form the references of this paper, do have existence in the sense of classical and functional analysis.

The Lay/Slavyanov paper [5] is yet another remarkable example of how the methods of applied mathematics lead to an inheritance of significant problems in classical and functional analysis which require sophisticated analytic methods to yield existence and/or uniqueness theorems.

The contents: Sections 2 to 8 are concerned with the properties of the Sturm-Liouville differential expression; Sections 9 and 10 with the properties of the associated differential operators; Section 11 with the spectral properties of these Heun differential operators; finally there is a Remarks section.

2. The Heun form of the differential equation

The Lay/Slavyanov case of the Heun differential equation, see (1) and (2), in the Heun form is, where $\lambda \in \mathbb{C}$,

$$\frac{d^2w(z)}{dz^2} + \left(\frac{c}{z} + \frac{d}{z-1} + \frac{e}{z+s}\right)\frac{dw(z)}{dz} + \frac{abz-\lambda}{z(z-1)(z+s)}w(z) = 0 \text{ for } z \in \mathbb{C}. \quad (8)$$

Recall that the parameters a, b, c, d, e, and s all belong to $\mathbb{R}$, and for the applications in [5] have to satisfy the conditions $s > 0$ and both (3) and (4). In this paper we are able to relax these conditions on the parameters; we work then with:

$$\begin{cases} (i) & a, b, e \in \mathbb{R} \\ (ii) & s \in (0, \infty) \\ (iii) & c, d \in (0, \infty) \end{cases} \quad (9)$$

and then the Fuchs condition

$$a + b + 1 = c + d + e. \quad (10)$$

The two conditions (ii) and (iii) of (9) are essential to the analysis that follows in this paper; condition (i) of (9) simply requires that a, b, e are all real, as in [5]. We have retained (10) since, when this condition holds the equation (8) is in the Heun class with four regular singularities in the complex plane (this condition (10) can be removed for the spectral analysis results that follow to be valid, but is retained here to keep within the Heun class of differential equations).

Note that when the conditions (3) and (4) of [5] hold then the conditions (9) and (10) are also satisfied, but not *vice versa*; thus all the spectral results of this paper hold for the differential equation given by (1) and (2). Further, we show that the structured boundary value problem defined within Hilbert function spaces, introduced in the following sections, is entirely equivalent to the boundary value problem represented by the equation (1) and (2), and the boundary conditions (7).

3. Frobenius solutions

If we apply the Frobenius series solution method to the Heun differential equation (8), see [4, Chapter XVI], at the two regular singularities at 0 and 1 then we obtain simple asymptotic properties of independent solutions of the equation. These asymptotic forms have branch points at the points 0 and 1 in $\mathbb{C}$; to avoid difficulties we cut the plane $\mathbb{C}$ from $-\infty$ on $\mathbb{R}$ to 0 and then again from 1 on $\mathbb{R}$ to $+\infty$.

Consider the regular singularity at 0; then a critical role is played by the parameter $c \in (0, \infty)$. If the Frobenius analysis is applied to the differential equation (8), in the neighbourhood of 0, then the indicial roots depend only upon the parameter c and are given by

$$\alpha = 0 \quad \text{and} \quad \alpha = 1 - c \text{ for all } c \in (0, \infty);$$

the double root for $c = 1$ introduces a logarithmic term into the series solutions. From these series solutions we obtain a basis of solutions $\{w_{0,1}, w_{0,2}\}$, in a cut neighbourhood of 0, with asymptotic forms

$$w_{0,1}(z) \sim 1 \text{ as } z \to 0 \tag{11}$$

$$w_{0,2}(z) \sim z^{1-c} \text{ as } z \to 0 \text{ for all } c \in (0,1) \cup (1,\infty) \tag{12}$$
$$w_{0,2}(z) \sim \log(z) \text{ as } z \to 0 \text{ for } c = 1. \tag{13}$$

For the regular singularity at 1 there are similar results, but now the parameter d plays a critical role to determine another basis $\{w_{1,1}, w_{1,2}\}$ with the asymptotic forms, in a cut neighbourhood of 1,

$$w_{1,1}(z) \sim 1 \text{ as } z \to 1 \tag{14}$$

$$w_{1,2}(z) \sim (z-1)^{1-d} \text{ as } z \to 1 \text{ for all } d \in (0,1) \cup (1,\infty) \tag{15}$$
$$w_{1,2}(z) \sim \log(z-1) \text{ as } z \to 1 \text{ for } d = 1. \tag{16}$$

We note that these dominant terms of the asymptotic form of these basis solutions are independent of the spectral parameter λ.

These asymptotic forms are used below to determine the endpoint classifications of the Sturm-Liouville form of the Heun differential equation (8).

4. The Sturm-Liouville form of the differential equation

We use L for Lebesgue integration, and the standard notations $L^1(0,1), L^1_{\text{loc}}(0,1)$ for integration spaces of complex-valued functions defined on the interval $(0,1)$

As in [5, Section 1] we now restrict the independent variable of equation (8) to lie in the interval $(0,1) \subset \mathbb{R} \subset \mathbb{C}$. Using standard methods this equation is transformed into Sturm-Liouville form, changing the independent variable $z \to x$, the dependent variable $w \to y$ and then restricting $x \in (0,1)$; compare with [8, Part A, Chapter 5, Section 5.2]. This process yields the ordinary linear differential equation, where the spectral parameter $\lambda \in \mathbb{C}$,

$$-(p(x)y'(x))' + q(x)y(x) = \lambda w(x)y(x) \text{ for all } x \in (0,1) \tag{17}$$

where for all $x \in (0,1)$ the coefficients p, q, w are defined by

$$\begin{cases} p(x) := x^c(1-x)^d(x+s)^e \\ q(x) := abx^c(1-x)^{d-1}(x+s)^{e-1} \\ w(x) := x^{c-1}(1-x)^{d-1}(x+s)^{e-1}. \end{cases} \tag{18}$$

We note that the coefficients p, q, w have the properties:

$$\begin{cases} (i) & p, q, w : [0,1] \to \mathbb{R} \\ (ii) & q, w \in C[0,1], i.e.\ q, w \in L^1(0,1) \\ (iii) & p(x), w(x) > 0 \text{ for all } x \in (0,1) \\ (iv) & 1/p \equiv p^{-1} \in L^1_{\text{loc}}(0,1) \\ (v) & p^{-1} \in L^1(0,1/2) \text{ if } c \in (0,1), \quad p^{-1} \notin L^1(0,1/2) \text{ if } c \in [1,\infty) \\ (vi) & p^{-1} \in L^1(1/2,1) \text{ if } d \in (0,1), \quad p^{-1} \notin L^1(1/2,1) \text{ if } d \in [1,\infty). \end{cases} \tag{19}$$

From these properties, and in terms of the definitions given in [6, Chapter V, Section 15.1], it follows that the differential equation (17) is regular at all points of the open interval $(0,1)$; regular at endpoint 0, respectively endpoint 1, if the parameter $c \in (0,1)$, respectively $d \in (0,1)$; otherwise the endpoints 0 and 1 are singular.

The linear ordinary differential equation (17) inherits asymptotic forms for its solutions from the Frobenius solutions and asymptotic forms of the Heun equation (8).

From the results (11), (12) and (13) there exists a basis $\{y_{0,1}, y_{0,2}\}$ for solutions for the differential equation (17) on the interval $(0,1)$ such that, for any $\lambda \in \mathbb{C}$,

$$y_{0,1}(x) \sim 1 \text{ as } x \to 0^+ \tag{20}$$

and

$$y_{0,2}(x) \sim x^{1-c} \text{ as } x \to 0^+ \text{ for all } c \in (0,1) \cup (1,\infty) \tag{21}$$

$$y_{0,2}(x) \sim \ln(x) \text{ as } x \to 0^+ \text{ for } c = 1. \tag{22}$$

Similarly, from the results (14), (15) and (16) there exists another basis $\{y_{0,1}, y_{0,2}\}$ for solutions for the differential equation (17) on the interval $(0,1)$ such that, for any $\lambda \in \mathbb{C}$,

$$y_{0,1}(x) \sim 1 \text{ as } x \to 1^- \tag{23}$$

and

$$y_{0,2}(x) \sim x^{1-d} \text{ as } x \to 1^- \text{ for all } d \in (0,1) \cup (1,\infty) \tag{24}$$

$$y_{0,2}(x) \sim \ln(x) \text{ as } x \to 1^- \text{ for } d = 1. \tag{25}$$

5. The differential expression M

The letters AC denote absolute continuity (Lebesgue); the term $AC_{\text{loc}}(0,1)$ denotes the collection of complex-valued functions defined on $(0,1)$ that are absolutely continuous on every compact interval $[\alpha, \beta] \subset (0,1)$.

We define the function space $L^2((0,1); w)$ as the collection of all Lebesgue measurable, complex-valued functions on the interval $(0,1)$ that satisfy

$$\int_0^1 w\,|f|^2 \equiv \int_0^1 w(x)\,|f(x)|^2\,dx < +\infty. \tag{26}$$

We use the same symbol $L^2((0,1);w)$ for the Hilbert space of such functions collected into equivalence classes, with inner product

$$(f,g)_w := \int_0^1 wf\overline{g} \text{ for all } f,g \in L^2((0,1);w). \tag{27}$$

We define the differential expression $M : D(M) \subset AC_{\text{loc}}(0,1) \to L^1_{\text{loc}}(0,1)$ by

$$D(M) := \{f : (0,1) \to \mathbb{C} : f, pf' \in AC_{\text{loc}}(0,1)\}$$

and

$$M[f] := -(pf')' + qf \text{ for all } f \in D(M). \tag{28}$$

Note that the differential equation (17) can now be written in the form

$$M[y] = \lambda wy \text{ on } (0,1) \tag{29}$$

and that any solution y satisfies $y \in D(M)$.

We can now define, see[6, Chapter V, Section 17.1], the maximal domain (linear manifold) $\Delta(c,d)$ generated by M in $L^2((0,1);w)$; this domain depends upon all the parameters a,b,c,d,e,s but is critically dependent upon the two parameters c and d and hence the notation $\Delta(c,d)$ is used in all following sections.

$$\Delta(c,d) := \{f \in D(M) : f, w^{-1}M[f] \in L^2((0,1);w)\}. \tag{30}$$

We note that from the condition $pf' \in AC_{\text{loc}}(0,1)$ and the properties of the coefficient p, we obtain, for all $f \in D(M)$,

$$f' \in AC_{\text{loc}}(0,1). \tag{31}$$

6. Endpoint classification of M in $L^2((0,1);w)$

For the purpose of defining self-adjoint differential operators in the Hilbert space $L^2((0,1);w)$, generated by the differential expression M, it is necessary to classify both of the endpoints $0,1$ of M in $L^2((0,1);w)$. For the general theory of classification of endpoints of arbitrary Lagrange symmetric (formally self-adjoint) differential expressions, see the results in [6, Chapter V, Section 17]; for a more explanatory and detailed form for second-order Sturm-Liouville expressions, see [1, Section 4].

The endpoints $0,1$ of M are classified, disjointly, as either *regular* or *limit-circle* or *limit-point* in $L^2((0,1);w)$; for many purposes the regular case classification can be regarded, without loss of generality, as a special case of the limit-circle classification; the limit-point classification is distinct from the limit-circle case, and hence from the regular case.

Note that the regular cases $c,d \in (0,1)$ do not occur in the Lay/Slavyanov paper [5], since therein the parameters c,d are restricted to $c,d \in [1,\infty)$ and this restriction excludes the regular cases.

In the regular/limit-circle case for the endpoint 0 of M, the differential equation (17) has, for all $\lambda \in \mathbb{C}$, all solutions in the space $L^2((0,1/2);w)$, *i.e.* for any solution $y(\cdot)$ of the equation

$$\int_0^{1/2} w(x)\,|y(x)|^2\,dx < +\infty. \tag{32}$$

Similarly, in the regular/limit-circle case for the endpoint 1 of M, the differential equation (17) has, for all $\lambda \in \mathbb{C}$, all solutions in the space $L^2((1/2,1);w)$, *i.e.* for any solution $y(\cdot)$ of the equation

$$\int_{1/2}^{1} w(x)\,|y(x)|^2\,dx < +\infty. \tag{33}$$

In the limit-point case for the endpoint 0 of M, the differential equation (17) has for some $\lambda \in \mathbb{C}$ a solution $y(\cdot) \notin L^2((0,1/2);w)$, *i.e.*

$$\int_0^{1/2} w(x)\,|y(x)|^2\,dx = +\infty. \tag{34}$$

Similarly, in the limit-point case for the endpoint 1 of M, the differential equation (17) has for some $\lambda \in \mathbb{C}$ a solution $y(\cdot) \notin L^2((1/2,1);w)$, *i.e.*

$$\int_{1/2}^{1} w(x)\,|y(x)|^2\,dx = +\infty. \tag{35}$$

We can now state

Theorem 2. *The differential expression M has the following endpoint classifications in the space $L^2((0,1);w)$:*

(1) *The endpoint* 0 *is:*
- (*i*) *regular for all $c \in (0,1)$*
- (*ii*) *limit-circle for all $c \in [1,2)$*
- (*iii*) *limit-point for all $c \in [2,\infty)$*

(2) *The endpoint* 1 *is:*
- (*i*) *regular for all $d \in (0,1)$*
- (*ii*) *limit-circle for all $d \in [1,2)$*
- (*iii*) *limit-point for all $d \in [2,\infty)$.*

Proof. These results follow from the classification criteria (32) to (35), given in terms of the solutions of the differential equation (17), the asymptotic forms of the solutions near 0^+ and 1^- given in (20) to (25), and the explicit form of the weight w given in (18).

In the regular case both $c, d \in (0,1)$, all solutions y of the differential equation (17) are not only in the space $L^2((0,1);w)$ but, for any solution y, both the solution y and the quasi-derivative py' are in the space $AC[0,1]$. □

7. The symplectic form for M

The symplectic form for the differential expression M, defined in Section 5, is obtained from the classical Green's formula for M; here, it is sufficient to define the mapping $[\cdot,\cdot] : \Delta(c,d) \times \Delta(c,d) \times (0,1) \to \mathbb{C}$ by

$$[f,g](x) := f(x).(p\overline{g}')(x) - (pf')(x).\overline{g}(x) \text{ for all } f,g \in \Delta(c,d) \text{ and all } x \in (0,1). \tag{36}$$

For this definition of the symplectic form see [6, Chapter V, Section 15.3] and [1, Section 5, (5.3)].

Theorem 3.

(1) *For all $c,d \in (0,\infty)$ and for all $f,g \in \Delta(c,d)$ the limits*

$$\lim_{x\to 0^+} [f,g](x) \qquad \lim_{x\to 1^-} [f,g](x) \tag{37}$$

exist and are finite in $\mathbb{C}$.

(2) *The differential expression M is limit-point at 0^+, i.e. $c \in [2,\infty)$, if and only if*

$$\lim_{x\to 0^+} [f,g](x) = 0 \text{ for all } f,g \in \Delta(c,d). \tag{38}$$

The differential expression M is limit-point at 1^-, i.e. $d \in [2,\infty)$, if and only if

$$\lim_{x\to 1^-} [f,g](x) = 0 \text{ for all } f,g \in \Delta(c,d). \tag{39}$$

(3) *The differential expression M is regular/limit-circle at 0^+, i.e. $c \in (0,2)$, if and only if there exists a pair $f,g \in \Delta(c,d)$ such that*

$$\lim_{x\to 0^+} [f,g](x) \neq 0. \tag{40}$$

The differential expression M is regular/limit-circle at 1^-, i.e. $d \in (0,2)$, if and only if there exists a pair $f,g \in \Delta(c,d)$ such that

$$\lim_{x\to 1^-} [f,g](x) \neq 0. \tag{41}$$

Proof. For the proof of these known results see [6, Chapter V, Section 18.3] and [1, Section 5]. □

8. Some properties of the linear manifold $\Delta(c,d)$

We have

Lemma 4. *For all $c,d \in (0,\infty)$ the following properties hold:*

(*i*) *Let 1 represent the unit function on the interval $[0,1]$; then $1 \in \Delta(c,d)$.*

(*ii*) *For all $f \in \Delta(c,d)$ the functions $M[f], qf, (pf')'$, defined on the open interval $(0,1)$, all belong to the space $L^1(0,1)$.*

(*iii*) *For all $f \in \Delta(c,d)$ the function $pf' \in AC[0,1]$.*

Proof. (*i*) From the definition (18) we note that $w \in L^1(0,1)$; from this it follows that $1 \in L^2((0,1);w)$. Now $w^{-1}M[1] = w^{-1}q$ and so

$$\int_0^1 w\left(w^{-1}|M[1]|\right)^2 = \int_0^1 w\left(w^{-1}q\right)^2 = \int_0^1 w^{-1}q^2 < +\infty$$

since, on inspection of the definitions in (18), the function $w^{-1}q^2$ is bounded on $(0,1)$. Hence $w^{-1}M[1] \in L^2((0,1);w)$ and $1 \in \Delta(c,d)$.

(*ii*) We have

$$\int_0^1 |M[f]| = \int_0^1 w^{1/2}w^{-1/2}|M[f]| \le \left\{\int_0^1 w \int_0^1 w\left(w^{-1}|M[f]|\right)^2\right\}^{1/2} < +\infty.$$

and so $M[f] \in L^1(0,1)$.

Also

$$\int_0^1 |qf| = \int_0^1 w^{-1/2}|q|\,w^{1/2}|f| \le \left\{\int_0^1 w^{-1}q^2 \int_0^1 w|f|^2\right\}^{1/2} < +\infty.$$

and so $qf \in L^1(0,1)$.

Finally from the definition (28) of the differential expression M it follows that $(pf')' \in L^1(0,1)$.

(*iii*) We know from the definition of the domain, see Section 5, that $pf' \in AC_{\text{loc}}(0,1)$. From (*ii*) above we have $(pf')' \in L^1(0,1)$ and so

$$(pf')(x) = (pf')(1/2) + \int_{1/2}^x (pf')' \text{ for all } x \in (0,1);$$

this gives $pf' \in AC[0,1]$ if we define, respectively,

$$(pf')(0^+,1^-) = (pf')(1/2) + \lim_{x \to 0^+,1^-} \int_{1/2}^x (pf')'.$$

□

9. The self-adjoint operator $T_{c,d}$

We now define the self-adjoint operator

$$T_{c,d} : D(T_{c,d}) \subset L^2((0,1);w) \to L^2((0,1);w);$$

it is this operator, and its properties developed below, that provide the functional analytic background for the boundary value problem given in Section 1 of the Lay/Slavyanov paper [5]. However, in defining the domain $D(T_{c,d})$ it is necessary to take into account the application of structured boundary conditions at the endpoints of the differential expression M as given in the following definition.

Definition 5. *Define the operator*

$$T_{c,d} : D(T_{c,d}) \subset L^2((0,1);w) \to L^2((0,1);w)$$

for all $c, d \in (0,\infty)$

$$\begin{cases} D(T_{c,d}) & := \Delta(c,d) \text{ if both } c,d \in [2,\infty) \\ & := \{f \in \Delta(c,d) : \lim\limits_{x \to 0^+} [f,1](x) = 0\} \text{ if } c \in (0,2) \text{ and } d \in [2,\infty) \\ & := \{f \in \Delta(c,d) : \lim\limits_{x \to 1^-} [f,1](x) = 0\} \text{ if } c \in [2,\infty) \text{ and } d \in (0,2) \\ & := \{f \in \Delta(c,d) : \lim\limits_{x \to 0^+} [f,1](x) = \lim\limits_{x \to 1^-} [f,1](x) = 0\} \\ & \qquad \text{if both } c,d \in (0,2) \end{cases} \tag{42}$$

and

$$T_{c,d}f := w^{-1}M[f] \text{ for all } f \in D(T_{c,d}). \tag{43}$$

This definition leads to

Theorem 6. *The linear operator* $T_{c,d}$ *with domain* $D(T_{c,d})$ *defined by* (42) *and* (43) *is self-adjoint in the Hilbert function space* $L^2((0,1);w)$.

Proof. The proof of this theorem follows from the now established analysis for linear ordinary boundary value problems, given in [6, Chapter V, Section 18] and [1, Section 5]. □

Remark 7.

(1) *The definition of the domain* $D(T_{c,d})$ *reflects upon the classification of the differential expression* M *in* $L^2((0,1);w)$ *of the endpoints* 0 *and* 1*; see the Theorems* 2 *and* 3 *for comparison.*

(2) *For uniformity of treatment in the regular/limit-circle cases we have adopted the same boundary condition at the endpoints* 0 *and* 1*, i.e.* $\lim_{x\to 0^+}[f,1](x) = 0$ *and* $\lim_{x\to 1^-}[f,1](x) = 0$.

(3) *For any endpoint in the limit-point case, i.e.* $c, d \in [2,\infty)$*, no boundary conditions are required; in particular this phenomenon is to be seen in the case when both* 0 *and* 1 *are in the limit-point case.*

(4) *For any endpoint in the limit-circle case, i.e.* $c, d \in [1,2)$*, the boundary condition applied to the elements of the linear manifold* $\Delta(c,d)$ *to give the domain* $D(T_{c,d})$*, e.g.* $\lim_{x\to 0^+}[f,1](x) = 0$*, is the so-called principal or Friedrichs boundary condition; see* [7, Section 1].

(5) *For any regular endpoint, say* 0 *when* $c \in (0,1)$*, the boundary condition* $\lim_{x\to 0^+}[f,1](x) = 0$ *is not the Friedrichs condition but leads to the so-called Neumann boundary condition* $(pf')(0) = 0$*; for the Friedrichs condition take* $\lim_{x\to 0^+}[f, x^{1-c}](x) = 0$ *which is equivalent to the so-called Dirichlet boundary condition* $y(0) = 0$.

(6) *The principal or Friedrichs boundary condition is frequently the boundary condition required in analysis concerned with, or derived from, boundary value problems in applied mathematics and mathematical physics; this reflection is certainly to be seen in the work of Lay and Slavyanov; see* [5].

10. Some properties of the domain $D(T_{c,d})$

The operator domain $D(T_{c,d})$ for the Heun problem, under consideration in this paper, possesses certain special properties now given in

Theorem 8. *For all $f, g \in D(T_{c,d})$ the following properties hold:*

(1) $\lim_{x \to 0+} (pf')(x) = \lim_{x \to 1-} (pf')(x) = 0.$

(2) $p^{1/2} f' \in L^2(0,1)$.

(3) $\lim_{x \to 0+} (pf'g)(x) = \lim_{x \to 1-} (pf'g)(x) = 0.$

(4) *Let both $c, d \in (0,2)$; then:*

(i) *$f' \in L^1(0,1)$ and $f \in AC[0,1]$ for all $f \in D(T_{c,d})$*

(ii) *let $f \in \Delta(c,d)$; then $f \in D(T_{c,d})$ if and only if*

$$f \text{ is bounded on the interval } [0,1]. \tag{44}$$

Proof. We prove these results for the endpoint 0; there is a similar proof for the endpoint 1.

(1) If $c \in (0,2)$ then the regular/limit-circle cases hold and the required result follows from the boundary condition (42) to determine the domain $D(T_{c,d})$ in these cases.

If $c \in [2,\infty)$ then the limit-point case holds; since $1 \in \Delta(c,d)$, from (i) of Lemma 4, and $f \in D(T_{c,d}) = \Delta(c,d)$ the required result follows from (38) of Theorem 3.

(2) Let $c \in (0,\infty)$ and let $f \in D(T_{c,d})$; then $f \in \Delta(c,d)$ and from (ii) of Lemma 4 we obtain the property $(pf')' \in L(0,1)$, and thus for all $(\xi, x) \subset (0,1)$

$$(pf')(x) = (pf')(\xi) + \int_\xi^x (pf')'(t)\,dt.$$

Now let $\xi \to 0^+$ and use result 1 above to yield, again calling on (ii) of Lemma 4, for all $x \in (0,1)$.

$$(pf')(x) = \int_0^x (pf')' = \int_0^x ((pf')' - qf) + \int_0^x qf = -\int_0^x M[f] + \int_0^x qf.$$

Hence, for all $x \in (0,1)$,

$$\begin{aligned}\left(p^{1/2} f'\right)(x) &= -p^{-1/2}(x) \int_0^x M[f] + p^{-1/2}(x) \int_0^x qf \\ &= -p^{-1/2}(x) \int_0^x w^{1/2} w^{-1/2} M[f] \qquad (45)\\ &\quad + p^{-1/2}(x) \int_0^x w^{-1/2} q w^{1/2} f. \qquad (46)\end{aligned}$$

We now use the standard little $o(1)$ and big-O notation for $x \in (0,1/2)$ and, from the explicit form of the coefficients p, q, w given in (18), obtain

$$\begin{gathered} p^{-1/2}(x) = O(x^{-c/2}) \qquad w^{1/2}(x) = O(x^{(c-1)/2}) \\ w^{-1/2}(x)q(x) = O(x^{-(c-1)/2}x^{c}). \end{gathered} \tag{47}$$

Next we note, from the definition (30), that

$$w^{-1/2}M[f] \in L^2(0,1) \quad \text{and} \quad w^{1/2}f \in L^2(0,1). \tag{48}$$

Taking these results together we obtain, on using the Cauchy-Schwarz integral inequality and (48),

$$\begin{aligned} &\left| -p^{-1/2}(x) \int_0^x w^{1/2} w^{-1/2} M[f](t) \right| \\ &= O\left(x^{-c/2} \left\{ \int_0^x t^{c-1}\, dt \int_0^x \left| w^{-1/2} M[f] \right|^2 \right\}^{1/2} \right) \\ &= O\left(\left\{ \int_0^x \left| w^{-1/2} M[f] \right|^2 \right\}^{1/2} \right) \\ &= o(1) \quad \text{as} \quad x \to 0^+, \end{aligned} \tag{49}$$

and

$$\begin{aligned} &\left| p^{-1/2}(x) \int_0^x w^{-1/2} q w^{1/2} f \right| \\ &= O\left(x^{-c/2} \left\{ \int_o^x t^{(c+1)/2}\, dt \int_0^x \left| w^{1/2} f \right|^2 \right\}^{1/2} \right) \\ &= O\left(\left\{ \int_0^x \left| w^{1/2} f \right|^2 \right\}^{1/2} \right) \\ &= o(1) \quad \text{as} \quad x \to 0^+. \end{aligned} \tag{50}$$

Thus from the equation (46) for the term $p^{1/2}f'$ and the two limit expressions (49) and (50), it follows that, for any $f \in D(T_{c,d})$, we have the result

$$\lim_{x \to 0+} \left(p^{1/2} f' \right)(x) = 0 \tag{51}$$

as required.

From similar analysis we can establish the corresponding result for the endpoint 1^-, *i.e.* for any $f \in D(T_{c,d})$ we have the result

$$\lim_{x \to 1-} \left(p^{1/2} f' \right)(x) = 0. \tag{52}$$

(Note that there is one difference between the proofs of (51) and (52); this involves the coefficient $q(x) = abx^c(1-x)^{d-1}(x+s)^{e-1}$; however the analysis given above for the proof of (51) continues to hold, notwithstanding the difference between the factors x^c and $(1-x)^{d-1}$, to give the proof of (52).)

From the results (51) and (52), together with the properties $p^{1/2} \in C[0,1]$ and $f' \in AC_{\text{loc}}(0,1)$ (see (31)), it follows that

$$p^{1/2}f' \text{ is bounded on the interval } (0,1) \text{ for each } f \in D(T_{c,d}) \tag{53}$$

and so

$$p^{1/2}f' \in L^2(0,1) \text{ for all } f \in D(T_{c,d}) \tag{54}$$

as required.

Note that in (53), we have established more than stated in item 2 of Theorem 8; however the notation of item 2 is retained as stated for a reason given below.

(3) To prove item 3 for the endpoint 0, for $c \in (0,\infty)$, it is sufficient to argue for real-valued functions $f, g \in D(T_{c,d})$, since then general functions of $D(T_{c,d})$ can be represented in terms of such real-valued elements.

We have then the identity

$$\begin{aligned}\int_x^{1/2} \{pf'g' + qfg\} &= [pf'g]_x^{1/2} - \int_x^{1/2} \{(pf')'g + qfg\} \\ &= [pf'g]_x^{1/2} + \int_x^{1/2} M[f]g,\end{aligned}$$

i.e.

$$\begin{aligned}\int_x^{1/2} \left\{p^{1/2}f'p^{1/2}g' + (q/w)w^{1/2}fw^{1/2}g\right\} &= [pf'g]_x^{1/2} \\ &\quad + \int_x^{1/2} w(w^{-1}M[f])g,\end{aligned} \tag{55}$$

valid for all $x \in (0,1/2]$. Since by inspection, the term $q/w = O(1)$ on the interval $(0,1/2]$ the integrand on the left-hand side is $L^1(0,1)$, using (54) and $f \in L^2((0,1);w)$; thus the left-hand side tends to a finite limit as $x \to 0^+$. The integrand on the right-hand side is $L^1(0,1)$ since both $w^{-1}M[g]$ and $g \in L^2((0,1);w)$. Thus we have the result

$$\lim_{x \to 0+} (pf'g)(x) = 2k \text{ (say)}, \tag{56}$$

where $k \in \mathbb{R}$. Now, if $k \neq 0$ then, near to 0^+, we have $p|f'g| > |k|$; thus, since from (i) of Theorem 8 $\lim_{x\to 0+}(pf')(x) = 0$, we have $\lim_{x \to 0+}|g(x)| = +\infty$; without loss of generality we can assume that $\lim_{x\to 0+} g(x) = +\infty$. Hence for some $\xi \in (0,1/2)$

$$(p|f'g'|)(x) > |k| \left|\frac{g'(x)}{g(x)}\right| \text{ for all } x \in (0,\xi),$$

and integrating gives

$$\begin{aligned}\int_x^{\xi} p|f'g'| &> |k| \int_x^{\xi} \left|\frac{g'(t)}{g(t)}\right| dt \\ &\geq |k| \left|\int_x^{\xi} \frac{g'(t)}{g(t)} dt\right| = |k| \left|[\ln(g(t))]_x^{\xi}\right| \text{ for all } x \in (0,\xi).\end{aligned}$$

As $x \to 0^+$ the left-hand side of this last result tends to a finite limit, from (54), but the right-hand side tends to $+\infty$; this contradiction implies that k, defined in (56), must satisfy $k = 0$.

Thus

$$\lim_{x \to 0+} (pf'g)(x) = 0 \text{ for all } f, g \in D(T_{c,d}). \tag{57}$$

There is a similar proof to give

$$\lim_{x \to 1-} (pf'g)(x) = 0 \text{ for all } f, g \in D(T_{c,d}).$$

Combining the above results, in particular (55), we obtain the following identity, valid for all $c, d \in (0, \infty)$,

$$\int_0^1 \{pf'\overline{g}' + qf\overline{g}\} = \int_0^1 ww^{-1}M[f]\overline{g} \text{ for all } f, g \in D(T_{c,d}). \tag{58}$$

(4) To prove (i) and (ii) of item 4 of this Theorem we restrict both $c, d \in (0, 2)$.

To prove (i).

Assume that $c \in (0, 2)$. Let $f \in D(T_{c,d})$; then from the proof of item 2 of this Theorem, see above, we have

$$(pf')(x) = \int_0^x (pf')' \text{ for all } x \in (0, 1). \tag{59}$$

From (48) we have $w^{-1/2}M[f] \in L^2(0, 1)$ for all $f \in \Delta(c, d)$; also

$$\left|w^{-1/2}qf\right| \le \left|w^{-1}qw^{1/2}f\right| = O\left(\left|w^{1/2}f\right|\right) \text{ on } (0, 1)$$

so that

$$w^{-1/2}qf \in L^2(0, 1) \text{ for all } f \in \Delta(c, d); \tag{60}$$

hence, from the definition of the differential expression M we obtain

$$w^{-1/2}(pf')' \in L^2(0, 1) \text{ for all } f \in \Delta(c, d). \tag{61}$$

Thus, from (59) to (61) we obtain, for $f \in D(T_{c,d})$,

$$\begin{aligned} |f'(x)| &\le \frac{1}{p(x)} \int_0^x w^{1/2}w^{-1/2} \left|(pf')'\right| \\ &\le \frac{1}{p(x)} \left\{ \int_0^x w \int_0^x \left|w^{-1/2}(pf')'\right|^2 \right\}^{1/2} \\ &= O\left(x^{-c} \left\{ \int_0^x t^{c-1}\,dt \right\}^{1/2} \right) = O\left(x^{-c/2}\right) \text{ for all } x \in (0, 1/2). \end{aligned}$$

This last result implies that, for all $c \in (0, 2)$

$$f' \in L^1(0, 1/2) \text{ for all } f \in D(T_{c,d}). \tag{62}$$

There is a similar proof, for the case when $d \in (0, 2)$ to give the corresponding result for the endpoint 1^-, *i.e.*

$$f' \in L^1(1/2, 1) \text{ for all } f \in D(T_{c,d}). \tag{63}$$

Taken together these results imply that, again noting the essential restriction of $c, d \in (0,2)$,

$$f' \in L^1(0,1) \text{ for all } f \in D(T_{c,d}). \tag{64}$$

If now we define, respectively,

$$f(0^+, 1^-) := f(1/2) + \lim_{x \to 0^+, 1^-} \int_{1/2}^{x} f'(t)\, dt$$

then it is clear that $f \in AC[0,1]$; this proves (i) of item 4.

Proof of (ii). We note that (64) yields the required result that, for all $c, d \in (0,2)$,

$$f \in D(T_{c,d}) \text{ implies that } f \text{ is bounded on } (0,1). \tag{65}$$

To prove the converse of this result, again taking $c \in (0,2)$, we have to show that if $f \in \Delta(c,d)$ and f is bounded on the interval $(0,1/2)$, then $f \in D(T_{c,d})$, *i.e.* that $\lim_{x \to 0^+}[f,1](x) = 0$. With this hypothesis assume that $p^{1/2} f' \notin L^2(0,1/2)$; then from the identity (55), with $g = f$, it follows that $\lim_{x \to 0^+}(pf'f)(x) = -\infty$; thus for some interval $(0,\xi) \subset (0,1/2)$ we can assume, without loss of generality, that for a positive number K

$$(pf'f)(x) < -K, \quad f(x) < 0, \quad (pf')(x) > 0 \text{ for all } x \in (0,\xi). \tag{66}$$

Hence

$$(-f(x))\,|(pf')'|\,(x) > K \frac{|(pf')'|\,(x)}{(pf')\,(x)} \text{ for all } x \in (0,\xi).$$

Thus

$$\int_x^{\xi} w^{1/2}\,|f|\,.w^{-1/2}\,|(pf')'| \geq K \int_x^{1/2} \frac{|(pf')'|}{pf'} \geq K[\ln(pf')]_x^{\xi} \text{ for all } x \in (0,\xi). \tag{67}$$

Since $p(x) > 0$ for all $x \in (0,\xi)$ we have $f'(x) > 0$ for all $x \in (0,\xi)$, and so $f(x)$ is negative and decreasing as $x \to 0^+$; since f is bounded on $(0,1)$ it follows that f tends to a finite limit $-k$, say, where $k > 0$. From the result that $\lim_{x \to 0^+}(pf'f)(x) = -\infty$ it then follows that

$$\lim_{x \to 0^+} (pf')(x) = +\infty. \tag{68}$$

Now from (61) and $f \in L^2((0,1);w)$ the left-hand side of (67) is bounded as $x \to 0^+$; from (68) the right-side tends to $+\infty$ as $x \to 0^+$; this is a contradiction from which the escape is to accept that $p^{1/2} f' \in L^2(0,1)$, for all $f \in D(T_{c,d})$.With this last result established, as in (54) we can follow the argument in the proof of item 3, as given above to obtain the result that if $f \in \Delta(c,d)$ and f is bounded on $(0,1)$, and if $g \in D(T_{c,d})$, then

$$\lim_{x \to 0^+} (pf'g)\,(x) = 0;$$

in this last result we can substitute $g = 1 \in D(T_{c,d})$ to give $\lim_{x \to 0^+} (pf')\,(x) = 0$, *i.e.*

$$\lim_{x \to 0^+} [f,1](x) = 0.$$

Let $d \in (0,2)$; then a similar argument proves the corresponding result for the endpoint 1^-, *i.e.*

$$\lim_{x \to 1^-} [f,1](x) = 0.$$

Thus we have shown that if $f \in \Delta(c,d)$ and f is bounded on $(0,1)$ then $f \in D(T_{c,d})$.

These results prove that (ii) of item 4 is established.

This concludes the proof of item 4 and the proof of Theorem 8. □

Remark 9.

(1) *The boundedness result for the domain $D(T_{c,d})$, of item 4 of Theorem 8, when $c, d \in (0,2)$, does not hold for this operator domain when c and/or $d \in [2,\infty)$. For example, if $c > 2$ and $d \in [2,\infty)$, and if f is defined by $f(x) := \ln(x)$ for all $x \in (0,1)$, then a computation shows that $f \in D(T_{c,d})$; however, f is unbounded near the endpoint 0^+. There is a more protracted example to give this result when $c = 2$.*

(2) *In the case when $c \in (0,2)$ and $d \in [2,\infty)$ the boundedness property item 4 for the domain $D(T_{c,d})$ holds at the endpoint 0^+ but not at the endpoint 1^-; similarly when $c \in [2,\infty)$ and $d \in (0,2)$.*

(3) *The result represented by item 2 is often called the Dirichlet property of the domain $D(T_{c,d})$.*

(4) *The result represented by item 3 is often called the strong limit-point property of the domain $D(T_{c,d})$.*

Corollary 10. *For all $c, d \in (0,\infty)$ the self-adjoint operator $T_{c,d}$ is bounded below in the Hilbert function space $L^2((0,1);w)$.*

If the non-negative number $K(c,d) \in [0,\infty)$ is defined by

$$K(c,d) := \sup\{|w^{-1}(x)q(x)| : x \in (0,1)\} \tag{69}$$

then, for all $c, d \in (0,\infty)$,

$$(T_{c,d}f, f)_w \geq -K(c,d)(f,f)_w \text{ for all } f \in D(T_{c,d}). \tag{70}$$

If the parameters a, b of the Heun differential equation (2) are non-negative then

$$(T_{c,d}f, f)_w \geq 0 \text{ for all } f \in D(T_{c,d}). \tag{71}$$

If both the parameters a, b are positive then

$$(T_{c,d}f, f)_w > 0 \text{ for all } f \in D(T_{c,d}) \text{ but } f \neq 0.$$

Proof. From the identity (58), valid for all $c, d \in (0,\infty)$, with $g = f$ we obtain

$$(w^{-1}M[f], f)_w = \int_0^1 \left\{p\,|f'|^2 + q\,|f|^2\right\} \text{ for all } f \in D(T_{c,d}).$$

Then from Definition 5 of the operator $T_{c,d}$ it follows that

$$\begin{aligned}(T_{c,d}f,f)_w &= (w^{-1}M[f],f)_w \\ &= \int_0^1 \left\{p\,|f'|^2 + q\,|f|^2\right\} \\ &\geq \int_0^1 p\,|f'|^2 - \int_0^1 \left|w^{-1}q\right| w\,|f|^2 \\ &\geq -K(c,d)\int_0^1 w\,|f|^2 = -K(c,d)(f,f)_w \text{ for all } f \in D(T_{c,d}).\end{aligned}$$

If the parameters a, b are non-negative then $q(x) \geq 0$ for all $x \in (0,1)$ and, from above,

$$(T_{c,d}f,f)_w = \int_0^1 \left\{p\,|f'|^2 + q\,|f|^2\right\} \geq 0 \text{ for all } f \in D(T_{c,d}). \tag{72}$$

If both a, b are positive then $q(x) > 0$ for all $x \in (0,1)$, and since $p(x) > 0$ for all $x \in (0,1)$, it follows from (72) that

$$(T_{c,d}f,f)_w = \int_0^1 \left\{p\,|f'|^2 + q\,|f|^2\right\} > 0 \text{ for all } f \in D(T_{c,d})$$

unless, firstly, $f'(x) = 0$ for all $x \in (0,1)$ and, secondly, $f(x) = 0$ for all $x \in (0,1)$. □

11. Spectral properties of the operator $T_{c,d}$

Let $\mathbb{N}_0$ denote the set of non-negative integers $\{0,1,2,\ldots\}$.

The spectral properties of the self-adjoint operator $T_{c,d}$ can now be stated in

Theorem 11. *For all $c, d \in (0,\infty)$ let the self-adjoint operator $T_{c,d}$ be defined as in Definition 5, noting the properties of $T_{c,d}$ given in Section 10. Then $T_{c,d}$ has the following spectral properties holding for all $c, d \in (0,\infty)$:*

1. *The spectrum $\sigma(T_{c,d})$ of $T_{c,d}$ is real, simple and discrete; that is the spectrum consists solely of real simple eigenvalues, say*

$$\sigma(T_{c,d}) = \left\{\lambda_n^{c,d} \in \mathbb{R} : n \in \mathbb{N}_0\right\}$$

with the properties, here $K(c,d)$ is the non-negative number defined in (69),

$$\begin{aligned}\lambda_0^{c,d} &\geq -K(c,d) \\ \lambda_n^{c,d} &< \lambda_{n+1}^{c,d} \text{ for all } n \in \mathbb{N}_0 \\ \lim_{n\to\infty} \lambda_n^{c,d} &= +\infty.\end{aligned}$$

The operator $T_{c,d}$ is bounded below in $L^2((0,1);w)$ with

$$(T_{c,d}f,f)_w \geq -K(c,d)(f,f)_w \text{ for all } f \in D(T_{c,d}).$$

Since each eigenvalue is simple the eigenspaces satisfy

$$\dim\left\{f \in D(T_{c,d}) : T_{c,d}f = \lambda_n^{c,d} f\right\} = 1 \textit{ for all } n \in \mathbb{N}_0.$$

2. *Let the eigenfunctions of* $T_{c,d}$ *be denoted by* $\left\{\psi_n^{c,d} : n \in \mathbb{N}_0\right\}$; *then*

$$T_{c,d}\psi_n^{c,d} = \lambda_n^{c,d}\psi_n^{c,d} \textit{ for all } n \in \mathbb{N}_0$$

and from the differential equation (29)

$$M[\psi_n^{c,d}] = \lambda_n^{c,d} w \psi_n^{c,d} \textit{ on } (0,1) \textit{ for all } n \in \mathbb{N}_0.$$

(*i*) *The eigenfunctions satisfy the structured boundary conditions, for all* $n \in \mathbb{N}_0$,

$$\begin{array}{ll} \text{if } c \in (0,2) & \text{then } \lim_{x \to 0^+} [\psi_n^{c,d}, 1](x) = 0 \\ \text{if } d \in (0,2) & \text{then } \lim_{x \to 1^-} [\psi_n^{c,d}, 1](x) = 0 \\ \text{if } c \in [2,\infty) & \text{then no boundary condition at } 0^+ \text{ is required} \\ \text{if } d \in [2,\infty) & \text{then no boundary condition at } 1^- \text{ is required.} \end{array}$$

(*ii*) *For all* $c, d \in (0,\infty)$ *the set of eigenfunctions* $\left\{\psi_n^{c,d} : n \in \mathbb{N}_0\right\}$ *is orthogonal and complete in* $L^2((0,1);w)$.

(*iii*) *For all* $c, d \in (0,\infty)$ *and for each* $n \in \mathbb{N}_0$ *the eigenfunction* $\psi_n^{c,d}$ *has exactly* n *zeros in the open interval* $(0,1)$.

(*iv*) *For all* $c, d \in (0,\infty)$ *and for each* $n \in \mathbb{N}_0$ *the eigenfunction* $\psi_n^{c,d}$, *defined on* $(0,1)$, *has the property that the two limits* $\lim_{x \to 0^+, 1^-} \psi_n^{c,d}(x)$ *exist and are finite; if these two limits are used to define* $\psi_n^{c,d}(0)$ *and* $\psi_n^{c,d}(1)$, *respectively, then for all* $c, d \in (0,\infty)$ *and for all* $n \in \mathbb{N}_0$, *the eigenfunction* $\psi_n^{c,d} \in AC[0,1]$.

Remark 12. *The results of this theorem provide justification for the statement of Theorem* 1, *made in the Introduction, Section* 1, *of this paper.*

Proof. The difficult part of the proof of this Theorem 11 is to establish that the spectrum $\sigma(T_{c,d})$ of $T_{c,d}$ is discrete; this part we give below.

Assuming that it is known that the spectrum is discrete then:

1. The proof that the spectrum is simple follows from the separated form of the boundary conditions used to determine the operator domain $D(T_{c,d})$ given in Definition 5. The other properties follow from the results obtained in Section 10.

2(*i*). This statement follows from again from the boundary conditions given in Definition 5; see [7]

2(*ii*). The results stated follow from known properties of self-adjoint operators, in Hilbert spaces, with discrete simple spectra; see [6, Chapter V].

2(*iii*). This is a known standard result for Sturm-Liouville differential operators that are bounded below and have a discrete spectrum; see [7].

2(*iv*). For an endpoint where c or $d \in (0,2)$, say $c \in (0,2)$, it follows from the analysis in the proof of 4(*i*) of Theorem 8 that the derivative $\psi_n^{c,d\,\prime} \in L^1(0,1/2)$; the absolute continuity property then follows to give $\psi_n^{c,d} \in AC[0,1/2]$.

For an endpoint where $c, d \in [2, \infty)$, say $c \in [2, \infty)$, the analysis in Section 10 is no longer applicable to give $f' \in L^1(0, 1/2)$ for all $f \in D(T_{c,d})$. However it is possible to prove that the eigenfunction $\psi_n^{c,d}$ has the property that the derivative $\psi_n^{c,d\,\prime} \in L^1(0, 1/2)$; this result follows from the fact that the solution $\psi_n^{c,d}$, in this the limit-point case, is only linearly dependent upon the solution $y_{0,1}$ of the differential equation (17) with asymptotic form $y_{0,1}(x) \sim 1$, see (20), as $x \to 0^+$; this asymptotic form can be extended to give information about the derivative of the solution.

Combining these results for all cases of $c, d \in (0, \infty)$ we obtain the required proof for $2(iv)$ of the Theorem.

It remains then to prove that the spectrum $\sigma(T_{c,d})$ is discrete for all values of $c, d \in (0, \infty)$; the proof given now provides an independent proof that the self-adjoint operator $T_{c,d}$ is bounded below in $L^2((0,1); w)$, see Corollary 10. □

Definition 13. *Let $T : D(T) \subseteq H \to H$ be a linear self-adjoint operator in the Hilbert space H; then T is said to have the property* BD *if the spectrum $\sigma(T)$ of T is bounded below on the real line $\mathbb{R} \subset \mathbb{C}$ and is discrete, i.e. the essential spectrum of T is empty.*

Notation 14. *If the operator T has the property* BD *then we write $T \in$ BD.*

Theorem 15. *For all $c, d \in (0, \infty)$ the self-adjoint operator $T_{c,d}$, see Definition 5, has the property $T_{c,d} \in$ BD.*

For details of the proof see the remaining results of this section.

Remark 16. *If both $c, d \in (0, 2)$ then the result that $T_{c,d} \in$ BD follows from the fact that both the endpoints 0 and 1 are either regular or limit-circle in $L^2((0,1); w)$, and the coefficient property $p(x) > 0$ for all $x \in (0, 1)$, see [6]. However we include these cases in the analysis that follows since the arguments apply to all the cases of $c, d \in (0, \infty)$.*

Note that the difficulties in proving the BD *property arise from the limit-point cases of the endpoints 0 and 1.*

Definition 17. *For all $c, d \in (0, \infty)$ the linear operators $T_{c,d}^0$ and $T_{c,d}^1$ are defined by*

$$D(T_{c,d}^0) := \{f \in D(T_{c,d}) : f \restriction_{(0,1/2]} \text{ and } f(1/2) = 0\} \tag{73}$$

$$D(T_{c,d}^1) := \{f \in D(T_{c,d}) : f \restriction_{[1/2,1)} \text{ and } f(1/2) = 0\} \tag{74}$$

and

$$T_{c,d}^0 f := w^{-1} M[f] \text{ for all } f \in D(T_{c,d}^0) \tag{75}$$

$$T_{c,d}^1 f := w^{-1} M[f] \text{ for all } f \in D(T_{c,d}^1). \tag{76}$$

Lemma 18. *The operators $T_{c,d}^0$ and $T_{c,d}^1$ are self-adjoint, respectively in the Hilbert spaces $L^2((0, 1/2); w)$ and $L^2((1/2, 1); w)$.*

Proof. This result follows from the now established analysis for linear ordinary boundary value problems, given in [6, Chapter V, Section 18] and [1, Section 5]. □

Lemma 19. *For all $c, d \in (0, \infty)$ the operator $T_{c,d} \in \mathsf{BD}$ if and only if both $T^0_{c,d} \in \mathsf{BD}$ and $T^1_{c,d} \in \mathsf{BD}$.*

Proof. See the results in [2, Pages 1483 and 1455]. □

Thus it is sufficient to prove that $T^0_{c,d} \in \mathsf{BD}$ since the proof that $T^1_{c,d} \in \mathsf{BD}$ follows along similar lines. Hence we restrict attention to the interval $(0, 1/2]$; further we now restrict the parameter λ to be real-valued.

The differential expression $M - \lambda w$ on $(0, 1/2]$ is said to be *oscillatory at* 0 if the endpoint 0 is an accumulation point of zeros of a non-trivial solution y of the differential equation $M[y] - \lambda w y = 0$ on $(0, 1/2]$; otherwise $M - \lambda w$ is *non-oscillatory at* 0. From the Sturm separation theorem this definition is independent of the particular solution y; further there can be no point of accumulation of zeros in $(0, 1/2]$ since for any solution y the zeros are isolated.

For a compact interval $[a, b] \subset (0, 1/2]$ the differential expression $M - \lambda w$ is said to be *oscillatory on* $[a, b]$ if there is a non-trivial solution of $M[y] - \lambda w y = 0$ with two zeros in $[a, b]$; otherwise $M - \lambda w$ is *non-oscillatory on* $[a, b]$.

Define the set of functions $A[a, b]$ by

$$A[a,b] := \left\{ f : [a,b] \to \mathbb{R} : f \in AC[a,b], f' \in L^2(a,b) \text{ and } f(a) = f(b) = 0 \right\}. \quad (77)$$

We now state two results from the paper [3] by Hinton and Lewis.

Theorem 20. *The differential expression $M - \lambda w$ is non-oscillatory on $[a, b] \subset (0, 1/2]$ if and only if for all $f \in A[a, b]$, with $f \neq 0$,*

$$\int_a^b \{pf'^2 + qf^2 - \lambda w f^2\} > 0. \quad (78)$$

Proof. For the proof of Theorem 20 see the results in [3, Theorem 0.2]. □

Lemma 21. *If $f \in A[a, b]$ and $f \equiv y$ is also a solution of the differential equation $M[y] - \lambda w y = 0$ on $[a, b]$, then*

$$\int_a^b \{pf'^2 + qf^2 - \lambda w f^2\} = 0. \quad (79)$$

Proof. This result follows on integration by parts on the term pf'^2. □

Theorem 22. *For all $c, d \in (0, \infty)$ the operator $T^0_{c,d} \in \mathsf{BD}$ if and only if for all $\lambda \in \mathbb{R}$ the differential expression $M - \lambda w$ is non-oscillatory at* 0.

Proof. For the proof of this Theorem see the results in [3, Theorem 0.1]. □

As a consequence of the results in the two theorems we have

Theorem 23. *The operator $T^0_{c,d} \in \mathsf{BD}$ if for each real number λ there exists a positive number $\delta \equiv \delta(\lambda)$ such that for all $[a, b] \subset (0, \delta)$ and then for all $f \in A[a, b]$, with $f \neq 0$, it follows that*

$$\int_a^b \{pf'^2 + qf^2 - \lambda w f^2\} > 0. \quad (80)$$

Proof. The proof of this theorem follows from the results in Theorems 20 and 22. For suppose that the conditions of this theorem hold but that, for some $\lambda \in \mathbb{R}$, $M - \lambda w$ is oscillatory at 0. Let y be a non-trivial solution of $M[y] - \lambda wy = 0$ on $(0, 1/2]$; let $\delta \in (0, 1/2]$ and take points a, b with $[a, b] \subset (0, \delta)$ such that $y(a) = y(b) = 0$. Then from Lemma 21 it follows that (80) does not hold. Thus $M - \lambda w$ is non-oscillatory at 0 and from Theorem 22 it follows that $T^0_{c,d} \in \mathsf{BD}$. □

We now prove that the conditions for Theorem 23 to hold are satisfied by the operator $T^0_{c,d}$. This next step requires the use of two Hardy type inequalities as given in

Lemma 24. *If $f \in A[a, b]$, with $f \neq 0$, then for all $\alpha \in \mathbb{R}$ but $\alpha \neq -1$*

$$\int_a^b x^\alpha f(x)^2\,dx < \frac{4}{(1+\alpha)^2} \int_a^b x^{\alpha+2} f'(x)^2\,dx \tag{81}$$

and

$$\int_a^b \frac{1}{x\,(\ln(x))^2} f(x)^2\,dx < 4 \int_a^b x f'(x)^2\,dx. \tag{82}$$

Proof. For the proof of these inequalities see [3, Section 2, Lemma 2.1]. □

From the definition of the three coefficients p, q, w in (18) we obtain the following bounds

$$p(x) \geq k_1 x^c \qquad |q(x)| \leq k_2 x^c \qquad w(x) \leq k_3 x^{c-1} \quad \text{for all } x \in (0, 1/2] \tag{83}$$

where the positive numbers k_1, k_2, k_3 are given by

$$k_1 = (1/2)^d s^e \qquad k_2 = |ab|\,(1+s)^{e-1} \qquad k_3 = (1+s)^{e-1}. \tag{84}$$

First, consider the case when the parameter $c \neq 1$. For $f \in A[a, b]$, with $f \neq 0$, the bounds (83) and (84) give

$$\int_a^b \{pf'^2 + qf^2 - \lambda w f^2\} \geq \int_a^b \{k_1 x^c f'^2 - k_2 x^c f^2 - k_3\,|\lambda|\, x^{c-1} f^2\}.$$

Applying the inequality (81) from Lemma 24, with $\alpha = c - 2$, we obtain

$$\begin{aligned} &\int_a^b \{pf'^2 + qf^2 - \lambda w f^2\} \\ &\qquad > \int_a^b \{(k_1(c-1)^2/4)x^{c-2} - k_2 x^c - k_3\,|\lambda|\, x^{c-1}\}\, f(x)^2\,dx \\ &\qquad = \int_a^b x^{c-2}\,\{(k_1(c-1)^2/4) - k_2 x^2 - k_3\,|\lambda|\, x\}\, f(x)^2\,dx. \end{aligned}$$

Clearly then, from these last results, there exists a positive number δ such that for all $x \in (0, \delta)$

$$\{(k_1(c-1)^2/4) - k_2 x^2 - k_3\,|\lambda|\, x\} > 0;$$

hence

$$\int_a^b \{pf'^2 + qf^2 - \lambda w f^2\} > 0.$$

Finally from the Theorems 20, 22 and 23 it now follows that $T^0_{c,d} \in \mathsf{BD}$, on using the result of Theorem 23.

Second, the proof for the case $c = 1$ is similar and follows from the use of the inequality (82) from Lemma 24. Following the reasoning given above we find that for $f \in A[a,b]$, with $f \neq 0$,

$$\int_a^b \{pf'^2 + qf^2 - \lambda w f^2\} > \int_a^b \frac{1}{x\,(\ln(x))^2}\{(k_1/4) - k_2 x^2\,(\ln(x))^2$$

$$-k_3\,|\lambda|\,x\,(\ln(x))^2\}f(x)^2 dx$$

and then the existence of the required positive number δ is clear; the proof is then completed as above.

This completes the proof that for all $c, d \in (0, \infty)$ the operator $T^0_{c,d} \in \mathsf{BD}$; there is a similar proof to give $T^1_{c,d} \in \mathsf{BD}$. From Lemma 19 it follows that the proof of Theorem 15 is now complete.

Finally, returning to the earlier part of this section it follows that the proof of Theorem 11 is now complete.

12. Remarks

1. The eigenfunctions $\{\psi_n^{c,d} : n \in \mathbb{N}_0\}$, see Theorem 11, are solutions of the Sturm-Liouville differential equation (17) and hence of the Heun equation (8). However we have not been able to classify these solutions as Heun functions; see the source book [8] on the general Heun differential equation, in particular [8, Chapter A, Section 5.2]; is it possible to classify the eigenfunctions $\{\psi_n^{c,d} : n \in \mathbb{N}_0\}$ within the Heun function classes I, II, III and IV relative to $0, 1$?
2. The Lay/Slavyanov paper [5], and in earlier contributions, is concerned with the properties of the boundary value problem determined by the differential equation (1) and the boundary conditions (7), when the parameter s tends to 0 through positive values and then its connection with the boundary value problem when $s = 0$. We hope to study this limiting process in the mode of strong resolvent convergence within the associated Hilbert function spaces.

Acknowledgement. Norrie Everitt thanks his colleague Sergei Slavyanov for help and advice in respect of the work of this paper, on the occasion of the Kovalevski Symposium held at the University of Stockholm in June 2000.

All four authors thank two referees for their careful scrutiny of the manuscript of this paper and for their comments which led to improvement in the presentation of the results.

References

[1] P.B. Bailey, W.N. Everitt and A. Zettl. 'The SLEIGN2 Sturm-Liouville code.' (To appear in *ACM Trans. Math. Software*; presently available as .pdf and .ps files on the web site "http://www.math.niu.edu/~zettl/SL2".)

[2] N. Dunford and J.T. Schwartz. *Linear operators* **II**: *spectral theory.* (Interscience, New York: 1963.)

[3] D.B. Hinton and R.T. Lewis. 'Singular differential operators with spectra discrete and bounded below.' *Proc. Roy. Soc. Edinburgh* A **84** (1979), 117–134.

[4] E.L. Ince. *Ordinary differential equations.* (Dover Publications Inc., New York: 1956.)

[5] W. Lay and S. Yu. Slavyanov. 'Heun's equation with nearby singularities.' *Proc. R. Soc. Lond.* A **455** (1999), 4347–4361.

[6] M.A. Naimark. *Linear differential operators* **II**. (Ungar Publishing Company, New York: 1968.)

[7] H.-D. Niessen and A. Zettl. 'Singular Sturm-Liouville problems: the Friedrichs extension and comparison of eigenvalues.' *Proc. London Math. Soc.* (3) **64** (1992), 545–578.

[8] A. Ronveaux. *Heun's differential equations.* (Oxford University Press: 1995.)

P.B. Bailey
c/o Department of Mathematical Sciences
Northern Illinois University
DeKalb, IL 60115-2888, USA
e-mail: `76021.3674@compuserve.com`

W.N. Everitt
School of Mathematics and Statistics
University of Birmingham
Edgbaston
Birmingham, England, UK
e-mail: `w.n.everitt@bham.ac.uk`

D.B. Hinton
Department of Mathematics
University of Tennessee
Knoxville, TN 37916, USA
e-mail: `hinton@math.utk.edu`

A. Zettl
Department of Mathematical Sciences
Northern Illinois University
DaKalb, IL 60115-2888, USA
e-mail: `zettl@math.niu.edu`

Operator Theory:
Advances and Applications, Vol. 132, 111–119

On the Approximation of the Solution of the Schrödinger Equation by Superpositions of Stationary Solutions

Johannes F. Brasche

Abstract. Let S be a symmetric operator with gap J. Suppose in addition that the deficiency indices of S are infinite, the Hamiltonian H is a self-adjoint extension of S and the support of the spectral measure $\mu_{f_0,H}$ of the initial state f_0 is a compact subset of J. Then there exist other self-adjoint extensions H_n of S and finite sums f_n of eigenvectors of H_n such that

$$e^{-itH_n} f_n \longrightarrow e^{-itH} f_0, \quad \text{as} \quad n \longrightarrow \infty,$$

locally uniformly in time. Upper estimates for the rate of convergence will be given.

1. Introduction

Obviously $f(t) = e^{-it\lambda} f_0$ is the solution of the Schrödinger equation

$$\begin{aligned} i\frac{d}{dt} f(t) &= Hf(t), \\ f(0) &= f_0, \end{aligned}$$

provided $Hf_0 = \lambda f_0$. Such solutions are called stationary. More generally it is trivial to solve the Schrödinger equation if the initial vector f_0 is a superposition of eigenvectors of H. Note that f_0 is a superposition of eigenvectors of the self-adjoint operator H if and only if f_0 belongs to the space $\mathcal{H}^{pp}(H)$ of vectors whose spectral measure (with respect to H) is a pure point measure.

Now let f_0 be any initial vector. Due to the continuous dependence on the initial conditions one might try to apply the following strategy in order to find the solution of the Schrödinger equation: One approximates f_0 by a sequence (f_n) in the space $\mathcal{H}^{pp}(H)$. Then the solutions $e^{-itH} f_n$ corresponding to the initial vectors f_n converge locally uniformly in t to the solution $e^{-itH} f_0$ corresponding to the initial vector f_0.

I would like to thank N. Elander and P. Kurasov for the organisation of a very pleasant and profitable conference. Financial support by the sponsors of the conference is gratefully acknowledged. I am grateful to H. Neidhardt and J. Weidmann for useful conversations. I would like to thank the anonymous referees for valuable suggestions and corrections.

As it is well known this strategy fails since the space $\mathcal{H}^{pp}(H)$ is closed and for $f_0 \notin \mathcal{H}^{pp}(H)$ the mentioned approximation is not possible.

While the above strategy to solve the Schrödinger equation fails a modification of it might be successful. One chooses self-adjoint operators H_n, $n \in \mathbb{N}$, and $f_n \in \mathcal{H}^{pp}(H_n)$ such that

$$e^{-itH_n} f_n \longrightarrow e^{-itH} f_0, \quad \text{as} \quad n \longrightarrow \infty,$$

locally uniformly in t.

We shall prove that this modified approach works if H is a self-adjoint extension of a densely defined symmetric operator S, the support of the spectral measure $\mu_{f_0,H}$ of f_0 with respect to H is a compact subset of a spectral gap J of S and for one and therefore every λ in J the dimension of the space of solutions of the eigenequation

$$S^* f = \lambda f$$

is infinite dimensional. Since J is a gap of S this last condition is equivalent to the fact that S has infinite deficiency indices.

More precisely we shall give a sequence (f_n) such that

$$e^{-itH_n} f_n \longrightarrow e^{-itH} f_0, \quad \text{as} \quad n \longrightarrow \infty,$$

locally uniformly in t and for every $n \in \mathbb{N}$ the vector f_n is the sum of finitely many, say $N(n)$, eigenvectors of another self-adjoint extension H_n of S.

We shall give upper estimates for the rate of convergence in terms of the numbers $N(n)$. Roughly speaking these estimates will depend on "how fast the spectral measure $\mu_{f_0,H}$ can be approximated by linear combinations of finitely many Dirac measures". Thus we are especially interested in the case when the measure $\mu_{f_0,H}$ is concentrated on a set with small Hausdorff-dimension. We refer to [6] and references therein for other results on the solution of the Schrödinger equation if the initial state has a continuous spectral measure concentrated on a set with small Hausdorff-dimension. We refer to [5] for the description of an important class of self-adjoint operators with continuous but zero-dimensional spectral measures and a discussion of the relation to the Anderson model.

Note that both H and H_n, $n \in \mathbb{N}$, are restrictions of the adjoint S^* of S. If S is a differential operator then this implies that both H and the H_n are described via the same differential expression but correspond to different choices of boundary conditions. Thus in many applications it is possible to approximate solutions of the Schrödinger equation by superpositions of stationary solutions via suitable variations of boundary conditions.

There is an intimate relationship between the mentioned result on the solution of the Schrödinger equation and a problem in spectral theory. K.O. Friedrichs and M. Stone resp. M.G. Krein have shown that the open interval $J = (a, b)$ is a gap of the symmetric operator S, i.e. there exists at least one self-adjoint extension H_∞

of S without spectrum in J, if and only if

$$\begin{aligned} (Sf,f) &\geq b \parallel f \parallel^2, &\text{if} \quad -\infty = a < b < \infty, \\ \parallel \left(S - \frac{a+b}{2}\right) f \parallel &\geq \frac{b-a}{2} \parallel f \parallel, &\text{if} \quad -\infty < a < b < \infty. \end{aligned}$$

In addition to the self-adjoint extensions of S preserving the gap J there might exist other self-adjoint extensions with some spectrum inside J and one might ask about which kinds of spectra inside J these other self-adjoint extensions can have. In 1947 M.G. Krein has given the complete answer to this question in the special case when the deficiency indices of S are finite. In 2000 I have given the complete answer in the general case [3].

A key in the proof has been the surprising observation that even certain vectors f in the domain $D(S^*)$ of S^* which cannot be represented as

$$f = \sum_{n=1}^{\infty} a_n e_n$$

for some orthonormal family of eigenvectors e_n of S^* can be approximated by finite sums of eigenvectors of S^*. This observation will also play a key role in the mentioned result on the approximative solution of the Schrödinger equation.

2. Approximate solution of the Schrödinger equation

In what follows let S be a symmetric operator in a complex Hilbert space $\mathcal{H}$. Suppose that the open interval J is a gap of S and for one and therefore every λ in J the space of solutions of the eigenequation

$$S^* f = \lambda f$$

is infinite-dimensional. Let H be a self-adjoint extension of S and f_0 a vector such that the support of the spectral measure $\mu_{f_0,H}$ of f_0 with respect to H is a compact subset of J. Here $E_H(\cdot)$ denotes the projection operator-valued measure associated to H and

$$\mu_{f,H}(\cdot) :=\parallel E_H(\cdot) f \parallel^2 .$$

The following lemma will play a keyrole in our proof of the mentioned result on the approximative solution of the Schrödinger equation.

Lemma 1. *Let $\lambda \in J$ and let P be the orthogonal projection onto the kernel $\mathcal{N}_\lambda$ of $S^* - \lambda$. Then for every h in the domain of the self-adjoint extension H of S*

$$\parallel h - Ph \parallel \leq \frac{1}{dist(\lambda, \partial J)} \sup_{\lambda' \in C_h} |\lambda - \lambda'| \parallel h \parallel$$

where C_h denotes the support of the spectral measure $\mu_{h,H}$ of h with respect to H.

Proof. We may assume that the operator S is closed. Then the range of $S-\lambda$ is a closed subspace of $\mathcal{H}$ and we have

$$(\operatorname{ran} P)^{\perp} = \mathcal{N}_{\bar\lambda}^{\perp} = \operatorname{ran}(S-\lambda).$$

We choose normalized vectors $e_1 \in (\operatorname{ran} P)^{\perp}$ and $e_2 \in \operatorname{ran} P$ such that

$$h = (e_1, h)\, e_1 + (e_2, h)\, e_2.$$

We have

$$\begin{aligned} \| (S^* - \lambda) h \|^2 &= \| (H-\lambda) h \|^2 \\ &= \int |\lambda' - \lambda|^2 \mu_{h,H}(d\lambda') \\ &\leq \sup_{\lambda' \in C_h} |\lambda - \lambda'|^2 \; \| h \|^2 \end{aligned}$$

since $\mu_{h,H}$ is supported by C_h.

We choose $g \in D(S)$ such that

$$e_1 = (S-\lambda)\, g.$$

We have

$$\| g \| \leq \| (S-\lambda)^{-1} \| \, \| e_1 \| \leq \frac{1}{\operatorname{dist}(\lambda, \partial J)}.$$

Thus

$$\begin{aligned} |(e_1, h)| &= |((S-\lambda)\, g, h)| \\ &= |(g, (S^* - \lambda)\, h)| \\ &\leq \frac{1}{\operatorname{dist}(\lambda, \partial J)} \; \| (S^* - \lambda) h \| \\ &\leq \frac{1}{\operatorname{dist}(\lambda, \partial J)} \sup_{\lambda' \in C_h} |\lambda - \lambda'| \; \| h \| . \end{aligned}$$

Since $h - Ph = (e_1, h)\, e_1$ the assertion is proved. □

We shall use the following result from [1]:

Lemma 2. ([1], *Lemma* 2.2) *Let S be a symmetric operator in the Hilbert space $\mathcal{H}$. Suppose that the open interval J is a gap of S. Let $\mathcal{H}_0$ be a closed subspace of $\mathcal{H}$ and M a self-adjoint operator in the Hilbert space $\mathcal{H}_0$. Suppose that M is a restriction of the adjoint S^* of S and the spectrum of M is a subset of the gap J. Then the operator*

$$S_M := S^*_{\restriction D(S)+D(M)}, \tag{1}$$

i. e. the restriction S_M of S^* to the space

$$D(S) + D(M) := \{f + g : f \in D(S), g \in D(M)\},$$

can be represented in the form

$$S_M = M \oplus G_0 \tag{2}$$

for a unique symmetric operator G_0 in the Hilbert space $\mathcal{H}_0^{\perp}$. Moreover the gap J of S is also a gap of G_0.

By the following theorem, solutions of the Schrödinger equation

$$\begin{aligned} i\frac{d}{dt}f(t) &= Hf(t), \\ f(0) &= f_0, \end{aligned}$$

can be approximated by superpositions of stationary solutions corresponding to other self-adjoint extensions of S provided the support of the measure $\mu_{f_0,H}$ is a compact subset of a gap of S. The theorem gives an upper bound for the rate of convergence and the proof of the theorem a method to construct such approximate solutions.

Theorem 3. *Suppose that the support of the spectral measure $\mu_{f_0,H}$ of f_0 with respect to H is a compact subset of the gap J of S. Let $B_1, \ldots, B_N$ be pairwise disjoint Borel sets which cover the support of $\mu_{f_0,H}$. Let $\lambda_1, \ldots, \lambda_N$ be points in J such that*

$$\sup_{\lambda' \in B_j} |\lambda_j - \lambda'| \leq d, \quad j = 1, 2, \ldots, N,$$

for some constant d. Let D be the distance of the set $\{\lambda_1, \ldots, \lambda_N\}$ to the boundary of J. Then there exist an orthonormal family $(e_j)_{j=1}^N$ and a self-adjoint extension $\tilde{H}$ of S such that

(i) $\tilde{H}e_j = \lambda_j e_j, \quad j = 1, 2, \ldots, N,$ *and*

(ii) *for*

$$f := \sum_{j=1}^{N} \alpha_j e_j, \quad \alpha_j := \sqrt{\mu_{f_0,H}(B_j)}, \; j = 1, 2, \ldots, N, \tag{3}$$

the following estimate holds for all $t \in \mathbb{R}$:

$$\| e^{-it\tilde{H}} f - e^{-itH} f_0 \| \leq \sqrt{N}\, d \, \| f_0 \| \, \left(\frac{\sqrt{2}}{D} + |t|\right). \tag{4}$$

Proof. We may assume that $E_H(B_j)f_0 \neq 0$ for all j. Then the vectors

$$\tilde{e}_j := \frac{E_H(B_j)f_0}{\| E_H(B_j)f_0 \|}$$

form an orthonormal system and

$$f_0 = \sum_{j=1}^{N} \alpha_j \tilde{e}_j. \tag{5}$$

We shall apply Lemma 2 several times. First set

$$\mathcal{H}_0 := \operatorname{ran} E_H(B_2 \cup \cdots \cup B_N).$$

Without loss of generality we may assume that $B_1 \cup \cdots \cup B_N$ is a relatively compact subset of J. Then the space $\mathcal{H}_0$ is contained in the domain $D(H)$ of H,

$$M := H_{|\mathcal{H}_0}$$

is a self-adjoint operator in $\mathcal{H}_0$, $M \subset S^*$ and the spectrum of M is contained in J. By Lemma 2,

$$S \subset M \oplus G_0$$

for some symmetric operator G_0 in $\mathcal{H}_0^\perp$ and the gap J of S is also a gap of G_0.

Apparently

$$H = M \oplus G$$

for some self-adjoint extension G of G_0. It follows that $\tilde{e}_1 \in D(G)$ and $\mu_{\tilde{e}_1,G} = \mu_{\tilde{e}_1,H}$. Let $P : \mathcal{H}_0^\perp \longrightarrow \ker(G_0^* - \lambda_1)$ be the orthogonal projection onto the kernel of $G_0^* - \lambda_1$. By Lemma 1,

$$\| \tilde{e}_1 - P\tilde{e}_1 \| \leq \frac{d}{D}.$$

Thus there exists a normalized vector $e_1 \in \ker(G_0^* - \lambda_1)$ such that

$$\| e_1 - \tilde{e}_1 \| \leq \sqrt{2}\,\frac{d}{D}.$$

Note that $G_0^* \subset S^*$. Thus $S^* e_1 = \lambda_1 e_1$. Moreover, by construction, e_1 is orthogonal to the space $\text{ran}E_H(B_2 \cup \cdots \cup B_N)$.

Now we change the notation. We set

$$\mathcal{H}_0 := \text{span}\{e_1\} + \text{ran}E_H(B_3 \cup \cdots \cup B_N).$$

Obviously $\mathcal{H}_0 \subset D(S^*)$ and $M := S^*_{|\mathcal{H}_0}$ is a self-adjoint operator in $\mathcal{H}_0$, $M \subset S^*$ and the spectrum of M is contained in J. By Lemma 2,

$$S \subset M \oplus G_0$$

for some symmetric operator G_0 in $\mathcal{H}_0^\perp$ and J is also a gap of G_0.

Apparently G_0 has a self-adjoint extension G such that $\text{ran}E_H(B_2) \subset D(G)$ and $Gh = Hh$ for all $h \in \text{ran}E_H(B_2)$. In particular, $\mu_{\tilde{e}_2,G} = \mu_{\tilde{e}_2,H}$. By applying Lemma 1 in the same way as before, we can show that there exists a normalized vector e_2 in the kernel of $S^* - \lambda_2$ such that

$$\| e_2 - \tilde{e}_2 \| \leq \sqrt{2}\,\frac{d}{D}$$

and e_2 is orthogonal to $\text{span}\{e_1\} + \text{ran}\,E_H(B_3 \cup \cdots \cup B_N)$.

Proceeding in this way, we get an orthonormal system $\{e_j\}_{j=1}^N$ such that

$$S^* e_j = \lambda_j e_j, \quad j = 1, \ldots, N,$$

and

$$\| e_j - \tilde{e}_j \| \leq \sqrt{2}\,\frac{d}{D}, \quad j = 1, \ldots, N. \tag{6}$$

Since $S^* e_j = \lambda_j e_j$ and the e_j are pairwise orthogonal, there exists a self-adjoint extension $\tilde{H}$ of S such that $\tilde{H} e_j = \lambda_j e_j$ for all j, cf., e.g., [1]. Since the measure $\mu_{\tilde{e}_j,H}$ is concentrated on B_j

$$\| e^{-itH}\tilde{e}_j - e^{-it\lambda_j}\tilde{e}_j \|^2 = \int | e^{-it\lambda} - e^{-it\lambda_j} |^2 \, \mu_{\tilde{e}_j,H}(d\lambda) \leq d^2 t^2. \tag{7}$$

By (6) and (7),

$$\| e^{-itH}\tilde{e}_j - e^{-it\lambda_j} e_j \| \leq d(\frac{\sqrt{2}}{D} + |\, t\,|). \tag{8}$$

By (3), (5) and (8),

$$\| e^{-itH} f_0 - e^{-it\tilde{H}} f \|^2 \leq \left(\sum_{j=1}^{N} \alpha_j^2 \right) \sum_{j=1}^{N} d^2 \, (\frac{\sqrt{2}}{D} + |\, t\,|)^2$$

and, by (3), the theorem is proved. □

Remark 4. *Generalizing the construction of the Cantor measure one gets for arbitrarily small $c > 0$ examples of continuous measures μ_c such that for each $n \in \mathbb{N}$ one can choose $N = 2^n$ and $d = c^n$ in the above theorem (with $\mu_{f_0,H} = \mu_c$). For c sufficiently small one gets a good appoximation of the solution of the Schrödinger equation by superpositions of stationary solutions.*
Obviously, the smaller c is, the smaller is the Hausdorff dimensionality of the measure μ_c. Recently a lot of work has been done in order to investigate Hamiltonians which have continuous spectral measures concentrated on sets of small Hausdorff dimension (often even Hausdorff dimension 0), cf. [5], [6],[8] *and references given therein.*

Remark 5. *In the above proof we have used the fact that the support of the spectral measure $\mu_{f_0,H}$ is contained in a gap of the symmetric operator S. It might be possible to weaken this hypothesis but one cannot completely omit it. E.g. let $\mathcal{H} = L^2(\mathbb{R}^d)$, $d > 1$. Let Γ be a closed subset of $\mathbb{R}$ such that Γ has Lebesgue measure zero and its complement $\mathbb{R}^d \setminus \Gamma$ is connected. Define the symmetric operator S in $L^2(\mathbb{R}^d)$ as follows:*

$$D(S) := C_0^\infty(\mathbb{R}^d \setminus \Gamma), \quad Sf := -\Delta f, \quad f \in D(S).$$

It is a well-known consequence of Kato's inequality that the adjoint S^* of S does not have nonnegative eigenvalues. Thus the method described in the proof of the above theorem cannot be applied if the support of the spectral measure $\mu_{f_0,H}$ contains positive real numbers. Note that the operator S has infinite deficiency indices if the set Γ is sufficiently big, e.g. if Γ has infinitely many points and its Hausdorff dimension is larger than $d-4$.

3. A result in Inverse Spectral Theory

Let A be a self-adjoint operator in a Hilbert space $\mathcal{H}$. It easily follows from the spectral theorem, that for every Borel set $B \subset \mathbb{R}$ we have

$$\mathcal{H} = \operatorname{ran} E_A(B) \oplus \operatorname{ran} E_A(\mathbb{R} \setminus B)$$

and there exist unique self-adjoint operators A_B in $\operatorname{ran}(E_A(B))$ and $A_{\mathbb{R}\setminus B}$ in $\operatorname{ran}(E_A(\mathbb{R} \setminus B))$ such that

$$A = A_B \oplus A_{\mathbb{R}\setminus B}.$$

Inside B the operators A and A_B have the same eigenvalues and for every eigenvalue $\lambda \in B$ of A the multiplicity mult (λ, A) of λ as an eigenvalue of A equals mult (λ, A_B).

For open sets J we have in addition that

$$\mu_{f,A}(B) = \mu_{f,A_J}(B)$$

for every Borel set $B \subset J$ and every $f \in \mathcal{H}$. In particular, we have

$$\sigma(A) \cap J = \sigma(A_J) \cap J, \ \sigma_{ac}(A) \cap J = \sigma_{ac}(A_J) \cap J, \ \sigma_{sc}(A) \cap J = \sigma_{sc}(A_J) \cap J,$$

and for every $\alpha \in [0, 1]$

$$\sigma_\alpha(A) \cap J = \sigma_\alpha(A_J) \cap J.$$

Here σ, σ_{ac}, σ_{sc} and σ_α denote the spectrum, the absolutely continuous spectrum, the singular continuous spectrum and the α-dimensional spectrum (cf. [5]), respectively.

Let S be a symmetric operator with deficiency indices (n, n). Suppose that the open interval J is a gap of S. It easily follows from von Neumann's extension theory that

$$\dim \operatorname{ran} E_A(J) \leq n$$

for every self-adjoint extension A of S; "dim" means dimension in the sense of Hilbert space theory, i.e. the cardinality of any orthonormal base. Up to unitary equivalence this is the only restriction for the operators A_J, A being a self-adjoint extension of S:

Theorem 6. ([3], *Theorem* 1) *Let S be a symmetric operator in the Hilbert space $\mathcal{H}$. Suppose that the open interval J is a gap of S and the deficiency indices of S equal (n, n). Let A^{aux} be any self-adjoint operator such that*

$$\dim \operatorname{ran} E_{A^{aux}}(J) \leq n.$$

Then there exists a self-adjoint extension A of S such that

$$A_J \simeq A_J^{aux},$$

i.e. $A_J = U^{-1} A_J^{aux} U$ for some unitary operator U.

Remark 7. *In particular, A and A^{aux} have the same eigenvalues inside J and for every eigenvalue $\lambda \in J$ of A we have*

$$mult(\lambda, A) = mult(\lambda, A^{aux}).$$

Moreover

$$\sigma(A) \cap J = \sigma(A^{aux}) \cap J, \ \sigma_{ac}(A) \cap J = \sigma_{ac}(A^{aux}) \cap J, \ \sigma_{sc}(A) \cap J = \sigma_{sc}(A^{aux}) \cap J,$$

and for every $\alpha \in [0, 1]$

$$\sigma_\alpha(A) \cap J = \sigma_\alpha(A^{aux}) \cap J.$$

Remark 8. *The theorem had been formulated as a conjecture in* [1].

Remark 9. *In the special case when the deficiency index n is finite the theorem has already been proved by M.G. Krein* ([7]).

Let S be a symmetric operator with infinite deficiency indices and J a gap of S. Let μ be a finite measure with compact support inside J. Let A^{aux} be the operator of multiplication by the independent variable in the Hilbert space $L^2(\mathbb{R},\mu)$. Then $A_J^{aux} = A^{aux}$ and, by the above theorem, there exist a self-adjoint extension A of S and a unitary transformation U such that $A_J = U^{-1}A^{aux}U$.

Let $f := U^{-1}1$, 1 being (the μ-equivalence class of) the function which equals 1 everywhere. Since the support of μ is compact and A a restriction of S^*, the vector f belongs to the domain of S^{*k} for every k and

$$(S^{*k}f, S^{*j}f) = \int \lambda^{k+j}\mu(d\lambda), \quad k,j = 0,1,2,\ldots \tag{9}$$

One of the key problems in the proof of the above theorem has been to show the existence of a vector f satisfying these equations (9). This could be done by a construction which is similar to the one in the proof of Theorem 3 but more complicated. Once this problem was solved the proof of the theorem could be completed by applying ideas and results from [1], [2] and [4], cf. [3] for the details.

References

[1] S. Albeverio, J. F. Brasche and H. Neidhardt: On inverse spectral theory for self-adjoint extensions, *J. Funct. Anal.* 154 (1998), 130–173.

[2] J. F. Brasche: Inverse spectral theory: Nowhere dense singular continuous spectra and Hausdorff dimension of spectra, *J. Op. Theory* 43 (2000), 145–169.

[3] J.F. Brasche: The spectra of the self-adjoint extensions of a symmetric operator S inside a gap of S. Preprint 2000:55 of the Department of Mathematics, Chalmers University of Technology and University of Göteborg. Submitted.

[4] J.F. Brasche, H. Neidhardt, J. Weidmann: On the point spectrum of self-adjoint extensions, *Math. Zeitschr.* 214 (1993), 343–355.

[5] R. del Rio, S. Jitomirskaja, Y. Last, B. Simon: Operators with singular continuous spectrum IV. Hausdorff dimensions, rank one perturbations, localization. *J. d'Analyse Mathématique 69* (1996), 153–200.

[6] A. Kiselev, Y. Last: Solutions, spectrum and dynamics for Schrödinger operators on infinite domains. *Duke Math. Journ.* 102, no. 1 (2000), 125–150.

[7] M.G. Krein: Theory of self-adjoint extensions of semi-bounded Hermitean operators and its applications. *Math. Sbornik* 20, no.3 (1947), 431–490.

[8] B. Simon: Operators with singular continuous spectrum: I. General operators. *Ann. of Math.* 141 (1995), 131–145.

Department of Mathematics
Chalmers University of Technology
and
University of Göteborg
41296 Göteborg, Sweden
e-mail: brasche@math.chalmers.se

Operator Theory:
Advances and Applications, Vol. 132, 121–130

Lyapunov Exponents in Continuum Bernoulli-Anderson Models

David Damanik, Robert Sims, and Günter Stolz

Abstract. We study one-dimensional, continuum Bernoulli-Anderson models with general single-site potentials and prove positivity of the Lyapunov exponent away from a discrete set of critical energies. The proof is based on Fürstenberg's Theorem. The set of critical energies is described explicitly in terms of the transmission and reflection coefficients for scattering at the single-site potential. In examples we discuss the asymptotic behavior of generalized eigenfunctions at critical energies.

1. The Main Result

We study Anderson-type random Schrödinger operators

$$H_\omega = -\frac{d^2}{dx^2} + \sum_{n\in\mathbb{Z}} q_n(\omega) f(x-n) \qquad (1)$$

in $L^2(\mathbb{R})$. The *single site potential* f is assumed to be real, supported in $[-1/2, 1/2]$, locally in L^1, and not identical to 0 (in L^1-sense). The *coupling constants* $q_n(\omega)$, $n \in \mathbb{Z}$, are independent, identically distributed Bernoulli random variables, i.e. they have distribution μ with $\operatorname{supp}\mu = \{0, 1\}$. Thus the random potential $V_\omega(x) = \sum_n q_n(\omega) f(x-n)$ is in $L^1_{loc,unif}(\mathbb{R})$ for all ω, which allows to associate a unique selfadjoint operator with H_ω that may equivalently be defined by form methods or in the sense of Sturm-Liouville theory.

Our interest in Bernoulli-Anderson models of this type arises from the fact that their spectral properties are not as well understood, by mathematically rigorous standards, as those of Anderson models with continuous (or absolutely continuous) distribution μ. Furthermore, these discrete distributions are physically more relevant as they are modeling the charge numbers of nuclei.

D.D. partially supported by the German Academic Exchange Service through HSP III (Postdoktoranden).
R.S. partially supported by NSF Grant DMS-9706076.
G.S. partially supported by NSF Grants DMS-9706076 and DMS-0070343.

One of the central objects in the study of one-dimensional random operators is the *Lyapunov exponent*. To define it for our model, let $g_E(n,\omega)$ be the transfer matrix of

$$-u'' + V_\omega u = Eu \tag{2}$$

from $n - 1/2$ to $n + 1/2$, i.e. for any solution of (2) one has

$$\begin{pmatrix} u(n+1/2) \\ u'(n+1/2) \end{pmatrix} = g_E(n,\omega) \begin{pmatrix} u(n-1/2) \\ u'(n-1/2) \end{pmatrix}.$$

Also, for $n \in \mathbb{N}$ let $U_E(n,\omega) = g_E(n,\omega)\dots g_E(1,\omega)$, and define the Lyapunov exponent at E by

$$\gamma(E) = \lim_{n\to\infty} \frac{1}{n}\mathbb{E}(\log \|U_E(n,\omega)\|), \tag{3}$$

where $\mathbb{E}$ denotes expectation with respect to ω. Existence of $\gamma(E)$ follows from the subadditive ergodic theorem, e.g. [5]. Since $\|U_E\| \geq 1$, we have $\gamma(E) \geq 0$, and it can also be seen that one gets the same $\gamma(E)$ if transfer matrices on the negative half line are used analogously in (3).

For many applications it is crucial to know at which energies one has that $\gamma(E) > 0$, which corresponds to exponential growth (or decay) of the solutions to (2). Our main goal here is to show that under the above assumptions this holds for all but a discrete set M of energies E, and to explicitly describe M in terms of the *transmission* and *reflection coefficients* for scattering at the single site potential f. To define them, let $k \in \mathbb{C} \setminus \{0\}$ and $u_+(x,k)$ be the *Jost solution* of

$$-u'' + fu = k^2 u, \tag{4}$$

i.e. the solution satisfying

$$u_+(x,k) = \begin{cases} e^{ikx} & \text{for } x \leq -1/2, \\ a(k)e^{ikx} + b(k)e^{-ikx} & \text{for } x \geq 1/2. \end{cases} \tag{5}$$

Since e^{ikx} and e^{-ikx} are linearly independent solutions of $-u'' = k^2u$, this defines $a(k)$ and $b(k)$ uniquely. They are related to the transmission and reflection coefficients used in physics by $a = t^{-1}$ and $b = rt^{-1}$. In particular, vanishing of b is equivalent to vanishing of r. For real k we use constancy of the Wronskian to get

$$|a(k)|^2 - |b(k)|^2 = 1, \tag{6}$$

corresponding to the familiar $|r|^2 + |t|^2 = 1$.

It can also be seen from (5) that

$$\begin{pmatrix} a(k) \\ b(k) \end{pmatrix} = \frac{1}{2ik}e^{-ik/2} \begin{pmatrix} ik & 1 \\ ike^{ik} & -e^{ik} \end{pmatrix} \begin{pmatrix} u_+(1/2,k) \\ u_+'(1/2,k) \end{pmatrix}.$$

Since u_+ is a solution of (4), a linear differential equation which is analytic in k, and satisfies the analytic initial condition $(u_+(-1/2), u_+'(-1/2)) = (1, ik)$, we see that $u_+(1/2,k)$ and $u'_+(1/2,k)$ are entire in k. Thus $a(k)$ and $b(k)$ are analytic in $\mathbb{C}\setminus\{0\}$ with a possible pole at $k=0$. Neither $a(k)$ nor $b(k)$ vanish identically. For $a(k)$ this is a trivial consequence of (6). For $b(k)$ it follows from the fact that a compactly supported $f\neq 0$ cannot be *reflectionless* (i.e. a *soliton*), as follows from inverse scattering theory, e.g. [7], see also [14]. Thus the roots of $a(k)$ and $b(k)$ cannot accumulate, neither at (the pole) 0 nor away from 0.

We are now ready to state our main result:

Theorem 1. *Let $E\in\mathbb{R}\setminus M$, where the set of critical energies M is given by*

$$\begin{aligned} M \;&:=\; \{\left(\frac{n\pi}{2}\right)^2 : n\in\mathbb{Z}\}\cup\{E=k^2 : k>0, b(k)=0\} && (7)\\ &\cup\{-\alpha^2<0 : a(i\alpha)a(-i\alpha)b(i\alpha)b(-i\alpha)=0\}, && (8)\end{aligned}$$

then $\gamma(E)>0$.

Note that the critical set M is the union of the discrete set $\{(n\pi/2)^2 : n\in\mathbb{Z}\}$ and all real numbers in $\{k^2 : k\in\mathbb{C}\setminus\{0\}, a(k)=0 \text{ or } b(k)=0\}$ (this uses that for real k, $a(k)\neq 0$ and $b(-k)=\overline{b(k)}$). Therefore M is discrete.

Our main tool for proving Theorem 1 in Section 2 will be Fürstenberg's Theorem, which has been used extensively in proofs of positivity of γ for discrete one-dimensional random operators, see [12, Sec.14A] for a summary. For continuum models, the use of Fürstenberg's Theorem until recently seems to have been restricted to the special cases $f=\chi_{[-1/2,1/2]}$ and $f=\delta_0$, a δ-point-interaction [1, 8]. The first to have used Fürstenberg's Theorem systematically for continuum Anderson models with general classes of single sites f has been Kostrykin and Schrader [9]. While they state their general results for absolutely continuous distribution μ, they point out that most of their ideas can also be applied to discrete distributions. Our work here can be seen as an implementation of this fact with, as we feel, a minimal amount of technical effort.

A nice feature of using Fürstenberg's Theorem in the proof of Theorem 1 is that once we know that the Theorem holds under the assumption that $\operatorname{supp}\mu=\{0,1\}$, then it is immediately clear that the Theorem extends to the case where $\{0,1\}\subset\operatorname{supp}\mu$. In fact, for larger support the critical set M should be smaller. Kostrykin and Schrader note in [10] that M should be empty if $\operatorname{supp}\mu$ has at least one non-isolated point, but there doesn't seem to be a proof of this yet. While Theorem 1 does not state that $\gamma(E)=0$ for all $E\in M$ (which in fact is not generally true), we will demonstrate in Section 3 that for Bernoulli-Anderson models one can indeed find many critical energies with $\gamma(E)=0$. In the discrete case this has already been observed for the so-called *dimer model*, see [2]. In particular, Section 3 discusses the example $f=\lambda\chi_{[-1/2,1/2]}$, $\lambda\in\mathbb{R}$ a constant, which leads to two different types of critical energies. They can be classified by the asymptotics of solutions of (2). In some cases they are of plane wave type, in particular bounded, while other critical energies lead to solutions which grow like $\exp\sqrt{x}$, due to a connection with random walks.

In [6] we use a result much like Theorem 1 to prove exponential and dynamical localization for continuum Bernoulli-Anderson models, where in addition we build on ideas which were developed for discrete models in [4]. Discreteness of the critical set M is crucial to this approach. In particular, it shows that the result found in Theorem 1 is superior to the well-known results of Kotani theory which prove that $\gamma(E) > 0$ for almost every E in a class of models containing ours. The methods of [6] also allow to extend Theorem 1 to the case where the support of μ is $\{a, b\}$, where $a \neq b$ are arbitrary real numbers. This requires considerably more technical effort, due to our use of scattering theory at a periodic background, and thereby somewhat obscures the rather simple ideas which we present here in Section 2.

2. Positivity of the Lyapunov Exponent

To prove positivity of γ, we need to understand properties of the transfer matrices. The transfer matrix from $-1/2$ to $1/2$ of

$$-u'' + fu = Eu, \tag{9}$$

is the matrix, $g(E)$, for which

$$\begin{pmatrix} u(1/2) \\ u'(1/2) \end{pmatrix} = g(E) \begin{pmatrix} u(-1/2) \\ u'(-1/2) \end{pmatrix} \tag{10}$$

for any solution u of (9). By $g_0(E)$ we denote the corresponding transfer matrix of $-u'' = Eu$. Set $G(E)$ to be the closed subgroup of $SL(2, \mathbb{R})$ generated by $g_0(E)$ and $g(E)$. Let $P(\mathbb{R}^2)$ be the projective space, i.e. the set of the directions in $\mathbb{R}^2$ and $\overline{v}$ be the direction of $v \in \mathbb{R}^2 \setminus \{0\}$. Note that $SL(2, \mathbb{R})$ acts on $P(\mathbb{R}^2)$ by $g\overline{v} = \overline{gv}$. We say that $G \subset SL(2, \mathbb{R})$ is strongly irreducible if and only if there is no finite G-invariant set in $P(\mathbb{R}^2)$.

In order to prove Theorem 1 we will consider energies $E > 0$ and $E < 0$ separately. Note that $E = 0$ is contained in M, so we don't need to consider it.

The general approach is as follows: We will first prove that $G(E)$ is not compact by showing that a sequence of elements has unbounded norm. This argument will be valid for all $E \in \mathbb{R} \setminus M'$, where

$$\begin{aligned} M' &= \{(n\pi)^2 : n \in \mathbb{Z}\} \cup \{E = k^2 : k > 0, b(k) = 0\} \\ &\quad \cup \{-\alpha^2 < 0 : a(i\alpha)a(-i\alpha)b(i\alpha)b(-i\alpha) = 0\}. \end{aligned}$$

Once non-compact, then the group G is known to be strongly irreducible if and only if for each $\overline{v} \in P(\mathbb{R}^2)$,

$$\#\{g\overline{v} : g \in G\} \geq 3, \tag{11}$$

see [3]. In order to prove that this condition is also satisfied we will in addition have to exclude k's which are odd multiples of $\pi/2$. We then use Fürstenberg's Theorem which states, in our context, that if the group $G(E)$ is not compact and strongly irreducible then $\gamma(E) > 0$, see also [3].

We start with positive energies, i.e. $E = k^2$, $k > 0$, and express the transfer matrices over $[-1/2, 1/2]$ in terms of the Jost solutions (5), at $\pm k$. We have

$$g(E) = \begin{pmatrix} u_N(1/2,k) & u_D(1/2,k) \\ u_N'(1/2,k) & u_D'(1/2,k) \end{pmatrix}$$

where u_N and u_D are the solutions of (4) satisfying $u_N(-1/2,k) = u_D'(-1/2,k) = 1$, $u_N'(-1/2,k) = u_D(-1/2,k) = 0$. Writing u_N and u_D as linear combinations of $u_+(x,k)$ and $u_-(x,k) = \overline{u_+(x,k)}$ and setting $z_\pm(k) = a(k)e^{ik} \pm b(k)$, we see that

$$g(k^2) = \begin{pmatrix} \mathrm{Re}[z_+(k)] & \frac{1}{k}\mathrm{Im}[z_+(k)] \\ -k\mathrm{Im}[z_-(k)] & \mathrm{Re}[z_-(k)] \end{pmatrix}. \tag{12}$$

Clearly,

$$g_0(k^2) = \begin{pmatrix} \cos k & \frac{1}{k}\sin k \\ -k\sin k & \cos k \end{pmatrix}. \tag{13}$$

Note for $k > 0$,

$$\begin{pmatrix} 1 & 0 \\ 0 & \frac{1}{k} \end{pmatrix}\begin{pmatrix} a & \frac{1}{k}b \\ kc & d \end{pmatrix}\begin{pmatrix} 1 & 0 \\ 0 & k \end{pmatrix} = \begin{pmatrix} a & b \\ c & d \end{pmatrix}. \tag{14}$$

Thus proving non-compactness for $G(k^2)$ is equivalent to proving non-compactness for the group $\tilde{G}(k^2)$ conjugate to $G(k^2)$ via (14).

If $b(k) = 0$ we have that $|a(k)| = 1$, and hence, we may write $a(k) = e^{i\phi}$. By (12), one has then that

$$\tilde{g}(k^2) = \begin{pmatrix} \cos(k+\phi) & \sin(k+\phi) \\ -\sin(k+\phi) & \cos(k+\phi) \end{pmatrix},$$

which is a rotation. The same holds for $g_0(k^2)$ by (13). Thus $\tilde{G}(k^2)$ is a group of rotations and thereby compact. This not only excludes an application of Fürstenberg, but shows $\gamma(k^2) = 0$ as a direct consequence of the definition (3).

Now assume $b(k) \neq 0$. Set $a(k) = Ae^{i\alpha}$, $b(k) = Be^{i\beta}$, and see that (6) implies $A^2 - B^2 = 1$. It follows then that $A > B > 0$, $A + B > 1$, and $A - B < 1$. With this notation, one may recalculate

$$\tilde{g}(k^2) = A\begin{pmatrix} \cos(\varphi) & -\sin(\varphi) \\ \sin(\varphi) & \cos(\varphi) \end{pmatrix} + B\begin{pmatrix} \cos(\psi) & -\sin(\psi) \\ -\sin(\psi) & -\cos(\psi) \end{pmatrix},$$

where we have set $\varphi := -k - \alpha$ and $\psi := -\beta$.

We wish to show that a sequence of elements in $\tilde{G}(k^2)$ has unbounded norm. To do so, consider an arbitrary unit vector

$$v(\theta) := \begin{pmatrix} \cos(\theta) \\ \sin(\theta) \end{pmatrix}.$$

The relation

$$\tilde{g}(k^2)v(\theta) = Av(\varphi + \theta) + Bv(-\psi - \theta)$$

shows that the image of the unit circle under $\tilde{g}$ is an ellipse, with the choices $\eta := -\frac{1}{2}(\varphi+\psi)$ and $\eta' := \frac{\pi}{2} - \frac{1}{2}(\varphi+\psi)$ identifying the semi-major and semi-minor axes as follows:

$$\tilde{g}(k^2)v(\eta) = (A+B)v(\tfrac{\varphi-\psi}{2}) \quad \text{and} \quad \tilde{g}(k^2)v(\eta') = (A-B)v(\tfrac{\pi}{2} + \tfrac{\varphi-\psi}{2}) \ .$$

Defining $R(\theta) := \|\tilde{g}(k^2)v(\theta)\|^2$, a short calculation, using $A^2 - B^2 = 1$, yields

$$R(\theta) - 1 = 2B\,[B + A\cos(2\theta + \varphi + \psi)] \,.$$

As a consequence, we see that $R(\theta) - 1$ is π-periodic and has exactly two roots in $[0, \pi)$. In particular, $R(\theta) = 1$ if and only if $\cos(2\theta + \varphi + \psi) = -\frac{B}{A}$, which shows that the distance between the zeros of $R(\theta) - 1$ is not equal to $\frac{\pi}{2}$: recall $B \neq 0$ (and $B \neq A$). Similarly, $R(\theta) > 1$ if and only if $\cos(2\theta + \varphi + \psi) > -\frac{B}{A}$ and hence $|\{\theta \in [0,\pi) : R(\theta) > 1\}| > \frac{\pi}{2}$. As a result, there exists a compact interval K (not necessarily in $[0, \pi)$) with $|K| > \frac{\pi}{2}$ and $\|\tilde{g}(k^2)v(\theta)\| > c > 1$ for all $\theta \in K$.

Now, applying $\tilde{g}$ once to $v(\eta)$, η as above, produces a new vector with norm greater than one, but the direction, initially η, is possibly altered. A vector with this new direction may not increase in norm by directly applying $\tilde{g}$ again. However, as long as k is not an integer multiple of π, then finitely many applications of $\tilde{g}_0(k^2)$, which is rotation by k, produces a vector with direction in K. Once in K, a direct application of $\tilde{g}$ does increase the norm size uniformly by $c > 1$, as indicated above. In this manner, we can produce a sequence of elements in $\tilde{G}(k^2)$ with unbounded norm. Thus $\tilde{G}(k^2)$ is non-compact if $b(k) \neq 0$ and k is not an integer multiple of π.

To complete the proof of Theorem 1 for positive energies, we note that if in addition k is not an integer multiple of $\pi/2$, then the free transfer matrix produces three distinct elements in projective space, i.e. (11) is satisfied and Fürstenberg's Theorem implies $\gamma(k^2) > 0$.

We finally turn to energies $E < 0$. In this case $G(E)$ is non-compact for every $E < 0$, since the free transfer matrix

$$g_0(E) = \begin{pmatrix} \cosh(\alpha) & \frac{1}{\alpha}\sinh(\alpha) \\ \alpha\sinh(\alpha) & \cosh(\alpha) \end{pmatrix}$$

has unbounded powers, where $\alpha = \sqrt{|E|}$. It remains to check (11), which we have to do for $E \in (-\infty, 0) \setminus \{-\alpha^2 : a(i\alpha)a(-i\alpha)b(i\alpha)b(-i\alpha) = 0\}$.

Let $E = -\alpha^2$ such that $a(\pm i\alpha) \neq 0$ and $b(\pm i\alpha) \neq 0$. By (5), this is equivalent to

$$g(E)\overline{v}_\pm \notin \{\overline{v}_+, \overline{v}_-\}, \tag{22}$$

where $\overline{v}_\pm$ are the directions of $v_\pm := (1, \pm\alpha)^t$. If $\overline{v} \notin \{\overline{v}_+, \overline{v}_-\}$, then $\#\{g_0(E)^n\overline{v} : n \in \mathbb{Z}\} = \infty$. If, on the other hand, $\overline{v} \in \{\overline{v}_+, \overline{v}_-\}$, then we use (22) to conclude that an initial application of $g(E)$ followed by iteration of $g_0(E)$ gives an infinite orbit. This shows (11) and completes the proof of Theorem 1.

3. Critical energies

In this section we discuss the appearance of critical energies in concrete examples. We thereby illustrate the following points: (i) critical energies with vanishing Lyapunov exponent do indeed exist, (ii) the structure and "size" of the set of energies with $b(k) = 0$ depends strongly on the concrete example, and (iii) at critical energies at least two different types of non-exponential asymptotics of solutions can be observed.

3.1. Zero reflection at degenerate gaps

Our first general observation is that if H^f_{per} is the Schrödinger operator with 1-periodic potential V satisfying $V(x) = f(x)$, $-1/2 \le x \le 1/2$ and H^f_{per} has a degenerate gap at energy $E = k^2$, i.e. either all solutions of (4) satisfy periodic boundary conditions on $[-1/2, 1/2]$ or all solutions satisfy anti-periodic boundary conditions on $[-1/2, 1/2]$, then $g(k^2) = I$ or $g(k^2) = -I$. By (12) this implies $(a(k), b(k)) = (1, 0)$ or $(-1, 0)$, respectively. Conversely, $(a(k), b(k)) = (\pm 1, 0)$ necessarily requires a degenerate gap. Thus, the possibility of degenerate gaps in periodic potentials is one mechanism which leads to reflectionless energies. Whether all reflectionless energies arise in this way is equivalent to deciding if $b(k) = 0$ necessarily leads to $a(k) = \pm 1$. In Example 2 below we will see that this is not the case.

3.2. Example 1

Consider first $f := \lambda\chi_{[-1/2,1/2]}$ with $\lambda \in \mathbb{R}$. For $E = k^2 > \max\{0, \lambda\}$ we get with $\alpha := \sqrt{E - \lambda}$

$$a(k)e^{ik} = \cos\alpha + i\frac{2k^2 - \lambda}{2k\alpha}\sin\alpha$$

$$b(k) = \frac{i\lambda}{2k\alpha}\sin\alpha$$

$$g_0(k^2) = \begin{pmatrix} \cos k & \frac{1}{k}\sin k \\ -k\sin k & \cos k \end{pmatrix}$$

$$g(k^2) = \begin{pmatrix} \cos\alpha & \frac{1}{\alpha}\sin\alpha \\ -\alpha\sin\alpha & \cos\alpha \end{pmatrix}$$

We can now distinguish between the following types of critical energies:

Type 1a: k^2 such that $b(k) = 0$: This happens for $E = k^2 = (n\pi)^2 + \lambda$, $n \in \mathbb{N}$. In this case we have $g(k^2) = \pm I$, and it easily follows that the group $\langle g_0, g\rangle$ is bounded. This means that solutions to (2) are bounded and that $\gamma(k^2) = 0$.

Type 1b: $k = n\pi$, $n \in \mathbb{N}$: This means $g_0(k^2) = \pm I$, and again $\langle g_0, g\rangle$ bounded, $\gamma(k^2) = 0$. This type is "dual" to Type 1, since $k^2 = n^2\pi^2$ can be viewed as the energies where $-d^2/dx^2 + \lambda - \lambda\chi_{[-1/2,1/2]}$ is reflectionless with respect to $-d^2/dx^2 + \lambda$.

Type 2: $E = k^2$, where $k = (2n-1)\pi/2$, $n \in \mathbb{N}$. This leads to a critical energy with a rather different asymptotic behavior of solutions if also $\alpha = (2m-1)\pi/2$, $m \in \mathbb{N}$, that is, for specific values of $\lambda = k^2 - \alpha^2 = \pi^2(n-m)(n+m-1)$. In this case we have

$$g_0(k^2) = \pm \begin{pmatrix} 0 & -1/k \\ k & 0 \end{pmatrix}, \quad g(k^2) = \pm \begin{pmatrix} 0 & -1/\alpha \\ \alpha & 0 \end{pmatrix}.$$

Let $h_n(\omega) := g_{2n+1}(\omega) g_{2n}(\omega)$, where $g_n(\omega)$ is the transfer matrix of H_ω from $n-1/2$ to $n+1/2$. Then

$$h_n(\omega) = \begin{cases} -I & \text{with probability } p^2+q^2, \\ \pm \begin{pmatrix} k/\alpha & 0 \\ 0 & \alpha/k \end{pmatrix} & \text{with probability } pq, \\ \pm \begin{pmatrix} \alpha/k & 0 \\ 0 & k/\alpha \end{pmatrix} & \text{with probability } qp. \end{cases}$$

If $h_{n,1}$ and $h_{n,2}$ are the diagonal entries of h_n, then the sums of $\log |h_{n,i}|$, $i = 1,2$, give a symmetric random walk (observing that $(p^2+q^2)\log 1 + pq \log(k/\alpha) + qp \log(\alpha/k) = 0$). Thus it follows that for every $\varepsilon > 0$ and almost surely

$$\limsup_{n\to\infty} n^{-1/2-\varepsilon} \log \| \prod_{k=1}^n h_k(\omega) \| = 0 \quad \text{and} \quad \limsup_{n\to\infty} n^{-1/2+\varepsilon} \log \| \prod_{k=1}^n h_k(\omega) \| = \infty,$$

see e.g. [13]. The interpretation of this is that the hull of the (oscillatory) solutions of (2) asymptotically grows like $\exp c\sqrt{x}$ or decays like $\exp(-c\sqrt{x})$ for some $c > 0$, where a generalization of the Ruelle-Osceledec Theorem from [11, Section 8] is used.

3.3. Example 2

Consider $f := \lambda(\chi_{[-1/2,0]} - \chi_{(0,1/2]})$ with $\lambda > 0$. For $E = k^2 > \lambda$, we may again explicitly calculate the transfer matrix

$$g(E) = \begin{pmatrix} g_{11} & g_{12} \\ g_{21} & g_{22} \end{pmatrix},$$

and see that,

$$\begin{aligned} g_{11} &= \cos(\alpha_-/2)\cos(\alpha_+/2) - \frac{\alpha_-}{\alpha_+}\sin(\alpha_-/2)\sin(\alpha_+/2), \\ g_{12} &= \frac{1}{\alpha_-}\sin(\alpha_-/2)\cos(\alpha_+/2) + \frac{1}{\alpha_+}\cos(\alpha_-/2)\sin(\alpha_+/2), \\ g_{21} &= -\alpha_+\cos(\alpha_-/2)\sin(\alpha_+/2) - \alpha_-\sin(\alpha_-/2)\cos(\alpha_+/2), \\ g_{22} &= -\frac{\alpha_+}{\alpha_-}\sin(\alpha_-/2)\sin(\alpha_+/2) + \cos(\alpha_-/2)\cos(\alpha_+/2), \end{aligned}$$

for $\alpha_\pm := \sqrt{E \pm \lambda}$. From this and the boundary conditions of the Jost solution, i.e.

$$g \begin{pmatrix} e^{-ik/2} \\ ike^{-ik/2} \end{pmatrix} = \begin{pmatrix} a(k)e^{ik/2} + b(k)e^{-ik/2} \\ ika(k)e^{ik/2} - ikb(k)e^{-ik/2} \end{pmatrix},$$

one can show that

$$b(k) = 0 \text{ if and only if } \sin(\alpha_+/2) = 0 = \sin(\alpha_-/2).$$

Therefore, if we fix some $\lambda > 0$, then any $E = k^2 > \lambda$ for which $b(k) = 0$ must satisfy

$$\frac{\sqrt{E+\lambda}}{2} = n\pi, \text{ and } \frac{\sqrt{E-\lambda}}{2} = m\pi,$$

for some $n, m \in \mathbb{N}$. Thus λ must have the form

$$\lambda = 2\pi^2(n^2 - m^2), \text{ with } n, m \in \mathbb{N}, \tag{23}$$

and the corresponding reflectionless E's are then

$$E = 2\pi^2(n^2 + m^2). \tag{24}$$

From this we conclude that for most values of λ (e.g., Lebesgue almost every λ), no reflectionless energy exists above λ. On the other hand, every λ satisfying (23) has only finitely many reflectionless energies. In fact, if $N := \frac{\lambda}{2\pi^2} \in \mathbb{N}$, then $N = n^2 - m^2$ for only finitely many pairs $(n, m) \in \mathbb{N}^2$. But this finite number can be arbitrarily large, depending on λ: To find at least j exceptional energies, choose

$$\frac{\lambda_j}{2\pi^2} = N_j = 2^{j+1}(2^{j-1} + 1)$$

and see that for $0 \leq \ell < j$ one may choose

$$n_\ell := \left(2^\ell(2^{j-1} + 1) + 2^{j-\ell-1}\right) \qquad \text{and} \qquad m_\ell := \left(2^\ell(2^{j-1} + 1) - 2^{j-\ell-1}\right)$$

satisfying $n_\ell^2 - m_\ell^2 = N_j$. Proof: Take $a := 2^\ell(2^{j-1} + 1)$ and $b := 2^{j-\ell-1}$ and see that $(a+b)^2 - (a-b)^2 = 4ab$. Therefore,

$$4ab = 2^2 \cdot 2^\ell(2^{j-1} + 1) \cdot 2^{j-\ell-1} = 2^{j+1}(2^{j-1} + 1) = N_j.$$

Recall from our remarks in 3.1 that in some cases, the condition $b(k) = 0$ implies that $a(k) = \pm 1$. To see that this is not always the case, we calculate $a(k)$ in this example for such values of k. For λ as in (23) and E as in (24), one can show that $a(k) = e^{-2\pi i\sqrt{n^2+m^2}}$ if $n+m$ is even and $a(k) = -e^{-2\pi i\sqrt{n^2+m^2}}$ if $n+m$ is odd.

References

[1] M. Benderskii and L. Pastur, On the asymptotics of the solutions of second-order equations with random coefficients (in Russian). *Teoria Funkcii, Func. Anal. i Priloz.* (Kharkov University) **N 22** (1975), 3–14

[2] S. de Bièvre and F. Germinet, Dynamical Localization for the Random Dimer Schrödinger Operator. *J. Stat. Phys.* **98** (2000), 1135–1148

[3] P. Bougerol and J. Lacroix, *Products of random matrices with applications to Schrödinger operators.* Birkhäuser, Boston–Stuttgart (1985)

[4] R. Carmona, A. Klein, and F. Martinelli, Anderson localization for Bernoulli and other singular potentials. *Commun. Math. Phys.* **108** (1987), 41–66

[5] R. Carmona and J. Lacroix, *Spectral theory of random Schrödinger operators.* Birkhäuser, Basel–Berlin (1990)

[6] D. Damanik, R. Sims and G. Stolz, Localization for one-dimensional, continuum, Bernoulli-Anderson models. Preprint 2000, mp-arc/00-404

[7] P. Deift and E. Trubowitz, Inverse scattering on the line. *Commun. Pure Appl. Math.* **32** (1979), 121–251

[8] K. Ishii, Localization of eigenstates and transport phenomena in one-dimensional disordered systems. *Progress Theor. Phys. Suppl.* **53** (1973), 77–118

[9] V. Kostrykin and R. Schrader, Scattering theory approach to random Schrödinger operators in one-dimension. *Rev. Math. Phys.* **11** (1999), 187–242

[10] V. Kostrykin and R. Schrader, Global bounds for the Lyapunov exponent and the integrated density of states of random Schrödinger operators in one dimension. Preprint 2000, mp-arc/00-226

[11] Y. Last and B. Simon, Eigenfunctions, transfer matrixes, and absolutely continuous spectrum of one-dimensional Schrödinger operators. *Invent. math.* **135** (1999), 329–367

[12] L. Pastur and A. Figotin, *Spectra of Random and Almost-Periodic Operators.* Springer Verlag, Berlin-Heidelberg-New York (1992)

[13] Yu. V. Prohorov and Yu. A. Rozanov, *Probability Theory.* Springer Verlag, Berlin-Heidelberg-New York (1969)

[14] R. Sims and G. Stolz, Localization in one-dimensional random media: a scattering theoretic approach. *Commun. Math. Phys.* **213** (2000), 575–597

David Damanik
Department of Mathematics
University of California at Irvine
CA 92697-3875, USA
e-mail: damanik@math.uci.edu

Robert Sims
Department of Mathematics
University of Alabama at Birmingham
AL 35294-1170, USA
e-mail: sims@math.uab.edu

Günter Stolz
Department of Mathematics
University of Alabama at Birmingham
AL 35294-1170, USA
e-mail: stolz@math.uab.edu

Operator Theory:
Advances and Applications, Vol. 132, 131–140

Families of Spectral Measures with Mixed Types

R. del Rio, S. Fuentes, and A. Poltoratski

Abstract. Consider a family of Sturm-Liouville operators H_θ on the half-axis defined as

$$H_\theta u = -u'' + q(x)u \qquad 0 \le x < \infty$$

with the boundary condition

$$u(0)\cos\theta + u'(0)\sin\theta = 0 \qquad 0 \le \theta < \pi$$

and the limit point case at infinity. We show that it is possible for all H_θ to have dense absolutely continuous and dense singular spectrum. The construction is based on integral representations of Pick functions in the upper half-plane. We also discuss applications to the Krein spectral shift.

1. Introduction

This note studies the interplay between various types of spectra under small perturbations or a change of the boundary condition.

Let A_0 be a cyclic self-adjoint operator, φ its cyclic vector and μ the corresponding spectral measure. Denote by A_λ the rank one perturbations of A_0:

$$A_\lambda = A + \lambda(\cdot, \varphi)\varphi, \quad \lambda \in \mathbb{R}.$$

Let μ_λ be the spectral measures of A_λ corresponding to φ. We study the following general problem: What can happen to μ_λ when the parameter (the coupling constant) λ is changing?

It is well known that the same problem can be formulated in terms of Sturm–Liouville operators on the half-axis. Let the operator H_θ be defined as

$$H_\theta u = -u'' + q(x)u \qquad 0 \le x < \infty$$

where q is real valued, locally integrable function defined in $[0,\infty)$). The domain is restricted by the boundary condition at zero

$$u(0)\cos\theta + u'(0)\sin\theta = 0 \qquad 0 \le \theta < \pi.$$

We assume that the limit point case occurs at infinity.

R. dR. was partially supported by projects IN-102998 PAPIIT-UNAM and 27487E CONACyT.
S.F. was supported by DGEP-UNAM.
A.P. was supported in part by the NSF grant DMS-9970151.

For each θ one can define the so-called Weyl-m function $m_\theta(z)$ which is analytic and has positive imaginary part in the upper half-plane. The imaginary part of $m_\theta(z)$ has the following integral representation

$$Im\ m_\theta(z) = \int_{\mathbb{R}} \frac{y}{(t-x)^2+y^2} d\rho_\theta(t) \qquad z = x+iy$$

The measures ρ_θ are called the Weyl spectral measures of the boundary value problem. The following relation is satisfied:

$$m_\theta(z) = \frac{-\sin(\theta-\beta) + m_\beta(z)\cos(\theta-\beta)}{\cos(\theta-\beta) + m_\beta(z)\sin(\theta-\beta)}, \tag{1}$$

see [4] and [9]. In these settings our general question transforms into: How does the spectrum of H_θ (the measure ρ_θ) depend on the parameter θ in the boundary condition? The same question can also be reformulated in terms of self-adjoint extensions of a symmetric operator with deficiency indices (1,1) or discrete Schrödinger operators. For more on these connections see [8] and [18].

In this paper we focus on the correlations between the absolutely continuous and singular parts of the spectrum under the change of the coupling constant λ (the boundary parameter θ). By the Kato-Rosenblum theorem the absolutely continuous spectrum is invariant under such small perturbations. At the same time, examples obtained by many researchers in recent years show that the singular spectrum can be extremely unstable under the change of the coupling constant λ in particular in "mixed" situations when both singular and absolutely continuous spectrum are present. In most cases, the singular spectrum of A_λ is located in the gaps of the absolutely continuous spectrum. With the change of λ the isolated eigenvalues of A_λ move inside their respective gaps and often dissappear when they hit the edge of the gap. But what happens if there are no gaps in the absolutely continuous spectrum? It seems that in this case there is no space for the singular spectrum. Even if some of A_λ's have nontrivial singular components, squeezed somewhere in the midst of the absolutely continuous spectrum, they must be easily destroyed by the change of λ. In particular, it seems unlikely that all operators can have dense absolutely continuous and dense singular spectrum. All the examples we know seem to support this intuitive argument: the examples of Naboko [12] and Simon [19] exhibit singular spectra only for a set of boundary conditions of Lebesgue measure zero (in θ), and the examples of Remling [16] do not have dense singular spectra. In [5] such a coexistence of spectra is shown for a set of positive measure in the coupling constant but not for all coupling constants. Also, in [3] it is shown that two other types of spectrum, the pure point and continuous, cannot coexist for all coupling constants if the spectrum of A_0 is dense. Other papers where coexistence is studied are [1] and [10].

In this paper we show that, despite all the evidence mentioned above, it is possible for *all* A_λ's to have everywhere dense absolutely continuous *and* everywhere dense singular spectrum. The corresponding example is constructed in Section 3.

In Section 2 we develop our machinery. It is based on the integral representations of analytic Pick functions in the upper half-plane. Our main tool is Lemma 2, which reveals the relations between the families of Pick functions appearing in Perturbation Theory. In addition to the main example, in Section 3 we explain how one can use Lemma 2 to obtain the singular components of the spectral measures μ_λ directly from the corresponding Krein function.

2. Preliminaries

Every Pick function $F(z)$ (an analytic function which takes the upper half-plane into itself, also known as Nevanlinna or Herglotz function) has an integral representation of the form

$$F(z) = a + bz + \int_{\mathbb{R}} \left[\frac{1}{t-z} - \frac{t}{1+t^2}\right] d\mu(t) \tag{2}$$

where $a, b \in \mathbb{R}, b \geq 0$ and μ is a non-negative Borel measure which satisfies

$$\mid \mu \mid := \int_{\mathbb{R}} \frac{d\mu(t)}{1+t^2} < \infty. \tag{3}$$

Conversely, any function of this form is analytic and takes the upper half-plane into itself, see [7] and [17]. The integral on the right-hand side of (2) is the Cauchy integral of μ in the upper half-plane. We will denote it by $K\mu$. We will also denote by $P\mu$ the Poisson integral of μ:

$$P\mu(x+iy) = ImK\mu(x+iy) = \int_{\mathbb{R}} \frac{y}{(x-t)^2+y^2} d\mu(t).$$

The Poisson integral is a so-called approximative identity: its kernel $\frac{y}{(x-\epsilon)^2+y^2}$ is positive, tends to zero uniformally outside of any neighborhood of x as $y \to 0$ and its L^1-norm is constant. This implies the following version of the Lebesgue Theorem (about the Lebesgue points of a summable function), see [14]. The measure $f\mu$ is defined for any measurable set A as

$$f\mu(A) := \int_A f(x) d\mu(x)$$

Lemma 1. *If μ is a complex Borel measure on $\mathbb{R}$ such that $|\mu|$ satisfies (3) and $f \in L^1(\mu)$ then*

$$\lim_{y \downarrow 0} \frac{Pf\mu(x+iy)}{P\mu(x+iy)} = f(x)$$

for μ-a. e. x.

Corollary 1. *If μ and ν are complex Borel measures on $\mathbb{R}$ such that $|\mu|$ and $|\nu|$ satisfy* (3) *and $\mu = f\nu + \eta$, where $f \in L^1(\nu)$ and $\eta \perp \nu$ (η and ν are mutually singular), then*

$$\lim_{y\downarrow 0} \frac{P\mu(x+iy)}{P\nu(x+iy)} = f(x)$$

for ν-a. e. x.

Proof. Since

$$\lim_{y\downarrow 0} \frac{P\mu(x+iy)}{P\nu(x+iy)} = \lim_{y\downarrow 0} \frac{Pf\nu(x+iy)}{P\nu(x+iy)} + \lim_{y\downarrow 0} \frac{P\eta(x+iy)}{P\nu(x+iy)}$$

by Lemma 1 it is enough to show that the last summand tends to 0 ν^s-a. e. Consider $f \in L^1(\nu+\eta)$ defined as 1 η-a.e. and as 0 ν-a.e. Then by Lemma 1

$$\lim_{y\downarrow 0} \frac{P\eta(x+iy)}{P(\nu+\eta)(x+iy)} = \lim_{y\downarrow 0} \frac{Pf(\nu+\eta)(x+iy)}{P(\nu+\eta)(x+iy)} = 0$$

ν-a.e. Therefore

$$\left[\frac{P\eta(x+iy)}{P(\nu+\eta)(x+iy)}\right]^{-1} = 1 + \frac{P\nu(x+iy)}{P(\eta)(x+iy)} \to \infty$$

ν-a.e. and we obtain our statement. □

To construct our main example we will also need the following lemma. Consider the family of Pick functions

$$f_\theta(z) := \frac{\cos\theta + z\sin\theta}{\sin\theta - z\cos\theta}, \quad \theta \in \mathbb{R}. \tag{4}$$

Let $L(z)$ be another Pick function such that $0 \le ImL(z) \le \pi$. For any $\alpha \in \mathbb{R}$ denote

$$M_\alpha(z) := f_\alpha(L(z)) \text{ and } N_\alpha(z) := f_\alpha(\exp L(z)).$$

Both $M_\alpha(z)$ and $N_\alpha(z)$ are Pick functions admitting representations similar to (2). We denote by μ_α and ν_α the measures appearing in the representations for M_α and N_α respectively. Then the singular parts of these measures, ν_α^s and μ_β^s, enjoy the following relation:

Lemma 2. *Let ν_α and μ_β be as above. Define the function $\alpha(\beta)$ as*

$$\alpha(\beta) = tg^{-1}(\exp tg\beta), \quad \beta \in (-\pi/2, \pi/2)$$

then

$$\mu_\beta^s = \alpha'(\beta)\, \nu_{\alpha(\beta)}^s.$$

Proof. Let us first show that

$$\lim_{y\downarrow 0} \frac{ImM_\beta(x+iy)}{ImN_\alpha(x+iy)} = \alpha'(\beta)$$

for μ_β^s a.e. x.

From the definition of M_β and N_α we have $(z = x + iy)$

$$\lim_{y\downarrow 0} \frac{ImM_\beta(z)}{ImN_\alpha(z)} = \lim_{y\downarrow 0} \frac{ImL(z)}{Im \exp L(z)} \left| \frac{\sin\alpha - \exp L(z)\cos\alpha}{\sin\beta - L(z)\cos\beta} \right|^2 .$$

From the definition of α

$$\frac{\sin\alpha - \exp L(z)\cos\alpha}{\sin\beta - L(z)\cos\beta} = \frac{\cos\alpha}{\cos\beta} \cdot \frac{\exp(tg\beta) - \exp L(z)}{tg\beta - L(z)}. \tag{5}$$

It is well known that for μ_β^s-a.e. x the Cauchy integral of μ_β at $x + i\varepsilon$ tends to infinity as $\varepsilon \to 0$. Therefore (by (2)) $M_\beta(x+i\varepsilon) \overset{\varepsilon\downarrow 0}{\longrightarrow} \infty$. The formula for f_β and the definition of M_β now imply that for μ_β^s- a.e. x $L(x + i\varepsilon) \overset{\varepsilon\downarrow 0}{\longrightarrow} tg\beta$.

Hence the expression in the right-hand side of (5) tends to

$$\frac{\cos\alpha}{\cos\beta} \quad \exp tg\beta \quad \text{when } \varepsilon \downarrow 0$$

and we get

$$\left| \frac{\sin\alpha - \exp L(z)\cos\alpha}{\sin\beta - L(z)\cos\beta} \right|^2 \overset{\varepsilon\downarrow 0}{\longrightarrow} \left(\frac{\cos\alpha}{\cos\beta} \right)^2 (\exp tg\beta)^2$$

for μ_β^s *a.e.* x.

Now if $L = a(z) + ib(z)$ then

$$\frac{ImL(z)}{Im \exp L(z)} = \frac{b(z)}{e^{a(z)} \sin b(z)}.$$

For μ_β^s *a.e.* x $\quad b \overset{\varepsilon\downarrow 0}{\longrightarrow} 0$ and therefore $\dfrac{b}{\sin b} \overset{\varepsilon\downarrow 0}{\longrightarrow} 1$. Since $e^{a+ib} \overset{\varepsilon\downarrow 0}{\longrightarrow} tg\alpha$ then $e^a \overset{\varepsilon\downarrow 0}{\longrightarrow} tg\alpha$. Hence

$$\frac{ImL}{Im \exp L} \overset{\varepsilon\downarrow 0}{\longrightarrow} \frac{1}{tg\alpha} \quad \text{for } \mu_\beta^s \text{ a.e. } x.$$

Therefore we obtain

$$\lim_{y\downarrow 0} \frac{ImM_\beta(z)}{ImN_\alpha(z)} = \left(\frac{\cos\alpha}{\cos\beta} \right)^2 (tg\alpha) = \alpha'(\beta)$$

for μ_β^s *a.e.* x. Since the Poisson integral $P\mu_\beta(x+i\varepsilon)$ tends to infinity at μ_β^s *a.e.* x, the last equation implies

$$\lim_{y\downarrow 0} \frac{ImM_\beta(x+iy)}{ImN_\alpha(x+iy)} = \lim_{y\downarrow 0} \frac{ImK\mu_\beta(x+iy) + C_1 y}{ImK\nu_\alpha(x+iy) + C_2 y} = \lim_{y\downarrow 0} \frac{P_{\mu_\beta}(z)}{P\nu_\alpha(z)} = \alpha'(\beta) \tag{6}$$

for μ_β^s *a.e.* x. If $\mu_\beta = f\nu_\alpha + \eta$, where $f \in L^1(\nu_\alpha)$ and $\eta \perp \nu_\alpha$, then Corollary 1 and (6) imply $f = \alpha'(\beta) > 0$ μ_β^s-a.e. Thus $\mu_\beta^s << \nu_\alpha^s$. In the same way one can show that

$$\lim_{y\downarrow 0} \frac{P_{\nu_\alpha}(z)}{P\mu_\beta(z)} = \frac{1}{\alpha'(\beta)}$$

for ν_α^s *a.e.* x and therefore $\nu_\alpha^s << \mu_\beta^s$. Hence $f\nu_\alpha^s = \mu_\beta^s$. Again by (6) and Lemma 1, $f \equiv \alpha'(\beta)$. □

Remark 1. In the definition of the function N_α in the above lemma instead of $\exp(z)$ one can use any other function analytic in the neighborhood of the strip $S = \{0 < Imz < \pi\}$ which takes S to the upper half-plane and $\mathbb{R}$ to $\mathbb{R}$. Such functions arise in many other problems related to Perturbation Theory. The definition of $\alpha(\beta)$ in the statement would have to be changed accordingly.

3. The Krein function and coexistence of spectra

We now reveal the meaning of Lemma 2 from the point of view of Perturbation Theory and Mathematical Physics.

First, let us notice that Lemma 2 allows one to see Cauchy integrals of spectral measures (resolvent functions) directly from the corresponding Krein function. Let us consider the following example.

Let A_0 be a self-adjoint operator, φ its cyclic vector and μ the corresponding spectral measure. Once again, denote by A_λ the rank one perturbations:

$$A_\lambda = A + \lambda(\cdot, \varphi)\varphi, \quad \lambda \in \mathbb{R}. \tag{7}$$

Then there exists a function u on $\mathbb{R}$ satisfying

$$u(x) = \arg(1 + K\mu(x + i0)) \quad \text{for } a.e.x \tag{8}$$

where arg stands for the principal branch of argument taking values in $(-\pi; \pi]$. The function u is called the Krein spectral shift for the perturbation problem (A_0, A_1).

To apply Lemma 2 notice that (8) implies that

$$1 + K\mu = \exp(Ku + c)$$

for some real c. For any $\alpha \in (-\pi/2, \pi/2]$ denote by μ_α and u_α the measures corresponding to the Pick functions $f_\alpha(1 + K\mu)$ and $f_\alpha(Ku + c)$ respectively. Lemma 2 immediately gives us the singular components of the spectral measures μ_λ:

Theorem 1.

$$u^s_\beta = \alpha'(\beta)\, \mu^s_{\alpha(\beta)} \tag{9}$$

where $tg\alpha = \exp tg\beta$

A similar result can be formulated for self-adjoint extensions, Sturm-Liouville operators on the half-axis, discrete Schrödinger operators etc.

Theorem 1 implies relations such as

$$\int_0^{\pi/2} u^s_\beta(A) d\,\beta = \int_{\pi/4}^{\pi/2} \mu^s_\alpha(A)\, d\alpha.$$

For more about the relation between the measures μ_λ and the Krein function see [15].

Next, we will show how to apply Lemma 2 to construct families of Sturm-Liouville operators (rank one perturbations, etc.) of mixed spectral types. As was announced in the introduction, in our example all the operators will have dense singular and dense absolutely continuous spectrum on an interval, regardless of the boundary condition.

Example 1. Consider a fixed interval I and take a set $E \subset I$ such that for every subinterval $J \subset I$ we have

$$0 < |E \cap J| < |J|$$

Such sets can easily be constructed, see [15] or [6, Examples 4 and 5].

Let

$$u(x) = \begin{cases} 1 & x \in E \\ 0 & x \in E^c \end{cases}$$

define

$$d\,\mu_{\pi/2} = u\,dx$$

inside I and set $\mu_{\pi/2}$ outside I such that (3) holds, and the necessary decay conditions required by the Gelfand-Levitan inverse theorem are satisfied, [13], the measure $\mu_{\pi/2}$ will be the spectral measure of a Sturm-Liouville operator $H_{\pi/2}$ (defined as in the introduction). Let μ_β be the spectral measures of H_β, μ_β^s and μ_β^{ac} stand for their singular and absolutely continuous components correspondingly. We claim:

a) $\mu_\beta^s(J) > 0$, for every subinterval $J \subset I \quad \beta \in (-\pi/2, \pi/2)$
b) $\mu_\beta^{ac}(J) > 0$, for every subinterval $J \subset I \quad \beta \in (-\pi/2, \pi/2]$

Proof. a) Let

$$K\nu_{\pi/2}(z) := \exp(K\mu_{\pi/2}(z)). \tag{10}$$

Then

$$u(x) = \frac{1}{\pi} \arg K\nu_{\pi/2}(x + i0) \quad \text{for } a.e. x \in I$$

and using the definition of u it follows that

$$Im K\nu_{\pi/2}(x + i0) = 0 \quad \text{for } a.e. x \in I.$$

Since the support of the absolutely continuous part of ν_α is the set

$$\{x / Im K\nu_{\pi/2}(x + i0) > 0\}$$

(see [11]), ν_α is purely singular in I for every $\alpha \in (-\pi/2, \pi/2)$.

Given an interval $J \subset I$ assume that $\nu_{\pi/2}(J) = 0$. Then $K\nu_{\pi/2}(z)$ can be extended analytically across J and from (10) the same follows for $K\mu_{\pi/2}(z)$. Since $\mu_{\pi/2} > 0$, this implies $\mu_{\pi/2}(J) = 0$, which contradicts the construction of $\mu_{\pi/2}$. Hence $\nu_{\pi/2}(J) > 0$ for every $J \subset I$. From this we obtain $\nu_\alpha^s(J) > 0$ for every $\alpha \in (-\pi/2, \pi/2)$ (see, for instance, [9, p. 38, Theorem 2.52]).

Now to obtain a) we just recall that from Theorem 1 we have

$$\mu_\beta^s(F) = \alpha'(\beta)\,\nu_\alpha^s(F)$$

when $\alpha(\beta) = tg^{-1}(\exp tg\beta)$ for every Borel set F. $\beta \in (-\pi/2, \pi/2), \alpha \in (0, \pi/2)$

b) Follows from the well-known stability of the absolutely continuous part of the spectrum (see [18, Theorem 2.1]) since $\mu_{\pi/2}$ is a.c. by construction. □

Remark 2. Note that in our construction the absolutely continuous spectra is recurrent ([2]).

Remark 3. In [6] five examples are given of families of measures $\{\mu_\beta\}$ where $d\mu = \chi_B(x)dx$, B is Lebesgue measurable set and

$$\chi_B(x) = \begin{cases} 1 & x \in B \\ 0 & x \notin B \end{cases} .$$

The occurrence of the singular spectrum embedded in the a.c. spectrum is only shown for a set of $\beta's$ of positive Lebesgue measure. The construction above proves in [6, examples 4 and 5], coexistence for all $\beta's$ with the exception of one $(\pi/2)$.

To construct a family of measures such that a) holds for all $\beta \in (-\pi/2, \pi/2]$ observe that

$$f_\theta(z) = i\frac{-e^{2i\theta} + \varphi(z)}{-e^{2i\theta} - \varphi(z)} \text{ where } \varphi(z) = \frac{z-i}{z+i}.$$

Also,

$$f_\theta(z) = f_{\theta+\pi}(z)$$

and

$$i\frac{-e^{2i\theta} + \varphi^2(z)}{-e^{2i\theta} - \varphi^2(z)} = \frac{1}{2} f_{\frac{\theta}{2}+\frac{\pi}{4}}(z) + \frac{1}{2} f_{\frac{\theta}{2}-\frac{\pi}{4}}(z)$$

Denote by $\tilde{\mu}_\beta$ the measure corresponding to the Pick function

$$i\frac{-e^{2i\beta} + \varphi^2(K\mu_{\pi/2}(z))}{-e^{2i\beta} + \varphi^2(K\mu_{\pi/2}(z))}.$$

Then

$$\tilde{\mu}_\beta = \frac{1}{2}\mu_{\frac{\beta}{2}+\frac{\pi}{4}} + \frac{1}{2}\mu_{\frac{\beta}{2}-\frac{\pi}{4}}. \tag{11}$$

The measures $\tilde{\mu}_\beta$ satisfy the decay condition and can be realized as spectral measures for a family of Sturm-Liouville operators. Also, from (11) it follows that $\tilde{\mu}_\beta$ have dense singular and absolutely continuous parts on I for every $\beta \in (-\pi/2, \pi/2]$.

Remark 4. If we multiply the measure $\mu_{\pi/2}$ used in Example 1 by a constant less than 1, then we get singular components only for some of the coupling constants β. More precisely, let $H_{\pi/2}$ be the Sturm-Liouville operator whose spectral measure is defined as $\gamma_{\pi/2} = \alpha\mu_{\pi/2} = \alpha u(x)dx$ where u is as in Example 1 and $0 < \alpha < 1$. Let γ_α be the spectral measures of H_α's. Then using the same methods as in Example 1 one can prove the following claim:

For the family of measures γ_α we have $\gamma_\alpha^s(J) > 0$ for every subinterval $J \subset I$ if $\alpha \in (0, \pi/2)$. If $\alpha \in (-\pi/2, 0]$ then γ_α is absolutely continuous.

References

[1] S. Albeverio, J. Brasche and H. Neidhardt, *On Inverse Spectral theory for Self-Adjoint Extensions: Mixed Types of Spectra*, Journal of Funct. Analysis, Vol. 154 No. 1 pp. 130–173, Apr. 1998.

[2] J. Avron and J. Simon, *Transient and recurrent spectrum*, J. Fund. Anal., 43 pp. 1–31 (1981).

[3] R. del Rio, N. Makarov, B. Simon *Operators with singular spectrum II: Rank one operators*, Comm. Math. Phys., vol. 165, 1994, pp. 59–67.

[4] R. del Rio and A. Poltoratski, *Spectral measures and category*, Operator Theory: Advances and Applications, Vol. 108 pp. 149–159, 1999.

[5] R. del Rio and B. Simon, *Point spectrum and mixed spectral types for rank one perturbations*, Proceedings of the American Mathematical Society, Vol. 125 No. 12 pp. 3593–3599, december 1997.

[6] R. del Rio, B. Simon and G. Stolz, *Stability of spectral types for Sturm-Liouville operators*, Mathematical Research Letters 1, pp. 437–450 (1994).

[7] W.F. Donoghue Jr., *Monotore Matrix Functions and Analytic Continuation*, Springer-Verlag, 1974.

[8] W.F. Donoghue Jr., *On the Perturbation of Spectra*, Communications on Pure and Applied Mathematics, Vol. XVIII pp. 559–579 (1965).

[9] M.S.P. Eastham and H. Kalf, *Schrödinger-type operators with continuous spectra*, Pitman Advanced Publishing Program, 1982.

[10] A.L. Figotin and L.A. Pastur, *A Schrödinger operator with a nonlocal potential whose a.c. and point spectra coexist*, Comm. Math. Phys., 130 pp. 357–380, 1990.

[11] D.J. Gilbert and D.B. Pearson, *On Subordinacy and Analysis of the Spectrum of One-Dimensional Schrödinger Operators*, Journal of Mathematical Analysis and Applications Vol. 128 No. 1 pp. 30–56, november 15,1987.

[12] S.N. Naboko, *Dense point spectra of Schrödinger and Dirac operators*, Theor. Math., 68 pp. 18–28 (1986).

[13] M. A. Naimark, *Linear Differential Operators, Part II: Linear Differential Operators in Hilbert Space*, George G. Harrap & Co., Ltd., 1968.

[14] A.Poltoratski, *On the boundary behavior of pseudocontinuable functions*, St. Petersburg Math. J., vol. 5, 1994, pp. 389–406

[15] A. Poltoratski, *The Kreĭn spectral shift and rank one perturbations of spectra*, St. Petersburg Math. J., Vol. 10 No. 5 pp. 833–859 (1999).

[16] C. Remling, *Embedded singular continuos spectrum for one-dimensional Schrödinger operators*, Trans Amer. Math. Soc., 351 No. 6 pp. 2479–2497 (1999).

[17] M. Rosenblum and J. Rovnyak, *Topics in Hardy Classes and Univalent Functions*, Birkhäuser, 1994.

[18] B. Simon, *Spectral analysis of rank one perturbations and applications*, CRM Proceedings and Lecture Notes Vol. 8 pp. 109–149, 1995.

[19] B. Simon, *Some Schrödinger Operators with dense point spectrum*, Proc. Amer. Math. Soc., 1998.

R. del Rio
IIMAS-UNAM
Circuito Escolar
Ciudad Universitaria
04510, México, D.F., México.
e-mail: delrio@servidor.unam.mx

S. Fuentes
Cajamarca 104
Col. Las Americas
Naucalpan de Juarez
Edo. de Mexico, 53040 Mexico
e-mail: sfuentesm@terra.com.mx

A. Poltoratski
Department of Mathematics
College Station
Texas 77843–3368, USA
e-mail: alexei.poltoratski@math.tamu.edu

Operator Theory:
Advances and Applications, Vol. 132, 141–181

Singular Point-like Perturbations of the Laguerre Operator in a Pontryagin Space

Aad Dijksma and Yuri Shondin

Abstract. The spectral problem for the Laguerre equation on $(0,\infty)$ with real parameter α in the case $0<|\alpha|<1$ is closely related to the Nevanlinna function

$$Q_\alpha(z) = -\pi\Gamma(-z)/(\sin\pi\alpha)\Gamma(-z-\alpha).$$

If $|\alpha|>1$ and $|\alpha|\neq 2,3,\ldots$, this function belongs to the generalized Nevanlinna class N_m, $m=[\frac{|\alpha|+1}{2}]$. A natural question appears: to what spectral problem does this function correspond? For $\alpha<-1$, $\alpha\neq -2,-3,\ldots$, an answer was given by Derkach [D]. He obtained an operator representation for the function $m_\alpha(z) = -Q_\alpha(-z)/\Gamma^2(1+\alpha)$ in terms of a self-adjoint operator in a Pontryagin space and an interpretation of $m_\alpha(z)$ as the Titchmarsh–Weyl function of some boundary value problem related to the Laguerre equation. That an indefinite metric was needed was made clear earlier by Morton and Krall [MK]. In this note for $\alpha>1$, $\alpha\neq 2,3,\ldots$ we answer this and related questions by using Pontryagin space operator realizations of suitable singular point-like perturbations of the Laguerre operator. We describe the operator models for $Q_\alpha(z)$ and compare them with the models for $-\alpha$. Also we discuss the spectral properties of the self-adjoint linear relations in the representation of the functions $Q_\alpha(z)$ and $-Q_\alpha(z)^{-1}$. Finally, we describe the connection between the self-adjoint linear relations in the representations of $Q_\alpha(z)$ and $Q_{-\alpha}(z+\alpha)$ and show that this connection can be viewed as an operator implementation of the Kummer transform for confluent hypergeometric functions.

0. Introduction

For $\alpha \in \mathbf{R}\setminus\mathbf{Z}$, we consider the differential expression

$$\ell_\alpha = -x\frac{d^2}{dx^2} - (1+\alpha-x)\frac{d}{dx}, \quad x\in\mathbf{R}^+, \tag{0.1}$$

in the Hilbert space $\mathcal{H}_0^\alpha = L^2(\mathbf{R}^+, w_\alpha)$ with weight function $w_\alpha(x) = x^\alpha e^{-x}$. Here $\mathbf{R}^+$ and $\mathbf{R}^-$ denote the intervals $(0,\infty)$ and $(-\infty,0)$. The differential expression ℓ_α is formally symmetric in $\mathcal{H}_0^\alpha$. It differs from the standard Laguerre differential

The research of Yuri Shondin was supported by the Netherlands Organization for Scientific Research NWO (NB 047-008-008) and the Russian RFBR grant (N 0001-00544).

expression only by the factor -1. We prefer this choice of ℓ_α because now the operator realizations of ℓ_α in the space $\mathcal{H}_0^\alpha$ which we consider in this note are nonnegative. In this paper we shall use that a general solution of the Kummer equation

$$xf'' + (1+\alpha-x)f' - zf = 0 \tag{0.2}$$

can be written in the form

$$f(x) = c_1\Phi(z, 1+\alpha; x) + c_2\Psi(z, 1+\alpha; x),$$

where c_1, c_2 are complex numbers and $\Phi(a, c; x)$ and $\Psi(a, c; x)$ are the confluent hypergeometric functions for $c \notin \mathbf{Z}$ given by [1]

$$\Phi(a, c; x) = \sum_{n=0}^{\infty} \frac{(a)_n}{(c)_n} \frac{x^n}{n!}, \tag{0.3}$$

$$\Psi(a, c; x) = \frac{\Gamma(1-c)}{\Gamma(a-c+1)}\Phi(a, c; x) + \frac{\Gamma(c-1)}{\Gamma(a)} x^{1-c}\,\Phi(a-c+1, 2-c; x). \tag{0.4}$$

Thus, in our notation, the solutions $f(x)$ of $\ell_\alpha f = zf$ are linear combinations of $\Phi(-z, 1+\alpha; x)$ and $\Psi(-z, 1+\alpha; x)$.

Note that the unitary transformation $U : \mathcal{H}_0^\alpha \to L^2(\mathbf{R}^+)$ given by

$$\begin{aligned} U : f(x) &\mapsto \widehat{f}(y) = (2y)^{1/2} w_\alpha^{1/2}(y^2) f(y^2), \\ U^{-1} : \widehat{f}(y) &\mapsto f(x) = 2^{-1/2} x^{-1/4} w_\alpha^{-1/2}(x) \widehat{f}(\sqrt{x}), \end{aligned}$$

transforms the expression ℓ_α into the expression $\frac{1}{4}h_\alpha$ in $L^2(\mathbf{R}^+)$, where

$$h_\alpha = -\frac{d^2}{dy^2} + y^2 + \frac{\alpha^2 - 1/4}{y^2} - 2(\alpha+1).$$

The expression h_α and the self-adjoint extensions of the minimal operator associated with h_α in $L^2(\mathbf{R}^+)$ have found applications in quantum mechanics; see [C, La, F]. The formula for h_α suggests that the spectral properties of operators associated with ℓ_α are closely related to the spectral properties of the operators associated with $\ell_{-\alpha}$. In this note we shall investigate this (and more) by studying in detail the function

$$Q_\alpha(z) := -\frac{\pi}{\sin \pi\alpha} \frac{\Gamma(-z)}{\Gamma(-z-\alpha)} \tag{0.5}$$

which is defined for all $\alpha \in \mathbf{R} \setminus \mathbf{Z}$ and $z \in \mathbf{C} \setminus \mathbf{Z}^+$. It satisfies the simple relation

$$Q_{-\alpha}(z) = -\frac{\pi^2}{\sin^2 \pi\alpha} Q_\alpha(z-\alpha)^{-1}, \tag{0.6}$$

which we call the Kummer symmetry, and it has the following series expansion, where $\sigma_\alpha(k) = \Gamma(\alpha+k)/k!$.

---

[1] See, for example, [AS, E]. We use the notation as in [E]; in [AS] instead of Φ and Ψ the symbols M and U are used.

Lemma 0.1.

(i) *For* $\alpha \in \mathbf{R}^- \setminus \{-1, -2, -3, \ldots\}$,

$$Q_\alpha(z) = \sum_{n=0}^{\infty} \frac{\sigma_{\alpha+1}(n)}{n - z}. \tag{0.7}$$

(ii) *For* $m < \alpha < m + 1$, $m = 0, 1, 2, \ldots$,

$$Q_\alpha(z) = (z + \alpha)(z + \alpha - 1) \cdots (z + \alpha - m) \sum_{n=0}^{\infty} \frac{\sigma_{\alpha-m}(n)}{n - z}. \tag{0.8}$$

The series converge uniformly on every compact set in $\mathbf{C} \setminus \{0, -1, -2, \ldots\}$.

Proof. We start with the following expansion

$$\frac{\Gamma(-z)\Gamma(a+1)}{\Gamma(-z+a)} = \sum_{n=0}^{\infty} \frac{(-1)^n a(a-1)(a-2)\cdots(a-n)}{n!\,(n-z)},$$

which holds for $a > 0$ and $z \neq 0, 1, \ldots$ (see [[WW], Ch.12, Example 9]). We multiply both sides of this equality by $\Gamma(-a)$ assuming $a \notin \mathbf{Z}^+$. The identity

$$\Gamma(x)\Gamma(1-x) = \frac{\pi}{\sin \pi x}$$

and the chain equality

$$\Gamma(x + n + 1) = x(x+1)\ldots(x+n)\Gamma(x)$$

applied for $x = -a$ and the substitution $a = -\alpha$ yield (0.7).

In the case $0 < \alpha < 1$, we write $\Gamma(-z - \alpha) = (-z - \alpha)^{-1}\Gamma(-z + 1 - \alpha)$ and obtain $Q_\alpha(z) = (z + \alpha)Q_{\alpha-1}(z)$. Now (ii) follows from (i). In the case $m < \alpha < m + 1$, by using the chain equality, we find again $Q_\alpha(z) = (z + \alpha)\cdots(z + \alpha - m + 1)Q_{\alpha-m}(z)$. Then (0.8) follows by applying the foregoing case. □

From the formulas in the lemma it follows that for $0 < |\alpha| < 1$, $Q_\alpha(z)$ is a Nevanlinna function. By definition, a function $Q(z)$ is called a Nevanlinna function if it is defined and holomorphic on $\mathbf{C} \setminus \mathbf{R}$ and if $\operatorname{Im} Q(z)/\operatorname{Im} z \geq 0$ and $Q(z^*) = Q(z)^*$, $z \in \mathbf{C} \setminus \mathbf{R}$. We denote the set of Nevanlinna functions by N_0.

As is known (see, for example, [D]) in the case $0 < |\alpha| < 1$ the minimal operator S_α in $\mathcal{H}_0^\alpha$ associated with the expression ℓ_α is symmetric and has defect numbers $(1,1)$. The set of all self-adjoint extensions A_α^t, $t \in \mathbf{R} \cup \{\infty\}$, of S_α can be described via their resolvents by M.G. Krein's resolvent formula

$$(A_\alpha^t - z)^{-1} = (A_\alpha^\infty - z)^{-1} - \frac{\langle \cdot, \varphi_\alpha(z^*)\rangle}{Q_\alpha(z) + t}\varphi_\alpha(z), \tag{0.9}$$

where $\langle \cdot, \cdot \rangle$ denotes the inner product of $\mathcal{H}_0^\alpha$, the function $\varphi_\alpha(z) \equiv \varphi_\alpha(\cdot, z)$ is given by

$$\varphi_\alpha(x, z) := \Gamma(-z)\Psi(-z, 1 + \alpha; x), \quad x \in \mathbf{R}^+, \tag{0.10}$$

and A_α^∞ is the self-adjoint extension of S_α uniquely determined by:

$(\mathrm{a})_\infty$ its spectrum is pure point spectrum and consists of all nonpositive integers, and

$(\mathrm{b})_\infty$ the generalized Laguerre polynomials (sometimes called Laguerre–Sonine or Laguerre–Chebyshev polynomials)

$$L_k^\alpha(x) = \frac{1}{k!}\mathrm{e}^x x^{-\alpha}\left(\mathrm{e}^{-x}x^{k+\alpha}\right)^{(k)}$$

are the eigenfunctions corresponding to the eigenvalues $k = 0, 1, 2, \ldots$.

In this context the function $Q_\alpha(z)$ is called the Q-function associated with S_α and A_α^∞ (see [KL1]): $Q_\alpha(z)$ is a solution (uniquely determined up to a real constant) of the defining relation for Q-functions

$$\frac{Q_\alpha(z) - Q_\alpha(\zeta)^*}{z - \zeta^*} = \langle \varphi_\alpha(z), \varphi_\alpha(\zeta)\rangle. \tag{0.11}$$

Here the function φ_α can be written as

$$\varphi_\alpha(z) = (1 + (z-\mu)(A_\alpha^\infty - z)^{-1})\varphi_\alpha(\mu),$$

where μ is a fixed point from the resolvent set $\rho(A_\alpha^\infty)$ of A_α^∞ and $\varphi_\alpha(\mu) = \varphi_\alpha(\cdot, \mu)$. Since for all $z \in \rho(A_0)$, $\varphi_\alpha(z)$ belongs to $\ker(S_\alpha^* - z)$, the function φ_α will be called the defect function for S and A_α^∞. The relation (0.11) corresponds to the integral formula

$$\Gamma(a)\Gamma(b)\int_0^\infty \Psi(a, 1+\alpha; x)\Psi(b, 1+\alpha; x)\, x^\alpha \mathrm{e}^{-x}dx \tag{0.12}$$
$$= \frac{\pi}{\sin\pi\alpha}(a-b)^{-1}\left(\frac{\Gamma(a)}{\Gamma(a-\alpha)} - \frac{\Gamma(b)}{\Gamma(b-\alpha)}\right),$$

which can easily be derived from [[E], Ch.6, p.6.15, Formula (24)] by using the representation [[AS], Formula (13.1.3)] for the function Ψ. Note that (0.11) also implies that $Q_\alpha(z) \in N_0$.

Besides the self-adjoint extension A_α^∞ determined by the conditions $(\mathrm{a})_\infty$ and $(\mathrm{b})_\infty$ above, also the self-adjoint extension A_α^0 of S_α (corresponding to $t = 0$ in Krein's formula (0.9)) plays an important role in the sequel. It is characterized by the following properties:

$(\mathrm{a})_0$ its spectrum consists of the points $k - \alpha$, $k = 0, 1, \ldots$, and

$(\mathrm{b})_0$ the functions $\Psi(-k+\alpha, 1+\alpha; x)$ are the corresponding eigenfunctions.

The two extensions A_α^∞ and A_α^0 are extremal in the sense that if $0 < \alpha < 1$ then A_α^∞ is the Friedrichs extension and A_α^0 is the Krein extension of S_α, and if $-1 < \alpha < 0$ then it is the other way around: A_α^∞ is the Krein extension and A_α^0 is the Friedrichs extension of S_α (see [D, KO]) .

These facts no longer hold when $|\alpha| > 1$, $\alpha \neq \pm 2, \pm 3, \ldots$.

In this paper we focus to a large extend on the case $\alpha > 1$, $\alpha \neq 2, 3, \ldots$. The main deviations from the case described above are:

(a) The minimal operator in $\mathcal{H}_0^\alpha$ which corresponds to ℓ_α is not symmetric with defect numbers $(1,1)$, but symmetric with defect numbers $(0,0)$, that is, self-adjoint. Throughout the text we denote this operator by L_α. It is a nonnegative operator.

(b) The functions $\varphi_\alpha(x,z) = \Gamma(-z)\Psi(-z, 1+\alpha; x)$, $z \in \rho(L_\alpha)$, have a singular behavior when $x \to 0$: $\varphi_\alpha(x,z) \sim \Gamma(\alpha)x^{-\alpha}$ and $\varphi_\alpha(x,z) \notin \mathcal{H}_0^\alpha$. Hence the integral on the left-hand side of (0.12) diverges and the formulas (0.11) and (0.9) loose their meaning.

(c) As we shall see below, the function $Q_\alpha(z)$ is not a Nevanlinna function but a generalized Nevanlinna function belonging to the class N_m with $m = [\frac{\alpha+1}{2}]$ negative squares. Here and further, for a positive number a, $[a]$ stands for the integer part of a and $\{a\} = a - [a]$.

So for the case $\alpha > 1$, $\alpha \neq 2, 3, \ldots$, natural problems appear and in this note we will try to solve them:

1) Find a self-adjoint operator (or a linear relation) for which the functions $\Psi(-n+\alpha, 1+\alpha; x)$, $n = 0, 1, \ldots$, can be interpreted as eigenfunctions.
2) Derive an operator representation for $Q_\alpha(z)$. Since it belongs to the class N_m this involves self-adjoint operators or linear relations in a Pontryagin space with negative index m.
3) Find new interpretations for the formulas (0.9) and (0.11). For example, how should one define $\varphi_\alpha(z)$ so that (0.11) makes sense?

These problems are addressed in Subsection 2.3. To obtain their solutions we use the method of realizations of singular perturbations of self-adjoint operators developed in [S1, DT, S2] and further in [DLSZ]. In [DS] we applied this method to the Bessel operator with index ν. There we posed and solved problems similar to 1), 2), and 3) in case $\nu > 1$. To indicate the direction in which we look for solutions to the problems 1), 2), and 3), we again consider the case $0 < |\alpha| < 1$ and look at the self-adjoint extension A_α^t whose resolvent is described by Krein's formula (0.9). It can be viewed as the realization of a singular point-like perturbation of operator L_α in $\mathcal{H}_0^\alpha$:

$$A_\alpha^t = L_\alpha + t^{-1}\langle \cdot, \chi\rangle\chi, \quad t \in \mathbf{R} \cup \{\infty\}. \tag{0.13}$$

Here χ is the generalized element

$$\chi = (L_\alpha - \mu)\,\Gamma(-\mu)\Psi(-\mu, 1+\alpha; x), \tag{0.14}$$

where now μ is some fixed point in $\mathbf{R}^-$. Since $0 < |\alpha| < 1$, χ belongs to the negative scale space $\mathcal{H}_{-1}^\alpha$ associated with L_α. (We recall the definition of scale spaces in the next section.) The equality

$$(\ell_\alpha - \mu)\Psi(-\mu, 1+\alpha; x) = 0,\ x \in \mathbf{R}^+,$$

shows that the support of χ is concentrated just in the one point $x = 0$. It is the reason why A_α^t in (0.13) is called a "point-like perturbation" of L_α. The monographs [AGHH] and [AK] develop an extensive theory for such perturbations and provide ample motivations. In this case the theory is that the A_α^t's are the canonical self-adjoint extensions of the one-dimensional restriction S_0 of L_α:

$$\operatorname{dom} S_0 = \{u \in \operatorname{dom} L_\alpha | \langle u, \chi \rangle = 0\}, \qquad S_0 u = L_\alpha u, \ u \in \operatorname{dom} S_0.$$

In the case $\alpha > 1$, the generalized element χ in (0.14) belongs to scale space $\mathcal{H}_{-m-1}^\alpha \setminus \mathcal{H}_{-m}^\alpha$ with $m = [\frac{\alpha+1}{2}]$ and the formal perturbation (0.13) becomes more singular. Following the cited papers we "lift" the operator L_α to a self-adjoint relation A_α^∞ in a suitable Pontryagin space Π_m^α with negative index m which extends $\mathcal{H}_0^\alpha$. The formal expressions A_α^t, $t \in \mathbf{R} \cup \{\infty\}$, defined by (0.13) are interpreted as the canonical self-adjoint extensions of a one-dimensional restriction of A_α^∞. There is lot of freedom in the construction of this model. For a suitable choice of the "free" parameters, the function $Q_\alpha(z)$ is the Q-function associated with the lifting of L_α and its one-dimensional restriction, and in this model the function $\varphi_\alpha(z)$ can be redefined and (0.9) has a meaning.

In Subsection 2.1 we describe the realization of (0.13) in a model case where L_α is replaced by the operator $\widehat{A}_0$ in $l^2(\mathbf{Z}^+)$ of multiplication by the discrete independent variable n, say, and χ is given by the sequence

$$\widehat{\chi}_\alpha(n) = (\Gamma(\alpha + n + 1)/n!)^{1/2}.$$

This model is called the Friedrichs model; in Subsection 2.2 we specialize the parameters and call the model the Friedrichs-Laguerre model. The realization in the Pontryagin space Π_m^α associated with L_α can be obtained from this model by using the Laguerre transform; this realization is called the Laguerre model. We describe properties of two extensions A_α^∞ and A_α^0 and show that in the decomposition

$$\Pi_m^\alpha = \mathcal{H}_0^\alpha \oplus (\mathcal{L}^\alpha \dot{+} \mathcal{M}^\alpha)$$

$\mathcal{L}^\alpha$ is the root subspace of A_α^∞ corresponding to its eigenvalue ∞ and $\mathcal{M}^\alpha$ is the span of the first m eigenvectors of A_α^0. We have that $P_{\mathcal{H}_0^\alpha}(A_\alpha^\infty - z)^{-1}|_{\mathcal{H}_0^\alpha} = (L_\alpha - z)^{-1}$, that is, the left-hand side is a resolvent operator on $\mathcal{H}_0^\alpha$, whereas $P_{\mathcal{H}_0^\alpha}(A_\alpha^0 - z)^{-1}|_{\mathcal{H}_0^\alpha}$ is a generalized resolvent, but not a resolvent operator.

In Section 3 we describe the point-like perturbation of the Laguerre operator in the case $\alpha < -1$. Following the line of Section 2 we first describe the realization of (0.13) in a model case where L_α is replaced by the operator, now denoted by $\widehat{A}_\alpha^\infty$, of multiplication by the discrete independent variable n in $l^2(\mathbf{Z}^+)$ and $\widehat{\chi}$ is given by the sequence $\widehat{\chi}_\alpha(n) = s_n \left(|\Gamma(\alpha + n + 1)|/n!\right)^{1/2}$ (see (3.1) and (3.2)) which belongs to $\operatorname{dom}(\widehat{A}_\alpha^\infty)^{m-1}$, $m = [\frac{|\alpha|+1}{2}]$. The new feature is that we equip $l^2(\mathbf{Z}^+)$ with an indefinite inner product which makes it a Pontryagin space Π_m^α with negative index m. The analog of the perturbation (0.13):

$$\widehat{A}_\alpha^t = \widehat{A}_\alpha^\infty + t^{-1}\langle \cdot, \widehat{\chi} \rangle \widehat{\chi}$$

is now a rank one perturbation and the operator $\widehat{A}_\alpha^t$ is well defined for each $t \neq 0$, and so is, via Krein's formula, $\widehat{A}_\alpha^0$; it is a self-adjoint linear relation in Π_m. The realization in the Pontryagin space Π_m^α associated with L_α for $\alpha < -1$ can be obtained from this model by using the Laguerre transform. In Section 3 we do the reverse of Section 2: In Subsection 3.3 we look for a decomposition

$$\Pi_m^\alpha = \mathcal{H}_0^\alpha \oplus (\mathcal{L}^\alpha \dot{+} \mathcal{M}^\alpha), \tag{0.15}$$

such that $\mathcal{H}_0^\alpha$ is a Hilbert space, $\mathcal{L}^\alpha$ is the root subspace for $\widehat{A}_\alpha^0$ at ∞ and $\mathcal{M}^\alpha$ is the span of the first m eigenvectors of $\widehat{A}_\alpha^\infty$, $P_{\mathcal{H}_0^\alpha}(A_\alpha^0 - z)^{-1}|_{\mathcal{H}_0^\alpha}$ is a resolvent operator on $\mathcal{H}_0^\alpha$, whereas $P_{\mathcal{H}_0^\alpha}(A_\alpha^\infty - z)^{-1}|_{\mathcal{H}_0^\alpha}$ is a generalized resolvent, but not a resolvent operator.

Morton and Krall in [MK] observed that in the case $-n-1 < \alpha < -n$, $n = 1, 2, \ldots$, the generalized Laguerre polynomials $L_k^\alpha(x)$ are orthogonal relative to indefinite inner product

$$\langle u, v \rangle = \int_0^\infty x^\alpha \left(\mathrm{e}^{-x} u v^* - \sum_{j=0}^{n-1} \left(\mathrm{e}^{-x} u v^*\right)^{(j)}(0) \frac{x^j}{j!} \right) dx. \tag{0.16}$$

In the subsequent papers [K1, K2] the following problem was studied: Find an appropriate operator treatment of the expression $(-\ell_\alpha)$ in a space which can be associated with the indefinite inner product (0.16), rather than in H_0^α. A complete solution of this problem was given recently by Derkach in [D]. Using extension theory in Pontryagin spaces he describes a self-adjoint operator A in a Pontryagin space Π_κ with negative index $\kappa = [\frac{n+1}{2}]$ for which the generalized Laguerre polynomials $L_k^\alpha(x)$ are the eigenfunctions. Moreover, he gives an operator representation for the function

$$m_\alpha(z) = -Q^\alpha(-z)/\Gamma^2(1+\alpha)$$

in terms of a self-adjoint operator in a Pontryagin space and an interpretation of $m_\alpha(z)$ as the Titchmarsh–Weyl function of some boundary value problem associated with the Laguerre equation. His method is related to the paper [JLT] of Jonas, Langer, and Textorius, in which general models for cyclic self-adjoint operators in a Pontryagin space are developed. In Section 3 we show that our model is isomorphic to the realization obtained by Derkach. In fact, we prove that the space $\mathcal{H}_0^\alpha$ can be identified with the space $L^2(\mathbf{R}^+, w_\alpha)$ and that under this identification $P_{\mathcal{H}_0^\alpha}(A_\alpha^0 - z)^{-1}|_{\mathcal{H}_0^\alpha}$ coincides with the resolvent $(L_\alpha - z)^{-1}$ of the minimal Laguerre operator. Moreover, the operator A obtained by Derkach corresponds to our operator $\widehat{A}_\alpha^\infty$.

Besides this introduction the paper consists of four sections and the references. In Section 1 we review with a slight modification the construction of the operators A^t in Π_m. We specialize it in Section 2 to the point-like perturbation of the Laguerre operator in the case $\alpha > 1$. In Section 3 we study the point-like perturbation of the Laguerre operator in the case $\alpha < -1$. Section 4 concerns the Kummer symmetry (0.6). There we also describe a unitary correspondence between the realizations for α and $-\alpha$ and in particular, between the linear relations and operators $A_{\pm\alpha}^\infty$ and $A_{\mp\alpha}^0$ which appear in the operator representations of the Q-functions $Q_\alpha(z)$ and $Q_{-\alpha}(z+\alpha)$.

1. A Pontryagin space realization of singular perturbations of a nonnegative operator

Consider an unbounded nonnegative self-adjoint operator A_0 in a Hilbert space $\mathcal{H}_0$ with inner product $\langle\cdot,\cdot\rangle_0$, a nonnegative integer m, and the corresponding scale of Hilbert spaces

$$\mathcal{H}_{m+1} \hookrightarrow \mathcal{H}_m \hookrightarrow \cdots \hookrightarrow \mathcal{H}_1 \hookrightarrow \mathcal{H}_0 \hookrightarrow \mathcal{H}_{-1} \hookrightarrow \cdots \hookrightarrow \mathcal{H}_{-m} \hookrightarrow \mathcal{H}_{-m-1}, \tag{1.1}$$

where the inclusion mappings $\hookrightarrow$ are contractions with a dense range. Recall (see [B]) that for $j = 1, 2, \ldots, m+1$, $\mathcal{H}_j$ is the Hilbert space $\operatorname{dom} A_0^j$ equipped with the inner product

$$\langle u, v\rangle_j = \langle (A_0+1)^j u, (A_0+1)^j v\rangle_0, \quad u, v \in \mathcal{H}_j, \tag{1.2}$$

and that $\mathcal{H}_{-j}$ is the Hilbert space completion of $\mathcal{H}_0$ with respect to the norm

$$\|f\|_{-j} = \sup_{0\neq u\in\mathcal{H}_j} \frac{\langle f, u\rangle_0}{\|u\|_j}, \quad f \in \mathcal{H}_0.$$

The scale (1.1) defines a rigging of $\mathcal{H}_0$ in the sense that for $j = 1, 2, \ldots, m+1$, the inner product $\langle f, u\rangle_0$ on $\mathcal{H}_0$ (viewed as a sesquilinear form on $\mathcal{H}_0 \times \mathcal{H}_0$) can be extended/restricted to a bounded sesquilinear form on $\mathcal{H}_{-j}\times\mathcal{H}_j$: for each $f \in \mathcal{H}_{-j}$ there is a sequence $f_n \in \mathcal{H}_0$ converging to f in the norm of $\mathcal{H}_{-j}$ such that for all $u \in \mathcal{H}_j$ the limit

$$\langle f, u\rangle_0 := \lim_{n\to\infty} \langle f_n, u\rangle_0$$

exists (and is independent of the sequence f_n) and satisfies

$$\|\langle f, u\rangle_0\| \leq \|f\|_{-j}\|u\|_j, \quad f \in \mathcal{H}_{-j},\ u \in \mathcal{H}_j.$$

Thus, alternatively, via the mapping $u \in \mathcal{H}_j \mapsto \langle f, u\rangle_0$, $f \in \mathcal{H}_{-j}$ can be viewed as a continuous antilinear functional on $\mathcal{H}_j$, $j = 1, 2, \ldots, m+1$. We set $\langle u, f\rangle_0 = \langle f, u\rangle_0^*$. By induction the resolvent $R_0(z) = (A_0 - z)^{-1}$ can be extended to a mapping from $\mathcal{H}_{-j-1}$ to $\mathcal{H}_{-j}$ via

$$\langle R_0(z) f, u\rangle_0 = \langle f, R_0(z^*)u\rangle_0, \quad f \in \mathcal{H}_{-j-1},\ u \in \mathcal{H}_j.$$

This extended resolvent is continuous, satisfies the resolvent identity, and (1.2) holds true for negative j's also. The spaces $\mathcal{H}_{j+1/2}$, $j = 0, \pm1, \pm2, \ldots$ are defined similarly.

In this section we describe the construction of an operator realization of the singular perturbations of A_0 formally represented by the formal expression (compare with (0.13))

$$A^t = A_0 + t^{-1}\langle\cdot,\chi\rangle_0\chi, \tag{1.3}$$

where $\chi \in \mathcal{H}_{-m-1}\setminus\mathcal{H}_{-m}$. Throughout the sequel we shall assume the minimality condition

$$\overline{\operatorname{span}}\{(A_0 - z)^{-1}(A_0+1)^{-m}\chi \mid z \in \rho(A_0)\} = \mathcal{H}_0. \tag{1.4}$$

The construction is a slight modification of the one in [S1, DT, S2](see also [DLSZ]): Now we use $m+1$ auxiliary level points $\mu_j \in \mathbf{R}^-$ (see below) instead of one single one. This allows us to be more flexible when we apply the model to the Laguerre operator. Earlier in [DS] a two-point modification was applied to the realization problem for point-like perturbations of the Bessel operator. The realization method we use now is an adaptation of the Λ-representation method in [DLSZ].

When $\chi \in \mathcal{H}_0$ or when $\chi \in \mathcal{H}_{-1}$ (the case $m = 0$) the formal expression (1.3) is realizable in the Hilbert space $\mathcal{H}_0$. Indeed, in the first case it can be viewed as a rank one perturbation and in the second case it is interpreted (see, for example, [AGHH] and [AK]) as the canonical self-adjoint extension of the one-dimensional S_0 restriction of A_0:

$$S_0 = A_0|_{\operatorname{dom} S_0}, \quad \operatorname{dom} S_0 = \{u \in \operatorname{dom} A_0 | \langle u, \chi \rangle_0 = 0\}.$$

Evidently, S_0 is symmetric, has defect indices $1, 1$, and its adjoint is given by

$$S_0^* = A_0 \dot{+} \operatorname{span}\{\{(A_0 - \mu)^{-1}\chi, \mu(A_0 - \mu)^{-1}\chi\}\},$$

where μ is a point from $\rho(A_0)$ and the sum is a direct sum in $\mathcal{H}_0^2$. Here and elsewhere in the sequel when it is convenient we use graph notation.

1.1. The realization of (1.3) when $\chi \in \mathcal{H}_{-m-1} \setminus \mathcal{H}_{-m}$, $m \geq 1$. In this case S_0 becomes essentially self-adjoint in $\mathcal{H}_0$ and Berezin [BE] was the first to propose that non-trivial perturbations of A_0 generated by χ should be realized in an indefinite metric space. To construct such a realization one associates with A_0 and χ a self-adjoint relation A^∞ in a Pontryagin space Π_m with (negative) index m and considers a suitable one-dimensional restriction S of A^∞. The realization or the interpretation of (1.3) is then defined as a corresponding representative from the family of all canonical self-adjoint extensions A^t of S. The construction will be given below in three steps: First we build the space Π_m, then the relation A^∞, and then we define S and determine the extensions A^t. For proofs of the statements we refer to the proofs of analogous results in the previously cited papers.

The construction of the realization admits some freedom: First we select m different points $\mu_j \in \mathbf{R}^-$ on the negative half-axis and form the Λ-set $\Lambda = \{\mu_1, \mu_2, \ldots, \mu_m\}$. Evidently, $\Lambda \subset \rho(A_0)$. With Λ we define the monomials

$$b_0(z) = 1, \; b_j(z) = (z - \mu_1)(z - \mu_2) \cdots (z - \mu_j), \; j = 1, 2, \ldots, m. \tag{1.5}$$

We also choose a point $\mu_0 \in \mathbf{R}^-$ and $2m - 1$ real numbers $g_2, g_3, \ldots, g_{2m}$. Later we also fix a real number g_1 (see after (1.12)) and if need be a real number g_{2m+1} (see the case $(c)_{ii}$ below). These are all the parameters that play a role in the construction below. Later in the case of the Laguerre operator we shall make suitable choices for them to obtain the right model for the Q-function.

The Pontryagin space Π_m.
With the Λ-set and the point μ_0 we associate $2m + 1$ elements

$$\varepsilon_j \in \mathcal{H}_{-m-1+j} \setminus \mathcal{H}_{-m+j},$$

$j = 1, 2, \ldots, 2m+1$:

$$\begin{aligned} \varepsilon_j &= b_j(A_0)^{-1}\chi, \quad j = 1, 2, \ldots, m, \\ \varepsilon_{j+m} &= (A_0 - \mu_0)^{-j} b_m(A_0)^{-1}\chi, \quad j = 1, 2, \ldots, m+1. \end{aligned}$$

With the first $2m$ ε_j's we form the linear space

$$\mathcal{P}_m = \mathcal{H}_m \dot{+} \operatorname{span}\{\varepsilon_1, \varepsilon_2, \ldots, \varepsilon_{2m}\}.$$

Here the sum is a direct sum, and the elements of $\mathcal{P}_m$ are written in the form

$$f = f_m + \sum_{j=1}^{2m} F_j \varepsilon_j, \quad f_m \in \mathcal{H}_m, \ (F_j)_{j=1}^{2m} \in \mathbf{C}^{2m}.$$

For example, $(A_0 - z)^{-1}\chi \in \mathcal{P}_m$ and it has the decomposition

$$(A_0 - z)^{-1}\chi = \sum_{j=1}^{m} b_{j-1}(z)\varepsilon_j + b_m(z)(A_0 - z)^{-1}\varepsilon_m. \tag{1.6}$$

On $\mathcal{P}_m$ we define the inner product

$$\langle f, f' \rangle = \langle f_m, f'_m \rangle_0 + \sum_{j=1}^{2m} (F_j \langle \varepsilon_j, f'_m \rangle_0 + F_j'^{*} \langle f_m, \varepsilon_j \rangle_0) + \sum_{j,k=1}^{2m} g_{j,k} F_k F_j'^{*}, \tag{1.7}$$

where real numbers $g_{j,k}$ have been substituted for the formal quantities $\langle \varepsilon_k, \varepsilon_j \rangle_0$, $j, k = 1, \ldots, 2m$, and these numbers are defined as follows.

(a) If $j + k \geq 2m + 2$, then $g_{j,k} = \langle \varepsilon_k, \varepsilon_j \rangle_0$; these numbers are well defined in the rigging.

(b) If $j, k = 1, \ldots, m$, then first we define

$$\begin{aligned} g_{1,k} &= g_{k,1} = g_{k+1}, \quad k = 1, 2, \ldots, m, \\ g_{j,m} &= g_{m,j} = g_{j+m}, \quad j = 2, 3, \ldots, m, \end{aligned}$$

and then we determine the other $g_{j,k}$'s by the recurrence relations

$$g_{j,k} = g_{k,j}, \quad (\mu_k - \mu_j) g_{j,k} = g_{j-1,k} - g_{j,k-1}.$$

(c) If $1 \leq j \leq m$, $m + 1 \leq k \leq 2m$ and $j + k < 2m + 2$, we first consider two cases.

(c)$_i$] If $\chi \in \mathcal{H}_{-m-1/2} \setminus \mathcal{H}_{-m}$, then $g_{m+1,m} = g_{m+1,m} = g_{2m+1}$, where

$$g_{2m+1} = \langle (A_0 - \mu_0)^{-1/2}\varepsilon_m, (A_0 - \mu_0)^{-1/2}\varepsilon_m \rangle_0,$$

which is well defined since $(A_0 - \mu_0)^{-1/2}\varepsilon_m \in \mathcal{H}_0$.

(c)$_{ii}$ If $\chi \in \mathcal{H}_{-m-1} \setminus \mathcal{H}_{-m-1/2}$, we choose a real number g_{2m+1} and define

$$g_{m+1,m} = g_{m,m+1} = g_{2m+1}.$$

The numbers $g_{j,k}$ for the remaining values of j and k are defined by the rule

$$g_{j,m+n} = g_{m+n,j} = \sum_{l=n}^{m-j+1} A_l g_{m,j+l} + \sum_{i=2}^{n} B_l \langle \varepsilon_m, \varepsilon_{m+i} \rangle_0, \quad n+j \leq m+1,$$

where the real numbers $A_l = A_l(\Lambda)$ and $B_l = B_l(\Lambda)$ are uniquely determined from the expansion

$$\begin{aligned}(z-\mu_0)^{-n} b_j(z)^{-1} &= \sum_{l=n}^{m-j} A_l b_{j+l}(z)^{-1} + A_{m-j+1}(z-\mu_0)^{-1} b_m(z)^{-1} \\ &\quad + \sum_{i=2}^{n} B_l (z-\mu_0)^{-m} b_m(z)^{-1}, \quad n+j \leq m+1.\end{aligned}$$

The **G**-space (see [AI]) Π_m is defined as the space $\mathcal{H}_0 \oplus \mathbf{C}^m \oplus \mathbf{C}^m$ whose elements are column vectors of the form

$$f = \text{column}\left(f_0, \widetilde{F}, F\right), \quad f_0 \in \mathcal{H}_0, \ \widetilde{F} \in \mathbf{C}^m, \ F \in \mathbf{C}^m,$$

and whose inner product on Π_m is the **G**-inner product $\langle \cdot, \cdot \rangle = \langle \cdot, \mathbf{G} \cdot \rangle_{\mathcal{H}_0 \oplus \mathbf{C}^m \oplus \mathbf{C}^m}$, were the Gram operator **G** is of the form

$$\mathbf{G} = \begin{pmatrix} 1_0 & 0 \\ 0 & G_{2m} \end{pmatrix}, \qquad G_{2m} = \begin{pmatrix} 0 & 1_m \\ 1_m & G \end{pmatrix}. \tag{1.8}$$

Here 1_0 and 1_m are the identity operators in $\mathcal{H}_0$ and $\mathbf{C}^m$, and $G = (g_{j,k})_{j,k=1}^m$. Evidently, Π_m is a Pontryagin space of index m.

The mapping τ_χ which is defined by the rule

$$\tau_\chi\Big(f_m + \sum_{j=1}^{2m} F_j \varepsilon_j\Big) = \begin{pmatrix} f_m + \sum_{k=m+1}^{2m} F_k \varepsilon_k \\ \left(\langle f_m, \varepsilon_j \rangle_0 + \sum_{k=m+1}^{2m} g_{j,k} F_k\right)_{j=1}^m \\ (F_j)_{j=1}^m \end{pmatrix} \tag{1.9}$$

maps $\mathcal{P}_m$ isometrically in Π_m and has a dense range. Hence $\mathcal{P}_m$ is a pre-Pontryagin space and Π_m is its Pontryagin space completion.

The self-adjoint linear relation A^∞ in Π_m.

The operator A_0 is naturally embedded in $\mathcal{P}_m$ as the operator $\overset{\circ}{A}$ defined on the domain

$$\operatorname{dom}(\overset{\circ}{A}) = \{ f \in \mathcal{P}_m | f_m = f_{m+1} + F_{2m+1} \varepsilon_{2m+1}, f_{m+1} \in \mathcal{H}_{m+1}, F_{2m+1} \in \mathbf{C}, F_1 = 0 \}$$

by

$$\overset{\circ}{A} f = A_0 f_{m+1} + \mu_0 F_{2m+1} \varepsilon_{2m+1} + \sum_{j=1}^{m} (F_{j+1} + \mu_j F_j) \varepsilon_j + \sum_{j=1}^{m} (F_{j+m+1} + \mu_0 F_{j+m}) \varepsilon_{j+m}.$$

Let A^∞ be the closure in the Pontryagin space Π_m of $\tau_\chi \overset{\circ}{A} \tau_\chi^{-1}$; the latter operator is an isometric copy of $\overset{\circ}{A}$ in Π_m under action of τ_χ. It can be shown that A^∞ is a self-adjoint linear relation in Π_m and its multivalued part, that is, its eigenspace corresponding to the eigenvalue ∞, is the subspace

$$A^\infty(0) = \text{span}\,\{\text{column}\,(0, e_1, 0)\}\,,$$

where e_j is the j-th standard unit vector in $\mathbf{C}^m$, $j = 1, \ldots, m$. Its root subspace at ∞ is the neutral subspace

$$\mathcal{L} = \{0\} \oplus \mathbf{C}^m \oplus \{0\}.$$

The point ∞ is the only critical of nonpositive type of A^∞ and it has multiplicity m; see [DLSZ], Theorems 5.7 and 5.8. The relation A^∞ can be considered as a kind of "lifting" of A_0 from $\mathcal{H}_0$ to Π_m; in particular, $\rho(A^\infty) = \rho(A_0)$ and $P_{\mathcal{H}_0}(A^\infty - z)^{-1}|_{\mathcal{H}_0} = (A_0 - z)^{-1}$, $z \in \rho(A_0)$.

The self-adjoint operators A^t in Π_m.[2]
These operators are defined as the canonical self-adjoint extensions of a one-dimensional restriction S of A^∞. S is chosen such that the defect subspace $\ker\,(S^* - \mu_1) = \text{ran}\,(S - \mu_1)^\perp$ is the one-dimensional space generated by the vector $\tau_\chi\varepsilon_1 =$ column $(0, 0, e_1)$. Hence S and its adjoint S^* in Π_m are given by

$$S = A^\infty \cap \{\{\tau_\chi\varepsilon_1, \mu_1\tau_\chi\varepsilon_1\}\}^*, \quad S^* = A^\infty \dotplus \text{span}\,\{\{\tau_\chi\varepsilon_1, \mu_1\tau_\chi\varepsilon_1\}\}.$$

The canonical self-adjoint extension A^t of S in Π_m is a restriction of S^* and can be written in the form

$$\begin{aligned} A^t \;=\; & \{\{f_\infty + \varphi(\mu_1)F_1, f'_\infty + \mu_1\varphi(\mu_1)F_1\} \in S^* \\ & \qquad \mid \{f_\infty, f'_\infty\} \in A^\infty,\, F_1 \in \mathbf{C},\, \langle\langle (f'_\infty - \mu_1 f_\infty), \varphi(\mu_1)\rangle\rangle = -tF_1\}, \end{aligned}$$

where

$$\varphi(\mu_1) := \tau_\chi\varepsilon_1, \qquad \varphi(\mu_1) \in \ker\,(S^* - \mu_1).$$

The parameter $t \in \mathbf{R} \cup \{\infty\}$ distinguishes between the different canonical self-adjoint extensions of S, and every canonical self-adjoint extension of S coincides with one of the A^t's. A^∞ is the unique multivalued canonical extension that is, linear relation rather than operator. The family of operators A^t, $t \in \mathbf{R} \cup \{\infty\}$, in Π_m or equivalently, the triple $\{\Pi_m, S, A^\infty\}$ represents and will be termed the operator realization of the formal expression (1.3) corresponding to the parameter set $\{\Lambda, \mu_0, \{g_k\}_{k=2}^l\}$, where $l = 2m$ if $\chi \in \mathcal{H}_{-m-1/2} \setminus \mathcal{H}_{-m}$ and $l = 2m+1$ if $\chi \in \mathcal{H}_{-m-1} \setminus \mathcal{H}_{-m-1/2}$. The triple uniquely determines the extensions A^t of S and the extensions uniquely determine the triple.

1.2. Krein's formula for resolvents. Now we describe Krein's formula for the resolvents $R^t(z) = (A^t - z)^{-1}$ of A^t. We define the function $\varphi(z)$ by

$$\varphi(z) = \tau_\chi(A_0 - z)^{-1}\chi = (I + (z - \mu_1)R^\infty(z))\varphi(\mu_1), \quad z \in \rho(A_0). \tag{1.10}$$

[2]Occasionally in the sequel the term operator should be interpreted so that it includes the possibility that the operator is a linear relation (or "multivalued operator").

It satisfies the relation

$$\varphi(z)-\varphi(\zeta)=(z-\zeta)R^{\infty}(z)\varphi(\zeta), \quad z,\zeta\in\rho(A_0). \tag{1.11}$$

Evidently, $\varphi(z)\in\ker(S^*-z)\subset\Pi_m$ and hence $\varphi(z)$ is the defect function associated with the symmetric operator S and its self-adjoint extension A^{∞} in Π_m. By definition, the Q-function $Q(z)$ related to S and A^{∞} is the solution of the equation

$$\frac{Q(z)-Q(\zeta)^*}{z-\zeta^*}=\langle\varphi(z),\varphi(\zeta)\rangle. \tag{1.12}$$

This equation determines $Q(z)$ up to a real constant. We put $Q(\mu_1)=: g_1$ and obtain

$$Q(z)=(z-\mu_1)\langle\varphi(z),\varphi(\mu_1)\rangle+g_1=b_m^2(z)Q_0(z)+p_{2m-1}(z), \tag{1.13}$$

where

$$Q_0(z)=(z-\mu_0)\langle R_0(z)\varepsilon_m,\varepsilon_{m+1}\rangle_0+g_{2m+1}, \tag{1.14}$$

and

$$p_{2m-1}(z)=\sum_{j=2}^{m}b_m(z)b_{j-1}(z)g_{j+m}+\sum_{j=1}^{m}b_1(z)b_{j-1}(z)g_{j+1}+g_1. \tag{1.15}$$

Note that $p_{2m-1}(z)$ is a polynomial of degree at most $2m-1$, which is self-adjoint, that is, $p_{2m-1}(z^*)^*=p_{2m-1}(z)$.

Krein's formula for the resolvent of A^t takes the following form (see, for example, [DLSZ])

$$R^t(z)=R^{\infty}(z)-\frac{\langle\cdot,\varphi(z^*)\rangle}{Q(z)-Q(\mu_1)+t}\varphi(z), \quad z\in\rho(A_0)\cap\rho(A^t). \tag{1.16}$$

Because of (1.10), the first equality in (1.13) takes the form

$$Q(z)=(z-\mu_1)\langle(1+(z-\mu_1)R^{\infty}(z))\varphi(\mu_1),\varphi(\mu_1)\rangle+g_1. \tag{1.17}$$

This identity is called the operator representation of $Q(z)$. The function $Q(z)$ belongs to the generalized Nevanlinna class N_m of functions with m negative squares. We recall the definition from [KL3]: A function $Q(z)$ belongs to N_m if

(1) $Q(z)$ is meromorphic on $\mathbf{C}\setminus\mathbf{R}$ (we denote by $\rho(Q)$ the set of $z\in\mathbf{C}\setminus\mathbf{R}$ in which Q is holomorphic),
(2) $Q(z^*)=Q(z)^*$ for all $z\in\rho(Q)$, and
(3) the kernel

$$K_Q(z,\zeta)=\frac{Q(z)-Q(\zeta)^*}{z-\zeta^*}$$

has m negative squares on $\rho(Q)$. This means that all hermitian matrices of the form $(K_Q(z_j,z_k))_{j,k=1}^n$, where n is an arbitrary integer ≥ 1 and the z_j's are arbitrary points in $\rho(Q)$, have at most and at least one of them has exactly m negative eigenvalues counted with multiplicity.

For $m=0$, the class N_0 coincides with the class of Nevanlinna functions defined in the introduction. Thus, for example, the function $Q_0(z)$ in (1.14) belongs to N_0, and $Q_0(\mu_0)=g_{2m+1}$.

In the realization problem for singular perturbations a special subclass N_m^∞ of holomorphic functions in the class N_m appears naturally; see [DLSZ]. We recall its definition: Let

$$N_0^\infty = \{Q_0(z) \in N_0 \mid \lim_{y\to\infty} y \operatorname{Im} Q_0(iy) = \infty, \quad \lim_{y\to\infty} y^{-1} \operatorname{Im} Q_0(iy) = 0\}.$$

For a nonpositive integer m, N_m^∞ is the set of all functions $Q(z)$ which admit the representation

$$Q(z) = (z^2+1)^m\, Q_0(z) + p_{2m-1}(z), \tag{1.18}$$

where $Q_0(z) \in N_0^\infty$ and $p_{2m-1}(z)$ is a self-adjoint polynomial of degree at most $2m-1$. It follows easily, that $N_m^\infty \subset N_m$, and that for a function $Q(z) \in N_m^\infty$ the point ∞ is a generalized pole of $Q(z)$ of nonpositive type of multiplicity m (see [KL3, L]). Therefore, for $Q(z) \in N_m^\infty$ the total number of its zeros in $\mathbf{C}^+$ and its generalized zeros of nonpositive type on $\mathbf{R}$ equals m; here *total* means, that these zeros are counted according to their multiplicities. In the sequel A_0 is nonnegative and then the Q-functions encountered in the realization problem form a subclass N_{m+}^∞ of N_m^∞. To define this subclass we note that if $Q \in N_m^\infty$ has the representation (1.18) and if $\mathbf{R}^- \subset \rho(Q)$ then the limit

$$\gamma := \lim_{x\to-\infty} \frac{Q(x)}{x^{2m}} = \lim_{x\to-\infty} Q_0(x)$$

exists and $\gamma \in [-\infty, \infty)$. This follows from the integral representation for $Q_0 \in N_0^\infty$. The Q-functions arising from our model with $A_0 \geq 0$ satisfy the properties $\rho(Q) \supset \mathbf{R}^-$ and

$$\lim_{x\to-\infty} \frac{Q(x)}{x^{2m}} = \lim_{x\to-\infty} Q_0(x) = \begin{cases} 0, & \text{if } \chi \in \mathcal{H}_{-m-1/2} \setminus \mathcal{H}_{-m}, \\ -\infty, & \text{if } \chi \in \mathcal{H}_{-m-1} \setminus \mathcal{H}_{-m-1/2}. \end{cases} \tag{1.19}$$

We refer to [[DS], p. 58–59] where this limit is shown for a similar realization. Hence $Q(z)$ belongs to the class

$$N_{m+}^\infty := \{Q(z) \in N_m^\infty \mid \mathbf{R}^- \subset \rho(Q), \lim_{x\to-\infty} x^{-2m} Q(x) = 0 \text{ or } -\infty\}.$$

Since $b_m(A_0)R_0(-1)^m R_0(z)\varepsilon_m = R_0(z)R_0(-1)^m\chi$ and $b_m(A_0)R_0(-1)^m$ is boundedly invertible, the minimality condition (1.4) is equivalent to the condition $\mathcal{H}_0 = \overline{\operatorname{span}}\,\{R_0(z)\varepsilon_m \mid z \in \rho(A_0)\}$, and this in turn is equivalent to the minimality of the operator representation (1.17) of $Q(z)$ in the sense that

$$\Pi_m = \overline{\operatorname{span}}\,\{\varphi(z) \mid z \in \rho(A^\infty)\}. \tag{1.20}$$

For a proof of this equivalence in a similar realization we refer to [[DLSZ], Proposition 3.5 and Theorem 5.8].

The triple $\{\Pi_m, S, A^\infty\}$, the Q-function in the class N_{m+}^∞, and the defect function $\varphi(z)$ uniquely determine each other.

We shall use the following geometric results from [DLSZ]. From (1.6) we obtain the expansion

$$\varphi(z) = \tau_\chi (A_0 - z)^{-1}\chi = \sum_{j=1}^{m} b_{j-1}(z) u_j + b_m(z)\phi(z), \tag{1.21}$$

where $u_j \in \Pi_m$, $j = 1, \dots, m$, and the holomorphic function $\phi(z)$ on $\mathbf{C} \setminus \mathbf{R}^+$ are defined by $u_1 = \tau_\chi \varepsilon_1 = \varphi(\mu_1)$,

$$u_j = \tau_\chi \varepsilon_j = \prod_{l=2}^{j} (A^\infty - \mu_l)^{-1} u_1, \quad \phi(z) = \tau_\chi (A_0 - z)^{-1} \varepsilon_m = (A^\infty - z)^{-1} u_m. \tag{1.22}$$

The following formula shows how to calculate the matrix $G = (g_{jk})$ in (1.8) from $\varphi(z)$

$$g_{jk} = \langle u_k, u_j \rangle = \langle \frac{\varphi(z) - \sum_{l=1}^{k-1} b_{l-1}(z) u_l}{b_{k-1}(z)}, \frac{\varphi(\zeta) - \sum_{l=1}^{j-1} b_{l-1}(\zeta) u_l}{b_{j-1}(\zeta)} \rangle \Bigg|_{z=\mu_k, \zeta=\mu_j}. \tag{1.23}$$

The elements $\{u_j\}_{j=1}^m$ form a basis for the space $\mathcal{M} = \{0\} \oplus \{0\} \oplus \mathbf{C}^m$. Let $\{w_j\}_{j=1}^m$ be a basis of $\mathcal{L} = \ker R^\infty(z)^m = \{0\} \oplus \mathbf{C}^m \oplus \{0\}$ which is biorthonormal to $\{u_j\}_{j=1}^m$ in the $\mathbf{G}$-inner product. With respect to the decomposition

$$\Pi_m = \mathcal{H}_0 \oplus (\mathcal{L} \dot{+} \mathcal{M}) \tag{1.24}$$

and the bases of $\mathcal{L}$ and $\mathcal{M}$, A^∞ has the matrix representation

$$A^\infty = \left\{ \left\{ \begin{pmatrix} h + c_{m+1}\varepsilon_{m+1} \\ \begin{pmatrix} b_1 \\ (b_j)_{j=2}^m \end{pmatrix} \\ (c_j)_{j=1}^m \end{pmatrix}, \begin{pmatrix} A_0 h + \mu_0 c_{m+1} \varepsilon_{m+1} \\ \begin{pmatrix} \mu_1 b_1 + b_0 \\ (\mu_j b_j + b_{j-1} - g_{jm} c_{m+1})_{j=2}^m \end{pmatrix} \\ (\mu_j c_j + c_{j+1})_{j=1}^m \end{pmatrix} \right\} \;\middle|\; \begin{array}{l} h \in \operatorname{dom} A_0, \ c_1 = 0, \\ b_m = \langle (A_0 - \mu_0) h, \varepsilon_{m+1} \rangle_0 + g_{2m+1} c_{m+1}, \\ c_{j+1}, b_{j-1} \in \mathbf{C}, \\ j = 1, \dots, m \end{array} \right\}. \tag{1.25}$$

For $z \in \rho(A)$ the defect function $\varphi(z) \in \ker(S^* - z)$ is represented as

$$\varphi(z) = \left(I + (z - \mu_1)(A^\infty - z)^{-1}\right) \varphi(\mu_1) = \begin{pmatrix} b_m(z)\varphi_0(z) \\ (b_m(z) q_j(z))_{j=1}^m \\ (b_{j-1}(z))_{j=1}^m \end{pmatrix}, \tag{1.26}$$

where

$$q_j(z) = (b_j(z))^{-1} \left(b_m(z) Q_0(z) + \sum_{k=j+1}^{m} b_{k-1}(z) g_{k,m} \right). \tag{1.27}$$

The adjoint linear relation S^* becomes

$$S^* = A^\infty \dot{+} \operatorname{span}\{\{u_1, \mu_1 u_1\}\} = S \dot{+} \{0 \oplus A^\infty(0)\} \dot{+} \operatorname{span}\{\{u_1, \mu_1 u_1\}\}.$$

With the decomposition (1.24) of Π_m the canonical self-adjoint extensions A^t of S have the direct sum representation

$$A^t = S \dot{+} \operatorname{span}\{\{u_1, \mu_1 u_1 - (t + Q(\mu_1))w_1\}\}, \quad t \in \mathbf{R}. \tag{1.28}$$

1.3. Krein's formula for generalized resolvents. There is an alternative interpretation of the operators A^t: A^t is a noncanonical self-adjoint extension with exit to Π_m of the one-dimensional restriction S_0 of A_0 in $\mathcal{H}_0$ to the domain

$$\operatorname{dom}(S_0) = \{f \in \operatorname{dom}(A_0) | \langle f, \varepsilon_m \rangle_0 = 0\}. \tag{1.29}$$

Clearly, S_0 is symmetric, has defect indices $1, 1$, and its adjoint is given by

$$S_0^* = A_0 \dot{+} \operatorname{span}\{\{\varepsilon_{m+1}, \mu_0 \varepsilon_{m+1}\}\}.$$

The extension A^t is determined by (that is, can be recovered up to unitary equivalence from) its generalized resolvent $P_{\mathcal{H}_0} R^t(z) |_{\mathcal{H}_0}$; see, for example, [DLS]. Krein's formula for generalized resolvents in this case reads as follows:

$$P_{\mathcal{H}_0} R^t(z) |_{\mathcal{H}_0} = R_0(z) - \frac{\langle \cdot, \varphi_0(z^*) \rangle_0}{Q_0(z) + t(z)} \varphi_0(z), \quad z \in \rho(A_0) \cap \rho(A^t).$$

Here

(i) the function $Q_0(z)$ is given by (1.14) and is the Q-function associated with S_0 and its self-adjoint extension A_0,

(ii) $\varphi_0(z) \in \ker(S_0^* - z) \subset \mathcal{H}_0$ is the defect function

$$\varphi_0(z) := (A_0 - z)^{-1} \varepsilon_m = (I + (z - \mu_0)(A_0 - z)^{-1}) \varepsilon_{m+1},$$

and

(iii) $t(z)$ is the rational N_m-function

$$t(z) = \frac{p_{2m-1}(z) - Q(\mu_1) + t}{b_m^2(z)}.$$

Note that $\varphi_0(z) = P_{\mathcal{H}_0} \phi(z)$, where $\phi(z)$ is from (1.22).

2. Point-like perturbation of the Laguerre operator in the case $\alpha > 1$

In this section we apply our realization model for singular perturbations to a self-adjoint operator with a discrete spectrum. This operator is a multiplication operator and the model will be called the Friedrichs model, the F-model for short. We specialize this model by choosing the free parameters such that it can be applied to the Laguerre operator. The resulting model will be called the Friedrichs–Laguerre model, or the FL-model. Finally using the unitary Laguerre transformation we describe the Laguerre model associated with ℓ_α, abbreviated as the L-model. We stipulate that in this section we assume that $\alpha > 1$ and $\alpha \neq 2, 3, \ldots$, unless indicated otherwise. Spaces, operators and other notions depend on α, but to avoid cumbersome notations we do not write $\mathcal{H}_0^\alpha, A_{0\alpha}$ and so on but simply $\mathcal{H}_0, A_0$ and so on. We set $m = [\frac{\alpha+1}{2}]$.

2.1. The F-model. In this subsection we study the realization of the singular perturbation (1.3) of A_0 where A_0 is a multiplication operator. We take for $\mathcal{H}_0$ the Hilbert space $\widehat{\mathcal{H}}_0 = l^2(\mathbf{Z}^+)$ of square summable sequences on $\mathbf{Z}^+$. We denote the elements of this space by symbols like h, $h(n)$. For A_0 we take the operator $\widehat{A}_0$ which acts in $\widehat{\mathcal{H}}_0$ as multiplication by the independent variable n: $\widehat{A}_0 h(n) = nh(n)$. Evidently, $\widehat{A}_0 \geq 0$. The scale spaces associated with $\widehat{A}_0$ are denoted by $\widehat{\mathcal{H}}_j$, $j = 0, \pm 1, \ldots$. We take for χ the sequence

$$\widehat{\chi}(n) = \widehat{\chi}_\alpha(n) = \left(\frac{\Gamma(\alpha + n + 1)}{n!} \right)^{1/2}. \tag{2.1}$$

Its action as a generalized element is that of an antilinear functional $\widehat{\chi}$ on the space of fast decreasing sequences:

$$\langle \widehat{\chi}, h \rangle_0 = \sum_0^\infty h(n)^* \widehat{\chi}(n).$$

According to Stirling's formula $\widehat{\chi}(n) \sim n^{\alpha/2}$, $n \to \infty$. Hence $\widehat{\chi} \in \widehat{\mathcal{H}}_{-m-1} \setminus \widehat{\mathcal{H}}_{-m}$. This inclusion can be specified further:

$$\begin{array}{ll} \widehat{\chi} \in \widehat{\mathcal{H}}_{-m-\frac{1}{2}} \setminus \widehat{\mathcal{H}}_{-m}, & \text{if } [\alpha] = 2m - 1, \\ \widehat{\chi} \in \widehat{\mathcal{H}}_{-m-1} \setminus \widehat{\mathcal{H}}_{-m-\frac{1}{2}}, & \text{if } [\alpha] = 2m. \end{array}$$

Since $\widehat{\chi}(n) \neq 0$, $n = 0, 1, \ldots$, the self-adjoint operator $\widehat{A}_0$ is cyclic (or an operator with simple spectrum) with generating element $(n+1)^{-m-1}\widehat{\chi}(n)$. This property implies that $\widehat{A}_0$ and $\widehat{\chi}$ satisfy the condition (1.4), which guarantees the minimality condition (1.20).

The realization of $\widehat{A}^t = \widehat{A}_0 + t^{-1}\langle \cdot, \widehat{\chi} \rangle \widehat{\chi}$ corresponding to the parameter set $\{\Lambda, \mu_0, \{g_k\}_{k=2}^{[\alpha]+1}\}$ will be denoted by $\{\widehat{\Pi}_m, \widehat{S}, \widehat{A}^\infty\}$ and called the Friedrichs model. The embedding (1.9) in this model will be denoted by $\widehat{\tau} = \widehat{\tau}_\alpha$. The analog of the function in (1.10) is given by

$$\begin{aligned} \widehat{\varphi}(z) &= (1 + (z - \mu_1)(\widehat{A}^\infty - z)^{-1})\widehat{\tau}\varepsilon_1 \\ &= \widehat{\tau}(1 + (z - \mu_1)(\widehat{A}_0 - z)^{-1})\varepsilon_1 = \widehat{\tau}\left(\frac{\widehat{\chi}(n)}{n - z} \right). \end{aligned}$$

The Q-function $\widehat{Q}(z)$ in this model is determined by the expression (1.13) but with $\widehat{\varphi}(z)$ instead of $\varphi(z)$. It will be calculated explicitly in the next proposition. The canonical self-adjoint extensions $\widehat{A}^t$ of $\widehat{S}$ and their resolvents are described by (1.28) and by Krein's formula (1.16) with S, φ, and Q replaced by $\widehat{S}$, $\widehat{\varphi}$ and $\widehat{Q}$.

Proposition 2.1. *The Q-function $\widehat{Q}(z)$ in the F-model belongs to the class N_{m+}^{∞} and for $z \in \mathbf{C}^+$, it takes the form*

$$\widehat{Q}(z) = \begin{cases} -\dfrac{\pi}{\sin \pi\alpha} \dfrac{\Gamma(-z)}{\Gamma(-z-\alpha)} + \widehat{p}_{[\alpha]}(z), & \text{if } \alpha \neq 0,1,2,\ldots, \\ (-1)^{k+1} \dfrac{\Gamma(-z)\psi(-z-k)}{\Gamma(-z-k)} + \widehat{p}_k(z), & \text{if } \alpha = k = 0,1,2,\ldots, \end{cases} \tag{2.2}$$

where $\widehat{p}_n(z)$ is a polynomial of degree at most n with real coefficients and $\psi(z) = \frac{d}{dz}\ln\Gamma(z)$ is the psi-function.

Proof. That $\widehat{Q}(z) \in N_m^{\infty}$ is a direct consequence of the representation (1.13) and of the fact that $\widehat{\chi}_\alpha(n)$ in (2.1) belongs to $\mathcal{H}_{-m-1} \setminus \mathcal{H}_{-m}$. From (1.13) we see that $\widehat{Q}(z)$ is analytic on $\mathbf{C} \setminus \overline{\mathbf{R}^+}$ and the following limits hold:

$$\begin{aligned} &\lim_{x\to-\infty} x^{-2m}\widehat{Q}(x) = 0, && \text{if } 2(m-1) < \alpha < 2m+1, \\ &\lim_{x\to-\infty} x^{-2m}\widehat{Q}(x) = -\infty, && \text{if } 2m \leq \alpha < 2m+1. \end{aligned}$$

Hence $\widehat{Q}(z) \in N_{m+}^{\infty}$. To calculate $\widehat{Q}(z)$ we actually substitute $\widehat{\chi}_\alpha = \widehat{\chi}_\alpha(n)$ in (1.13) and obtain

$$\widehat{Q}(z) = b_m^2(z)(z-\mu_0)\sum_{n=0}^{\infty} \frac{\Gamma(\alpha+n+1)}{n!(n-z)(n-\mu_0)b_m^2(n)} + b_m^2(z)g_{2m+1} + p_{2m-1}(z).$$

We apply the method of analytic continuation in α (see [GS]) to the function

$$F(\alpha, z) = \sum_{n=0}^{\infty} \frac{\Gamma(\alpha+n+1)}{n!(n-z)}.$$

This function is well defined and analytic in α and z with $-1 < \operatorname{Re}\alpha < 0$ and $z \in \mathbf{C} \setminus \mathbf{Z}^+$. For these values of α and z the series on the right-hand side equals (see Lemma 0.1)

$$F(\alpha, z) = -\frac{\pi}{\sin \pi\alpha} \frac{\Gamma(-z)}{\Gamma(-z-\alpha)}. \tag{2.3}$$

The function on the right-hand side is analytic in $\alpha \in \mathbf{C} \setminus \{0, \pm 1, \pm 2, \ldots\}$ and $z \in \mathbf{C} \setminus \mathbf{Z}^+$. Iterating the resolvent identity we expand $F(\alpha, z)$ and obtain

$$\begin{aligned} F(\alpha, z) &= (z-\mu_1)\sum_{n=0}^{\infty} \frac{\Gamma(\alpha+n+1)}{n!(n-z)(n-\mu_1)} + a_1(\alpha) \\ &= (z-\mu_0)b_m^2(z)\sum_{n=0}^{\infty} \frac{\Gamma(\alpha+n+1)}{n!(n-z)(n-\mu_0)b_m^2(n)} + \\ &+ \sum_{j=1}^{m} a_{j+m+1}(\alpha)b_j(z)b_m(z) + \sum_{j=1}^{m} a_{j+1}(\alpha)b_j(z)b_1(z) + a_1(\alpha). \end{aligned} \tag{2.4}$$

Here the infinite series is well defined for $-1 < \operatorname{Re}\alpha < 2m+1$ and the functions $a_j(\alpha)$ are meromorphic in α. Comparing the formulas for $F(\alpha, z)$ and $\widehat{Q}(z)$ we observe that the difference of these functions is a polynomial in z of degree at most $2m$. The leading coefficient of this polynomial can determined from the limit behavior of $x^{-2m}\widehat{Q}(x)$ for $x \to -\infty$, and this leads to the representation (2.2) for $\widehat{Q}(z)$.

In the case $\alpha = k$, $k = 0, 1, 2, \ldots$, every point $\alpha = k$ is a simple pole of $F(\alpha, z)$:

$$F(\alpha, z) = (-1)^{k+1}\frac{\Gamma(-z)}{\Gamma(-z-k)}\frac{1}{\alpha - k} + (-1)^{k+1}\frac{\Gamma(-z)\psi(-z-k)}{\Gamma(-z-k)} + \cdots,$$

where the dots stand for a sum in positive powers of $\alpha - k$. The finite sum on the right-hand side of (2.4) and the function $F(\alpha, z)$ have a simple pole at $\alpha = k$ and have the same residue. The difference between the two functions is regular at $\alpha = k$ and, up to a polynomial in z with real coefficients and of degree at most $2m$, coincides with the function $\widehat{Q}(z)$. The behavior of $\widehat{Q}(z)$ near infinity, as in the case of a noninteger α, implies that $\widehat{Q}(z)$ is of the form (2.2). □

2.2. The FL-model. In this subsection we make a specific choice of the elements in the parameter set $\{\Lambda, \mu_0, \{g_k\}_{k=2}^{[\alpha]+1}\}$ of the F-model. We take as Λ-set the set

$$\Lambda = \Lambda_\alpha = \{\mu_1, \mu_2, \ldots, \mu_m\}, \quad \mu_j = -\alpha + j - 1, \; j = 1, 2, \ldots, m, \tag{2.5}$$

so the corresponding monomials $b_j(z)$ are

$$b_0(z) = 1, \quad b_j(z) = \prod_{k=1}^{j}(z + \alpha - k + 1), \; j = 1, 2, \ldots, m. \tag{2.6}$$

We set $\mu_0 = -\alpha + m$. Finally we choose the numbers g_k, $k = 2, \ldots, [\alpha] + 1$, such that the Q-function $\widehat{Q}(z)$ in Proposition 2.1 for $z \in \mathbf{R}^-$ coincides with the function $Q_\alpha(z)$ in (0.5). From now on we write again $Q(z)$ for $\widehat{Q}(z) = Q_\alpha(z)$. With this particular choice of the parameters the F-model will be called the FL-model, where FL stands for Friedrichs-Laguerre. The self-adjoint extensions $\widehat{A}^t$, $t \in \mathbf{R} \cup \{\infty\}$, of $\widehat{S}$ in this model are given by

$$\begin{aligned}\widehat{A}^t \;&=\; \{\{f_\infty + \widehat{\varphi}(-\alpha)F_1, f'_\infty - \alpha\widehat{\varphi}(-\alpha)F_1\} \in \widehat{S}^* \\ &\qquad\qquad \mid \{f_\infty, f'_\infty\} \in A^\infty, \; F_1 \in \mathbf{C}, \; \langle (f'_\infty + \alpha f_\infty), \widehat{\varphi}(-\alpha)\rangle = -tF_1\},\end{aligned}$$

where the adjoint of $\widehat{S}$ is the linear relation

$$\widehat{S}^* = \widehat{A}^\infty \dotplus \operatorname{span}\{(\widehat{\varphi}(-\alpha), -\alpha\widehat{\varphi}(-\alpha))\}.$$

In the FL-model the matrix elements $g_{j,k}$ of the Gram matrix G follow from (1.23) and a simple calculation gives

$$g_{j,k} = \frac{\Gamma(\alpha - j + 1)\Gamma(\alpha - k + 1)}{(j-1)!(k-1)!}\Gamma(j + k - \alpha - 1). \tag{2.7}$$

Hence

$$\begin{aligned} g_{1+j} &= g_{1,j} \\ &= \frac{1}{(j-1)!}\Gamma(\alpha-j+1)\Gamma(\alpha)\Gamma(j+\alpha), \\ g_{m+j} &= g_{m,j} \\ &= \frac{\Gamma(\alpha-j+1)\Gamma(\alpha+m-1)}{(j-1)!(m-1)!}\Gamma(j+\alpha+m-1), \quad j=1,\dots,m. \end{aligned} \tag{2.8}$$

In the following proposition we list some important properties of the Q-function $Q(z)$. By (0.5), its zeros are the points $z=-\alpha,-\alpha+1,\dots$. Continuing earlier notations (see (2.5)) we denote the first $2m$ of them by μ_j; hence $\mu_j=-\alpha+j-1$, $j=1,\dots,2m$.

Proposition 2.2.

(a) *The function $Q(z)=Q_\alpha(z)$ in (0.5) for $\alpha>1$ and $\alpha\neq 2,3,\dots$ has $m=[\frac{\alpha+1}{2}]$ simple generalized zeros of nonpositive type. More specifically: If $[\alpha]$ is even, then these are the points $\mu_{2j}=2j-1-\alpha$, $j=1,\dots,m$, and if $[\alpha]$ is odd, then these are the points $\mu_{2j-1}=2(j-1)-\alpha$, $j=1,\dots,m$.*

(b) *Furthermore, $Q(z)$ admits the representation*

$$Q(z)=b_m^2(z)\left(Q_0(z)+t(z)\right), \tag{2.9}$$

where (i) $t(z)$ is the rational function given by

$$t(z)=\sum_{j=1}^{m}\frac{t_j}{j-1-\alpha-z}, \quad t_j=(-1)^{[\alpha]+j}\frac{\pi}{\sin\pi\{\alpha\}}\frac{\Gamma(\alpha-j+1)}{(j-1)!\,((m-j)!)^2}, \tag{2.10}$$

and belongs to the class N_κ with $\kappa=[\frac{m}{2}]$, if $[\alpha]$ is even, and with $\kappa=[\frac{m+1}{2}]$, if $[\alpha]$ is odd, and (ii) the function $Q_0(z)$ has the form

$$Q_0(z)=(z+\alpha-m)\sum_{n=0}^{\infty}\frac{\widehat{\chi}^2(n)}{b_m^2(n)(n+\alpha-m)(n-z)}+\sum_{j=1}^{m}\frac{t_j}{m+1-j}, \tag{2.11}$$

hence $Q_0(z)\in N_0^\infty$ and $\rho(Q_0)=\rho(Q)$.

Proof. From (0.5) it follows that all zeros of $Q_\alpha(z)$ are simple and given by $\mu_j:=\alpha+j-1, j=1,2,\dots$. Moreover,

$$\lim_{z\to\mu_j}\frac{Q_\alpha(z)}{z-\mu_j}=(-1)^{[\alpha]+j+1}\frac{\pi}{\sin\pi\{\alpha\}}(j-1)!\Gamma(\alpha-j+1).$$

By a criteria of Langer [L] the point μ_j is a generalized zero of nonpositive type of $Q_\alpha(z)$ if and only if $[\alpha]+j+1$ is odd and $j\le[\alpha]+1$. This implies the first statement in (a).

Consider the function $\widehat{Q}_\alpha(z)=b_m^2(z)^{-1}Q_\alpha(z)$. Now (0.5) and

$$\frac{\Gamma(-z)}{\Gamma(-z-\alpha)}=(-1)^k(z+\alpha)\dots(z+\alpha-k+1)\frac{\Gamma(-z)}{\Gamma(-z-\alpha+k)}$$

with $k = 2m$ yield

$$\widehat{Q}_\alpha(z) = r(z)Q_{\alpha-2m}(z),$$

where $r(z) := b_m(z)^{-1}\prod_{j=m}^{2m-1}(z+\alpha-j)$. As $-1 < \alpha - 2m < 1$ and $\alpha - 2m \neq 0$, Lemma 0.1 implies $Q_{\alpha-2m}(z) \in N_{0+}^{\infty}$. The rational function $r(z)$ has positive values at the points $z = n$, $n = 0, 1, \ldots$, which are simple poles of $Q_{\alpha-2m}(z)$. The function $\widehat{Q}_\alpha(z)$ has additionally simple poles at points μ_j, $j = 1, 2, \ldots, m$, with the residues

$$-t_j := \operatorname{res}_{z=\mu_j} \widehat{Q}_\alpha(z) = (-1)^{[\alpha]+j+1}\frac{\pi}{\sin\pi\{\alpha\}}\frac{\Gamma(\alpha-j+1)}{(j-1)![(m-j)!]^2}.$$

The numbers t_j appear in the definition (2.10) of the function $t(z)$. There are $[\frac{m}{2}]$ positive and $[\frac{m+1}{2}]$ negative numbers in the set $\{t_j\}_{j=1}^m$, if $[\alpha]$ is even, and vice versa, if $[\alpha]$ is odd. Hence the function $t(z)$ belongs to the class N_κ with κ as in the proposition.

The function

$$Q_0(z) := \widehat{Q}_\alpha(z) - t(z) = r(z)Q_{\alpha-2m}(z) - t(z)$$

has simple poles at the points $z = n$, $n = 0, 1, \ldots$, and $\rho(Q_0) = \rho(Q_\alpha) = \rho(Q_{\alpha-2m})$. Also, by using either (0.7) or (0.8) for $Q_{\alpha-2m}(z)$ we find

$$\begin{aligned} -\operatorname{res}_{z=n} Q_0(z) &= -\operatorname{res}_{z=n} r(z)Q^{(\alpha-2m)}(z) = r(n)\frac{\Gamma(\alpha-2m+1+n)}{n!} \\ &= \frac{\Gamma(\alpha+n+1)}{n!b^2(n)} = \frac{\widehat{\chi}^2(n)}{b^2(n)}. \end{aligned}$$

Hence $Q_0(z)$ is a Nevanlinna function. The asymptotic behavior of $Q_0(z)$ when $\operatorname{Im} z \to \pm\infty$ is the same as the behavior of the function $Q_{\alpha-2m}(z) \in N_0^\infty$. Therefore $Q_0(z) \in N_0^\infty$. From this and the equality $Q_\alpha(-\alpha+m) = 0$ we conclude that $Q_0(z)$ has the representation (2.9). □

2.3. The L-model. In this subsection we describe the realization of the point-like perturbation $A^t = L_\alpha + t^{-1}\langle\cdot,\chi\rangle\chi$, where L_α is the self-adjoint operator associated with the differential expression ℓ_α in (0.1) in the Hilbert space $\mathcal{H}_0^\alpha = L^2(\mathbf{R}^+, w_\alpha)$ with $w_\alpha(x) = x^\alpha e^{-x}$, and χ is given by (0.14) with $\mu = \mu_0 = -\alpha+m$. This realization is called the Laguerre model or L-model. It is obtained from the general construction in Section 1 by taking $A_0 = L_\alpha$, $\mathcal{H}_0 = \mathcal{H}_0^\alpha$ and χ as above. It can be constructed more explicitly via the F- and FL-models and a unitary transformation $\mathsf{L}_\alpha : \mathcal{H}_0 \to \widehat{H}_0 = l^2(\mathbf{Z}^+)$ which makes the multiplication operator in these models the spectral representation of the unperturbed operator L_α. If $\mathsf{L}_\alpha : f \mapsto \widehat{f}$, this transformation is defined by

$$\widehat{f}(n) = (\mathsf{L}_\alpha f)(n) = \int_0^\infty \widetilde{L}_n^{(\alpha)}(x)f(x)\,x^\alpha e^{-x}dx,$$

where $\widetilde{L}_n^{(\alpha)}(x)$ is the normalized generalized Laguerre polynomial:

$$\widetilde{L}_n^{\alpha}(x) = N(\alpha,n)\, L_n^{\alpha}(x), \qquad N(\alpha,n) := \left(\frac{n!}{\Gamma(\alpha+n+1)}\right)^{1/2}.$$

The mapping L_α is unitary, since these polynomials form an orthonormal basis of $\mathcal{H}_0$ and it transforms A_0 to the operator $\widehat{A}_0$ of multiplication by n in $\widehat{\mathcal{H}}_0$, because $\widetilde{L}_n^{\alpha}(x)$ is the eigenfunction of L_α with eigenvalue n. The mapping L_α can be extended to a mapping from the scale space $\mathcal{H}_j$ associated with the Laguerre operator to the scale space $\widehat{\mathcal{H}}_j$ associated with the multiplication operator via the formula

$$\langle \mathsf{L}_\alpha \xi(n), \widehat{u}(n)\rangle_0 = \langle \xi, \mathsf{L}_\alpha^* \widehat{u}\rangle_0 = \langle \xi, \mathsf{L}_\alpha^{-1}\widehat{u}\rangle_0,$$

for all $\widehat{u}(n) \in \widehat{\mathcal{H}}_j$, $\xi \in \mathcal{H}_{-j}$, $j = 0, 1, 2, \ldots$.
With $\varphi(x,z)$ as in (0.10):

$$\varphi(x,z) \equiv \varphi_\alpha(x,z) = \Gamma(-z)\Psi(-z, 1+\alpha; x)$$

(note that the right-hand side is well defined for $\alpha \geq 1$, $z \in \mathbf{C} \setminus \mathbf{Z}^+$) and the generalized element χ as in (0.14) the action of L_α on the main objects are listed in the following table:

$$\begin{aligned} f(x) \in \mathcal{H}_0 = L^2(\mathbf{R}^+, w_\alpha) &\xrightarrow{\mathsf{L}_\alpha} \widehat{f}(n) \in \widehat{\mathcal{H}}_0 = l^2(\mathbf{Z}^+), \\ A_0 = L_\alpha &\xrightarrow{\mathsf{L}_\alpha} \text{multiplication by } n, \\ \varphi(x,z) \in \mathcal{H}_{-m} &\xrightarrow{\mathsf{L}_\alpha} \widehat{\varphi}(n) = \left(\frac{\Gamma(\alpha+n+1)}{n!}\right)^{1/2} \frac{1}{n-z} \in \widehat{\mathcal{H}}_{-m}, \\ \chi(x) \in \mathcal{H}_{-m-1} &\xrightarrow{\mathsf{L}_\alpha} \widehat{\chi}(n) = \left(\frac{\Gamma(\alpha+n+1)}{n!}\right)^{1/2} \in \widehat{\mathcal{H}}_{-m-1}. \end{aligned}$$

The fourth correspondence in the table follows from the third since, in the sense of distributions, $\chi(x) = (L_\alpha - \mu)\varphi(x,\mu)$, $\mu \in \mathbf{R}^-$. The third correspondence in the table follows from the formula

$$\Gamma(b)\int_0^\infty \Phi(a, 1+\alpha; x)\Psi(b, 1+\alpha; x)\, x^\alpha e^{-x} dx = \frac{\Gamma(1+\alpha)}{a-b} \tag{2.12}$$

after multiplication by $\left(\frac{\Gamma(\alpha+n+1)}{n!}\right)^{1/2}$. The formula (2.12) is obtained easily from the more general formula [[E], ch.6, p.6.15, formula (24)]. It now easily follows that $\chi \in \mathcal{H}_{-m-1} \setminus \mathcal{H}_{-m}$ with $m = [\frac{\alpha+1}{2}]$, more specifically, $\chi \in \mathcal{H}_{-m-1} \setminus \mathcal{H}_{-m-1/2}$ if $[\alpha] = 2m$ and $\chi \in \mathcal{H}_{-m-1/2} \setminus \mathcal{H}_{-m}$ if $[\alpha] = 2m-1$.

The F-model provides a realization $\{\widehat{\Pi}_m, \widehat{A}^\infty, \widehat{S}\}$ for $\widehat{A}_0 + t^{-1}\langle\,\cdot\,, \widehat{\chi}\rangle\widehat{\chi}$. One can repeat the construction of Section 1 to obtain a realization $\{\Pi_m, A^\infty, S\}$ for $A_0 + t^{-1}\langle\cdot, \chi\rangle\chi$. It is evident that the two realizations coincide when the parameter sets are the same. In particular the Q-functions coincide and are described by the formula (2.2). In the pre-image of these models (that is before taking completion) the spaces $\widehat{\mathcal{P}}_m$ and $\mathcal{P}_m$ are defined differently: the ε_j's in $\widehat{\mathcal{P}}_m$ are generalized sequences $\widehat{\varepsilon}_j(n) \in \widehat{\mathcal{H}}_{-m-1+j}$, whereas in $\mathcal{P}_m$ they are generalized functions $\varepsilon_j(x) \in \mathcal{H}_{-m-1+j}$.

The Laguerre model for the expression $L_\alpha + t^{-1}\langle\,\cdot\,,\chi\rangle\chi$, or L-model for short, is the realization $\{\Pi_m, A^\infty, S\}$ relative to the parameter set $\{\Lambda_\alpha, m-\alpha, \{g_k\}_{k=2}^{[\alpha]+1}\}$ as in the FL-model (see (2.5) and (2.8)). Thus, by definition, the Q-function $Q(z)$ for the L-model is given by (0.5), that is, the operator representation of this function is provided by the L-model (but also the FL-model): In this model the linear relation A^∞ is described explicitly by (1.25) with the parameter set $\{\Lambda_\alpha, m-\alpha, \{g_k\}_{k=2}^{[\alpha]+1}\}$ and the symmetric operator S is given by

$$S = \{\{f, f'\} \in A^\infty \mid \langle (f' + \alpha f), \varphi(-\alpha)\rangle = 0\}, \tag{2.13}$$

where $\varphi(-\alpha) = \text{column}\,(0, 0, e_1)$ (the function $\varphi(z)$ is determined below). This is the solution to Problem 2) in the introduction.

In the L-model the analog of the expansion (1.21) has a clear interpretation in terms of functions:

$$\varphi(x,z) = \sum_{j=1}^{m} b_{j-1}(z)\varepsilon_j(x) + b_m(z)\varphi_0(x,z). \tag{2.14}$$

We calculate the functions $\varepsilon_j(x)$ and $\varphi_0(x,z)$ explicitly. To this end, for a function $f(x)$ on $\mathbf{R}^+$ which is k times differentiable at $x = 0$, we define the function $[f(x)]_k$ by

$$[f(x)]_k = f(x) - \sum_{j=0}^{k-1} \frac{x^j}{j!} f^{(j)}(0), \quad x \in \mathbf{R}^+. \tag{2.15}$$

From the formulas (0.3) and (0.4) we obtain the following expansion for the function $\varphi(x,z) = \Gamma(-z)\Psi(-z, 1+\alpha; x)$:

$$\begin{aligned}
\varphi(x,z) &= \Gamma(\alpha)x^{-\alpha} + (z+\alpha)\Gamma(\alpha-1)\frac{x^{-\alpha+1}}{1!} + (z+\alpha)(z+\alpha-1)\Gamma(\alpha-2)\frac{x^{-\alpha+2}}{2!} \\
&\quad + \cdots + (z+\alpha)(z+\alpha-1)\cdots(z+\alpha-m+1)\Gamma(\alpha-m+1)\frac{x^{-\alpha+m-1}}{(m-1)!} \\
&\quad + b_m(z)\Bigg((-1)^m \frac{\Gamma(-\alpha)\Gamma(-z)}{\Gamma(-z+\alpha+m)}\Phi(-z, 1+\alpha; x) \\
&\quad + \Gamma(\alpha-m)\frac{x^{-\alpha+m}}{m!} + (z+\alpha+m)\Gamma(\alpha-m-1)\frac{x^{-\alpha+m+1}}{(m+1)!} + \cdots\Bigg) \\
&= b_0(z)\Gamma(\alpha)x^{-\alpha} + b_1(z)\Gamma(\alpha-1)\frac{x^{-\alpha+1}}{1!} + \cdots + b_m(z)\Gamma(\alpha-m+1)\frac{x^{-\alpha+m-1}}{(m-1)!} \\
&\quad + x^{-\alpha}[x^\alpha \varphi(x,z)]_m.
\end{aligned}$$

Comparing this result with (2.14) we get

$$\varepsilon_j(x) = \Gamma(\alpha - j + 1)\frac{x^{-\alpha+j-1}}{(j-1)!}, \qquad b_m(z)\varphi_0(x,z) = x^{-\alpha}[x^\alpha\varphi(x,z)]_m.$$

Hence $\varphi_0(x,z) \sim x^{-\alpha+m}$ when $x \to 0$ and, therefore, $\varphi_0(x,z) \in \mathcal{H}_0$. The expression (1.26) for the defect function $\varphi(z)$ in the L-model becomes

$$\varphi(z) = \tau\varphi(\cdot,z) = \begin{pmatrix} \Gamma(-z)x^{-\alpha}[x^\alpha\Psi(-z,\alpha+1;x)]_m \\ (b_m(z)q_j(z))_{j=1}^m \\ (b_{j-1}(z))_{j=1}^m \end{pmatrix}, \tag{2.16}$$

where $\tau = \tau_\alpha$ is the embedding (1.9) for the L-model, $q_j(z)$ is determined by (1.27) with $\mu_j \in \Lambda_\alpha$ and $g_{j,k}$ is of the form (2.7). Thus we have the following solution of Problem 3) from the introduction: Krein's formula with $Q_\alpha(z)$ as in (0.5) and the function $\varphi(z)$ in (2.16) as $\varphi_\alpha(z)$ is the new version of (0.9) and (0.11) holds. The latter equality written out in full gives the following replacement of (0.12):

$$\begin{aligned}\Gamma(-z)\Gamma(-\zeta^*) &\int_0^\infty [x^\alpha\Psi(-z,1+\alpha;x)]_m[x^\alpha\Psi(-\zeta^*,1+\alpha;x)]_m\, x^{-\alpha}\mathrm{e}^{-x}dx \\ &+ \begin{pmatrix} (b_m(\zeta)q_j(\zeta))_{j=1}^m \\ (b_{j-1}(\zeta))_{j=1}^m \end{pmatrix}^* \begin{pmatrix} 0 & 1_m \\ 1_m & G \end{pmatrix} \begin{pmatrix} (b_m(z)q_j(z))_{j=1}^m \\ (b_{j-1}(z))_{j=1}^m \end{pmatrix} \\ &= \frac{\pi}{\sin\pi\alpha}(z-\zeta^*)^{-1}\left(\frac{\Gamma(-z)}{\Gamma(-z-\alpha)} - \frac{\Gamma(-\zeta^*)}{\Gamma(-\zeta^*-\alpha)}\right).\end{aligned}$$

Our results can be summed up in the following theorem.

Theorem 2.1. *Assume $\alpha > 1$, $\alpha \neq 2,3,\dots$ and let $m = [\frac{\alpha+1}{2}]$. The L-model of the formal point-like perturbation (0.13) of the Laguerre operator $A_0 = L_\alpha$ is the triple $\{\Pi_m, A^\infty, S\}$ relative to the parameter set $\{\Lambda_\alpha, m-\alpha, g_j\}$, where Λ_α is the set (2.5) and the numbers g_j, $j = 2,\dots,[\alpha]+1$ are given by (2.8):*

(a) *$\Pi_m = \mathcal{H}_0 \oplus \mathbf{C}^m \oplus \mathbf{C}^m$ is a Pontryagin space with negative index m and a $\mathbf{G}$-space with Gram operator (1.8) with $G = (g_{j,k})_{j,k=1}^m$ given by (2.7)*

(b) *The self-adjoint linear relation A^∞ is described by (1.25) and S is the one-dimensional restriction (2.13) of A^∞.*

(c) *The function (0.5) is the Q-function associated with S and A^∞.*

(d) *The canonical self-adjoint extension A^t, $t \in \mathbf{R}\cup\{\infty\}$ of the symmetric operator S is described by the Krein formula*

$$(A^t - z)^{-1} = (A^\infty - z)^{-1} - \frac{\langle\cdot,\varphi(z^*)\rangle}{Q(z)+t}\varphi(z),$$

where the defect function $\varphi(z)$ is given by (2.16).

(e) *The subspace $\mathcal{L} = \{0\}\oplus\mathbf{C}^m\oplus\{0\}$ is the root subspace of A^∞ at ∞ and the subspace $\mathcal{M} = \{0\}\oplus\{0\}\oplus\mathbf{C}^m$ coincides with the span of the eigenvectors of A^0 corresponding to its first m eigenvalues.*

We now come to Problem 1) of the Introduction. By [[DLSZ], Theorem 3.3], the spectrum $\sigma(A^t)$ of A^t, $t \in \mathbf{R}$, coincides with the set of zeros of the function $Q(z)+t$ and hence it is discrete and simple. If λ is a zero of $Q(z)+t$ then the corresponding normalized eigenvector is a multiple of $\varphi(\lambda)$. For the case $t=0$,

$$\sigma(A^0) = \{n-\alpha \mid n = 0,1,\dots\}.$$

By Proposition 2.2, the spectrum $\sigma^0(A^0)$ of nonpositive type of A^0 consists of simple eigenvalues:

$$\sigma^0(A^0) = \begin{cases} \{\mu_{2j} = 2j - 1 - \alpha \mid j = 1, \dots, m\}, & \text{if } [\alpha] \text{ is even,} \\ \{\mu_{2j-1} = 2(j-1) - \alpha \mid j = 1, \dots, m\}, & \text{if } [\alpha] \text{ is odd.} \end{cases}$$

The eigenvector of A^0 corresponding to $n - \alpha$ is of the form:

$$e_n^0 = \varphi(n-\alpha) = \tau\varphi(\cdot, n-\alpha) = \begin{pmatrix} (-1)^n n!\, x^{-\alpha}[L_n^{-\alpha}(x)]_m \\ (c(j,n,\alpha))_{j=1}^m \\ (b_{j-1}(n-\alpha))_{j=1}^m \end{pmatrix},$$

where $c(j,n,\alpha) = 0$ if $n \leq m-1$ and if $n \geq m$

$$c(j,n,\alpha) = -(n-j)! \sum_{i=2}^{j} \frac{n!}{(n-i)!(n-m)!} g_{im} - (n-j)! \sum_{i=1}^{m} \frac{n}{(n-i)!} g_{1i}.$$

The appearance of the generalized Laguerre polynomial in the first row of e_n^0 comes from

$$\varphi(x, n-\alpha) = \Gamma(\alpha) x^{-\alpha} \Phi(-n, 1-\alpha; x)$$

(see (0.4)) and the relation

$$\Phi(-n, 1-\alpha; x) = \frac{n!}{(1-\alpha)_n} L_n^{-\alpha}(x)$$

[see [AS], Table 13.6], which gives

$$\varphi(x, n-\alpha) = (-1)^n n! \Gamma(\alpha - n) x^{-\alpha} L_n^{-\alpha}(x). \tag{2.17}$$

It follows from (0.11) that

$$\langle e_k^0, e_n^0 \rangle = (-1)^{[\alpha]+n} \frac{\pi}{\sin \pi\{\alpha\}} \frac{\Gamma(\alpha - n)}{n!} \delta_{kn}.$$

Hence the normalized eigenvectors of A^0 form an orthogonal basis for the Pontryagin space Π_m.

As to the case $t = \infty$: $\sigma(A^\infty)$ coincides with the poles of the function $Q(z)$ and therefore,

$$\sigma(A^\infty) = \{n \mid n = 0, 1, \dots\} \cup \{\infty\};$$

we recall $\sigma^0(A^\infty) = \{\infty\}$. The orthonormalized eigenvectors corresponding to the finite eigenvalues can be written as

$$e_n^\infty = \tau \widetilde{L}_n^\alpha(x) = \begin{pmatrix} \widetilde{L}_n^\alpha(x) \\ N(\alpha, n)^{-1} \left(b_j(n)^{-1}\right)_{j=1}^m \\ 0 \end{pmatrix},$$

where $\widetilde{L}_n^\alpha(x)$ are the orthonormalized generalized Laguerre polynomials. Evidently, $P_{\mathcal{H}_0} e_n^\infty$ is an orthonormal basis in $\mathcal{H}_0$.

Unfortunately the spectral properties for A^t with $t \neq 0, \infty$ are not so explicit as in the case $t = 0, \infty$. We do not discuss them here. We end this subsection with a description of the compressions of the resolvents of A^∞ and A^0 to the Hilbert space $\mathcal{H}_0$.

Proposition 2.3. *If the L model is described by the triple* $\{\Pi_m, A^\infty, S\}$, $m = [\frac{\alpha+1}{2}]$, *then*

(i) *for* $z \in \mathbf{C}\backslash\mathbf{Z}^+$, $P_{\mathcal{H}_0}(A^\infty - z)^{-1}|_{\mathcal{H}_0} = (A_0 - z)^{-1}$, *where* $A_0 = L_\alpha$, *the minimal Laguerre operator, and*

(ii) *for* $z \in \mathbf{C}\setminus\{\mathbf{Z}^+ \cup \{\mathbf{Z}^+ - \alpha\}$,

$$P_{\mathcal{H}_0}(A^0 - z)^{-1}|_{\mathcal{H}_0} = R_0(z) - \frac{\langle\cdot, \varphi_0(z^*)\rangle_0}{Q_0(z) + t(z)}\varphi_0(z), \tag{2.18}$$

where the functions $Q_0(z)$ *and* $t(z)$ *are given by* (2.11) *and* (2.10). *This compressed resolvent is a generalized resolvent of the symmetric operator*

$$S_0 = \{\{f, f'\} \in A_0 \mid \langle (f' - \mu_0 f), \varphi_0(\mu_0)\rangle_0 = 0\}.$$

Proof. (i) is a consequence of two facts: $\mathcal{L}$ is A^∞-invariant subspace and that $\mathcal{M}$ is spanned by the special vectors $u_j, j = 1, 2, \ldots, m$. We refer to [[DLSZ], Lemma 5.6].

To prove (ii) we insert the expansion (1.21) of $\varphi(z)$ in the above Krein formula for $(A^t - z)^{-1}$ and then take the described compression to $\mathcal{H}_0$. If we factorize also $Q(z)$ in denominator of the same formula as in (2.9) then we come to the expression (2.18). The defect function $\varphi_0(z)$ in this formula satisfy both relations (1.11) and (1.12) with $Q(z) \to Q_0(z)$. Hence, by standard reasoning, this formula is a generalized resolvent of the described symmetric operator. □

3. Point-like perturbation of the Laguerre operator for $\alpha < -1$

The F-model in Subsection 2.1 also makes sense if $0 < |\alpha| < 1$: $\widehat{A}_0$ remains the operator of multiplication by n in $\widehat{\mathcal{H}}_0 = l^2(\mathbf{Z}^+)$ and $\widehat{\chi}(n) = (\Gamma(\alpha + n + 1)/n!)^{1/2}$ is well defined and belongs to $\widehat{\mathcal{H}}_{-1}$. Of course, $m = [\frac{\alpha+1}{2}] = 0$, $\widehat{\Pi}_0 = \widehat{\mathcal{H}}_0$, and the realizations $\widehat{A}^t$ of $\widehat{A}_0 + t^{-1}\langle\cdot, \widehat{\chi}\rangle\widehat{\chi}$, $t \in \mathbf{R} \cup \{\infty\}$, are the canonical self-adjoint extensions of the symmetric operator

$$\widehat{S} = \{\{f, f'\} \in \widehat{A}_0 \mid \langle f, \widehat{\chi}\rangle = 0\}.$$

The Q-function associated with $\widehat{S}$ and $\widehat{A}^\infty = \widehat{A}_0$ is the function $Q(z)$ in (0.5) and this function belongs to the class N_0. A subtle distinction between positive and negative α comes from the observation that $\widehat{\chi} \in \widehat{\mathcal{H}}_{-1}\backslash\widehat{\mathcal{H}}_{-1/2}$ if $0 < \alpha < 1$, whereas $\widehat{\chi} \in \widehat{\mathcal{H}}_{-1/2} \setminus \widehat{\mathcal{H}}_0$ if $-1 < \alpha < 0$. In the first case the usual singular perturbation theory applies as in the books [AGHH, AK] and $\widehat{A}_0$ is the Friedrichs extension of $\widehat{S}$. In the second case $\widehat{A}_0$ is the Krein extension of $\widehat{S}$. In this case $\widehat{A}^t$ also has an interpretation as the sum of the operators $\widehat{A}_0$ and $t^{-1}\langle\cdot, \widehat{\chi}\rangle\widehat{\chi}$, in a generalized sense, $t \in \mathbf{R} \cup \{\infty\}$; see [KJ, JL] and [AK].

The correspondence between the model for the abstract singular perturbation $A_0 + t^{-1}\langle\cdot,\chi\rangle\chi$ with $\chi \in \mathcal{H}_{-1} \setminus \mathcal{H}_{-1/2}$ and the model with $\chi \in \mathcal{H}_{-1/2} \setminus \mathcal{H}_0$ has been studied by many authors (see for example [GSi, KS, HS] and the books [AGHH, AK] and references therein). If $\alpha > 1$, the transition of the models in the previous section from α to models corresponding to $-\alpha$ is more delicate, and requires a different point of view which is presented in this section. From now on in this section we assume that $\alpha < -1$ and $\alpha \neq -2, -3, \ldots$ unless stated otherwise. Again we try to avoid writing α as a subscript to indicate the dependence on α. We set $m = [\frac{|\alpha|+1}{2}]$. In the first subsection we consider the analog of the FL-model, in the second the analog of the L-model.

3.1. The FL-model for $\alpha < -1$. We set $\widehat{\mathcal{H}}_0 = l^2(\mathbf{Z}^+)$ as in Subsection 2.1 and consider in this space the operator $\widehat{A}^\infty$ of multiplication by the independent variable: $\widehat{A}^\infty h(n) = nh(n)$. As before the Hilbert spaces $\widehat{\mathcal{H}}_j$, $j = 0, \pm 1, \ldots$, are the scale spaces corresponding to $\widehat{A}^\infty$. We set

$$\widehat{\chi}(n) = \widehat{\chi}_\alpha(n) = s_n \left(\frac{|\Gamma(\alpha + n + 1)|}{n!} \right)^{1/2}, \tag{3.1}$$

where the numbers s_n are given by

$$s_n = \begin{cases} (-1)^{[|\alpha|]+n}, & n = 0, 1, \ldots, [|\alpha|] - 1, \\ 1, & n = [|\alpha|], [|\alpha|] + 1, \ldots. \end{cases} \tag{3.2}$$

We observe that this sequence is fast decreasing and $\widehat{\chi} \in \mathcal{H}_{m-1/2} \setminus \mathcal{H}_m$ if $[|\alpha|] = 2m$ and $\widehat{\chi} \in \mathcal{H}_{m-1} \setminus \mathcal{H}_{m-1/2}$ if $[|\alpha|] = 2m - 1$. We denote by $\widehat{\Pi}_m = \widehat{\Pi}_m^\alpha$ the space $l^2(\mathbf{Z}^+)$ equipped with the indefinite inner product

$$\langle a, b \rangle = \sum_{n=0}^{\infty} s_n b(n)^* a(n). \tag{3.3}$$

Evidently, $\widehat{\Pi}_m$ is a Pontryagin space with m negative squares. We consider the self-adjoint operator $\widehat{A}^\infty$ in $\widehat{\Pi}_m$ which acts as multiplication by the independent variable: $\widehat{A}^\infty a(n) = na(n)$. Observe that $\sigma(\widehat{A}^\infty) = \mathbf{Z}^+$ and that $\widehat{A}^\infty$ is a cyclic operator with generating element $\widehat{\chi}$, or equivalently, $\widehat{\Pi}_m = \overline{\operatorname{span}}\,\{(\widehat{A}^\infty - z)^{-1}\widehat{\chi} \mid z \in \mathbf{C} \setminus \mathbf{Z}^+\}$.

The following rank one perturbation

$$\widehat{A}^t = \widehat{A}^\infty + t^{-1}\langle\cdot,\widehat{\chi}\rangle\widehat{\chi} \tag{3.4}$$

is a well-defined self-adjoint operator in the space $\widehat{\Pi}_m$ for $0 \neq t \in \mathbf{R} \cup \{\infty\}$. It is called the FL-model. For $z \in \mathbf{C} \setminus \mathbf{R}$ we define

$$\widehat{\varphi}(z) = (\widehat{A}^\infty - z)^{-1}\widehat{\chi}, \qquad \widehat{Q}(z) = \langle(\widehat{A}^\infty - z)^{-1}\widehat{\chi}, \widehat{\chi}\rangle.$$

Some properties of $\widehat{\chi}$, $\widehat{Q}(z)$, and $\widehat{A}^t$ are specified in the next theorem.

Theorem 3.1. *Assume* $\alpha < -1$, $\alpha \neq -2, -3, \ldots$.

(i) *The function* $\widehat{Q}(z)$ *is the* Q*-function associated with* $\widehat{A}^\infty$ *and the one-dimensional restriction*

$$S = \{\{f, f'\} \in \widehat{A}^\infty \mid \langle f, \widehat{\chi}\rangle = 0\},$$

and for $z \notin \mathbf{Z}^+$, $\widehat{Q}(z) = Q_\alpha(z)$ *given by* (0.5).

(ii) *The vector* $\widehat{\chi} \in \widehat{\Pi}_m$ *belongs to* $\operatorname{dom}(\widehat{A}^\infty)^k$, $k = 0, 1, 2, \ldots, m-1$ *and the subspace* $\mathcal{L} := \operatorname{span}\{(\widehat{A}^\infty)^k\widehat{\chi}, k = 0, 1, 2, \ldots, m-1\}$ *is a maximal neutral subspace in* $\widehat{\Pi}_m$.

(iii) *The resolvent of the self-adjoint operator* A^t, $0 \neq t \in \mathbf{R}$, *is expressed by the formula*

$$(\widehat{A}^t - z)^{-1} = (\widehat{A}^\infty - z)^{-1} - \frac{\langle \cdot, \widehat{\varphi}(z^*)\rangle}{Q(z) + t}\widehat{\varphi}(z). \tag{3.5}$$

When $t = 0$ *the right-hand side of this formula defines the resolvent of a self-adjoint linear relation* $\widehat{A}^0$ *in* $\widehat{\Pi}_m$ *whose root subspace at* ∞ *is* $\mathcal{L}$.

Proof. (i) is an easy exercise. (ii) By the Stirling formula the function $Q(z)$ behaves as $|z|^{-|\alpha|}$ when z approaches infinity inside the cone $C_\delta = \{\delta < \arg z < 2\pi - \delta\}$, $0 < \delta < \pi$. Hence

$$\langle(\widehat{A}^\infty - z)^{-1}\widehat{\chi}, \widehat{\chi}\rangle = Q(z) = O(|z|^{-\alpha}), \quad |z| \to \infty, \ z \in C_\delta.$$

This behavior and the inequality $|\alpha| - 1 < 2m < |\alpha| + 1$ imply by [[KL2], Satz 1.10] the inclusion $\widehat{\chi} \in \operatorname{dom}(\widehat{A}^\infty)^k$, $k = 0, 1, \ldots, m-1$ and the equalities:

$$\langle(\widehat{A}^\infty)^j\widehat{\chi}, \widehat{\chi}\rangle = 0, \quad j = 0, 1, \ldots, m-1;$$
$$\langle(\widehat{A}^\infty)^{m-1}\widehat{\chi}, (\widehat{A}^\infty)^k\widehat{\chi}\rangle = 0, \quad k = 1, \ldots, m-1.$$

(iii) If $0 \neq t \in \mathbf{R}$ then the (3.5) is the well-known expression for resolvent of rank one perturbation. We set $\widehat{R}^t(z) = (\widehat{A}^t - z)^{-1}$, $t \in \mathbf{R} \setminus \{0\}$. The right-hand side in (3.5) is continuous for all $t \in \mathbf{R}$. Taking $t = 0$ and $z \neq n$, $z \neq |\alpha| + n$, $n = 0, 1, \ldots$, we observe that $\widehat{R}^0(z)$ satisfies the resolvent identity and $\widehat{R}^0(z)^* = \widehat{R}^0(z^*)$. Hence it is resolvent of a self-adjoint linear relation (see [DSn]). We apply $\widehat{R}^0(z)$ to the vector $\widehat{\chi}$ and find

$$\begin{aligned}\widehat{R}^0(z)\widehat{\chi} &= \widehat{R}^\infty(z)\widehat{\chi} - \frac{\langle\widehat{\chi}, \widehat{\varphi}(z^*)\rangle}{Q(z)}\widehat{\varphi}(z) \\ &= \widehat{\varphi}(z) - \widehat{\varphi}(z) \equiv 0.\end{aligned}$$

Here we used the relations $\widehat{\varphi}(z) = (\widehat{A}^\infty - z)^{-1}\widehat{\chi}$ and $Q(z) = \langle(\widehat{A}^\infty - z)^{-1}\widehat{\chi}, \widehat{\chi}\rangle$. Hence $\widehat{\chi}$ belongs to the singular part of $\widehat{A}^0$. Next we calculate $\widehat{R}^0(z)(\widehat{A}^\infty)^k\widehat{\chi}$, $k = 1, 2, \ldots, m-1$. Observe that

$$\widehat{R}^\infty(z)(\widehat{A}^\infty)^k\widehat{\chi} = (\widehat{A}^\infty)^{k-1}\widehat{\chi} + z(\widehat{A}^\infty)^{k-3}\widehat{\chi} + \cdots + z^{k-1}\widehat{\chi} + z^k\widehat{\varphi}(z),$$

and hence for $k = 1, \ldots, m-1$,

$$\begin{aligned}\langle (\widehat{A}^\infty)^k \widehat{\chi}, \widehat{\varphi}(z^*) \rangle &= \langle \widehat{\chi}, (\widehat{A}^\infty)^k \widehat{\varphi}(z^*) \rangle = \langle R^\infty(z)(A^\infty)^k \widehat{\chi}, \widehat{\chi} \rangle \\ &= \sum_{j=0}^{k-1} z^j \langle (\widehat{A}^\infty)^j \widehat{\chi}, \widehat{\chi} \rangle + z^k \langle \widehat{\chi}, \widehat{\varphi}(z^*) \rangle = z^k Q(z).\end{aligned}$$

The last equality follows from that the subspace span $\{\widehat{\chi}, \widehat{A}^\infty \widehat{\chi}, \ldots, (\widehat{A}^\infty)^k \widehat{\chi}\}$ is neutral. Applying $R^0(z)$ to the vector $(\widehat{A}^\infty)^k \widehat{\chi}$ and by using (3.5) with $t = 0$ and the previous relations we get

$$\widehat{R}^0(z)(\widehat{A}^\infty)^k = (\widehat{A}^\infty)^{k-1}\widehat{\chi} + z(A^\infty)^{k-2}\widehat{\chi} + \cdots + z^{k-1}\widehat{\chi}.$$

Hence and from the previous equality $\widehat{R}^0(z)\widehat{\chi} = 0$ we have $R^0(z)^m\,(\widehat{A}^\infty)^k \widehat{\chi} = 0$ for $k = 0, 1, \ldots, m-1$. □

3.2. The L-model for $\alpha < -1$. Let $\mathcal{P}$ be the linear space spanned by the normalized generalized Laguerre polynomials

$$\widetilde{L}_n^\alpha(x) = N(\alpha, n) L_n^\alpha(x), \quad N(\alpha, n) = \left(\frac{n!}{|\Gamma(\alpha + n + 1)|} \right)^{1/2}, \quad n = 0, 1, \ldots,$$

and equipped with the inner product

$$\langle \widetilde{L}_j^\alpha, \widetilde{L}_k^\alpha \rangle = s_j\, \delta_{jk},$$

where the numbers s_j are defined by (3.2). The completion of $\mathcal{P}$ will be denoted by Π_m; evidently, it is a Pontryagin space with negative index m. The space Π_m defined here can be identified with the Pontryagin space introduced in [K1, D]: In these papers the space $\mathcal{P}$ was equipped with the inner product (0.16) and it was shown that the equalities

$$\langle L_j^\alpha, L_k^\alpha \rangle = \frac{\Gamma(j + \alpha + 1)}{j!}\, \delta_{jk} = s_j \frac{|\Gamma(j + \alpha + 1)|}{j!}\, \delta_{jk}$$

hold. In the sequel we assume that the two spaces coincide. Let $\widehat{e}_j = \{\delta_{jn}\}_{n=0}^\infty$, $j = 0, 1, \ldots$, where δ_{jn} is the Kronecker symbol, be the standard basis in $\widehat{\Pi}_m$. It satisfies

$$\langle \widehat{e}_j, \widehat{e}_k \rangle = s_j\, \delta_{jk}.$$

Hence the mapping : $\widetilde{L}_n^\alpha \mapsto \widehat{e}_n$ can be extended by linearity and continuity to an isomorphism $\mathsf{L}_\alpha : \Pi_m \to \widehat{\Pi}_m$ given by

$$\widehat{f}(n) = (\mathsf{L}_\alpha f)(n) = \langle \widetilde{L}_n^\alpha, f \rangle, \quad f \in \Pi_m.$$

Its inverse is given by

$$f = (\mathsf{L}_\alpha^{-1} \widehat{f}) = \sum_{n=0}^\infty s_n \widetilde{L}_n^\alpha(\cdot) \widehat{f}(n), \quad \widehat{f} \in \widehat{\Pi}_m.$$

For $f \in \widehat{\Pi}_m \cap C^\infty(\mathbf{R}^+)$, the series on the right-hand side converges uniformly on each compact subset of $\mathbf{R}^+$ and we have

$$f(x) = \sum_{n=0}^{\infty} s_n \widetilde{L}_n^\alpha(x)\widehat{f}(n), \quad x \in \mathbf{R}^+;$$

see [D]. Using this isomorphism we can transform the FL-model of the previous subsection which concerns sequence spaces to what we shall call the L-model which as we shall see concerns spaces of functions. In general for $f \in \Pi_m$, the series $\sum_{n=0}^{\infty} s_n \widetilde{L}_n^\alpha(x)\widehat{f}(n)$ does not converge pointwise, but has a meaning as a distribution, which we write as $f(x)$. We denote this correspondence by $\leftrightarrow$. For example, the sequence $\widehat{\chi}(n)$ in $\widehat{\Pi}_m$ is transformed into the element

$$\chi := (\mathsf{L}_\alpha^{-1}\widehat{\chi}) = \sum_{n=0}^{\infty} s_n \widetilde{L}_n^\alpha(\cdot)\widehat{\chi}(n) = \sum_{n=0}^{\infty} L_n^\alpha(\cdot),$$

where the series converge in Π_m, and $\chi \leftrightarrow \chi(x)$ where $\chi(x)$ represents the distribution

$$\chi(x) = \sum_{n=0}^{\infty} L_n^\alpha(x), \tag{3.6}$$

which is supported at 0, that is, $\chi(x) = 0$ for $x \in \mathbf{R}^+$. This can be seen by using the generating function for the Laguerre polynomials:

$$(1-z)^{-\alpha-1} \exp -\frac{zx}{1-z} = \sum_{n=0}^{\infty} L_n^\alpha(x) z^n, \quad |z| < 1. \tag{3.7}$$

Indeed, the series with $z = t \in (-1, 1)$ as $t \uparrow 1$ formally tends to the series on the right-hand side of (3.6) and for every $f \in C_0^\infty(\mathbf{R}^+)$ with support in some interval $[a, b] \subset \mathbf{R}^+$, we have

$$\lim_{t\uparrow 1}(1-t)^{-\alpha-1} \int_a^b f(x) \mathrm{e}^{-\frac{tx}{1-t}} x^\alpha \mathrm{e}^{-x} dx = 0.$$

We define the self-adjoint operator A^∞ on Π_m by

$$A^\infty = \mathsf{L}_\alpha^{-1}\widehat{A}^\infty \mathsf{L}_\alpha.$$

Clearly, $\sigma(A^\infty) = \mathbf{Z}^+$ and $\widetilde{L}_n^\alpha(x)$ is the normalized eigenfunction of A^∞ corresponding to the eigenvalue n. The one-dimensional perturbation

$$A^t = A^\infty + t^{-1}\chi\langle \cdot, \chi\rangle, \quad 0 \neq t \in \mathbf{R} \cup \{\infty\}, \tag{3.8}$$

of A^∞ will be called the L-model. It is isomorphic to the FL model (3.4). According to Theorem 3.1 the linear relation $A^0 = \mathsf{L}_\alpha^{-1}\widehat{A}^0\mathsf{L}_\alpha$ is well defined and self-adjoint. The resolvents of A^t are described by the formula (3.5) (of course without the "hats"). The Q-function for the model is the same as the Q-function for the FL-model and so coincides with $Q_\alpha(z)$ in (0.5). By (3.6),

$$\varphi(z) = (\mathsf{L}_\alpha^{-1}\widehat{\varphi}(z)) \longleftrightarrow \varphi(x,z) = \sum_{n=0}^{\infty} \frac{L_n^\alpha(x)}{n-z} = \Gamma(-z)\Psi(-z,1+\alpha;x).$$

3.3. The decomposition of Π_m for $\alpha < -1$. In the case $\alpha > 1$, the space $\mathcal{H}_0$ and the operator A_0 were given and used to construct the realization $\{\Pi_m, A^\infty, S\}$. Now in the case $\alpha < -1$ we have the realization and we want to construct the underlying Hilbert space $\mathcal{H}_0$ and operator A_0. We want to describe this Hilbert space as a space of functions and give a more concrete decomposition of the space Π_m. We begin by looking for the analog of the decomposition (1.24) of the space Π_m. For this we introduce the function

$$\widetilde{\varphi}(z) := -\frac{\pi}{\sin \pi\alpha} Q(z)^{-1}\varphi(z), \qquad z \neq \alpha + n,\, n = 0, 1, \ldots.$$

It has values in Π_m, is holomorphic at the points $z = n$, $n = 0, 1, \ldots$, and satisfies relations similar to (1.11) and (1.12) (see, for example, [[DLSZ], Section 7]):

$$\widetilde{\varphi}(z) = \left(1 + z(A^0 - z)^{-1}\right)\widetilde{\varphi}(0), \qquad \frac{\widetilde{Q}(z) - \widetilde{Q}(\zeta)^*}{z - w^*} = \langle \widetilde{\varphi}(z), \widetilde{\varphi}(\zeta)\rangle, \tag{3.9}$$

where

$$\widetilde{Q}(z) = \widetilde{Q}_\alpha(z) := -\frac{\pi^2}{\sin^2 \pi\alpha} Q_\alpha(z)^{-1} = Q_{-\alpha}(z + \alpha).$$

For the last equality we used (0.6).

As Λ-set we take

$$\Lambda_\alpha = \{\mu_1, \mu_2, \ldots, \mu_m\}, \quad \mu_j = j - 1, \quad j = 1, \ldots, m, \tag{3.10}$$

so that the corresponding monomials are given by

$$b_0(z) = 1, \quad b_j(z) = \prod_{k=0}^{j-1}(z - k), \quad j = 1, 2, \ldots, m,$$

and we set $\mu_0 = m$. The vectors u_j, $j = 1, 2, \ldots, m$, and the function $\widetilde{\phi}(z)$ are defined by an expansion of the type (1.21), namely

$$\widetilde{\varphi}(z) = \sum_{j=1}^{m} b_{j-1}(z)u_j + b_m(z)\widetilde{\phi}(z). \tag{3.11}$$

By using the first relation in (3.9) we find

$$u_1 = \widetilde{\varphi}(0), \quad u_j = (\prod_{k=1}^{j-1}(A^0 - k)^{-1})u_1, \quad \widetilde{\phi}(z) = (A^0 - z)^{-1}(\prod_{k=1}^{j-1}(A^0 - k)^{-1})u_1. \tag{3.12}$$

We form the subspace $\mathcal{M} = \operatorname{span}\{u_1, u_2, \ldots, u_m\}$. Note that $\mathcal{M}$ coincides with the span of the eigenvectors corresponding to the first m eigenvalues of A^∞. By [[DLSZ], Lemma 5.6] the subspaces $\mathcal{L} = \ker (A^0 - z)^{-m}$, which is independent of

$z \in \rho(A^0)$, and $\mathcal{M}$ are skewly linked (that is, form a dual pair) and Π_m admits the decomposition

$$\Pi_m = \mathcal{H}_0 \oplus \left(\mathcal{L} \dot{+} \mathcal{M}\right), \tag{3.13}$$

where $\mathcal{H}_0 := \Pi_m \ominus \left(\mathcal{L} \dot{+} \mathcal{M}\right)$ is a Hilbert subspace of Π_m. The decomposition (3.13) is the desired analog of (1.24) as we shall show below; see Theorem 3.2(iii).

Next we look for a Krein formula which expresses the resolvent of the operator A^t in terms of the resolvent of the linear relation A^0. For this we apply (3.5) without "hats" and obtain

$$(A^t - z)^{-1} - (A^0 - z)^{-1} = \frac{t\langle\,\cdot\,, \varphi(z^*)\rangle}{Q(z)(Q(z)+t)}\varphi(z).$$

Therefore the desired Krein formula reads

$$(A^t - z)^{-1} = (A^0 - z)^{-1} - \frac{\langle\,\cdot\,, \widetilde{\varphi}(z^*)\rangle}{Q_{-\alpha}(z+\alpha) - \pi^2/t\sin^2\pi\alpha}\widetilde{\varphi}(z). \tag{3.14}$$

This formula is similar to the formula in Theorem 2.1 for the resolvent of the operator A^t corresponding to the positive index $-\alpha$. It implies that A^t can be viewed as a canonical self-adjoint extension of the one-dimensional restriction S of A^0 defined by

$$S = \{\{f, f'\} \in A^0 \mid \langle (f' + \alpha f), \widetilde{\varphi}(-\alpha)\rangle = 0\}.$$

Similar to the case $\alpha > 1$ (see Proposition 2.3 (i)) the special decomposition (3.13) implies that the compressed resolvent $P_{\mathcal{H}_0}(A^0 - z)^{-1}\,|_{\mathcal{H}_0} = (A_0 - z)^{-1}$ is the resolvent of some self-adjoint operator A_0 in $\mathcal{H}_0$, the compressed resolvent $P_{\mathcal{H}_0}(A^\infty - z)^{-1}\,|_{\mathcal{H}_0}$ is a generalized resolvent of the one-dimensional restriction $\widetilde{S}_0$ of A_0:

$$\widetilde{S}_0 = \{\{f, f'\} \in A_0 \mid \langle (f' - \mu_0 f), \widetilde{\varphi}_0(\mu_0)\rangle_0 = 0\},$$

where $\widetilde{\varphi}_0(z) = P_{\mathcal{H}_0}\widetilde{\phi}(z)$. Moreover,

$$P_{\mathcal{H}_0}(A^\infty - z)^{-1}\,|_{\mathcal{H}_0} = (A_0 - z)^{-1} - \frac{\langle\,\cdot\,, \widetilde{\varphi}_0(z^*)\rangle_0}{Q_0(z+\alpha) + t(z+\alpha)}\widetilde{\varphi}_0(z), \tag{3.15}$$

where $Q_0(z)$ is given by (2.11) and $t(z)$ by (2.10).

Summing up we have the following theorem.

Theorem 3.2. *Assume $\alpha < -1$ and $\alpha \neq -2, -3, \ldots$ and set $m = [\frac{|\alpha|+1}{2}]$. The expression* (0.13) *admits a realization in the Pontryagin space Π_m with negative index m as a rank one perturbation of the self-adjoint operator A^∞ of the form* (3.8). *This realization has the following properties:*

(i) *A^t, $t \in \mathbf{R} \cup \{\infty\}$, is the canonical self-adjoint extension of the nondensely defined symmetric operator S. The function $Q_\alpha(z)$ in* (0.5) *is the Q-function associated with S and A^∞, and the function $Q_{-\alpha}(z+\alpha)$ is the Q-function associated with S and A^0.*

(ii) *The resolvent of A^t is described by formula* (3.5) *without "hats", and also by formula* (3.14).

(iii) *The space* Π_m *admits the decomposition* (3.13) *corresponding to the set* Λ_α *in* (3.10) *and* $\mu_0 = m$, *where* $\mathcal{H}_0$ *is the Hilbert subspace*

$$\mathcal{H}_0 = \overline{\operatorname{span}}\,\{\widetilde{\varphi}_0(z),\ z \in \mathbf{C} \setminus \{\mathbf{Z}^+ - \alpha\}\},$$

$\mathcal{L} = \ker(A^0 - z)^{-m}$, *and* $\mathcal{M}$ *is spanned by the eigenvectors of* A^∞ *corresponding to its first* m *eigenvalues.*

(iv) *For* $z \in \mathbf{C}\setminus\{\mathbf{Z}^+ - \alpha\}$, *the compression* $P_{\mathcal{H}_0}(A^0 - z)^{-1}\,|_{\mathcal{H}_0}$ *is the resolvent of a self-adjoint operator* A_0 *in* $\mathcal{H}_0$. *For* $z \in \mathbf{C}\setminus\{\mathbf{Z}^+ \cup \{\mathbf{Z}^+ - \alpha\}\}$, *the compressed resolvent* $P_{\mathcal{H}_0}(A^\infty - z)^{-1}\,|_{\mathcal{H}_0}$ *is a generalized resolvent of the symmetric operator* $\widetilde{S}_0$ *and it has the form* (3.15).

It follows from the formula (3.14) that the spectrum of the linear relation A^0 is simple, $\sigma(A^0) = \{n - \alpha,\ n = 0, 1, \ldots\} \cup \{\infty\}$, and, by [[DLSZ], Theorem 3.3], the $\varphi(n - \alpha)$ are its eigenvectors corresponding to the eigenvalues $n - \alpha$.

We use Theorem 3.2(iii) to describe the spaces Π_m and $\mathcal{H}_0$ as spaces of functionals. To this end, we return to the expansion (3.11). From the correspondence $\varphi(z) \longleftrightarrow \varphi(x, z) = \Gamma(-z)\Psi(-z, \alpha + 1; x)$ we get the correspondence

$$\widetilde{\varphi}(z) \longleftrightarrow \widetilde{\varphi}(x, z) = \Gamma(-z - \alpha)\Psi(-z, 1 + \alpha; x).$$

The analog of the expansion (1.6) for the $\widetilde{\varphi}(z, x)$ takes the form

$$\widetilde{\varphi}(x, z) = \sum_{j=1}^{m} b_{j-1}(z)\varepsilon_j(x) + b_m(z)\widetilde{\varphi}_0(x, z). \tag{3.16}$$

The reasoning used in the calculation of the functions $\varepsilon_j(x)$ and $\varphi_0(z, x)$ in the expansion (2.14) now leads to the formulas

$$\varepsilon_j(x) = \Gamma(-\alpha - j + 1)\,\frac{x^{j-1}}{(j-1)!}, \qquad b_m(z)\widetilde{\varphi}_0(x, z) = [\widetilde{\varphi}(x, z)]_m,$$

where the mapping $f(x) \to [f(x)]_m$ is defined by (2.15). Hence $\widetilde{\varphi}_0(x, z) = O(x^m)$ near $x = 0$. Comparing the expansions (3.16) and (3.11) we obtain the correspondence

$$u_j = \prod_{k=1}^{j-1}(A^0 - k)^{-1}u_1 \longleftrightarrow \varepsilon_j(x) = \Gamma(-\alpha - j + 1)\,\frac{x^{j-1}}{(j-1)!}, \quad j = 1, 2, \ldots, m.$$

We can extend the expansions of $\widetilde{\varphi}(z)$ and $\widetilde{\varphi}(x, z)$ in (3.11) and (3.16) with respect to the monomials $b_0 = 1$, $b_j(z) = \prod_{k=0}^{j-1}(z - k)$, $j = 1, 2, \ldots, r$, $r \geq m$ (and so include the first r zeros of $\widetilde{Q}(z)$). As a result we get the following extended correspondence

$$\prod_{k=1}^{j}(A^0 - k)^{-1}u_1 \longleftrightarrow \Gamma(-\alpha - j)\,\frac{x^j}{j!}, \quad j = 1, 2, \ldots, r. \tag{3.17}$$

Lemma 3.1. *The vectors*

$$w_j = \frac{\sin \pi|\alpha|}{\pi} b_{j-1}(A^\infty)\chi, \quad j = 1, 2, \ldots, m,$$

form a basis $\{w_j\}_{j=1}^m$ in the space $\mathcal{L}$ which is biorthonormal to the basis $\{u_j\}_{j=1}^m$ of $\mathcal{M}$. Every polynomial p with $p(x) = \sum_{j=0}^n p_j x^j$ considered as an element of Π_m can be represented with respect to decomposition (3.13) as the sum

$$p = \left([p]_m - \sum_{j=1}^{\min(m,n)} \widetilde{p}_j w_j \right) + \sum_{j=1}^{\min(m,n)} \widetilde{p}_j w_j + \sum_{j=1}^{m} \frac{(j-1)!\, p_{j-1}}{\Gamma(-\alpha - j + 1)} u_j, \qquad (3.18)$$

where $\widetilde{p}_j = \langle [p]_m, u_j \rangle$, $j = 1, 2, \ldots, m$. On the right-hand side the element inside the brackets belongs to the space $\mathcal{H}_0$, the second summand belongs to $\mathcal{L}$, and the last sum belongs to $\mathcal{M}$.

Proof. By Theorem 3.1(ii) the vectors w_j, $j = 1, 2, \ldots, m$, belong to $\mathcal{L}$. The biorthonormality is proved as follows. We have

$$\langle w_1, \widetilde{\varphi}(z) \rangle = \frac{1}{Q(z)} \langle \chi, \varphi(z) \rangle = 1.$$

Since $u_1 = \widetilde{\varphi}(0)$, this relation implies $\langle w_1, u_1 \rangle = 1$. It also implies the equalities $\langle w_1, u_j \rangle = 0$, $j = 2, 3, \ldots, m$. For $k > j$, the representation (3.12) of u_k implies

$$\langle w_j, u_k \rangle = \langle \prod_{i=1}^{k-1} (A^0 - i)^{-1} w_j, u_1 \rangle = 0.$$

By [[DLSZ], Theorem 3.3], $e_n^\infty = \widetilde{\varphi}(n)$ is the eigenvector of A^∞ corresponding to its eigenvalue n. Also, every u_j is a linear combination of the form $u_j = \sum_{n=0}^{j-1} c_n e_n^\infty$ in which $c_{j-1} = b_{j-1}(j-1)^{-1}$. Hence

$$\begin{aligned} \langle w_j, u_j \rangle &= \langle b_{j-1}(A^\infty) w_1, \sum_{n=0}^{j-1} c_n e_n^\infty \rangle = \sum_{n=0}^{j-1} b_{j-1}(n) c_n^* \langle w_j, e_n^\infty \rangle) \\ &= c_{j-1} b_{j-1}(j-1) \langle w_1, e_{j-1}^\infty \rangle = \langle w_1, \widetilde{\varphi}(j-1) \rangle = 1. \end{aligned}$$

This proves the first statement.

Now we prove the second statement. The correspondence (3.17) shows that each monomial x^r can represented as

$$x^r = \text{const.} \prod_{k=1}^{r} (A^0 - k)^{-1} u_1 \in \Pi_m.$$

Since $\mathcal{L}$ is the root subspace of A^0 at the eigenvalue ∞, $(A^0 - z)^{-m}\mathcal{L} = \{0\}$, and hence

$$\prod_{k=1}^{r} (A^0 - k)^{-1} w_j = 0, \quad r \geq m.$$

This implies

$$\langle x^r, w_j \rangle = \langle \prod_{k=1}^{n} (A^0 - k)^{-1} u_1, w_j \rangle = 0, \quad r \geq m.$$

Every polynomial p with $p(x) = \sum_{j=0}^{n} p_j x^j$ can be written in the form

$$p = [p]_m + \sum_{j=1}^{\min(m,n)} c_j(p)\varepsilon_j, \quad c_j(p) = \frac{(j-1)!\, p_{j-1}}{\Gamma(-\alpha - j + 1)}.$$

Consider p as element of Π_m and observe that the second summand belongs to the subspace $\mathcal{M}$. Since $[p]_m$, the first summand, is a finite linear combination of the monomials x^r with $r \geq m$, $\langle [p]_m, w_j \rangle = 0$, $j = 1, \ldots, m$, and therefore $[p]_m$ belongs to the space $\mathcal{H}_0 \oplus \mathcal{L}$. The decomposition (3.18) now readily follows. $\square$

The representation (3.18) corresponds to the decomposition (3.13) of Π_m. We denote by $\widetilde{\mathcal{H}}_0$ the quotient space $\left(\mathcal{H}_0 \dot{+} \mathcal{L}\right)/\mathcal{L}$, which is a Hilbert space because $\mathcal{L}$ is the isotropic part of $\mathcal{H}_0 \dot{+} \mathcal{L}$. Since $\mathcal{L} = \operatorname{span}\{\, w_j\}_{j=1}^{m}$ the elements $[p]_m - \sum_{j=1}^{m} \widetilde{p}_j w_j$ and $[p]_m$ belong to the same equivalence class in$\widetilde{\mathcal{H}}_0$. Therefore we may choose the polynomial $[p]_m$ as the representative of its equivalence class. By (0.16), the inner product $\langle \cdot, \cdot \rangle$ on the set of all polynomials of the form $[p]_m$ coincides with the inner product of the space $L^2(\mathbf{R}^+, w_\alpha)$. As the set of polynomials of the form $x^m q(x)$, where $q(x)$ is a polynomial, is dense in $L^2(\mathbf{R}^+, w_\alpha)$, the spaces $\mathcal{H}_0$, $\widetilde{\mathcal{H}}_0$, and $L^2(\mathbf{R}^+, w_\alpha)$ can be identified. More specifically: the mapping $[p]_m - \sum_{j=1}^{m} \widetilde{p}_j w_j \mapsto [p]_m$ can be extended by continuity to an isomorphism from $\mathcal{H}_0$ onto $L^2(\mathbf{R}^+, w_\alpha)$.

We can go one step further. Let $\Pi_m(G) = L^2(\mathbf{R}^+, w_\alpha) \oplus \mathbf{C}^m \oplus \mathbf{C}^m$ be a Pontryagin space, which is also a $\mathbf{G}$-space with the Gram operator of the form (1.8), where the entries of the matrix $G = (g_{j,k})_{j,k=1}^{m}$ are defined by the right-hand side of the formula (2.7), but now with α being negative. We define the embedding $\tau : \mathcal{P} \cap \Pi_m \to \Pi_m(G)$ by

$$\tau p = \tau_\alpha p := \operatorname{column}\left([p]_m, \left(\langle [p]_m, \varepsilon_j \rangle\right)_{j=1}^{m}, \left(\frac{(j-1)!\, p_{j-1}}{\Gamma(-\alpha - j + 1)} \right)_{j=1}^{m} \right) \tag{3.19}$$

and extend it to all of Π_m as follows. Let l_{nj} be the coefficients in the expansion

$$\widetilde{L}_n^\alpha(x) - [\widetilde{L}^\alpha]_m = \sum_{j=1}^{\min(m,n)} l_{nj}\varepsilon_j(x).$$

For an arbitrary element $f = \sum_{n=0}^{\infty} s_n f_n \widetilde{L}_n^\alpha$ of Π_m we set

$$\tau f = \tau_\alpha f = \begin{pmatrix} \sum_{n=m}^{\infty} s_n f_n [\widetilde{L}_n^\alpha]_m \\ \left(\sum_{n=m}^{\infty} s_n f_n \langle [\widetilde{L}_n^\alpha]_m, u_j \rangle \right)_{j=1}^{m} \\ \left(\sum_{n=0}^{j} s_n f_n l_{nj} \right)_{j=1}^{m} \end{pmatrix}. \tag{3.20}$$

The mapping $\tau : \Pi_m \to \Pi_m(G)$ so defined is an isomorphism and extends τ on $\mathcal{P}$. This proves the first two statements of the following proposition.

Proposition 3.1. *The space $\mathcal{H}_0$ can be identified with the space $L^2(\mathbf{R}^+, w_\alpha)$ and the mapping $\tau : \mathcal{P} \cap \Pi_m \to \Pi_m(G)$ in (3.19) can be extended by continuity to an isometry 3.20 from Π_m onto $\Pi_m(G)$. Moreover, if A_0 is as in Theorem 3.2(iv) then $\tau^{-1} A_0 \tau$ coincides with the minimal operator L_α associated with the Laguerre expression ℓ_α in $L^2(\mathbf{R}^+, w_\alpha)$, which is characterized by its spectrum $\sigma(A_0) = \{n - \alpha, n \in \mathbf{Z}^+\}$ and the corresponding eigenfunctions $x^{-\alpha} L_n^{-\alpha}(x)$.*

Proof. It remains to prove the last statement. We recall that $\sigma(A^0) = \{n-\alpha\} \cup \{\infty\}$ and that the function $\varphi(n - \alpha)$ is the eigenfunction corresponding to $n - \alpha$; see after Theorem 3.2. Since $\varphi(z) = \tau\varphi(\cdot, z)$, where $\varphi(x, z) = \Gamma(-z)\Psi(-z, \alpha + 1; x)$, and, by (2.17),

$$\varphi(x, n - \alpha) = (-1)^n n! \Gamma(\alpha - n)\, x^{-\alpha} L_n^{-\alpha}(x),$$

we conclude that $\varphi(n-\alpha) = \tau\varphi(\cdot, n-\alpha) \in L^2(\mathbf{R}^+, w_\alpha) \oplus \mathbf{C}^m \oplus \{0\}$. The subspace $\{0\} \oplus \mathbf{C}^m \oplus \{0\}$ of $\Pi_m(G)$ is invariant under the resolvent $(A^0 - z)^{-1}$. It is clear that $[x^{-\alpha} L_n^{-\alpha}(x)]_m = x^{-\alpha} L_n^{-\alpha}(x)$. Hence

$$(A_0 - z)^{-1} L_n^{-\alpha}(x) = P_{L^2}(A^0 - z)^{-1} x^{-\alpha} L_n^{-\alpha}(x) = (n - \alpha - z)^{-1} x^{-\alpha} L_n^{-\alpha}(x).$$

In the case $\alpha < -1$, the functions $x^{-\alpha} L_n^{-\alpha}(x)$ are the eigenfunctions of the minimal Laguerre operator and we have

$$(L_\alpha - z)^{-1} L_n^{-\alpha}(x) = (n - \alpha - z)^{-1} x^{-\alpha} L_n^{-\alpha}(x).$$

Hence $A_0 = L_\alpha$. □

The space $\Pi_m(G)$ and the embedding τ in Proposition 3.1 are similar to those described in [D]. They coincide if we use the monomials $\{x^{j-1}\}_{j=1}^m$ as a basis in $\mathcal{L}$ instead of our basis $\{\varepsilon_j(x)\}_{j=1}^m$ and the space $L^2(\mathbf{R}^+, w_{\alpha+2m})$ instead of $L^2(\mathbf{R}^+, w_\alpha)$, but simultaneously change the embedding (3.19) to the embedding

$$\tau_D p(x) = \text{column } \left(x^{-m}[p(x)]_m,\, (\langle [p(x)]_m, x^{j-1}\rangle)_{j=1}^m,\, (p_j)_{j=1}^m\right)$$

considered by Derkach.

We identify the vectors w_j with linear functionals on the space $\mathcal{P}$ of polynomials as follows: If $p(x) = \sum_{j=0} p_j x^j$ then the functional corresponding to w_j is defined by

$$w_j(p) = \langle p(\cdot), w_j \rangle = \frac{(j-1)!\, p_{j-1}}{\Gamma(-\alpha - j + 1)}.$$

This functional corresponds to the distributional series

$$w_j(x) = -\frac{\sin \pi\alpha}{\pi} \sum_{n=0}^{\infty} b_{j-1}(n) L_n^\alpha(x).$$

Consider the space $\mathcal{F}$ of functions $f = f_0 + \sum_{j=0}^{m-1} f_j \varepsilon_j$, where $f_0 \in L^2(\mathbf{R}^+, w_\alpha)$, that is, $\mathcal{F} = L^2(\mathbf{R}^+, w_\alpha) \dot{+} \operatorname{span}\{x^j, j = 0, 1, \ldots, m-1\}$, and the linear span of functionals $\mathcal{W} = \operatorname{span}\{w_j(x), j = 1, 2, \ldots, m\}$. Then Π_m can also be viewed as the direct sum space $\Pi_m = \mathcal{F} \dot{+} \mathcal{W}$, thus providing functional representation of Π_m.

4. Kummer symmetry

We assume in this section $\alpha > 0$, $\alpha \neq 1, 2, \ldots$, and describe a correspondence between the L-models for α and their analogs for $-\alpha$. It seems unavoidable that now we have to use the complex notation in which we add a subscript or superscript α to the operators, spaces and elements depending on α. For example, we write $\Pi_m^{\pm\alpha}$, $S_{\pm\alpha}$, $A_{\pm\alpha}^t$, $\varphi_{\pm\alpha}(z)$, and so on.

Recall that if $f_{\alpha,z}(x)$ is a solution of the Kummer equation (0.2) with parameters α and z, then $x^\alpha f_{\alpha,z}(x)$ is a solution of (0.2) with parameters $-\alpha, z-\alpha$, that is,

$$xf'' + (1-\alpha-x)f' - (z-\alpha)f = 0.$$

This property implies the Kummer transformation

$$\Psi(z, 1+\alpha; x) = x^{-\alpha}\,\Psi(z-\alpha, 1-\alpha; x). \tag{4.1}$$

First assume $0 < \alpha < 1$ and consider the unitary mapping

$$U_\alpha :\ L^2(\mathbf{R}^+, w_\alpha) \longrightarrow L^2(\mathbf{R}^+, w_{-\alpha}), \qquad (U_\alpha\ f)(x) = x^\alpha f(x).$$

Using (4.1) and the definition of the functions $\varphi_{\pm\alpha}(z)$ we obtain the relation

$$U_\alpha \varphi_\alpha(z) = \frac{\Gamma(-z)}{\Gamma(-z-\alpha)}\varphi_{-\alpha}(z+\alpha). \tag{4.2}$$

Hence U_α maps $\ker(S_\alpha^* - z)$ to $\ker(S_{-\alpha}^* - z - \alpha)$, and the following relations

$$(U_\alpha L_n^\alpha)(x) = \frac{(-1)^n}{n!}\,\Psi(-n-\alpha, 1-\alpha; x),$$
$$(U_\alpha \Psi(-n+\alpha, 1+\alpha; \cdot))(x) = (-1)^n n!\, L_n^{-\alpha}(x)$$

hold. These relations imply the operator equalities

$$\begin{aligned} U_\alpha\ S_\alpha\ U_\alpha^{-1} &= S_{-\alpha} - \alpha, \\ U_\alpha\ A_\alpha^\infty\ U_\alpha^{-1} &= A_{-\alpha}^0 - \alpha, \\ U_\alpha\ A_\alpha^0\ U_\alpha^{-1} &= A_{-\alpha}^\infty - \alpha. \end{aligned} \tag{4.3}$$

These equalities and the operator representations of the Q-functions $Q_{\pm\alpha}(z)$ imply the Kummer symmetry (0.6).

Next we define the unitary mapping $U_\alpha : \Pi_m^\alpha \to \Pi_m^{-\alpha}$ for $\alpha > 1$ by setting (in analogy with (4.2))

$$U_\alpha \widetilde{\varphi}_\alpha(z-\alpha) = \varphi_{-\alpha}(z) \tag{4.4}$$

or equivalently,

$$U_\alpha \varphi_\alpha(z-\alpha) = \widetilde{\varphi}_{-\alpha}(z), \tag{4.5}$$

where

$$\varphi_{\pm\alpha}(z) = \tau_{\pm\alpha}\Gamma(-z)\Psi(-z, 1\pm\alpha; \cdot)$$

has values in $\Pi_m^{\pm\alpha}$ (from the L-models) and

$$\widetilde{\varphi}_{\pm\alpha}(z) = \frac{\Gamma(-z\mp\alpha)}{\Gamma(-z)}\varphi_\alpha(z)$$

($\widetilde{\varphi}_{-\alpha}(z)$ has already been introduced in the previous section). The function $\widetilde{\varphi}_\alpha(z)$ can be written also as

$$\widetilde{\varphi}_\alpha(z) = -\frac{\pi}{\sin \pi\alpha} Q_\alpha(z)^{-1}\varphi_\alpha(z),$$

and hence it satisfies the relations (1.11) and (1.12) but with $Q_\alpha(z)$ replaced by

$$\widetilde{Q}_\alpha(z) = -\frac{\pi^2}{\sin^2 \pi\alpha} Q_\alpha(z)^{-1} = Q_{-\alpha}(z+\alpha).$$

Theorem 4.1. *For $\alpha > 1$ the linear mapping U_α defined by (4.4) can be extended by linearity and continuity to a unitary mapping from Π_m^α onto $\Pi_m^{-\alpha}$. This extended mapping maps $\mathcal{L}^\alpha$ onto $\mathcal{L}^{-\alpha}$, $\mathcal{M}^\alpha$ onto $\mathcal{M}^{-\alpha}$, and $\mathcal{H}_0^\alpha$ onto $\mathcal{H}_0^{-\alpha}$, where $\mathcal{L}^{\pm\alpha}$, $\mathcal{M}^{\pm\alpha}$, $\mathcal{H}_0^{\pm\alpha}$ are subspaces in the decompositions (1.24). Moreover, it intertwines the operators $A_\alpha^0 + \alpha$ and $A_{-\alpha}^\infty$, and the relations A_α^∞ and $A_{-\alpha}^0 - \alpha$, that is, the relations (4.3) also hold for $\alpha > 1$. Finally, for $t \neq 0, \infty$ it holds*

$$U_\alpha\ A_\alpha^t\ U_\alpha^{-1} = A_{-\alpha}^s - \alpha, \quad s = -t^{-1}\frac{\pi^2}{\sin^2 \pi\alpha}.$$

Proof. We assume that the minimality conditions (1.20) hold, that is,

$$\Pi_m^\alpha = \overline{\operatorname{span}}_{z-\alpha\in\mathbf{C}\setminus\mathbf{Z}}\{\varphi_\alpha(z)\} \text{ and } \Pi_m^{-\alpha} = \overline{\operatorname{span}}_{z\in\mathbf{C}\setminus\mathbf{Z}}\{\varphi_{-\alpha}(z)\}.$$

By the definitions of $\widetilde{\varphi}_{-\alpha}(z)$ and $\varphi_\alpha(z)$,

$$\begin{aligned}\langle \widetilde{\varphi}_{-\alpha}(z), \widetilde{\varphi}_{-\alpha}(\zeta)\rangle &= \frac{\widetilde{Q}_{-\alpha}(z) - \widetilde{Q}_{-\alpha}(\zeta)^*}{z-\zeta*}\\ &= \frac{Q_\alpha(z-\alpha) - Q_\alpha(\zeta-\alpha)^*}{z-\zeta*} = \langle \varphi_\alpha(z-\alpha), \varphi_\alpha(\zeta-\alpha)\rangle.\end{aligned}$$

It follows that if U_α is extended by linearity then it maps a dense set isometrically onto another dense set. This implies the first statement in the theorem.

Since $\mathcal{M}^\alpha = \operatorname{span}\{\varphi_\alpha(n-\alpha)\}_{n=0}^{m-1}$ and $\mathcal{M}^{-\alpha} = \operatorname{span}\{\widetilde{\varphi}_{-\alpha}(n)\}_{n=0}^{m-1}$, we have $U_\alpha\mathcal{M}^\alpha = \mathcal{M}^{-\alpha}$. Next, the vectors $w_j^{(-\alpha)}$ $j = 1, 2, \ldots$, which span the subspace $\mathcal{L}^{-\alpha}$, can be determined from the expansion of $\varphi_{-\alpha}(z)$ at infinity:

$$\varphi_{-\alpha}(z) = (A_{-\alpha}^0 - z)^{-1}\chi_- = \frac{\pi}{\sin \pi\alpha}\left(-\sum_{j=1}^m \frac{1}{b_j^{(-\alpha)}(z)} w_j^{(-\alpha)} + o(\frac{1}{z^m})\right).$$

Similarly (see [[DLSZ], Section 7.2]) the vectors $w_j^{(\alpha)}$, $j = 1, 2, \ldots$, which span the subspace $\mathcal{L}^\alpha$ and are biorthonormal to the vectors $u_j^{(\alpha)}$, can be determined from the expansion of $\widetilde{\varphi}_\alpha(z)$ at infinity:

$$\widetilde{\varphi}_\alpha(z) = \frac{\pi}{\sin \pi\alpha}\left(-\sum_{j=1}^m \frac{1}{b_j^{(\alpha)}(z)} w_j^{(\alpha)} + o(\frac{1}{z^m})\right).$$

Since $b_j^{(-\alpha)}(z) = b_j^{(\alpha)}(z-\alpha)$ and by (4.5), we have $U_\alpha \mathcal{L}^\alpha = \mathcal{L}^{-\alpha}$. By (1.24) for α and $-\alpha$, we also have $U_\alpha \mathcal{H}_0^\alpha = \mathcal{H}_0^{-\alpha}$. The last statement of the theorem can be proved by invoking Krein's formula (1.16) to the middle term in the expression $U_\alpha(A_\alpha^t - z)^{-1}U_\alpha^{-1}$ and compare the result with formula (3.14) for $(A_{-\alpha}^s - z)^{-1}$. □

Finally, we describe the action of the unitary mapping U_α^{-1} on the basis vectors $L_k^{-\alpha}(x)$ in $\Pi_m^{-\alpha}$ and the basis vectors $\widehat{e}_k$ in $\widehat{\Pi}_m^{-\alpha}$:

$$U_\alpha^{-1} L_k^{-\alpha}(x) = \frac{(-1)^k}{k!\Gamma(\alpha-k)}\varphi_\alpha(k-\alpha) = \tau_\alpha x^{-\alpha} L_n^{-\alpha}(x).$$

The translation of this relation in the L-models to the FL-models reads

$$\begin{aligned} U_\alpha^{-1}\widehat{e}_k &= \frac{(-1)^k}{k!\Gamma(k-\alpha)} N(-\alpha,k)\widehat{\varphi}_\alpha(k-\alpha) \\ &= \frac{(-1)^k}{k!\Gamma(k-\alpha)} N(-\alpha,k)\widehat{\tau}_\alpha\{\frac{\widehat{\chi}_\alpha(n)}{n-k+\alpha}\}_{n=0}^\infty, \end{aligned}$$

where $N(-\alpha,k)$ is the norm of $L_k^{-\alpha}(x)$.

References

[AS] M. Abramovitz and I.M. Stegan (ed.), *Handbook of mathematical functions with formulas, graphs and mathematical tables*, National Bureau of Standards, Mathematical Series 55, 4-th edition, Washington, 1965.

[AGHH] S. Albeverio, F. Gesztesy, R. Høegh–Krohn, and H. Holden, *Solvable models in quantum mechanics*, Springer–Verlag, Berlin, 1988.

[AI] T.Ya. Azizov and I.S. Iokhvidov, *Foundations of the theory of linear operators in spaces with an indefinite metric*, Nauka, Moscow, 1986 (Russian); English translation: *Linear operators in spaces with an indefinite metric*, Wiley, New York, 1989.

[AK] S. Albeverio and P. Kurasov, *Singular pertubations of differential operators*, London Math. Soc., Lecture Notes Series 271, Cambridge Univ. Press, 1999.

[B] Yu.M. Berezanskii, *Expansions in eigenfunctions of self-adjoint operators*, Transl. Amer. Math. Soc. 17, Providence, Rhode Island, 1968.

[BE] F.A. Berezin, *On the Lee model*, Matem. Sborn. **60** (1963), 425–453 (Russian).

[DL] K. Daho and H. Langer, *Sturm-Liouville operators with indefinite weight function*, Proc. Royal Soc. Edinburgh **78a** (1977), 161–191.

[C] F. Calogero, *Solution of a three body problem in one dimension*, J. Math. Phys. **10**(12) (1969), 2191–2196.

[D] V.A. Derkach, *On extensions of the Laguerre operator in indefinite inner product spaces*, Matem. Zametki **63**(4) (1998), 509–521 (Russian).

[DLS] A. Dijksma, H. Langer, and H.S.V. de Snoo, *Selfadjoint π_κ-extensions of symmetric subspaces: an abstract approach to boundary problems with spectral parameter in the boundary conditions*, Integral Equations Operator Theory **7** (1984), 459–515.

[DLSZ] A. Dijksma, H. Langer, Yu. Shondin, and C. Zeinstra, *Self-adjoint operators with inner singularities and Pontryagin spaces*, Operator Theory: Adv. Appl., vol. 118, Birkhäuser Verlag, Basel, 2000, 105–175.

[DS] A. Dijksma and Yu. Shondin, *Singular point-like perturbations of the Bessel operator in a Pontryagin space*, J. Diff. Equations **164** (2000), 49–91.

[DSn] A. Dijksma and H.S.V. de Snoo, *Symmetric and self-adjoint relations in Krein spaces* I, Operator Theory: Adv. Appl., vol. 24, Birkhäuser Verlag, Basel, 1987, 145–166.

[DT] J.F. van Diejen and A. Tip, *Scattering from generalized point interaction using self-adjoint extensions in Pontryagin spaces*, J. Math. Phys. **32**(3) (1991), 630–641.

[E] A. Erdélyi, *Higher transcendental functions*, vol. 1, Mcgraw-Hill, New York, 1953.

[F] J. Fuchs, *Physical state conditions and symmetry breaking in quantum mechanics*, J. Math. Phys. **27**(1) (1986), 349–353.

[GS] I.M. Gelfand and G.E. Schilow, *Verallgemeinerte Funktionen (Distributionen)* I, VEB Deutscher Verlag der Wissenschaften, Berlin, 1960.

[GSi] F. Gesztesy and B. Simon, *Rank one perturbation at infinite coupling*, J. Funct. Anal. **128** (1995), 245–252.

[HS] S. Hassi and H. de Snoo, *On rank one perturbations of self-adjoint operators*, Int. Eq. Oper. Theory **29** (1997), 288–300.

[JL] P. Jonas and H. Langer, *Some questions in the perturbation theory of J-nonnegative operators in Krein spaces*, Math. Nachr. **114** (1983), 205–226.

[JLT] P. Jonas, H. Langer, and B. Textorius, *Models and unitary equivalence of cyclic self-adjoint operators in Pontrjagin spaces*, Operator Theory: Adv. and Appl., vol. 59, Birkhäuser Verlag, Basel, 1992, 252–284.

[K1] A.M. Krall, *Laguerre polynomial expansions in indefinite inner product spaces*, J. Math. Anal. Appl., **70** (1979), 267–279.

[K2] A.M. Krall, *On boundary values for the Laguerre operator in indefinite inner product spaces*, J. Math. Anal. Appl., **85** (1982), 406–408.

[KJ] M.G.Krein and V.A. Javryan, *Spectral shift functions that arise in perturbations of a positive operator*, J. Oper.Theory **6** (1981), 155–191 (Russian).

[KL1] M.G. Krein and H. Langer, *Über die Q-Funktion eines π-hermiteschen Operators in Raume Π_κ*, Acta Sci. Math. (Szeged) **34** (1973), 191–230.

[KL2] M.G. Krein and H. Langer, *Über einige Fortzetzungsprobleme, die eng mit der Theorie hermitescher Operatoren im Raume Π_κ zusammenhängen. I. Einige Funktionenklassen und ihre Darstellungen*, Math. Nachr. **77** (1977), 187–236.

[KL3] M.G. Krein and H. Langer, *Some propositions on analytic matrix functions related to the theory of operators on the space Π_κ*, Acta Sci. Math. (Szeged) **43** (1981), 181–205.

[KO] M.G. Krein and I.E. Ovcharenko, *Inverse problems for Q-functions and resolvent matrices of positive hermitean operators*, Dokl. Akad. Nauk SSSR **242**(1978), 521–524 (Russian); English translation, Soviet Math. Dokl. **18**(1978.

[KS] A. Kiselev and B. Simon, *Rank one perturbations with infinitesimal coupling*, J. Funct. Anal. **130**(1995), 345–256.

[L] H. Langer, *A characterization of generalized zeros of negative type of functions of the class N_κ*, Operator Theory: Adv. Appl., vol. 17, Birkhäuser Verlag, Basel, 1986, 201–212.

[La] L. Lathouers, *The Hamiltonian $H = (-1/2)d^2/dx^2 + x^2/2 + \lambda/x^2$ reobserved*, J. Math. Phys. **16**(7) (1975), 1393–1395.

[MK] R.D. Morton and A.M. Krall, *Distributional weight functions for orthogonal polynomials*, SIAM J. Math. Anal. **2** (1978), 604–626.

[S1] Yu.G. Shondin, *Quantum-mechanical models in $\mathbf{R}^n$ associated with extension of the energy operator in a Pontryagin space*, Teor. Mat. Fiz. **74** (1988) 331–344 (Russian); English translation: Theor. Math. Phys. **74** (1988), 220–230.

[S2] Yu.G. Shondin, *Perturbation of elliptic operators supported on subsets of high codimension, and extension theory in indefinite metric spaces*, Seminars of St. Petersburg Math. Inst., vol. 222, Researches in linear operators and function theory, 23 (1995), 246–292 (Russian).

[WW] E.T. Whittaker and G.N. Watson, *A Course of modern analysis*, Cambridge Univ.Press, Cambridge, 1952.

Aad Dijksma
Department of Mathematics
University of Groningen
P. O. Box 800, 9700 AV Groningen
The Netherlands
e-mail: `a.dijksma@math.rug.nl`

Yuri Shondin
Department of Physics
Pedagogical State University
str. Ulyanova 1
Nizhny Novgorod 603600
Russia
e-mail: `shondin@shmath.nnov.ru`

Operator Theory:
Advances and Applications, Vol. 132, 183–198

Realizations of Herglotz-Nevanlinna Functions via F-systems

Seppo Hassi, Henk de Snoo, and Eduard Tsekanovskiĭ

Abstract. An extension of Brodskiĭ-Livšic operator systems, involving an unbounded main operator and an additional orthogonal projection, is studied. It leads to new types of representation and realization results for certain classes of Herglotz-Nevanlinna functions and for the associated transfer functions.

1. Introduction

An operator valued function $V(z)$ acting on a (finite-dimensional) Hilbert space $\mathfrak{E}$ belongs to the class of Herglotz-Nevanlinna matrix-functions if it is holomorphic on $\mathbb{C}\setminus\mathbb{R}$, if it is symmetric with respect to the real axis, i.e., $V(z)^* = V(\bar{z})$, $z \in \mathbb{C}\setminus\mathbb{R}$, and if it satisfies the positivity condition

$$\operatorname{Im} V(z) \geq 0, \quad z \in \mathbb{C}_+.$$

It is well known that matrix-valued Herglotz-Nevanlinna functions admit the following integral representation:

$$V(z) = Q + Lz + \int_{-\infty}^{+\infty} \left(\frac{1}{t-z} - \frac{t}{1+t^2} \right) d\Sigma(t), \quad z \in \mathbb{C}\setminus\mathbb{R}, \tag{1.1}$$

where $Q = Q^*$, $L \geq 0$, and $\Sigma(t)$ is a nondecreasing matrix-valued function on $\mathbb{R}$ with values in the class of nonnegative operators acting on $\mathfrak{E}$ such that

$$\int_{-\infty}^{+\infty} \frac{(d\Sigma(t)x, x)}{1+t^2} < \infty, \quad \text{for all } x \in \mathfrak{E}. \tag{1.2}$$

Special classes of Herglotz-Nevanlinna functions are known to have operator realizations, which have interpretations in system theory. Consider, for instance, a matrix-valued Herglotz-Nevanlinna function of the form

$$V(z) = \int_a^b \frac{d\Sigma(t)}{t-z}, \quad z \in \mathbb{C}\setminus\mathbb{R}, \tag{1.3}$$

The present research was supported by NATO (Collaborative Research Grant CRG 972210) and by the Academy of Finland (project 40362).

with $\Sigma(t)$ a nondecreasing matrix-valued function on the finite interval $(a,b) \subset \mathbb{R}$. Then $V(z)$ has an operator realization of the form

$$V(z) = K^*(D - zI)^{-1}K, \quad z \in \mathbb{C} \setminus \mathbb{R}, \tag{1.4}$$

where D is a bounded self-adjoint operator acting on a Hilbert space $\mathfrak{H}$ and K is a bounded invertible operator from the Hilbert space $\mathfrak{E}$ into $\mathfrak{H}$. Such realizations are due to M.S. Brodskiĭ and M.S. Livšic; they have been used in the theory of characteristic operator-valued functions as well as in system theory, cf. [4], [5], [10], [11]. The function $V(z)$ can be expressed via a fractional linear transform

$$V(z) = i[W_\Theta(z) - I][W_\Theta(z) + I]^{-1}J = i[W_\Theta(z) + I]^{-1}[W_\Theta(z) - I]J, \tag{1.5}$$

where the operator-valued function

$$W_\Theta(z) = I - 2iK^*(M - zI)^{-1}KJ \tag{1.6}$$

is the transfer (or characteristic) function of a linear stationary conservative dynamical system Θ of the form

$$\Theta = \begin{pmatrix} M & K & J \\ \mathfrak{H} & & \mathfrak{E} \end{pmatrix}, \quad \operatorname{Re} M = D, \quad \operatorname{Im} M = KJK^*. \tag{1.7}$$

The Brodskiĭ-Livšic representation theory was extended in [8], [9] to include Herglotz-Nevanlinna functions of the form (1.1) with compactly supported $\Sigma(t)$. In the present paper realization results for Herglotz-Nevanlinna functions of the form

$$V(z) = Q + Lz + \int_{-\infty}^{+\infty} \frac{d\Sigma(t)}{t - z}, \tag{1.8}$$

where $Q = Q^*$, $L \geq 0$, and $\Sigma(t)$ is a nondecreasing matrix-valued function on $\mathbb{R}$ such that

$$\int_{-\infty}^{+\infty} (d\Sigma(t)x, x) < \infty, \quad \text{for all } x \in \mathfrak{E}, \tag{1.9}$$

will be given. In particular, it is shown that each Herglotz-Nevanlinna function $V(z)$ of the form (1.8) satisfying (1.9) gives rise to an F-system of the form

$$\Theta_F = \begin{pmatrix} M & F & K & J \\ \mathfrak{H} & & & \mathfrak{E} \end{pmatrix}, \tag{1.10}$$

where the main operator M is now allowed to be unbounded. Moreover, the additional ingredient in (1.10) is the operator F which is an orthogonal projection in $\mathfrak{H}$, cf. [8], [9]. The system determined by (1.10) has properties, which are analogous to those of Brodskiĭ-Livšic systems (1.7). The function $V_{\Theta_F}(z)$ is obtained from (1.10) via an F-resolvent in the form

$$V_{\Theta_F}(z) = K^*(D - zF)^{-1}K, \quad z \in \mathbb{C} \setminus \mathbb{R}, \tag{1.11}$$

where $D = M_R = \frac{1}{2}(M + M^*)$ stands for the "real part" of the main operator M. The transfer (or characteristic) function $W_{\Theta_F}(z)$ associated to the F-system (1.10) is defined by

$$W_{\Theta_F}(z) = 1 - 2iK^*(M - zF)^{-1}KJ, \tag{1.12}$$

for all $z \in \mathbb{C}$ for which $M - zF$ is invertible. Again the functions $V_{\Theta_F}(z)$ and $W_{\Theta_F}(z)$ are related via (1.5). It is shown that the operators D and F in (1.11) can be selected so that they satisfy a certain commutativity condition precisely when the linear term in (1.8) is absent, i.e., $L = 0$.

The present paper forms a continuation of [8], [9]. The main results obtained in this paper concern transfer functions and their realizations in terms of F-systems of the form (1.10), where the main operator M in (1.10) is allowed to be unbounded. Different types of realization results for certain classes of Herglotz-Nevanlinna functions were recently obtained in [3]. Some other recent approaches to realization problems can be found in [1], [2], [6], and [7].

2. Some facts concerning F-systems

Let $\mathfrak{H}$ be a Hilbert space with inner product $(\cdot,\cdot)$. Let A be a closed linear operator in $\mathfrak{H}$ and let F be an orthogonal projection in $\mathfrak{H}$. Associated to the pair (A, F) is the resolvent set $\rho(A, F)$, i.e., the set of all $z \in \mathbb{C}$ for which $A - zF$ is boundedly invertible in $\mathfrak{H}$. The corresponding resolvent operator is defined as $(A - zF)^{-1}$, $z \in \rho(A)$. When the resolvent set $\rho(A, F)$ is non-empty, the following analog of the resolvent identity holds:

$$(A - zF)^{-1} - (A - wF)^{-1} = (z - w)(A - wF)^{-1}F(A - zF)^{-1}, \tag{2.1}$$

$z, w \in \rho(A, F)$. Note that if A is self-adjoint then $\rho(A, F)$ is symmetric with respect to the real axis. In the sequel an important role is played by the class of densely defined closed linear operators A in $\mathfrak{H}$ which are of the form

$$A = D + iR, \quad D = D^*, \quad R = R^* \in L(\mathfrak{H}). \tag{2.2}$$

Here D, and hence A, may be unbounded, but R is bounded. Therefore $A^* = D^* - iR^*$, which implies $2D = A + A^* = 2\mathrm{Re}\, A$ and $2iR = A - A^* = 2i\mathrm{Im}\, A$, so that

$$D = \mathrm{Re}\, A, \quad R = \mathrm{Im}\, A,$$

are the real part and the imaginary part of A, respectively. Below the real part D of A in (2.2) is often denoted by A_R.

Lemma 1. *Let $A = D + iR$ be a closed linear operator in the Hilbert space $\mathfrak{H}$ of the form (2.2), so that $D = D^*$ and $R = R^* \in L(\mathfrak{H})$. Let F be an orthogonal projection in $\mathfrak{H}$. Assume that $R \geq 0$ or that $R \leq 0$. Then:*

(i) $\ker(A - zF) = \ker(A^* - \bar{z}F) = \ker A \cap \ker F$ *for every $z \in \mathbb{C}_-$ or for every $z \in \mathbb{C}_+$, respectively;*

(ii) *if $z \in \rho(A, F)$ for some $z \in \mathbb{C}_-$ or for some $z \in \mathbb{C}_+$, then $z \in \rho(A, F)$ for every $z \in \mathbb{C}_-$ or for every $z \in \mathbb{C}_+$, respectively.*

Moreover, if A is self-adjoint or equivalently $R = 0$, then the statements (i) *and* (ii) *hold for all $z \in \mathbb{C}_- \cup \mathbb{C}_+$.*

Proof. It suffices to give a proof for the case $R \geq 0$. The case $R \leq 0$ can be dealt with in an analogous manner.

(i) Clearly, $\ker A \cap \ker F \subset \ker(A - zF)$ for any $z \in \mathbb{C}$. To see the converse, assume that $x \in \ker(A - zF)$, $z \in \mathbb{C}_-$. Then

$$0 = ((A - zF)x, x) = (Dx, x) + i(Rx, x) - z(Fx, x)$$

and, since $z \in \mathbb{C}_-$, one concludes that $(Rx, x) = (Fx, x) = 0$. Therefore, $Rx = Fx = 0$ and consequently $Ax = 0$, so that $x \in \ker A \cap \ker F$. Similarly one proves $\ker(A^* - \bar{z}F) = \ker A \cap \ker F$.

(ii) Assume that $z_0 \in \rho(A, F)$ for some $z_0 \in \mathbb{C}_-$, so that $\ker(A - z_0F) = \{0\}$ and $\operatorname{ran}(A - z_0F) = \mathfrak{H}$. It follows from (i) that $\ker(A - zF) = \{0\}$ for every $z \in \mathbb{C}_-$. It remains to show that $\operatorname{ran}(A - zF) = \mathfrak{H}$ for every $z \in \mathbb{C}_-$. Again by (i) $\ker(A^* - \bar{z}F) = \{0\}$, and thus it is enough to prove that $\operatorname{ran}(A - zF)$ is closed for every $z \in \mathbb{C}_-$; or equivalently, that for every $z \in \mathbb{C}_-$ there is a positive constant $c(z)$ such that $\|(A-zF)x\| \geq c(z)\|x\|$ for all $x \in \operatorname{dom} A$. Assume the converse, i.e., that there is $z \in \mathbb{C}_-$ such that $(A - zF)x_n \to 0$ for some sequence (x_n), $\|x_n\| = 1$, $x_n \in \operatorname{dom} A$. Then

$$((A - zF)x_n, x_n) = (Dx_n, x_n) + i(Rx_n, x_n) - z(Fx_n, x_n) \to 0 \tag{2.3}$$

implies that $(Fx_n, x_n) = \|Fx_n\|^2 \to 0$. Hence $Fx_n \to 0$, $Ax_n \to 0$, and consequently $(A - z_0F)x_n \to 0$, a contradiction to the assumption $z_0 \in \rho(A, F)$.

It is clear that if A is self-adjoint then the assertions (i) and (ii) are true for every $z \in \mathbb{C} \setminus \mathbb{R}$. □

Corollary 2. *Let $A = D + iR$ and F be as in Lemma 1, and assume that $R = R^* \in L(\mathfrak{H})$ satisfies $R \geq 0$ or $R \leq 0$, respectively. Then:*
(i) $\ker A \subset \ker A_R$;
(ii) $\rho(A_R, F) \neq \emptyset$ *implies* $\rho(A, F) \neq \emptyset$.

Proof. (i) Assume that $Ax = 0$ for some $x \in \operatorname{dom} A$. Then $((A_R + iR)x, x) = 0$, which implies $(A_Rx, x) = (Rx, x) = 0$. Consequently, $Rx = 0$ and hence also $A_Rx = 0$.

(ii) Assume that $\rho(A_R, F) \neq \emptyset$. Then, by Lemma 1, $\mathbb{C}_+ \cup \mathbb{C}_- \subset \rho(A_R, F)$. With $z \in \mathbb{C}_-$ part (i) and Lemma 1 imply $\ker(A - zF) = \ker A \cap \ker F \subset \ker A_R \cap \ker F = \{0\}$. To see that $z \in \rho(A, F)$ consider a sequence (x_n), $\|x_n\| = 1$, with $(A - zF)x_n \to 0$ as in the proof of Lemma 1. Then (2.3) shows that $Fx_n \to 0$ and $R^{\frac{1}{2}}x_n \to 0$. Therefore also $Rx_n \to 0$ and $A_Rx_n \to 0$. Consequently, $(A_R - zF)x_n \to 0$ and this implies (ii). □

Definition 3. *Let $\mathfrak{H}$ and $\mathfrak{E}$ be Hilbert spaces with $\mathfrak{E}$ finite-dimensional, and let M, F, K, and J be linear operators. The array*

$$\Theta_F = \begin{pmatrix} M & F & K & J \\ \mathfrak{H} & & & \mathfrak{E} \end{pmatrix} \tag{2.4}$$

is called an F-system (or an F-colligation) if:

(i) *M is of the form* (2.2);
(ii) $J = J^* = J^{-1} \in L(\mathfrak{E})$;
(iii) $\operatorname{Im} M = KJK^*$, *where* $K \in L(\mathfrak{E}, \mathfrak{H})$;
(iv) *F is an orthogonal projection in* $\mathfrak{H}$;
(v) *the resolvent sets* $\rho(M_R, F)$ *and* $\rho(M, F)$ *are nonempty.*

The system (2.4) is called a *scattering F-system* if in (ii) $J = I$, in which case the main operator M in (2.4) is dissipative: $\operatorname{Im} M \geq 0$, as follows from (iii). Moreover, in this case the first condition $\rho(M_R, F) \neq \emptyset$ in (v) implies the second condition $\rho(M, F) \neq \emptyset$ in (v), as follows from Corollary 2.

To each F-system (F-colligation) in Definition 3 one can associate the following function

$$V_{\Theta_F}(z) = K^*(M_R - zF)^{-1}K, \quad z \in \rho(M_R, F). \tag{2.5}$$

Proposition 4. *Let Θ_F be an F-system of the form* (2.4). *Then for all $z, w \in \rho(M_R, F)$, the function $V_{\Theta_F}(z)$ in* (2.5) *satisfies*

$$V_{\Theta_F}(z) - V_{\Theta_F}(\bar{w})^* = (z - \bar{w})K^*(M_R - zF)^{-1}F(M_R - \bar{w}F)^{-1}K. \tag{2.6}$$

If $\rho(M_R, F)$ is nonempty, then $\mathbb{C}_+ \cup \mathbb{C}_- \subset \rho(M_R, F)$ and $V_{\Theta_F}(z)$, $z \in \mathbb{C} \setminus \mathbb{R}$, is a Herglotz-Nevanlinna function.

Proof. For each $z, w \in \rho(M_R, F)$ the resolvent identity (2.1) reads as

$$(M_R - zF)^{-1} - (M_R - \bar{w}F)^{-1} = (z - \bar{w})(M_R - zF)^{-1}F(M_R - \bar{w}F)^{-1}. \tag{2.7}$$

The definition (2.5) implies (2.6). If $\rho(M_R, F)$ is nonempty, then according to Lemma 1, $\mathbb{C} \setminus \mathbb{R} \subset \rho(M_R, F)$. Clearly, $V_{\Theta_F}(z)^* = V_{\Theta_F}(\bar{z})$ when $z \in \rho(M_R, F)$. Moreover, it follows from (2.6) that $V_{\Theta_F}(z)$ is (locally) holomorphic on $\rho(M_R, F)$ and has a nonnegative imaginary in the upper halfplane, so that $V_{\Theta_F}(z)$ is a matrix-valued Herglotz-Nevanlinna function. □

To each F-system (F-colligation) in Definition 3 one can also associate the following transfer (or characteristic) function

$$W_{\Theta_F}(z) = I - 2iK^*(M - zF)^{-1}KJ, \quad z \in \rho(M, F). \tag{2.8}$$

Proposition 5. *Let Θ_F be an F-system of the form* (2.4). *Then for all $z, w \in \rho(M, F)$, the function $W_{\Theta_F}(z)$ in* (2.8) *satisfies*

$$W_{\Theta_F}(z)JW^*_{\Theta_F}(w) - J = 2i(\bar{w} - z)K^*(M - zF)^{-1}F(M^* - \bar{w}F)^{-1}K,$$
$$W^*_{\Theta_F}(w)JW_{\Theta_F}(z) - J = 2i(\bar{w} - z)JK^*(M^* - \bar{w}F)^{-1}F(M - zF)^{-1}KJ.$$

In particular, if $z \in \rho(M, F)$, then $W_{\Theta_F}(z)$ is J-unitary when $z \in \mathbb{R}$, J-expansive when $z \in \mathbb{C}_+$, and J-contractive when $z \in \mathbb{C}_-$.

Proof. The property (iii) in Definition 3 shows that for all $z, w \in \rho(M, F)$

$$
\begin{aligned}
&(M-zF)^{-1} - (M^* - \bar{w}F)^{-1} \\
&\quad =(M - zF)^{-1}[(M^* - \bar{w}F) - (M - zF)](M^* - \bar{w}F)^{-1} \\
&\quad =(z - \bar{w})(M - zF)^{-1}F(M^* - \bar{w}F)^{-1} - 2i(M - zF)^{-1}KJK^*(M^* - \bar{w}F)^{-1}.
\end{aligned}
$$

This identity together with (2.8) implies that

$$
\begin{aligned}
&W_{\Theta_F}(z)JW^*_{\Theta_F}(w) - J \\
&\qquad = [I - 2iK^*(M - zF)^{-1}KJ]J[I + 2iJK^*(M^* - \bar{w}F)^{-1}K] - J \\
&\qquad = 2i(\bar{w} - z)K^*(M - zF)^{-1}F(M^* - \bar{w}F)^{-1}K.
\end{aligned}
$$

This proves the first equality. Likewise, by using

$$
\begin{aligned}
(M - zF)^{-1} - (M^* - \bar{w}F)^{-1} = (z - \bar{w})&(M^* - \bar{w}F)^{-1}F(M - zF)^{-1} \\
&- 2i(M^* - \bar{w}F)^{-1}KJK^*(M - zF)^{-1},
\end{aligned}
$$

one proves the second identity. □

The function $V_{\Theta_F}(z)$ defined in (2.5) and the function $W_{\Theta_F}(z)$ defined in (2.8) are closely connected.

Proposition 6. *Let Θ_F be an F-system of the form (2.4). Then for all $z \in \rho(M_R, F) \cap \rho(M, F)$ the operators $I + iV_{\Theta_F}(z)J$ and $I + W_{\Theta_F}(z)$ are boundedly invertible. Moreover,*

$$
\begin{aligned}
V_{\Theta_F}(z) &= i[W_{\Theta_F}(z) - I][W_{\Theta_F}(z) + I]^{-1}J \\
&= i[W_{\Theta_F}(z) + I]^{-1}[W_{\Theta_F}(z) - I]J,
\end{aligned} \tag{2.9}
$$

and

$$
\begin{aligned}
W_{\Theta_F}(z) &= [I - iV_{\Theta_F}(z)J][I + iV_{\Theta_F}(z)J]^{-1} \\
&= [I + iV_{\Theta_F}(z)J]^{-1}[I - iV_{\Theta_F}(z)J].
\end{aligned} \tag{2.10}
$$

Proof. The following identity with $z \in \rho(M, F) \cap \rho(M_R, F)$

$$
(M_R - zF)^{-1} - (M - zF)^{-1} = i(M - zF)^{-1}\mathrm{Im}\, M(M_R - zF)^{-1},
$$

leads to

$$
K^*(M_R - zF)^{-1}K - K^*(M - zF)^{-1}K = iK^*(M - zF)^{-1}KJK^*(M_R - zF)^{-1}K.
$$

Now in view of (2.8) and (2.5), $2V_{\Theta_F}(z) + i(I - W_{\Theta_F}(z))J = (I - W_{\Theta_F}(z))V_{\Theta_F}(z)$, or equivalently,

$$
[I + W_{\Theta_F}(z)][I + iV_{\Theta_F}(z)J] = 2I. \tag{2.11}
$$

Similarly, the identity $(M_R - zF)^{-1} - (M - zF)^{-1} = i(M_R - zF)^{-1}\mathrm{Im}\, M(M - zF)^{-1}$ with $z \in \rho(M, F) \cap \rho(M_R, F)$ leads to

$$
[I + iV_{\Theta_F}(z)J][I + W_{\Theta_F}(z)] = 2I. \tag{2.12}
$$

The equalities (2.11) and (2.12) show that the operators are boundedly invertible and consequently one obtains (2.9) and (2.10). □

When M is a bounded operator and $F = I$, the F-system in Definition 3 reduces to the usual Brodskiĭ-Livšic operator system (colligation) as in [4]. When M is a bounded operator and F is any orthogonal projection, the F-system in Definition 3 reduces to the notion introduced in [8]. The present situation with M unbounded and F an orthogonal projection is of interest as it is phrased in pure Hilbert space terminology; it does not require any Hilbert space triplets as in [3].

If the system (2.4) is a scattering system ($J = I$), then it follows from (2.10) that $W_{\Theta_F}(z)$ is a Schur function; it is holomorphic on $\mathbb{C}_-$ and its values are contractions. In general, when (2.4) is not a scattering system, the domain of holomorphy of $W_{\Theta_F}(z)$ need not be $\mathbb{C}_-$. The realization of transfer functions in the case of a general directing operator $J = J^* = J^{-1}$ is more involved and details will be studied elsewhere.

3. Realizations of a class of Herglotz-Nevanlinna functions

Let A be a self-adjoint operator in a Hilbert space $\mathfrak{H}$, and let $\mathfrak{L}$ be a closed linear subspace of $\mathfrak{H}$. When A is bounded, the subspace $\mathfrak{L}$ is said to be invariant under A if $A\mathfrak{L} \subset \mathfrak{L}$. When A is possibly unbounded, the subspace $\mathfrak{L}$ is said to be invariant under A if for some point $z \in \mathbb{C} \setminus \mathbb{R}$,

$$(A - zI)^{-1}\mathfrak{L} \subset \mathfrak{L}. \tag{3.1}$$

Lemma 7. *Let $\mathfrak{L}$ be a closed linear subspace of $\mathfrak{H}$. If (3.1) holds for some $z \in \mathbb{C} \setminus \mathbb{R}$, then $\mathfrak{L}$ reduces $A = A^*$:*

$$A = A_1 \oplus A_2, \quad A_j = A_j^* \text{ in } \mathfrak{H}_j, \quad j = 1, 2, \quad \mathfrak{H}_1 = \mathfrak{L}, \quad \mathfrak{H}_2 = \mathfrak{H} \ominus \mathfrak{L}. \tag{3.2}$$

Proof. By means of the resolvent operator of A one can describe the graph of A for each $z \in \rho(A)$ as follows:

$$A = \{\, \{(A - zI)^{-1}h, (I + z(A - zI)^{-1})h\} : h \in \mathfrak{H} \,\}.$$

By assumption $(A - zI)^{-1}\mathfrak{L} \subset \mathfrak{L}$, and therefore also $(A - \bar{z}I)^{-1}\mathfrak{L}^{\perp} \subset \mathfrak{L}^{\perp}$. This gives rise to the following restrictions of A in $\mathfrak{H}_1$ and $\mathfrak{H}_2$, respectively:

$$A_1 = \{\, \{(A - zI)^{-1}h, (I + z(A - zI)^{-1})h\} : h \in \mathfrak{L} \,\},$$
$$A_2 = \{\, \{(A - \bar{z}I)^{-1}h, (I + \bar{z}(A - \bar{z}I)^{-1})h\} : h \in \mathfrak{H} \ominus \mathfrak{L} \,\}.$$

Now A_j as a restriction of A is symmetric, say, with defect numbers (n_j^+, n_j^-) in $\mathfrak{H}_j$, $j = 1, 2$. Since $A_1 \oplus A_2 = A$ and A is self-adjoint one concludes $(n_1^+ + n_2^+, n_1^- + n_2^-) = (0, 0)$. Consequently, A_1 and A_2 are (graphs of) self-adjoint operators. □

It was shown in Proposition 4 that the function $V_{\Theta_F}(z)$ associated to the F-system Θ_F in (2.4) is a Herglotz-Nevanlinna function, assuming that $\rho(M_R, F) \neq \emptyset$. However, more can be said when M_R and F have a "bounded noncommuting part" in the sense to be specified in the next theorem.

Theorem 8. *Let* Θ_F *be an* F*-system of the form* (2.4) *with* $\dim\ker F < \infty$. *Assume that* $\rho(M_R, F)$ *is nonempty, and let* $V_{\Theta_F}(z)$ *be defined by* (2.5). *Then the following statements hold.*

(i) *If there is an orthogonal projection* P *commuting with* F *and* $(M_R - zI)^{-1}$ *for some* $z \in \mathbb{C}\setminus\mathbb{R}$, *such that* $\operatorname{ran} P \subset \operatorname{ran} F$ *and the restriction* $M_R \restriction \ker P$ *is bounded, then* $V_{\Theta_F}(z)$ *is a matrix-valued Herglotz-Nevanlinna function with the integral representation*

$$V_{\Theta_F}(z) = Q + Lz + \int_{-\infty}^{+\infty} \frac{d\Sigma(t)}{t-z}, \quad z \in \rho(M, F), \tag{3.3}$$

where $Q = Q^*$, $L \geq 0$, *and* $\Sigma(t)$ *is a nondecreasing matrix-valued function on* $\mathbb{R}$ *satisfying* (1.9).

(ii) *If, in particular,* $F(M_R - zI)^{-1} = (M_R - zI)^{-1}F$ *for some* $z \in \mathbb{C}\setminus\mathbb{R}$, *then the integral representation of* $V(z)$ *is of the form* (3.3) *with* $L = 0$.

Proof. Let P be an orthogonal projection satisfying the assumptions in (i). Then by Lemma 7 M_R admits an orthogonal decomposition of the form (3.2), where $M_2 = M_R \restriction \ker P$ is bounded by the assumptions. Decomposing $K = K_1 \oplus K_2$ according to $\mathfrak{H} = \operatorname{ran} P \oplus \ker P$ one obtains

$$V_{\Theta_F}(z) = K_1^*(M_1 - zI)^{-1}K_1 + K_2^*(M_2 - zF_2)^{-1}K_2, \tag{3.4}$$

where $F_2 = (I-P)F$ is an orthogonal projection in $\ker P$. Now $\dim\ker F_2 < \infty$ and according to [9, Theorem 2.2] the second summand in (3.4) has the following integral representation

$$K_2^*(M_2 - zF_2)^{-1}K_2 = Q_2 + L_2 z + \int_a^b \frac{d\Sigma_2(t)}{t-z}, \quad a, b \in \mathbb{R}. \tag{3.5}$$

Clearly, the first summand in (3.4) has the integral representation

$$K_1^*(M_1 - zI)^{-1}K_1 = \int_{-\infty}^{+\infty} \frac{d\Sigma_1(t)}{t-z}, \quad \int_{-\infty}^{+\infty} (d\Sigma_1(t)x, x) < \infty, \quad x \in \mathfrak{E}. \tag{3.6}$$

Hence, (3.3) follows from (3.5) and (3.6) with $Q = Q_2$, $L = L_2$, and $\Sigma(t) = \Sigma_1(t) + \Sigma_2(t)$.

To prove (ii) observe that the decomposition $M_R = M_1 \oplus M_2$ for M_R in Lemma 7 gives (3.4) with $F_2 = (I-F)F = 0$. Since $\rho(M_R, F) \neq \emptyset$, the operator M_2 must be invertible, and therefore

$$(M_R - zF)^{-1} = (M_1 - zI)^{-1} \oplus M_2^{-1}$$

and

$$V_{\Theta_F}(z) = K_1^*(M_1 - zI)^{-1}K_1 + K_2^* M_2^{-1} K_2.$$

Now one obtains (3.3) with $Q = K_2^* M_2^{-1} K_2$ and $L = 0$. □

In the remaining part of this section the converse to Theorem 8 will be established.

Lemma 9. *Let Q and $L \geq 0$ be self-adjoint operators in $\mathfrak{E}$, $\dim \mathfrak{E} < \infty$. Then the following representations hold:*

$$Q = K_1^*(D_1 - zF_1)^{-1}K_1, \quad z \in \rho(D_1, F_1)\,(= \mathbb{C}), \tag{3.7}$$

$$zL = K_2^*(D_2 - zF_2)^{-1}K_2, \quad z \in \rho(D_2, F_2)\,(= \mathbb{C}), \tag{3.8}$$

where K_j is an injective operator from $\mathfrak{E}$ into a Hilbert space $\mathfrak{H}_j$, $\dim \mathfrak{H}_j < \infty$, D_j is a self-adjoint operator in $\mathfrak{H}_j$, and F_j is an orthogonal projection in $\mathfrak{H}_j$, $j = 1, 2$.

Proof. The representation (3.7) is straightforward to show if Q is invertible and one can take $\mathfrak{H}_1 = \mathfrak{E}$, $K = I$, $D = Q^{-1}$, $F = 0$. In the general case write $Q = Q_1 + Q_2$ with two invertible self-adjoint operators Q_1 and Q_2. Let $\mathfrak{H}_1 = \mathfrak{E} \oplus \mathfrak{E}$ and introduce in $\mathfrak{H}_1$ the following operators:

$$K_1 = \begin{pmatrix} I \\ I \end{pmatrix}, \quad D_1 = \begin{pmatrix} Q_1^{-1} & 0 \\ 0 & Q_2^{-1} \end{pmatrix}, \quad F_1 = 0.$$

Then K_1 is an injective operator from $\mathfrak{E}$ into $\mathfrak{H}_1 = \mathfrak{E} \oplus \mathfrak{E}$, D_1 is self-adjoint and invertible, and $F_1 = 0$ is an orthogonal projection in $\mathfrak{H}_1$. Clearly, $\rho(D_1, F_1) = \mathbb{C}$ and it is straightforward to check that $K_1^*(D_1 - zI)^{-1}K_1 = Q_1 + Q_2 = Q$. This proves (3.7).

Now consider the function zL. Let $\mathfrak{H}_2$ be the Hilbert space $\mathfrak{E} \oplus \mathfrak{E}$ and define in $\mathfrak{H}_2$ the following operators:

$$K_2 = \begin{pmatrix} L^{\frac{1}{2}} \\ kI \end{pmatrix}, \quad D_2 = \begin{pmatrix} 0 & iI \\ -iI & 0 \end{pmatrix}, \quad F_2 = \begin{pmatrix} 0 & 0 \\ 0 & I \end{pmatrix}, \tag{3.9}$$

where $k \in \mathbb{R}\backslash\{0\}$. Then K_2 is an injective operator from $\mathfrak{E}$ into $\mathfrak{H}_2$, D_2 is self-adjoint and F_2 is an orthogonal projection in $\mathfrak{H}_2$. Moreover, $\rho(D_2, F_2) = \mathbb{C}$, and it is easy to check that

$$K_2^*(D_2 - zF_2)^{-1}K_2 = zL + iL^{\frac{1}{2}}k - ikL^{\frac{1}{2}} = zL. \tag{3.10}$$

□

Theorem 10. *Each Herglotz-Nevanlinna function acting on a Hilbert space $\mathfrak{E}$, $n = \dim \mathfrak{E} < \infty$, of the form*

$$V(z) = Q + zL + \int_{-\infty}^{+\infty} \frac{d\Sigma(t)}{t - z} \tag{3.11}$$

where $Q = Q^*$, $L \geq 0$, and $\Sigma(t)$ is a nondecreasing matrix-valued function on $\mathbb{R}$ satisfying (1.9) can be represented in the form

$$V(z) = K^*(D - zF)^{-1}K, \quad z \in \mathbb{C} \setminus \mathbb{R} \subset \rho(D, F), \tag{3.12}$$

where K is an injective operator from $\mathfrak{E}$ into $\mathfrak{H}$, D is a self-adjoint operator in a Hilbert space $\mathfrak{H}$, and F is an orthogonal projection in $\mathfrak{H}$ with $\dim \ker F < \infty$. Moreover, if $L = 0$ then the operators D and F can be selected such that $(D - zI)^{-1}$, $z \in \mathbb{C} \setminus \mathbb{R}$, and F commute.

Proof. Consider the Hilbert space $\mathfrak{H}_3 = L^2(d\Sigma)$ of measurable vector functions $f = f(t)$ with

$$\|f\|^2 = \int_{-\infty}^{+\infty} (d\Sigma(t)f(t), f(t)) < \infty.$$

Define the vector functions $\omega_j(t) \equiv e_j$, $j = 1, \dots, n$, where e_j stand for the coordinate vectors in $\mathbb{C}^n$. The integrability property (1.9) guarantees that $\omega_j \in \mathfrak{H}_3$. Let $K_3 : \mathbb{C}^n \to \mathfrak{H}_3$ be the linear operator determined by $K_3 e_j = \omega_j(t)$, $j = 1, \dots, n$. Let $D_3 = t$ be the multiplication operator by the independent variable in $\mathfrak{H}_3$, and let $F_3 = I$. Then

$$K_3^*(D_3 - zF_3)^{-1}K_3 = \int_{\mathbb{R}} \frac{d\Sigma(t)}{t - z}.$$

Now let the Hilbert space $\mathfrak{H}$ be defined by $\mathfrak{H} = \mathfrak{H}_1 \oplus \mathfrak{H}_2 \oplus \mathfrak{H}_3$, where $\mathfrak{H}_1$ and $\mathfrak{H}_2$ are as in Lemma 9. In addition, using the operators introduced in the proof of Lemma 9, define

$$K = \begin{pmatrix} K_1 \\ K_2 \\ K_3 \end{pmatrix}, \quad D = \begin{pmatrix} D_1 & 0 & 0 \\ 0 & D_2 & 0 \\ 0 & 0 & D_3 \end{pmatrix}, \quad F = \begin{pmatrix} F_1 & 0 & 0 \\ 0 & F_2 & 0 \\ 0 & 0 & F_3 \end{pmatrix}. \tag{3.13}$$

Then K is an injective operator from $\mathfrak{E}$ into $\mathfrak{H}$, D is a self-adjoint operator and F is an orthogonal projection in $\mathfrak{H}$. Moreover, it is easy to check that (3.12) is satisfied. Observe, that if $L = 0$ then the definitions of K, D, and F can be reduced to $K^* = (K_1^*, K_3^*)$, $D = D_1 \oplus D_3$, and $F = F_1 \oplus F_3$, in which case the commutativity of D and F follows from $F_1 = 0$, $F_3 = I$. □

The next result shows that each Nevanlinna function of the form (3.3) can be realized via an F-colligation (2.4) with $J = I$, i.e., by means of a scattering F-system.

Theorem 11. *For each Herglotz-Nevanlinna function acting on a Hilbert space $\mathfrak{E}$, $\dim \mathfrak{E} < \infty$, of the form*

$$V(z) = Q + zL + \int_{-\infty}^{+\infty} \frac{d\Sigma(t)}{t - z}, \tag{3.14}$$

where $Q = Q^*$, $L \geq 0$, and $\Sigma(t)$ satisfies (1.9), there is a scattering F-system Θ_F of the form (2.4) with $\dim \ker F < \infty$ and the directing operator $J = I$, whose transfer function $W_{\Theta_F}(z)$ is defined and holomorphic on $\mathbb{C}_-$ and such that it determines $V(z)$ via

$$V(z) = i[W_{\Theta_F}(z) - I][W_{\Theta_F}(z) + I]^{-1}. \tag{3.15}$$

Proof. By Theorem 10 the function $V(z)$ can be represented via an F-resolvent in the form (3.12). Let $\mathfrak{H}$, K, D, and F be as in Theorem 10. Define in $\mathfrak{H}$ a dissipative operator M by $M = D + iKK^*$. Then $D = \operatorname{Re} M$ and by construction $\mathbb{C} \setminus \mathbb{R} \subset \rho(D, F)$. Now Corollary 2 shows that $\rho(M, F) \neq \emptyset$ and hence $\mathbb{C}_- \subset \rho(M, F)$, cf. Lemma 1. This gives rise to an F-colligation Θ_F satisfying all the properties required in Definition 3. Moreover, the characteristic function $W_{\Theta_F}(z)$

associated to Θ_F in (2.8) is holomorphic on $\mathbb{C}_-$. It remains to apply Proposition 6 to $V_{\Theta_F}(z) = V(z)$ and $W_{\Theta_F}(z)$ with $z \in \mathbb{C}_- \subset \rho(M,F) \cap \rho(D,F)$. □

4. Asymptotic properties of the transfer functions

In this section properties of transfer functions associated to F-systems of the form (2.4) will be studied. For this purpose an explicit construction for the F-system Θ_F in Theorem 11 will be presented. Here results are given for the case $J = I$. The case of a general directing operator $J = J^* = J^{-1}$ is more involved and differs essentially from the case of scattering systems. For instance, it may happen that the transfer function associated to a function $V(z)$ of the form (3.14) need not even be defined.

The next result shows that in the case of scattering F-systems, realizing functions of the form (3.14), the main operator M of the system Θ_F can be constructed in a specific form, which is convenient for studying the asymptotic properties of $W_{\Theta_F}(z)$.

Proposition 12. *Let $V(z)$ be a Herglotz-Nevanlinna function of the form* (3.14) *satisfying* (1.9). *Then there exists a realization for $V(z)$ via a scattering F-system Θ_F of the form* (2.4) *with* $\dim \ker F < \infty$ *and $J = I$, such that the main operator M admits an operator matrix representation with respect to $\mathfrak{H} = \ker F \oplus \operatorname{ran} F$ of the form $M = (M_{ij})_{i,j=1}^2$, in which $M_{11} = (I-F)M{\restriction}\ker F$ is invertible.*

Proof. Let Q be the matrix in (3.14). For every $k \in \mathbb{R}$, excluding a finite number of points, the matrix $Q - k^2 I_n$ is invertible. Let $\mathfrak{H}_1 = \mathbb{C}^{2n}$ and introduce in $\mathfrak{H}_1$ the following operators

$$K_1 = \begin{pmatrix} I_n \\ kI_n \end{pmatrix}, \quad D_1 = \begin{pmatrix} (Q-k^2I_n)^{-1} & 0 \\ 0 & I_n \end{pmatrix}, \quad F_1 = 0.$$

Let the matrix L in (3.14) be factorized by $L = SS^*$, $r = \operatorname{rank} L$, where S is an $n \times r$ matrix. Let $\mathfrak{H}_2 = \mathbb{C}^{2n}$ and introduce in $\mathfrak{H}_2$ the following operators

$$K_2 = \begin{pmatrix} S^* \\ 0 \end{pmatrix}, \quad D_2 = \begin{pmatrix} 0 & I_r \\ I_r & 0 \end{pmatrix}, \quad F_2 = \begin{pmatrix} 0 & 0 \\ 0 & I_r \end{pmatrix}.$$

Let $\mathfrak{H}_3$, K_3, D_3, and F_3 be as in the proof of Theorem 10. Introduce the Hilbert space $\mathfrak{H} = \mathfrak{H}_1 \oplus \mathfrak{H}_2 \oplus \mathfrak{H}_3$ and let the operators K, D, and F be defined by the orthogonal sums as in (3.13). It is easy to check that $K^*(D - zF)^{-1}K = V(z)$, see also the proof of Lemma 9 and Theorem 10. The operator $M = D + iKK^*$ has the following operator matrix representation:

$$\begin{pmatrix} (Q-k^2I_n)^{-1} + iI_n & ikI_n & iS & 0 & iK_3^* \\ ikI_n & (1+ik^2)I_n & ikS & 0 & ikK_3^* \\ iS^* & ikS^* & iS^*S & I_r & iS^*K_3^* \\ 0 & 0 & I_r & 0 & 0 \\ iK_3 & ikK_3 & iK_3S & 0 & D_3 + iK_3K_3^* \end{pmatrix}. \tag{4.1}$$

Decompose the operator M with respect to $\mathfrak{H} = \ker F \oplus \operatorname{ran} F$:

$$M = \begin{pmatrix} M_{11} & M_{12} \\ M_{21} & M_{22} \end{pmatrix},$$

so that

$$M_{11} = \begin{pmatrix} D_1 + iK_1K_1^* & i\widetilde{K} \\ i\widetilde{K}^* & iS^*S \end{pmatrix}, \quad \widetilde{K} = \begin{pmatrix} S \\ kS \end{pmatrix}.$$

If $L = 0$ the blocks involving K_2, D_2, F_2 are absent in (4.1) and $M_{11} = D_1 + iK_1K_1^*$ is invertible. Now assume that $L \neq 0$, in which case

$$\operatorname{Re} M_{11} = \begin{pmatrix} D_1 & 0 \\ 0 & 0 \end{pmatrix}, \quad \operatorname{Im} M_{11} = \begin{pmatrix} K_1K_1^* & \widetilde{K} \\ \widetilde{K}^* & S^*S \end{pmatrix} \geq 0.$$

Hence, if $M_{11}x = 0$, then $0 = (M_{11}x, x) = (\operatorname{Re} M_{11}x, x) + i(\operatorname{Im} M_{11}x, x)$, so that $\operatorname{Im} M_{11}x = 0$ and consequently $\operatorname{Re} M_{11}x = 0$. Since D_1 and S^*S are invertible, it follows that $x = 0$. Therefore $\ker M_{11} = \{0\}$, and M_{11} is invertible. □

The next result shows that the transfer function $W_{\Theta_F}(z)$ corresponding to a scattering F-system has a finite nontangential limit at $\pm i\infty$. First observe that for any operator $A = D + iR$ as in (2.2) the following identity

$$(A - z)^{-1} = (D - z - iR)^{-1} = (D - z)^{-1}\left(I - iR(D - z)^{-1}\right)^{-1},$$

shows that

$$(A - z)^{-1} \to 0 \text{ as } z \widehat{\to} \infty. \tag{4.2}$$

Proposition 13. *Let $W_{\Theta_F}(z)$ be the transfer function of a scattering F-system* (2.4) *with $J = I$. Then:*

(i) *$W_{\Theta_F}(z)$ is defined and holomorphic in a neighborhood of $\pm i\infty$;*

(ii) *if $D = \operatorname{Re} M$ is bounded, then $W_{\Theta_F}(z)$ is defined and holomorphic outside a neighborhood of the origin.*

Moreover, the following nontangential limit

$$W = \lim_{z \widehat{\to} \infty} W_{\Theta_F}(z) \tag{4.3}$$

exists as a unitary matrix.

Proof. It may be assumed that the corresponding scattering system (2.4) is constructed as in Proposition 12. Then the transfer function $W_{\Theta_F}(z)$ admits the following representation:

$$W_{\Theta_F}(z) = I - 2iK^* \begin{pmatrix} M_{11} & M_{12} \\ M_{21} & M_{22} - zI \end{pmatrix}^{-1} K, \tag{4.4}$$

where M_{11} is invertible. The Frobenius formula for the inverse of block operators shows that

$$\rho(M, F) = \rho(\widehat{M}_{22}), \qquad \widehat{M}_{22} := M_{22} - M_{21}M_{11}^{-1}M_{12},$$

and that $(M - zF)^{-1}$, $z \in \rho(M, F)$, has the form

$$\begin{pmatrix} M_{11}^{-1} + M_{11}^{-1}M_{12}(\widehat{M}_{22} - zI)^{-1}M_{21}M_{11}^{-1} & -M_{11}^{-1}M_{12}(\widehat{M}_{22} - zI)^{-1} \\ -(\widehat{M}_{22} - zI)^{-1}M_{21}M_{11}^{-1} & (\widehat{M}_{22} - zI)^{-1} \end{pmatrix}. \tag{4.5}$$

Hence, the statements (i) and (ii) are consequences of the realization of $W_{\Theta_F}(z)$ in (4.4). It follows from (4.2), that

$$(\widehat{M}_{22} - zI)^{-1} \to 0 \text{ as } z \widehat{\to} \infty. \tag{4.6}$$

Now (4.4), (4.5), and (4.6) imply that

$$W = \lim_{z \widehat{\to} +\infty} W_{\Theta_F}(z) = \lim_{z \widehat{\to} -\infty} W_{\Theta_F}(z),$$

exists. Furthermore, Proposition 5 with $w = \bar{z}$ yields

$$I = W_{\Theta_F}(z)W^*_{\Theta_F}(\bar{z}) = \lim_{z \widehat{\to} +\infty} W_{\Theta_F}(z)W^*_{\Theta_F}(\bar{z}) = WW^*,$$

and

$$I = W^*_{\Theta_F}(z)W_{\Theta_F}(\bar{z}) = \lim_{z \widehat{\to} +\infty} W^*_{\Theta_F}(z)W_{\Theta_F}(\bar{z}) = W^*W,$$

which shows that W is unitary. □

The proof of the above proposition shows that the limit W in (4.3) admits the representation

$$W = I - 2iK^* \begin{pmatrix} M_{11}^{-1} & 0 \\ 0 & 0 \end{pmatrix} K, \tag{4.7}$$

when the scattering F-system (2.4) is constructed as in Proposition 12. The following example shows that for $J \neq I$ the finite limit W as defined in Proposition 13 need not exist.

Example 14. *Let J and $V_k(z)$, $k \in \mathbb{R}$, be 2×2 matrices of the form*

$$V_k(z) = kI + zL, \quad L = \begin{pmatrix} 1 & 0 \\ 0 & 0 \end{pmatrix}, \quad J = \begin{pmatrix} 0 & 1 \\ 1 & 0 \end{pmatrix}.$$

Hence $V_k(z)$ is a Herglotz-Nevanlinna function and the corresponding transfer function $W_k(z)$ is given by

$$W_k(z) = \begin{pmatrix} \frac{2}{1+k(z+k)} - 1 & \frac{-2i(z+k)}{1+k(z+k)} \\ \frac{-2ik}{1+k(z+k)} & \frac{2}{1+k(z+k)} - 1 \end{pmatrix}.$$

Notice that for $k = 0$

$$V_0(z) = zL \text{ and } W_0(z) = \begin{pmatrix} 1 & -2iz \\ 0 & 1 \end{pmatrix},$$

and that

$$\lim_{k \to 0} V_k(z) = V_0(z) \text{ and } \lim_{k \to 0} W_k(z) = W_0(z).$$

Clearly, for $k \neq 0$ the nontangential limit W of $W_k(z)$ at ∞ exists and is J-unitary. However, for $k = 0$ this limit does not exist since the $(1,2)$-entry of $W_0(z)$ has a linear growth.

The operators in the next corollary are decomposed according to the decomposition $\mathfrak{E} = \operatorname{ran} L \oplus \ker L$ determined by the factor L of the linear term in $V(z)$. In particular, L can be rewritten as $L = L_1 \oplus 0$, where $L_1 \geq 0$ is invertible.

Corollary 15. *The unitary limit W in* (4.3) *has the following properties:*
(i) $W_{11} = -I$, $W_{21} = 0$, $W_{12} = 0$, *and W_{22} is unitary;*
(ii) *$W_{22} + I$ is invertible;*
(iii) *if L is invertible then the limit* (4.7) *equals $-I$;*
(iv) *if $L = 0$ then $-1 \notin \sigma(W)$;*
(v) *if $F = I$ then the limit* (4.7) *equals I.*

Proof. Proposition 6 gives the connection between the functions $V_{\Theta_F}(z)$ and $W_{\Theta_F}(z)$. In particular, dividing both sides of (2.11) by z and letting $z \widehat{\to} \infty$ gives

$$\begin{pmatrix} W_{11} + I & W_{12} \\ W_{21} & W_{22} + I \end{pmatrix} \begin{pmatrix} iL_1 & 0 \\ 0 & 0 \end{pmatrix} = 0.$$

Since L_1 is invertible, this implies that

$$W_{11} + I = 0, \quad W_{21} = 0.$$

Similarly, (2.12) leads to $W_{12} = 0$. Projecting the identity (2.11) from the right onto $\ker L$ and taking limits shows that

$$(W_{11}+I)iQ_{12}+W_{12}(I+iQ_{22}) = 0, \quad W_{21}iQ_{12}+(W_{22}+I)(I+iQ_{22}) = 2I. \quad (4.8)$$

Since $W_{11} + I = 0$, $W_{12} = 0$, $W_{21} = 0$ the second identity in (4.8) reduces to

$$(W_{22} + I)(I + iQ_{22}) = 2I.$$

Similarly, projecting (2.12) from the left onto $\ker L$ and taking limits leads to

$$(I + iQ_{22})(W_{22} + I) = 2I.$$

The last two identities show that $W_{22} + I$ is invertible and that W_{22} is a unitary operator in $\ker L$. This proves (i) and (ii), from which (iii) and (iv) obviously follow. Finally, if $F = I$ then $L = 0$, $Q = 0$ and, therefore, $W = I$. □

Let $W_{\Theta_F}(z)$ be the transfer function of a scattering F-system (2.4) with $J = I$. Then according to (4.3)

$$W_{\Theta_F}(z) = W + o(1), \quad z \widehat{\to} \infty.$$

This result can be made more precise: there is a square matrix R such that

$$W_{\Theta_F}(z) = W + \frac{R}{z} + o\left(\frac{1}{z}\right), \quad z \widehat{\to} \infty.$$

The matrix R can be expressed in terms of an operator representation of $W_{\Theta_F}(z)$.

Proposition 16. *Let $W_{\Theta_F}(z)$ be the transfer function of the scattering F-system* (2.4) *constructed as in Proposition* 12, *and let W be its limit in* (4.7). *Then*

$$\lim_{z \widehat{\to} \infty} z \frac{W_{\Theta_F}(z) - W}{2i} = K^* T_1 T_2 K,$$

where

$$T_1 = \begin{pmatrix} M_{11}^{-1}M_{12} \\ -I \end{pmatrix}, \quad T_2^\top = \begin{pmatrix} M_{21}M_{11}^{-1} \\ -I \end{pmatrix}. \tag{4.9}$$

Moreover, if the spectral measure of $V_{\Theta_F}(z)$ *has a compact support, then* $W_{\Theta_F}(z)$ *admits the following asymptotic expansion:*

$$W_{\Theta_F}(z) = W - 2iK^*T_1 \left(\sum_{j=0}^{\infty} \frac{\widehat{M}_{22}^j}{z^{j+1}} \right) T_2K, \quad |z| > r_\sigma(\widehat{M}_{22}), \tag{4.10}$$

where $r_\sigma(\widehat{M}_{22})$ *stands for the spectral radius of* $\widehat{M}_{22} = M_{22} - M_{21}M_{11}^{-1}M_{12}$.

Proof. The form of the limit $W = \lim_{z \hat{\to} \infty} W_{\Theta_F}(z)$ in (4.7) together with the representations of $W_{\Theta_F}(z)$ in (4.4) and $(M - zF)^{-1}$ in (4.5) imply

$$W_{\Theta_F}(z) - W = -2iK^*T_1(\widehat{M}_{22} - zI)^{-1}T_2K, \quad z \in \rho(M, F) = \rho(\widehat{M}_{22}), \tag{4.11}$$

where T_1 and T_2 are as defined in (4.9). Since $\lim_{z \hat{\to} \infty} z(\widehat{M}_{22} - zI)^{-1} = -I$, the first assertion follows from (4.11).

If the spectral measure of $V_{\Theta_F}(z)$ has a compact support, then the main operator M of Θ_F is bounded, cf. [9]. Consequently M_{22} and $\widehat{M}_{22}$ are also bounded operators, and therefore one obtains (4.10) by expanding the resolvent of $\widehat{M}_{22}$ as a geometric series in (4.11). □

References

[1] D. Alpay, A. Dijksma, J. Rovnyak, and H.S.V. de Snoo, *Schur functions, operator colligations, and reproducing kernel Pontryagin spaces*, Oper. Theory Adv. Appl., 96, Birkhäuser Verlag, Basel, 1997.

[2] D.Z. Arov and A.A. Nudel'man, "Passive linear stationary scattering systems with continuous time", Integral Equations Operator Theory, 24 (1996), 1–45.

[3] S.V. Belyi and E.R. Tsekanovskiĭ, "Realization theorems for operator-valued R-functions", Oper. Theory Adv. Appl., 98 (1997), 55–91.

[4] M.S. Brodskiĭ, *Triangular and Jordan representations of linear operators*, Moscow, Nauka, 1969 (Russian) (English translation: Transl. Math. Monographs, 32, Amer. Math. Soc., 1971).

[5] M.S. Brodskiĭ and M.S. Livšic, "Spectral analysis of non-self-adjoint operators and intermediate systems", Uspehi Mat. Nauk, 13, no. 1, 79, (1958), 3–85 (Russian) (English translation: Amer. Math. Soc. Transl., (2) 13 (1960), 265–346).

[6] F. Gesztesy, N.J. Kalton, K.A. Makarov, and E.R. Tsekanovskiĭ, "Some applications of operator-valued Herglotz functions", Oper. Theory Adv. Appl., **123** (2001), 271–321.

[7] F. Gesztesy and E.R. Tsekanovskiĭ, "On matrix-valued Herglotz functions", Math. Nachr., 218 (2000), 61–138.

[8] S. Hassi, H.S.V. de Snoo, and E.R. Tsekanovskiĭ, "An addendum to the multiplication and factorization theorems of Brodskiĭ-Livšic-Potapov", Applicable Analysis, **77** (2001), 125–133.

[9] S. Hassi, H.S.V. de Snoo, and E.R. Tsekanovskiĭ, "Commutative and noncommutative representations of matrix-valued Herglotz-Nevanlinna functions", Applicable Analysis, **77** (2001), 135–147.

[10] M.S. Livšic, "On the spectral decomposition of linear non-self-adjoint operators", Mat. Sb., 34, no. 76 (1954), 145–199 (Russian) (English translation: Amer. Math. Soc. Transl., (2) 5 (1957), 67–114).

[11] M.S. Livšic, *Operators, oscillations, waves*, Moscow, Nauka, 1966 (Russian) (English translation: Transl. Math. Monographs, 34, Amer. Math. Soc., 1973).

Seppo Hassi
Department of Mathematics and Statistics
University of Vaasa
PL 700, 65101 Vaasa
Finland
e-mail: `sha@uwasa.fi`

Henk de Snoo
Department of Mathematics
University of Groningen
Postbus 800
9700 AV Groningen
Nederland
e-mail: `desnoo@math.rug.nl`

Eduard Tsekanovskiĭ
Department of Mathematics
Niagara University
NY 14109, USA
e-mail: `tsekanov@niagara.edu`

Operator Theory:
Advances and Applications, Vol. 132, 199–206

Integral Equations of Relativistic Bound State Theory and Sturm-Liouville Problem

V. Kapshai

Abstract. Some relativistic integral equations of the three-dimensional single-time approach for bound state problem in quantum field theory are investigated. The relativistic potential is chosen as a local in the momentum Lobachevsky space one. Partial expansion of such three-dimensional equations is made. It is shown that the one-dimensional partial integral equations for considered potential can be reduced to Sturm-Liouville problems. The differential equations of such Sturm-Liouville problems in the momentum representation have the form of the Schrödinger equation in the coordinate representation.

1. Integral equations

Various covariant three-dimensional integral equations for the description of the elementary particle bound state problem have been introduced in quantum field theory [1, 2]. These equations for the relative movement wave function of two spinless particles are noted in the momentum representation (in the center of mass system) as follows:

$$G_{(j)}^{-1}\left(E_{iw}, E_p\right)\Psi\left(\vec{p}\right) = \frac{1}{(2\pi)^3}\int V\left(\vec{p},\vec{k};E_{iw}\right)\Psi\left(\vec{k}\right)\frac{md\vec{k}}{E_k}, \tag{1}$$

here $\Psi\left(\vec{p}\right)$ is the relativistic wave function, $E_k = \sqrt{\vec{k}^2+m^2}$, where m is the mass of each particle ($m_1 = m_2 = m$). $G_{(j)}\left(E_{iw}, E_p\right)$ are free Green functions, which for the Logunov-Tavkhelidze ($j = 1$) and Kadyshevsky ($j = 2$) equations have the form

$$G_{(1)}(E_{iw}, E_p) = \frac{1}{E_{iw}^2 - E_p^2}; \quad G_{(2)}(E_{iw}, E_p) = \frac{1}{E_p(2E_{iw} - 2E_p)}, \tag{2}$$

where $E_p = \sqrt{m^2+\vec{p}^2}$, and $V\left(\vec{p},\vec{k};E_{iw}\right)$ is the quasipotential. Procedures of quasipotential construction with the help of the two-time Green function ($j = 1$) or with the help of the scattering amplitude ($j = 2$) give a parametric dependence of the quasipotential on the full energy of the two particle system $2E_{iw}$ (the mass of a bound state), which is a parameter of equations (1). The energy E_{iw} should be less than m for bound states, but it is noted similarly to energies E_p and E_k, which are more than m obviously, because the parameterization E_{iw} (with an imaginary index) as

$$E_{iw} = \sqrt{m^2 + (iw)^2} = \sqrt{m^2 - w^2};\ 0 \leq w \leq m \tag{3}$$

is very convenient. We also consider the modified Logunov-Tavkhelidze ($j = 3$) and modified Kadyshevsky ($j = 4$) equations, where [3, 4]

$$G_{(3)}(E_{iw}, E_p) = \frac{1}{E_{iw}^2 - E_p^2} \cdot \frac{E_p}{m};\ G_{(4)}(E_{iw}, E_p) = \frac{1}{2E_{iw} - 2E_p} \cdot \frac{1}{m}. \tag{4}$$

For the relativistic equations Fourier analysis is of no use, since the equations in the coordinate representation are integro-differential (non-local) [5]. In the first place, relativistic quasipotentials $V\left(\vec{p}, \vec{k}; E_{iw}\right)$ are not local, secondly, equation (1) contain square roots like $E_p = \sqrt{\vec{p}^{\,2} + m^2}$.

2. Relativistic configurational representation

A 4-momentum $(p_0, \vec{p})$ of a relativistic physical particle of mass m belongs to the mass hyperboloid $p_0^2 - \vec{p}^{\,2} = m^2$, $(p_0 > 0)$. As a consequence the three-dimensional momentum space is to be the Lobachevsky space [4], the volume element of it is given as $md\vec{k}/E_k$. This fact allows to proceed in equation (1) with Green functions $G_{(1)} \ldots G_{(4)}$ into the relativistic configurational representation (RCR) [4, 6]. The transition is carried out with the help of expansions of all magnitudes of the theory (wave functions, quasipotentials, amplitudes) on the matrix elements of the irreducible unitary representations of the Lorentz group, being group of movements of the Lobachevsky space, that is on functions [7]

$$\eta(\vec{r}, \vec{p}) = \left(\frac{E_p - \vec{n}\vec{p}}{m}\right)^{-1-imr};\ \vec{r} = r\vec{n},\ 0 \leq r < \infty,\ m > 0. \tag{5}$$

These functions are the relativistic analog of the plane waves $\exp(i\vec{x}\vec{p})$ and pass in them in the non-relativistic limit

$$\lim_{m \to \infty} \eta(\vec{r}, \vec{p}) = \exp(i\vec{p}\vec{r}). \tag{6}$$

The transformations (direct and inverse) for the relativistic wave function look like

$$\Psi(\vec{r}) = \frac{1}{(2\pi)^3} \int \eta(\vec{r}, \vec{p})\, \Psi(\vec{p})\, \frac{md\vec{p}}{E_p};\ \Psi(\vec{p}) = \int \eta^*(\vec{r}, \vec{p})\, \Psi(\vec{r})\, d\vec{r}. \tag{7}$$

The equations for $\Psi(\vec{r})$ are the simplest ones in the RCR if $V\left(\vec{p},\vec{k};E\right)$ is local in the momentum Lobachevsky space [5], namely

$$V\left(\vec{p},\vec{k};E\right) = V\left(\left|\vec{\Delta}_{p,k}\right|;E\right) = \int \eta^*\left(\vec{r},\vec{\Delta}_{p,k}\right) V(r,E)\, d\vec{r}$$
$$= \int \eta^*(\vec{r},\vec{p})\, V(r,E)\, \eta\left(\vec{r},\vec{k}\right) d\vec{r}, \tag{8}$$

$$V(r,E) = \frac{1}{(2\pi)^3}\int \eta\left(\vec{r},\vec{\Delta}\right) V\left(\left|\vec{\Delta}\right|,E\right)\frac{m d\vec{\Delta}}{\sqrt{\vec{\Delta}^2+m^2}}. \tag{9}$$

In the latter formulas and below $\vec{\Delta}_{p,k}$ is the difference of two vectors in Lobachevsky space [4, 6]

$$\vec{\Delta}_{p,k} = \vec{p} - \frac{\vec{k}}{m}\left[E_p - \frac{\vec{p}\vec{k}}{E_k+m}\right] = \overrightarrow{L_k^{-1}p}, \tag{10}$$

where L_k^{-1} is a pure Lorentz transformation (boost), so that $L_k^{-1}(k_0,\vec{k}) = (m,\vec{0})$.

In the RCR equations (1) become finite-difference equations [4, 6]. Another approach is based on using the Green functions in the RCR [8, 9] when equations in the RCR are written as integral ones. It is necessary to note, that these finite-difference (and integral as well) equations are much more complicated than the appropriate non-relativistic differential Schrödinger equation. At the same time the knowledge of exact solutions of relativistic equations for the bound state theory is as important as the knowledge of exact solutions of the similar problems in non-relativistic quantum mechanics.

In this connection it would be attractive to consider such potentials for which exact solutions of equations (1) can be directly obtained in the momentum representation. An energy independent quasipotential was firstly considered in paper [10], in the RCR it has the simple form

$$V(r) = -\frac{g^2}{r^2}. \tag{11}$$

The behaviour $V(r)$ at $r \to 0$ is similar to the behaviour of the massless boson exchange quasipotential [4]

$$V(r) = -e^2 m\frac{\coth \pi m r}{r}, \tag{12}$$

where e is the coupling constant. At large distances (and in the non-relativistic limit) potential (11) takes the Coulomb form: $V(r)|_{r\to\infty} \cong -e^2 m/r$. At the same time the behaviour of solutions of the relativistic equations with quasipotentials (11) and (12) for weekly connected systems should be similar.

3. Partial expansion

In the momentum representation potential (11) according to (8) has the form

$$V(\vec{p},\vec{k};E_{iw}) = V\left(\left|\vec{\Delta}_{p,k}\right|\right) = -2\pi^2\frac{g^2}{\left|\vec{\Delta}_{p,k}\right|}. \quad (13)$$

It was shown in paper [10] that in the elementary case of spherically symmetrical wave functions $\Psi(\vec{p}) = \Psi(|\vec{p}|)$, which appropriate to zero angular momentum, integral equation (1) with Green functions $G_{(1)}$, $G_{(2)}$ and with potential (13) can be reduced to a differential one. Now we consider the general case of wave functions of any angular momentum, for which

$$\Psi(\vec{p}) = \Psi_l(p)Y_{lm}(\vec{n}_p); \qquad p = |\vec{p}| = \sqrt{\vec{p}^2}; \qquad \vec{n}_p = \vec{p}/p, \quad (14)$$

where $Y_{lm}(\vec{n}_p)$ are spherical harmonics. Decomposing the quasipotential V on spherical harmonics

$$\frac{1}{(2\pi)^3}V\left(\left|\vec{\Delta}_{p,k}\right|\right) = \sum_{n=0}^{\infty}\sum_{\mu=-n}^{n}\frac{1}{pk}V_n(p,k)Y_{n\mu}(\vec{n}_p)\,Y^*_{n\mu}(\vec{n}_k), \quad (15)$$

we have for the partial potentials $V_n(p,k)$ the expression

$$V_n(p,k) = \frac{pk}{(2\pi)^3}\int_{-1}^{1} V\left(\left|\vec{\Delta}_{p,k}\right|\right) P_n(\cos\theta_{p,k})\, d\cos\theta_{p,k}, \quad (16)$$

where $\theta_{p,k}$ is the angle between vectors $\vec{p}$ and $\vec{k}$, and P_n are Legendre polynomials.

Substituting (14) and (15) into equation (1) and entering functions

$$F_{(j)l}(p) = G^{-1}_{(j)}(E_{iw},E_p)\cdot p\Psi_l(p), \quad (17)$$

we obtain the following one-dimensional integral equation:

$$F_{(j)l}(p) = \int_0^{\infty} V_l(p,k)G_{(j)}(E_{iw},E_k)F_{(j)l}(k)m\,dk/E_k. \quad (18)$$

For a further investigation of this equation it is convenient to enter rapidities χ_p and χ_k, appropriated to momenta $\vec{p}$ and $\vec{k}$, which are determined by ratios

$$E_p = \sqrt{m^2+p^2} = m\cosh\chi_p; \qquad \vec{p} = \vec{n}_p m\sinh\chi_p, \quad (19)$$

and similarly for $\vec{k}$.

In spherically symmetrical case the potential $V_0(p,k)$ can be easily determined from (16) [10], it is given by

$$V_0(p,k) = -g^2m\left\{\theta(\chi_p-\chi_k)\chi_k + \theta(\chi_k-\chi_p)\chi_p\right\}, \quad (20)$$

where $\theta(\chi)$ is the Heaviside step-function. Generally for the partial potential appropriated to angular momentum l formulas (13) and (16) yield

$$V_l(p,k) = -\frac{g^2}{2} mpk \int_{-1}^{1} \frac{P_l(z)dz}{\sqrt{(E_pE_k - pkz)^2 - m^4}}, \tag{21}$$

where $z = \cos\theta_{p,k}$. After the substitution $m^2 \cosh x = E_pE_k - pkz$ this integral can be written in the form

$$V_l(p,k) = -\frac{g^2 m}{2} \int_{x_-}^{x_+} P_l\left(\coth\chi_p \coth\chi_k - \frac{\cosh x}{\sinh\chi_p \sinh\chi_k}\right) dx, \tag{22}$$

where the limits of integration are $x_\pm = |\chi_p \pm \chi_k|$. Finally for the partial potentials we have the expression

$$\begin{aligned} V_l(p,k) = & - g^2 m \{\theta(\chi_p - \chi_k) P_l(\coth\chi_p) Q_l(\coth\chi_k) \\ & + \theta(\chi_k - \chi_p) P_l(\coth\chi_k) Q_l(\coth\chi_p)\}, \end{aligned} \tag{23}$$

in which $P_l(Q_l)$ are the Legendre functions of the first (second) kind [11].

4. Sturm-Liouville problem

Note some properties of the functions

$$u_l^{(1)}(\chi) = P_l(\coth\chi); \qquad u_l^{(2)}(\chi) = Q_l(\coth\chi). \tag{24}$$

At first, both of them satisfy the following second-order differential equation

$$\left\{\frac{d^2}{d\chi^2} - \frac{l(l+1)}{\sinh^2\chi}\right\} u_l^{(l)}(\chi) = 0. \tag{25}$$

Secondly, their wronskian is

$$W\left(u_l^{(1)}(\chi), u_l^{(2)}(\chi)\right) = \begin{vmatrix} u_l^{(1)}(\chi) & u_l^{(2)}(\chi) \\ \frac{d}{d\chi}u_l^{(1)}(\chi) & \frac{d}{d\chi}u_l^{(2)}(\chi) \end{vmatrix} = 1. \tag{26}$$

Using the changing $F_{(j)l}(p) = f_{(j)l}(\chi_p)$ equation (18) with the partial potential (23) can now be presented as

$$\begin{aligned} f_{(j)l}(\chi_p) = \int_0^\infty & \left\{\theta(\chi_p - \chi_k) u_l^{(1)}(\chi_p) u_l^{(2)}(\chi_k) + \theta(\chi_k - \chi_p) u_l^{(1)}(\chi_k) u_l^{(2)}(\chi_p)\right\} \\ & \times \left(-g^2 m^2\right) G_{(j)}\left(E_{i\omega}, m\cosh\chi_k\right) f_{(j)l}(\chi_k) d\chi_k. \end{aligned} \tag{27}$$

It is not difficult to show that integral equation (27) is equivalent to Sturm-Liouville problem consisting of the differential equation

$$\left\{-\frac{d^2}{d\chi^2} + \frac{l(l+1)}{\sinh^2\chi} + g^2 m^2 G_{(j)}\left(E_{i\omega}, m\cosh\chi\right)\right\} f_{(j)l}(\chi) = 0 \tag{28}$$

and boundary conditions

$$\lim_{\chi\to 0} W\left(u_l^{(2)}(\chi), f_{(j)l}(\chi)\right) = 0,$$

$$\lim_{\chi\to \infty} W\left(u_l^{(1)}(\chi), f_{(j)l}(\chi)\right) = 0, \tag{29}$$

where W denotes wronskian.

Equation (28) written in the momentum representation (rapidity χ is connected to a momentum by a ratio such as (19)) is similar to the partial Schrödinger equation in the coordinate space, however does not coincide with it. The term $l(l+1)/\sinh^2\chi$ plays the role of a centrifugal potential in (28), while in Schrödinger equation the centrifugal potential looks like $l(l+1)/\chi^2$. They coincide at $\chi \to 0$ or at $l = 0$ only. The expression

$$V_{(j)}^{ef}(\chi) = g^2 m^2 G_{(j)}\left(E_{iw}, m\cosh\chi\right) \tag{30}$$

plays the role of an effective non-relativistic potential. The appropriate effective non-relativistic energy is equal to zero, that is possible only for some relation between parameters g^2 and E_{iw} of the effective potential.

Denoting

$$w = m\sin\xi; \qquad E_{iw} = m\cos\xi, \tag{31}$$

we obtain the following expressions for effective potentials in four considered equations:

$$V_{(1)}^{ef}(\chi) = \frac{g^2}{\cos^2\xi - \cosh^2\chi}; \qquad V_{(2)}^{ef}(\chi) = \frac{g^2}{2\cosh\chi(\cos\xi - \cosh\chi)};$$

$$V_{(3)}^{ef}(\chi) = \frac{g^2\cosh\chi}{\cos^2\xi - \cosh^2\chi}; \qquad V_{(4)}^{ef}(\chi) = \frac{g^2}{2\cos\xi - 2\cosh\chi}. \tag{32}$$

In some cases Sturm-Liouville problem (28), (29) obtained can be solved exactly. Let us consider, for example, the Logunov-Tavkhelidze ($j = 1$) and Kadyshevsky ($j = 2$) equations in so-called chiral limit [10], when $E_{iw} = 0$. Using for $j = 1, 2$ the notation $g^2 = j\nu(\nu+1)$ we obtain for functions $f_{(1)l}(\chi)$ and $f_{(2)l}(\chi)$ the identical equation

$$\left\{\frac{d^2}{d\chi^2} + \frac{l(l+1)}{\sinh^2\chi} + \frac{\nu(\nu+1)}{\cosh^2\chi}\right\} f_l(\chi) = 0. \tag{33}$$

Using the replacement $t = \cosh^2\chi$ and the substitution

$$f_l(\chi) = t^{\frac{\nu+1}{2}}(t-1)^{\frac{l+1}{2}} v_l(t), \tag{34}$$

we can find the general solution of equation (33) as

$$\begin{aligned} f_l(\chi) = {} & A(\cosh\chi)^{\nu+1}(\sinh\chi)^{l+1}\ {}_2F_1\left(\tfrac{l+\nu}{2}+1, \tfrac{l+\nu}{2}+1; \nu+\tfrac{3}{2}; \cosh^2\chi\right) + \\ & + B(\cosh\chi)^{-\nu}(\sinh\chi)^{l+1}\ {}_2F_1\left(\tfrac{l-\nu+1}{2}, \tfrac{l-\nu+1}{2}; \tfrac{1}{2}-\nu; \cosh^2\chi\right), \end{aligned} \tag{35}$$

where ${}_2F_1$ is the hypergeometric function [11], and A, B are arbitrary constants. Taking into account boundary conditions (29) we get as a result that $A = 0$ and

$$\frac{l-\nu+1}{2} = -n; \ n \in Z. \tag{36}$$

Thus for the case $E_{iw} = 0$ both solutions of the Sturm-Liouville problem (28), (29) and exact solutions of integral equation (27) for $j = 1, 2$ are only possible if

$$g^2 = j(2n+l+1)(2n+l+2); \ n \in Z. \tag{37}$$

They are given by

$$f_{nl}(\chi) = B(\cosh\chi)^{-(2n+l+1)}(\sinh\chi)^{l+1} \ {}_2F_1\left(-n, -n, -\frac{1}{2} - l - 2n, \cosh^2\chi\right). \tag{38}$$

Using (17) it is also easy to find relativistic wave functions in the momentum representation. Then from (7) one can determine wave functions in the RCR, especially for small values of l and n. Thereby solutions of the finite-difference equations mentioned as well as the integral equations in RCR can be found too. The constant B is not determined from the equations and can be found only from a normalization condition.

The author would like to thank T.A. Alferova and V.V. Kondratjuk for useful discussions.

References

[1] A.A. Logunov, A.N. Tavkhelidze, Quasi-Optical Approach in Quantum Field Theory, *Nuovo Cimento*, **V.29**, N2, 380–399, 1963.

[2] V.G. Kadyshevsky, Quasipotential type equation for the relativistic scattering amplitude, *Nucl.Phys.*, **V.B6**, N1, 125–148, 1968.

[3] V.A. Rizov, I.T. Todorov, Quasipotential approach to the problem of bound states in quantum electrodynamics, *Particles and Nucleus*, **V.6**, N3, 669–742, 1975.

[4] V.G. Kadyshevsky, R.M. Mir-Kasimov, N.B. Skachkov, Three-dimensional formulation of the relativistic two-body problem, *Particles and Nucleus*, **V.2**, N3, 635–690, 1972.

[5] A.A. Arhipov, V.I. Savrin, A method for solutions of the quasipotential equation, *Teret. Mat. Fiz.*, **V.53**, N3, 342–357, 1982.

[6] V.G. Kadyshevsky, R.M. Mir-Kasimov, N.B. Skachkov, Quasipotential Approach and the Expansion in Relativistic Spherical Functions, *Nuovo Cimento*, **V.55A**, N2, 233–257, 1968.

[7] I.S. Shapiro, The expansion of the wave function on the irreducible representations of the Lorentz group, *Docl. Akad. Nauk SSSR*, **V.106**, N4, 647–649. 1956.

[8] V.N. Kapshai, T.A. Alferova, Relativistic two-particle one-dimensional scattering problem for superposition of δ-potentials, *J. Phys. A: Math. Gen.*, **V.32**, 5329–5342, 1999.

[9] V.N. Kapshai, T.A. Alferova, Green Functions and Wave Functions of Relativistic Two-Particle Equations for States of Discrete and Continuous Spectrum, Proceed. of the Seventh Annual Seminar NPCS / Academy of Sciences of Belarus. Inst. of Phys., Minsk, 57–63, 1999.

[10] V.N. Kapshai, S.P. Kuleshov, N.B. Skachkov, On a class of exact solutions to quasipotential equations, *Teoret. Mat. Fiz.*, **V.55**, N3, 349–360, 1983.

[11] H. Bateman, A. Erdely, Higher transcendental functions, New York, V.1, 1953.

V. Kapshai
Department of Theoretical Physics
Gomel State University
Gomel, Belarus
e-mail: `kvn@gsu.unibel.by`

Operator Theory:
Advances and Applications, Vol. 132, 207–217

The Non-relativistic Scattering Problem for a Superposition of δ-potentials

V. Kapshai, T. Alferova, and N. Elander

Abstract. The two-body scattering problem is solved exactly for a single and a superposition of two δ-potentials. Conditions for the existence of bound states and resonances are investigated for these kinds of potentials.

1. Introduction

Models of point (or contact) interaction in non-relativistic quantum mechanics have recently been a popular topic (see monographs [1]–[4]) since using δ interactions makes these models exactly solvable. A δ interaction centered on a sphere of radius R in three dimensions, formally given by the Hamiltonian $H = -\Delta + \alpha\delta(|\mathbf{x}| - R)$, has also attracted a lot of attention. The first more complete analysis of δ-sphere interaction from physical point of view has been given by Romo [5] and Kok *et al.* [6]. A systematic mathematical investigation of such kinds interactions was given by Antoine *et al.* in [7]. A mathematical treatment of the Schrödinger equation for many sphere interactions, described by the Hamiltonian $H = -\Delta + \sum_{j=1}^{N} \alpha_j \delta(|\mathbf{x}| - R_j)$, can be found in the paper by Hounkonnou *et al* [8] and in references therein.

Two of the authors have recently studied one-dimensional relativistic integral equations for a superposition of two δ-potentials [9]. We shall not go into details concerning those quasipotential relativistic equations [10, 11]. Let us just note here that no differential equations can be derived in the relativistic configurational representation [12]. This implies that we need good methods for solving integral equations. Our immediate focus is thus to solve the three-dimensional **non-relativistic integral** and then to apply obtained methods to treat **relativistic integral** equations with superimposed δ sphere interactions.

The three-dimensional non-relativistic integral equation for the potential scattering partial wave function is well known [13] as

$$\psi_{l,p}(r) = \hat{j}_l(pr) + \int_0^\infty dr' G_{l,p}^{(0)}(r, r') U(r') \psi_{l,p}(r') . \tag{1}$$

The second author is supported by The Swedish Institute.

Here $U(r) = 2mV(r)$ where $V(r)$ is the interaction potential and m is reduced mass (here and further we use the atomic units $m = 1$). The Green's function $G^{(0)}_{l,p}(r,r')$ is expressed via Riccati-Bessel $\hat{j}_l(z)$ and Riccati-Hankel $\hat{h}^+_l(z)$ functions as

$$G^{(0)}_{l,p}(r,r') = -\frac{1}{p}\hat{j}_l(pr_<)\,\hat{h}^+_l(pr_>), \tag{2}$$

$r_<$ and $r_>$ are respectively the smaller and larger of r and r'.

When considering the bound states we make the change of variables

$$p = \sqrt{2mE} \to iw = i\sqrt{2m|E|}. \tag{3}$$

The non-relativistic integral equation for bound states thus has the form

$$\psi_{l,iw}(r) = \int_0^\infty dr' G^{(0)}_{l,iw}(r,r')\,U(r')\,\psi_{l,iw}(r'). \tag{4}$$

Resonances are one of the most interesting phenomena in scattering. Defined as purely outgoing states they play an important part of the description of reactions all the way from chemical, molecular, atomic nuclear to particle physics. A resonance is some sense a hybrid of bound state and a pure scattering state. It is localized as a bound state and is not integrable as pure scattering states. Experimentally one may associate the resonance conception with dramatic variation of the form of the total or differential cross section. One approach when investigating resonances is to examine the S-matrix for existence of poles in the lower half plane $\{\mathrm{Im}\, p < 0\}$. Usually we look for poles of the S-matrix as zeros of the Jost function but we can obtain the same result by considering the partial-wave amplitude $f_l(p)$

$$f_l(p) = \frac{1}{p}e^{i\tilde{\delta}_l(p)}, \tag{5}$$

$$S_l(p) = e^{2i\tilde{\delta}_l(p)} = \frac{f_l(p)}{f^*_l(p)}, \tag{6}$$

where $\tilde{\delta}_l(p)$ is the phase shift. The integral representation for the exact amplitude (5) is

$$f_l(p) = -\frac{1}{p^2}\int_0^\infty dr\ \hat{j}_l(pr)\,U(r)\,\psi_{l,p}(r). \tag{7}$$

2. Bound states, resonances and poles of the S-matrix for a single δ potential

Let us consider equation (1) with a single δ-potential

$$U(r) = \frac{U_0}{a}\delta(r-a);\ (U_0, a) \in \mathrm{R}, a > 0 \tag{8}$$

describing δ interactions centered on a sphere of radius $a > 0$. In this case the solution of (1) can be given as

$$\psi_{l,p}(r) = \hat{j}_l(pr) + G^{(0)}_{l,p}(r,a)\,U_0 \frac{\hat{j}_l(pa)}{a - G^{(0)}_{l,p}(a,a)\,U_0}. \tag{9}$$

Note that the radial wave function (9) obtained in this way satisfies the radial Schrödinger equation

$$\left[\frac{d^2}{dr^2} - \frac{l(l+1)}{r^2} + E\right]\psi_{l,p}(r) = \frac{U_0}{a}\delta(r-a)\psi_{l,p}(r) \tag{10}$$

with the boundary conditions:

$$\psi_{l,p}(0) = 0, \tag{11}$$

$$\psi_{l,p}(r)|_{r\to\infty} \to \frac{i}{2}\left[\hat{h}^-_l(pr) - S_l(p)\hat{h}^+_l(pr)\right]. \tag{12}$$

Integration of eq. (10) across the delta-function wall, $\psi_{l,p}(r)$ being continuous, yields

$$\lim_{\varepsilon\to 0+} \left.\frac{d\psi_{l,p}(r)}{dr}\right|_{a-\varepsilon}^{a+\varepsilon} = U_0\psi_{l,p}(a). \tag{13}$$

If ε is a small positive number we have

$$\psi_{l,p}(a+\varepsilon) = \psi_{l,p}(a-\varepsilon) \equiv \psi_{l,p}(a). \tag{14}$$

The partial wave amplitude (7) is then written as

$$f_l(p) = -\frac{\hat{j}_l(pa)^2}{p^2} \cdot \frac{U_0}{a - G^{(0)}_{l,p}(a,a)\,U_0}. \tag{15}$$

Considering eq.(6) and (15) we can see that the partial wave S-matrix has poles either as zeros of the denominator or poles of the numerator. We only consider those S-matrix poles which are zeros of the denominator. While the partial wave S-matrix has the form

$$S_l(p) = \frac{pa + U_0\hat{h}^-_l(pa)\,\hat{j}_l(pa)}{pa + U_0\hat{h}^+_l(pa)\,\hat{j}_l(pa)}. \tag{16}$$

In order to determine the resonances we thus only consider the condition

$$pa + U_0\hat{h}^+_l(pa)\,\hat{j}_l(pa) = 0. \tag{17}$$

Since the Riccati-Bessel function $\hat{j}_l(pa)$ can be expressed in the Riccati-Hankel functions $\hat{h}^\pm_l(pa)$ as

$$\hat{j}_l(pa) = \frac{i}{2}\left[\hat{h}^-_l(pa) - \hat{h}^+_l(pa)\right] \tag{18}$$

we can rewrite (17) in the form

$$2pa + iU_0\left[\hat{h}^-_l(pa) - \hat{h}^+_l(pa)\right]\hat{h}^+_l(pa) = 0. \tag{19}$$

2.1. Zero angular momentum ($l = 0$)

If the angular momentum is equal to zero the condition (19) is given by

$$2pa + iU_0\left[1 - \exp\left(2ipa\right)\right] = 0. \tag{20}$$

Now we can analytically continue the latter result into the complex plane p as

$$2pa = Z = x + iy;\ (x, y) \in \mathrm{R}. \tag{21}$$

Condition (20) is satisfied when both the real and imaginary parts are equal to zero as in

$$\begin{cases} x + U_0 e^{-y} \sin x = 0 \\ y + U_0\left(1 - e^{-y}\cos x\right) = 0. \end{cases} \tag{22}$$

Let x in (22) be equal to zero and $y > 0$ (p is purely imaginary), then bound states exist if U_0 satisfies

$$U_0 = -\frac{y}{1 - \exp\left(-y\right)} \leq -1. \tag{23}$$

If $U_0 > 0$ and $x \neq 0$ there are no poles in the upper half of the p-plane.

For $|U_0| \to \infty$ ($|U_0| \gg 1$, i.e. a very strong repulsive or a strong attractive potential) the system (22) can in this case be simplified to

$$\begin{cases} \sin x \cong -\frac{1}{U_0} x \\ \cos x \cong 1. \end{cases} \tag{24}$$

2.2. Non-zero angular momentum ($l = 1, 2$)

We can here analogously obtain the corresponding results for the non-zero angular momentum. However the analog of the condition (20) is much more complicated. Limiting ourselves to $l = 1$ with the notation (21) the analogue of eq. (20) is

$$Z + iU_0\left[1 + \frac{4}{Z^2} + \left(1 + \frac{2i}{Z}\right)^2 e^{iZ}\right] = 0. \tag{25}$$

Bound states furthermore exist if the parameter U_0 satisfies the condition

$$U_0 = -\frac{y^3}{y^2 - 4 + \left(y + 2\right)^2 \exp\left(-y\right)} \leq -3. \tag{26}$$

For $l = 2$ the condition for S-matrix poles are obtainable from

$$Z + iU_0\left[1 + \frac{12}{Z^2} + \frac{144}{Z^4} - \left(1 + \frac{6i}{Z} - \frac{12}{Z^2}\right)^2 e^{iZ}\right] = 0 \tag{27}$$

while the condition for the existence of bound states is

$$U_0 = -\frac{y^5}{y^4 - 12y^2 + 144 - \left(y^2 + 6y + 12\right)^2 \exp\left(-y\right)} \leq -5. \tag{28}$$

In Figure 1 we present results of numerical calculations of the system of the three transcendental equations (22), (25), (27) in the case of single δ-potential (8). As above discussed [5]–[7] and illustrated in figures there is an infinite number

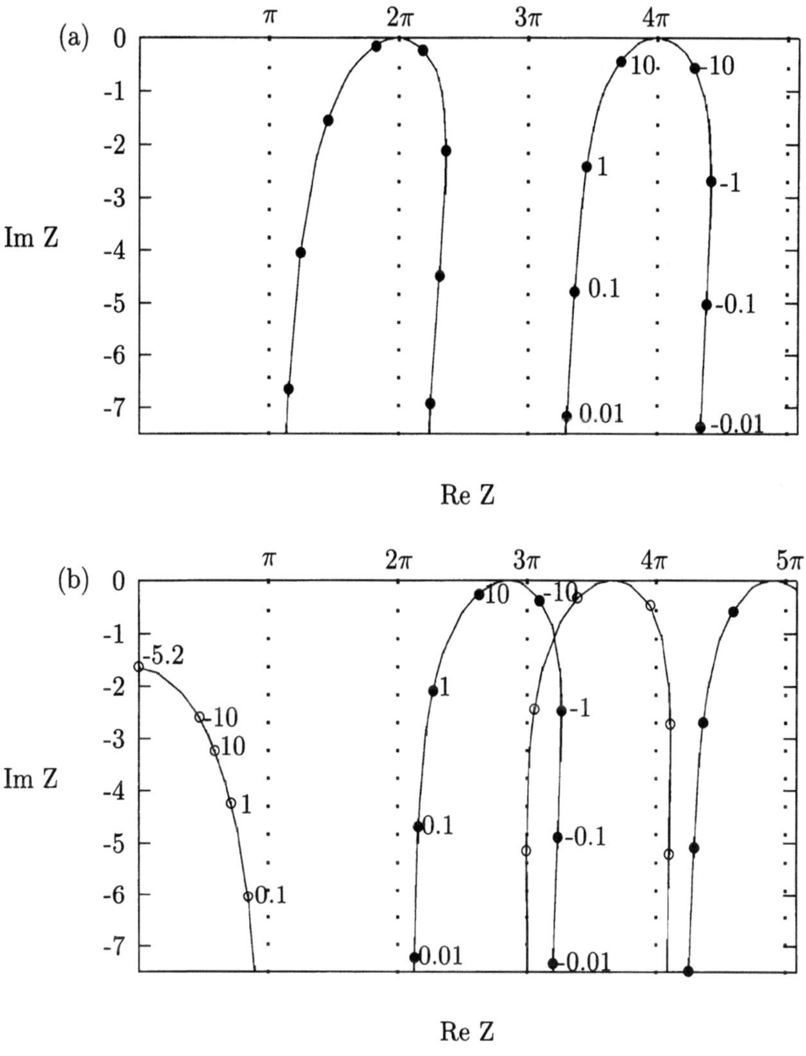

FIGURE 1. *Trajectories of poles of the partial wave S-matrix in the complex plane $Z = 2pa$ for a single δ-potential with $a = 1$. The numbers along the curves correspond to the value of the potential strength U_0:* **(a)** *$l = 0$;* **(b)** *$l = 1$ ($\bullet$) and $l = 2$ ($\circ$).*

of resonances off the imaginary axis for both $Re\ p > 0$ and $Re\ p < 0$ (we present results only for $Re\ p > 0$):

(i) for a weak repulsive potential the poles located quite far from the real axis (physical region) and can be found close the limiting points

$$2p_n a = (2n + l - 1)\pi - i\infty,\ n = 1, 2, \ldots .$$

In addition there is an exception for $l = 2$. The poles along the first trajectory locate close to the limit $2pa = \pi - i\infty$;

(ii) if the value U_0 increases then $y = \text{Im}\ Z$ decreases implying that the partial wave S-matrix has poles close to the real Z-axis. From (24) it is easy to find the limiting points in the case of $l = 0$ as

$$2p_n a = 2\pi n,\ n = 1, 2, \ldots$$

In the case of $l = 2$ poles of the first trajectory for a very strong potential ($|U_0| >> 1$) locate near the limiting point ($Re\ Z \cong 1.732$, $Im\ Z \cong -3$);

(iii) for a weak attractive potential the poles move to the right and down and their trajectories asymptotically approach the points

$$2p_n a = (2n + l)\pi - i\infty,\ n = 1, 2, \ldots$$

For $l = 2$ poles of the first trajectory move from the point ($Re\ Z \cong 1.732$, $Im\ Z \cong -3$) to the left and up toward the imaginary axis. For $U_0 = -5.195$ the pole locate at the point ($Re\ Z \cong 0$, $Im\ Z \cong -1.65$) this corresponds to the virtual state.

3. Superposition of two δ-potentials

Let us now consider a superposition of two δ-potentials (U_1, U_2, a, a_1, a_2 are real) such that

$$U(r) = \frac{U_1}{a}\delta(r - a_1) + \frac{U_2}{a}\delta(r - a_2). \tag{29}$$

Combining eq.(1) and (29) we obtain the wave function

$$\psi_{l,p}(r) = \hat{j}_l(pr) + \sum_{s=1}^{2} G_{l,p}^{(0)}(r, a_s)\frac{U_s}{a}\psi_{l,p}(a_s). \tag{30}$$

The constants $\psi_{l,p}(a_s)$ can be found in matrix form as

$$\psi_{l,p}(a_s) = \left[M_{s,k}^{(l)}(p)\right]^{-1}\hat{j}_l(pa_k),\ (k = 1, 2), \tag{31}$$

where

$$M_{s,k}^{(l)}(p) = \left[\delta_{s,k} - G_{l,p}^{(0)}(a_s, a_k)\frac{U_s}{a}\right]. \tag{32}$$

The partial wave amplitude is now defined as

$$f_l(p) = -\frac{1}{p^2}\sum_{s,k=1}^{2}\hat{j}_l(pa_s)\frac{U_s}{a}\left[M_{s,k}^{(l)}(p)\right]^{-1}\hat{j}_l(pa_k). \tag{33}$$

This yields the partial wave S-matrix

$$S_l(p) = \frac{\sum_{s,k=1}^{2}\hat{j}_l(pa_s)U_s\left[M_{s,k}^{(l)}(p)\right]^{-1}\hat{j}_l(pa_k)}{\sum_{s,k=1}^{2}\hat{j}_l(pa_s)U_s\left[M_{s,k}^{*(l)}(p)\right]^{-1}\hat{j}_l(pa_k)}. \tag{34}$$

The S-matrix has poles if condition (35) is satisfied

$$\det\left[M^{(l)}(p)\right] = \prod_{s=1}^{2}\left[a - G_{l,p}^{(0)}(a_s,a_s)\,U_s\right] - \left[G_{l,p}^{(0)}(a_1,a_2)\right]^2 U_1U_2 = 0. \quad (35)$$

Using Hankel functions eq. (35) can be written to

$$\begin{aligned} &\left\{2pa + i\left[\hat{h}_l^-(pa_1) - \hat{h}_l^+(pa_1)\right]\hat{h}_l^+(pa_1)\,U_1\right\}\\ \times\ &\left\{2pa + i\left[\hat{h}_l^-(pa_2) - \hat{h}_l^+(pa_2)\right]\hat{h}_l^+(pa_2)\,U_2\right\} \qquad (36)\\ +\ &\left\{\left[\hat{h}_l^-(pa_1) - \hat{h}_l^+(pa_1)\right]\hat{h}_l^+(pa_2)\right\}^2 U_1U_2 = 0. \end{aligned}$$

Bound states exist in this case if condition (36) is restricted to $p = iw$ ($w \in R$), i.e. the S-matrix only has poles on the imaginary axis in the upper half p-plane.

Using the eq. (21) and taking into account that $a_1 = \mu_1 a$, $a_2 = \mu_2 a$ (μ_1, μ_1 are real) we can rewrite (36) for $l = 0$ as

$$\begin{aligned} &Z^2 + iZ\left[1 - e^{i\mu_1 Z}\right]U_1 + iZ\left[1 - e^{i\mu_2 Z}\right]U_2\\ &-\left[1 - e^{i\mu_1 Z} + e^{i\mu_2 Z} - e^{i(\mu_2-\mu_1)Z}\right]U_1U_2 = 0. \qquad (37) \end{aligned}$$

For $l = 1$ we obtain

$$\begin{aligned} &\left\{Z + iU_1\left[1 + \left(\frac{2}{\mu_1 Z}\right)^2 + \left(1 + \frac{2i}{\mu_1 Z}\right)^2 \exp(i\mu_1 Z)\right]\right\}\\ \times&\left\{Z + iU_2\left[1 + \left(\frac{2}{\mu_2 Z}\right)^2 + \left(1 + \frac{2i}{\mu_2 Z}\right)^2 \exp(i\mu_2 Z)\right]\right\}\\ +&\left\{\left[1 + \frac{2i}{Z}\left(\frac{1}{\mu_2} - \frac{1}{\mu_1}\right) + \frac{4}{\mu_1\mu_2 Z^2}\right]\exp\left[\frac{i}{2}(\mu_2-\mu_1)Z\right]\right.\\ +&\left.\left[1 + \frac{2i}{Z}\left(\frac{1}{\mu_2} + \frac{1}{\mu_1}\right) - \frac{4}{\mu_1\mu_2 Z^2}\right]\exp\left[\frac{i}{2}(\mu_2+\mu_1)Z\right]\right\}^2 U_1U_2 = 0. \qquad (38) \end{aligned}$$

The numerical solutions of equations (36) for $l = 0, 1, 2$ and for equal values of parameters U_1 and U_2 ($U_1 = U_2 = U_0$) are given in Figure 2–Figure 4. Two δ-potentials produce new poles. For a strong potential ($|U_0| >> 1$) they look like double poles. Comparing locations of the poles for a single δ-potential with ones for a superposition of two δ-potentials (U_0 is fixed and the same in the both cases) we note that in the second case the poles locate two times closer to the real axis.

In Figure 2–Figure 4 one can see that

(i) for a weak repulsive potential ($U_0 > 0$, $U_0 << 1$) the poles move close to the limiting points

$$2p_n a = (2n + l - 1)\frac{\pi}{2} - i\infty,\ n = 1, 2, \ldots;$$

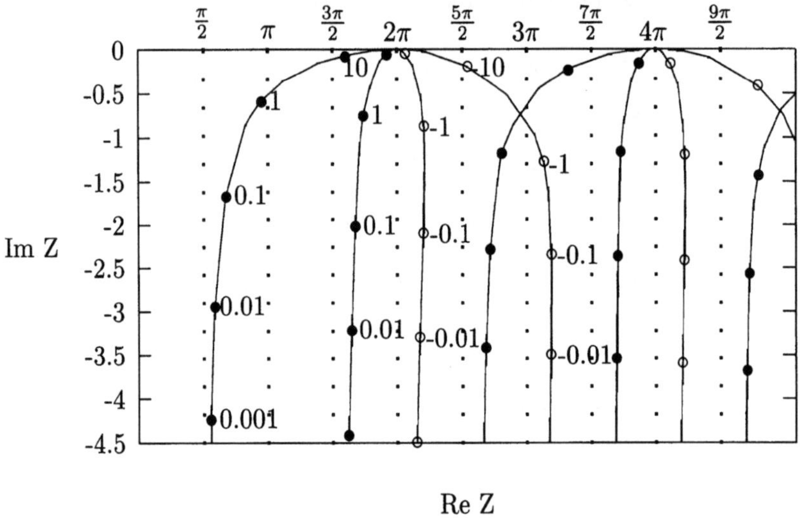

FIGURE 2. *Trajectories of poles of the partial wave S_0-matrix in the complex plane $Z = 2pa$ in the case of a superposition of two δ-potentials ($U_1 = U_2 = U_0$), $\mu_1 = 1$, $\mu_2 = 2$, $a = 1$. The numbers along the curves correspond to the value of the potential strength U_0.*

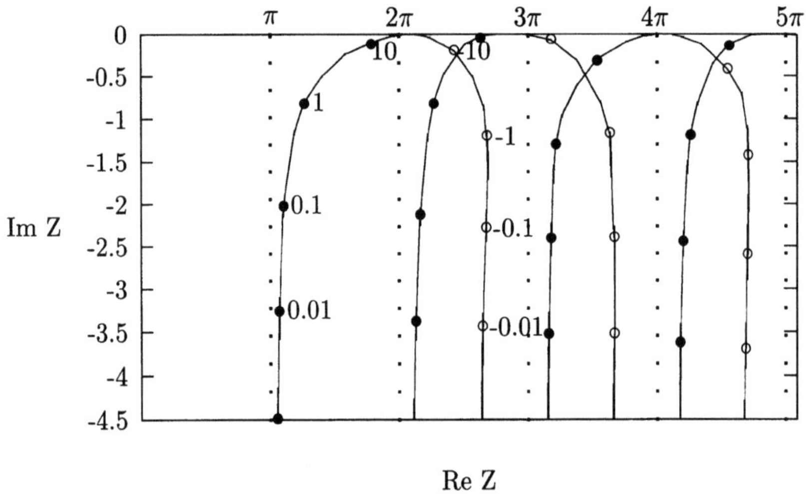

FIGURE 3. *Same as Figure 2, for $l = 1$.*

in the case when $l = 2$ there is another trajectory for poles being close to the limit $2pa = \frac{\pi}{2} - i\infty$;

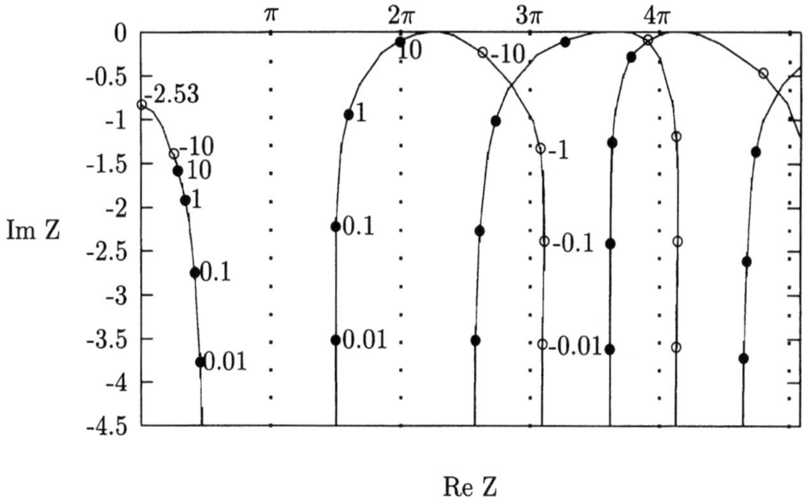

FIGURE 4. *Same as Figure 2, for $l = 2$.*

(ii) for a strong potential the double poles of S_0-matrix for both repulsive and attractive potentials move to the limiting points

$$2p_n a = 2\pi n, \; n = 1, 2, \ldots$$

like for a single δ-potential (see Figure 1(a)), for $l > 0$ it is not too easy to determine the locations of poles. In the case $l = 2$ the poles along the first trajectory do not move to the real axis but to the point (*Re* $Z \cong 0.866$, *Im* $Z \cong -1.5$);

(iii) for a weak attractive potential ($U_0 < 0$, $|U_0| << 1$) the poles reach the limit

$$2p_n a = (2n + 2 + l)\frac{\pi}{2} - i\infty, \; n = 1, 2, \ldots,$$

except for $l = 2$ since there are poles along the first trajectory which move from the point (*Re* $Z \cong 0.866$, *Im* $Z \cong -1.5$) to the left and up towards the imaginary axis and with $U_0 \cong -2.526$ reach the point (*Re* $Z \cong 0$, *Im* $Z \cong -0.841$).

4. Conclusion

We have here demonstrated how the solution of the commonly used integral equations (1), (4) and (7) are drastically simplified by using superposed δ-potentials into simple analytically expressions like eq. (15) and (16) and generalizations (33), (34). They are evaluated by inserting appropriate Riccati-Bessel or Riccati-Hankel functions. Conditions for the existence of partial wave S-matrix poles and thereby resonances are obtained.

The strength of the present approach is twofold. First, it attacks the potential scattering problem from a completely new point of view, avoiding any kind of numerical integration, yet yielding a scattering wave function. Secondly, the present method can be useful when solving three-dimensional relativistic scattering problem for the partial wave function where only the integral equation and no differential equation is obtainable.

References

[1] A.I. Baz, Y.B. Zeldovich and A.M. Perelomov, Scattering, Reactions and Decay in non-relativistic Quantum Mechanics, Moscow: Nauka (In Russian), 1971.

[2] Yu.N. Demkov and V.N. Ostrovsky, Zero-range potentials and their applications in atomic physics, Plenum Press, New York and London, 1988.

[3] S. Albeverio, F. Gesztesy, R. Høegh-Krohn and H. Holden, Solvable models in quantum mechanics, Springer, 1988.

[4] S. Albeverio and P. Kurasov, Singular perturbations of differential operators, London Mathem. Soc. Lecture Notes 271: Cambridge Univ. Press, 2000.

[5] W.J. Romo, Poles of the S-Matrix for a Complex Surface Delta Function Potential, *Can. J. Phys.* **52**, 1974, 1603–1614.

[6] L.P. Kok, J.W. de Maag and H.H. Brouwer, Formulas for the δ-shell-plus-Coulomb potential for all partial waves, *Phys. Rev. C* **26**(6), 1982, 2381–2396.

[7] J-P. Antoine, F. Gesztesy and J. Shabani, Exactly solvable models of sphere interactions in quantum mechanics, *J. Phys. A: Math. Gen.* **20**(12), 1987, 3687–3712.

[8] M.N. Hounkonnou, M. Hounkpe, J. Shabani, Scattering theory for finitely many sphere interactions supported by concentric spheres, *J. Math. Phys.* **38**, 1997, 2832–2850.

[9] V.N. Kapshai and T.A. Alferova, Relativistic two-particle one-dimensional scattering problem for superposition of δ-potentials, *J. Phys. A: Math. Gen.* **32**, 1999, 5329–5342.

[10] A.A. Logunov and A.N. Tavkhelidze, Quasi-Optical Approach in Quantum Field Theory, *Nuovo Cimento* **29**, 1963, 380–399.

[11] V.G. Kadyshevsky, Quasipotential type equation for the relativistic scattering amplitude, *Nucl. Phys.* **B6**, 1968, 125–148.

[12] V.N. Kapshai and N.B. Skachkov, Covariant two-particle wave functions for model quasipotentials that admit exact solutions. II. Solutions in the relativistic configurational representation, *Sov. Journ. Theor. Math. Phys.* **55** N1, 1983, 330–337.

[13] R.J. Taylor, Scattering theory, New York, London, Sydney, Toronto: John Wiley & Sons, Inc, 1972.

V. Kapshai
Department of Physics
Gomel State University
Gomel 246699, Belarus
e-mail: kvn@gsu.unibel.by

T. Alferova
Department of Physics
Stockholm University
Stockholm Center for Physics, Astronomy and Biotechnology
SE-106 91 Stockholm, Sweden
e-mail: alferova@physto.se

N. Elander
Department of Physics
Stockholm University
Stockholm Center for Physics, Astronomy and Biotechnology
SE-106 91 Stockholm, Sweden
e-mail: Nils.Elander@physto.se

Operator Theory:
Advances and Applications, Vol. 132, 219–231

On the Periodic Magnetic Schrödinger Operator in R^d. Eigenvalues and Model Functions

Yu.E. Karpeshina

Abstract. An asymptotic formula for a rich family of eigenvalues of the periodic magnetic Schrödinger operator is constructed in the high-energy region. Model functions, which solve the equation for eigenfunctions approximately are described explicitly. It is shown that the system of model functions is almost orthogonal and form a rich subspace in $L_2(R^d)$.

1. Introduction

We consider the operator

$$H = \sum_{j=1}^{d} \left(\frac{1}{i} \frac{\partial}{\partial x_j} + a_j(x) \right)^2 + \alpha a_0(x) \tag{1}$$

in $L_2(R^d)$, where $a_0(x)$, $a_1(x), \ldots, a_d(x)$ are real periodic potentials, α is a real parameter. We assume that the potentials have the same periods $\vec{b}_1, \ldots, \vec{b}_d$. For the sake of simplicity, we consider the periods to be orthogonal, however all the results are also valid for nonorthogonal periods. Without loss of generality we assume that the periods are directed along the coordinate axes and

$$\int_Q a_s(x)dx = 0, \quad s = 0, 1, \ldots, d, \quad Q = [0, b_1] \times \cdots \times [0, b_d].$$

We use the Fourier representation of the potentials:

$$a_s(x) = \sum_{m \in Z^d,\ m \neq 0} a_{sm} \exp i(\vec{p}_m(0), x), \quad s = 0, 1, \ldots, d,$$

where $(\cdot, \cdot)$ is the inner product in R^d and $\vec{p}_m(0)$ is a vector of the dual lattice:

$$\vec{p}_m(0) = 2\pi(m_1 b_1^{-1}, \ldots, m_d b_d^{-1}). \tag{2}$$

We use the notation $\vec{A}(x) = (a_0(x), a_1(x), \ldots, a_d(x))$. It is known (see e.g. [ReSi]) that spectral analysis of H can be reduced to analysis of a family of operators $H(t)$, $t \in K$, where K is the elementary cell of the dual lattice,

$$K = [0, 2\pi b_1^{-1}) \times \cdots \times [0, 2\pi b_d^{-1}).$$

Research partially supported by USNSF Grant DMS-9803498.

The vector t is called the *quasimomentum*. The operator $H(t)$, $t \in K$, acts in $L_2(Q)$. It is described by the formula (1) and the quasiperiodic conditions:

$$u(b_1, x_2, \dots, x_d) = \exp(it_1 b_1) u(0, x_2, \dots, x_d),$$

$$\dots\dots$$

$$u(x_1, \dots, x_{d-1}, b_d) = \exp(it_d b_d) u(x_1, \dots, x_{d-1}, 0). \tag{3}$$

The first derivatives with respect to $x_1, \dots, x_d$ have to satisfy the analogous conditions. Each operator $H(t)$, $t \in K$, has a discrete semibounded below spectrum $\Lambda(t)$:

$$\Lambda(t) = \cup_{n=1}^{\infty} \lambda_n(t), \ \lambda_n(t) \to_{n \to \infty} \infty.$$

The spectrum Λ of the operator H is the union of the spectra of the operators $H(t)$: $\Lambda = \cup_{t \in K} \Lambda(t) = \cup_{n \in N, t \in K} \lambda_n(t)$. The functions $\lambda_n(t)$ are continuous, so Λ has a band structure:

$$\Lambda = \cup_{n=1}^{\infty} [q_n, Q_n], \ q_n = \min_{t \in K} \lambda_n(t), \ Q_n = \max_{t \in K} \lambda_n(t).$$

It is proven in [BiSu, Sob] that the spectrum of operator H is absolutely continuous, i.e. $[q_n, Q_n]$ cannot degenerate to a point.

The eigenfunctions of $H(t)$ and H are simply related. Extending all the eigenfunctions of the operators $H(t)$ quasiperiodically (see (3)) to R^d, we obtain a complete system of generalized eigenfunctions of H.

In the case of $\vec{A} = 0$, the eigenvalues and eigenfunctions of the corresponding operators $H_0(t)$, $t \in K$, are naturally indexed by the points of Z^d: $\Psi_m^0(t, x) = \exp i(\vec{p}_m(t), x)$, $\lambda_m^0(t) = p_m^2(t)$, $m \in Z^d$, here and below $\vec{p}_m(t) = \vec{p}_m(0) + t$, $\vec{p}_m(0)$ being given by (2), and $p_m^2(t) = |\vec{p}_m(t)|^2$.

The goal of this paper is to construct an asymptotic formula in the high energy region for a rich set of eigenvalues of the operators $H(t)$, $t \in K$, and to construct explicitly model functions, which solve the equation for eigenfunctions approximately. We will show that the system of model functions is almost orthogonal and form a rich subspace in $L_2(R^d)$.

We consider the case of $\vec{A} = (a_0(x), a_1(x), \dots, a_d(x))$ being a trigonometric polynomial of the "length" R_0:

$$a_s(x) = \sum_{m \in Z^d, 0 < |m| \leq R_0} a_{sm} \exp i(\vec{p}_m(0), x), \quad s = 0, 1, \dots, d. \tag{4}$$

It turns out that the model functions of $H(t)$, $t \in K$, are not close to the unperturbed eigenfunctions $\exp i(\vec{p}_m(t), x)$. This makes the magnetic case different from the Schrödinger operator with only an electric potential ($a_s(x) = 0, s \geq 1$), where perturbed eigenfunctions are mostly close to unperturbed ones (see [K]). The new phenomena is due to the fact that the perturbation by a magnetic field is much stronger than the perturbation by an electric field only and, therefore, leads to a more complicated diffraction picture. The construction of the model functions includes the construction of approximate solutions of the Eikonal equation by perturbative methods.

Let us introduce some notations. First,

$$\vec{a}(x) = (a_1(x), \dots, a_d(x)).$$

Let $C^\infty(Q)$ be the class of infinitely differentiable functions periodic on Q (all the derivatives are also periodic.) Let $\|\cdot\|_\nu$, $\nu = 0, 1, \dots$ be the family of norms in $C_\infty(Q)$ defined as follows. For any $f \in C^\infty(Q)$,

$$\|f\|_\nu = |f_0| + \sum_{r \in Z^d \setminus \{0\}} |f_r||r|^\nu,$$

f_r being the Fourier coefficients of $f(x)$. If $\vec{f}(x)$ is a vector function: $f : Q \to \mathbf{C}^N$, $\vec{f}(x) = (f_1(x), \dots, f_N(x))$, then

$$\|\vec{f}\|_\nu = \sum_{s=1}^{N} \|f_s\|_\nu.$$

Obviously, $\|D^{\vec{\nu}} \vec{f}\|_0 \le \|\vec{f}\|_\nu$, $\vec{\nu} = (\nu_1, \dots, \nu_d)$, $\nu = \nu_1 + \dots + \nu_d$.

We denote by $\|L\|_*$ the norm of a matrix L:

$$2\|L\|_* = \sup_i \sum_j |L_{ij}| + \sup_j \sum_i |L_{ij}|. \tag{5}$$

Last, let $|Q| = b_1 \cdot \dots \cdot b_d$, $b_0 = \max\{2\pi/b_1, \dots, 2\pi/b_d, 1\}$. Further, $(\cdot, \cdot)$ is the scalar product in R^d, $< \cdot, \cdot >$ is the scalar product in $L_2(Q)$.

We denote by C an absolute constant independent of the parameters of H, by C_b a constant defined only by $b_1 \dots b_d$, by $C_{b,\vec{A}}$ a constant defined only by $b_1, \dots, b_d, \vec{A}$, etc.

2. Construction of Model Functions and an Asymptotic Formula for Eigenvalues

In this section we will prove that the model functions are almost orthogonal (Theorem 2) and form a rich subspace in $L_2(Q)$ (Theorem 3).

Let k be a large parameter and $Z_0(k, \beta, \varepsilon)$ be the set of $j \in Z^d$ satisfying the conditions:

(1) $$k/4 < p_j(t) < 7k/4. \tag{6}$$

(2) For any $q \in Z^d : 0 < |q| \le k^\varepsilon$:

$$|(\vec{p}_j(t), \vec{p}_q(0))| > k^{1/2+\beta}, \tag{7}$$

ε, β, being parameters, $0 < \beta < 1/2$, $0 < \varepsilon d < 1/2 - \beta$.

Clearly, $Z_0(k, \beta, \varepsilon)$ is a rich set in the sense that

$$\lim_{k \to \infty} \frac{\#Z_0(k, \beta, \varepsilon)}{\#S(k)} = 1, \tag{8}$$

here $\#Z_0(k,\beta,\varepsilon)$ and $\#S(k)$ are the number of points j in $Z_0(k,\beta,\varepsilon)$ and in the set $S(k)=\{j : k/4 < p_j(t) < 7k/4\}$, respectively. The goal of this section is to construct the model functions and formula for perturbed eigenvalues when $j\in Z_0(k,\beta,\varepsilon)$. The construction of model functions follows immediately (see (9)–(17)). The formula for eigenvalues is given in Theorem 1 at the end of this section.

If $j\in Z_0(k,\beta,\varepsilon)$, we define the model function Ψ_j by the formula

$$\Psi_j(x)=\frac{1}{|Q|^{1/2}}e^{i(\vec p_j(t),x)+S_j(x)}. \tag{9}$$

with some periodic function $S_j(x)$. We want Ψ_j to satisfy the equation $H(t)\Psi_j=\lambda_j\Psi_j$ as precise as possible, λ_j being a constant. This consideration gives us an asymptotic equation to define the periodic function $S_j(x)$:

$$2i\,(\vec p_j(t),\nabla S_j(x))+\Delta S_j(x)+(\nabla S_j(x),\nabla S_j(x))+2i\,(\nabla S_j(x),\vec a(x))$$

$$=2\,(\vec p_j(t),\vec a(x))+(\vec a(x),\vec a(x))+\alpha a_0(x)-i\mathrm{div}\vec a(x)+C_{j,t,N}+\epsilon_N(x), \tag{10}$$

$$\|\epsilon_N(x)\|_0<ck^{1-2(N+2)\beta}, \tag{11}$$

where $C_{j,t,N}$ is a constant, N is a fixed natural number. The choice of N is defined by the accuracy required from the approximation: $N=0$ corresponds to a simpler formula for $S_j(x)$ (see (13)), while a larger N requires additional terms in the formula for $S_j(x)$ (see (12)), which provide a higher accuracy of the approximation.

To satisfy this equation we define $S_j(x)$ by the formula:

$$S_j(x)=\sum_{n=0}^{N+1}S_j^{(n)}(x), \tag{12}$$

where

$$S_j^{(0)}(x)=-\sum_{r\in Z^d\setminus\{0\}}\frac{2(\vec p_j(t),\vec a_r)+(\alpha a_0-i\mathrm{div}\vec a)_r}{p_{j+r}^2(t)-p_j^2(t)}e^{i(\vec p_r(0),x)}, \tag{13}$$

$\vec a_r$, a_{0r} being the Fourier coefficients of $\vec a(x)$ and $a_0(x)$, respectively. Next,

$$S_j^{(1)}(x)=\sum_{r\in Z^d\setminus\{0\}}\frac{\tilde a_{jr}^{(1)}}{p_{j+r}^2(t)-p_j^2(t)}e^{i(\vec p_r(0),x)}, \tag{14}$$

where $\tilde a_{jr}^{(1)}$ are the Fourier coefficients of the function

$$\tilde a_j^{(1)}(x)=2i\left(\nabla S_j^{(0)}(x),\vec a(x)\right)+\left(\nabla S_j^{(0)}(x),\nabla S_j^{(0)}(x)\right)-(\vec a(x),\vec a(x)). \tag{15}$$

The functions $S_j^{(n)}$, $n\geq 2$, are defined by the recurrent formula:

$$S_j^{(n)}(x)=\sum_{r\in Z^d\setminus\{0\}}\frac{\tilde a_{jr}^{(n)}}{p_{j+r}^2(t)-p_j^2(t)}, \tag{16}$$

where $a_{jr}^{(n)}$ are the Fourier coefficients of the function

$$\tilde{a}_j^{(n)} = 2i\left(\nabla S_j^{(n-1)}, \vec{a}\right) + 2\left(\nabla S_j^{(n-1)}, \sum_{s=0}^{n-2} \nabla S_j^{(s)}\right)$$
$$+ (\nabla S_j^{(n-1)}, \nabla S_j^{(n-1)}). \tag{17}$$

Lemma 1. *Suppose $0 < \beta < 1/2$, $0 < \varepsilon d < 1/2 - \beta$, $j \in Z_0(k, \beta, \varepsilon)$, the parameter α in the definition of H satisfies the estimate $|\alpha| \leq k$, $\vec{A}$ being a trigonometric polynomial of the length R_0 (see (4)) and $N \in \mathbf{N}$. Then, the functions $S_j^{(n)}(x)$, n=0,1,...,N+1, and $\tilde{a}_j^{(n)}(x)$, $n = 1, \ldots, N+1$, have the following properties:*

$$S_{jr}^{(n)} = \tilde{a}_{jr}^{(n)} = 0 \quad \text{if } |r| > 2^{n+1}R_0, \tag{18}$$

$$\left\|D^{\vec{\nu}} S_j^{(n)}\right\|_0 \leq c_{n,\nu} \left\|\vec{A}\right\|_{\nu+n}^{n+1} k^{1/2-(1+2n)\beta}, \tag{19}$$

$$\left\|D^{\vec{\nu}} \Re S_j^{(n)}\right\|_0 \leq c_{n,\nu} \left\|\vec{A}\right\|_{\nu+n+2}^{n+1} k^{-2\beta(n+1)}, \tag{20}$$

$$\left\|D^{\vec{\nu}} \left(S_j^{(n)} - S_m^{(n)}\right)\right\|_0 \leq c_{n,\nu} \left\|\vec{A}\right\|_{\nu+n}^{n+1} |m-j| k^{-2(n+1)\beta}, \tag{21}$$

$$\text{if } m \in Z_0(k, \beta, \varepsilon),$$

$$\left\|D^{\vec{\nu}} \tilde{a}_j^{(n)}\right\|_0 \leq c_{n,\nu} \left\|\vec{A}\right\|_{\nu+n}^{n+1} k^{1-2n\beta}, \tag{22}$$

$$\left\|D^{\vec{\nu}} \tilde{\Im} a_j^{(n)}\right\|_0 \leq c_{n,\nu} \left\|\vec{A}\right\|_{\nu+n}^{n+1} k^{1/2-2n\beta}, \tag{23}$$

$$\left\|D^{\vec{\nu}} \left(\tilde{a}_j^{(n)} - \tilde{a}_m^{(n)}\right)\right\|_0 \leq c_{n,\nu} \left\|\vec{A}\right\|_{\nu+n}^{n+1} |m-j| k^{1/2-(2n+1)\beta}, \tag{24}$$

$$\text{if } m \in Z_0(k, \beta, \varepsilon),$$

where $c_{n,\nu} = c^{(n+1)(n+1+\nu)}$. The estimates (19)–(24) hold for any order ν of derivatives and k large enough, namely $k : k^{\varepsilon} > 2^{N+2} R_0 b_0^2$, $k^{2\beta} > c_{n,\nu} \|\vec{A}\|_{\nu+N}$.

Corollary 1.

$$\left\|D^{\vec{\nu}} S_j\right\|_0 \leq c \left\|\vec{A}\right\|_{\nu} k^{1/2-\beta}, \tag{25}$$

$$\left\|D^{\vec{\nu}} \Re S_j\right\|_0 \leq c \left\|\vec{A}\right\|_{\nu+2} k^{-2\beta}, \tag{26}$$

$$\left\|e^{S_j}\right\|_0 < 1 + \left\|\vec{A}\right\|_2 k^{-2\beta}, \tag{27}$$

$$\left\|D^{\vec{\nu}} (S_j - S_m)\right\|_0 \leq c \left\|\vec{A}\right\|_{\nu} |m-j| k^{-2\beta}. \tag{28}$$

Proof. By (4), $\vec{A}_r = 0$ when $|r| > R_0$. From the definition of $S_j^{(0)}$ we get $S_{jr}^{(0)} = 0$ when $|r| > R_0$. It follows now from (15) that $\tilde{a}_{jr}^{(1)} = 0$ when $|r| > 2R_0$. Using (16) we immediately obtain $S_{jr}^{(1)} = 0$ when $|r| > 2R_0$. Further, using (17) and (16), we obtain (18) for all n by mathematical induction.

Next we prove the pair of estimates (19), (22) by mathematical induction. Using (7) and the inequality $2^{N+2}R_0 < k^{\varepsilon}$, we get

$$\left|p_{j+r}^2(t) - p_j^2(t)\right| > k^{1/2+\beta} \quad \text{for all } r \in Z^d, 0 < |r| \le 2^{N+2}R_0. \tag{29}$$

Using the definition (13) of $S_j^{(0)}(x)$, inequality (29) and the estimates $|\alpha| \le k$, $\|\mathrm{div}\, \vec{a}\|_0 \le \|\vec{A}\|_1$, we easily get (19) for $n = 0$. Next, we get (22) for $\tilde{a}_j^{(1)}$. In fact, it is easy to see that for any $\vec{f}, \vec{g} : Q \to R^d$ the following definition holds:

$$\left\| D^{\vec{\nu}}\left(\vec{f}(x), \vec{g}(x)\right)\right\|_0 \le 2^{\nu}\|\vec{f}\|_{\nu}\|\vec{g}\|_{\nu}. \tag{30}$$

Applying (30) in (15) we obtain:

$$\|D^{\vec{\nu}}\tilde{a}_j^1(x)\|_0 \le 2^{\nu}\left(2\|\nabla S_j^0(x)\|_{\nu}\|\vec{a}(x)\|_{\nu} + \|\nabla S_j^{(0)}(x)\|_{\nu}^2 + \|\vec{a}(x)\|_{\nu}^2\right).$$

Using already proven estimate (19) for $S_j^{(0)}$, we get (22) for $\tilde{a}_j^1$. It follows from (16) and (29) that for all $\nu \ge 0$ and $n = 1, \dots, N+1$,

$$\|S_j^{(n)}(x)\|_{\nu} \le \|\tilde{a}_j^{(n)}(x)\|_{\nu} k^{-1/2-\beta}. \tag{31}$$

Formulae (17) and (30) yield:

$$\|D^{\vec{\nu}}\tilde{a}_j^{(n)}(x)\|_0 \le \|\tilde{a}_j^{(n)}(x)\|_{\nu}$$

$$\le 2^{\nu}\left(\|\nabla S_j^{(n-1)}(x)\|_{\nu}\|\vec{a}(x)\|_{\nu} + 2\|\nabla S_j^{(n-1)}(x)\|_{\nu}\sum_{s=0}^{n-1}\|\nabla S_j^{(s)}(x)\|_{\nu}\right).$$

Using (31), we obtain

$$\|\tilde{a}_j^{(n)}(x)\|_{\nu} \le 2^{\nu}k^{-1/2-\beta}\|\tilde{a}_j^{(n-1)}(x)\|_{\nu+1}$$

$$\times\left(\|\vec{a}(x)\|_{\nu} + 2\|\nabla S_j^{(0)}(x)\|_{\nu} + 2k^{-1/2-\beta}\sum_{s=1}^{n-1}\|\tilde{a}_j^{(s)}(x)\|_{\nu+1}\right).$$

Suppose (22) holds for $s = 1, \dots, n-1$. Then, taking into account the estimate $2^{N+\nu}\|\vec{A}\|_{N+\nu}k^{-2\beta} < 1$ and inequality (19) for $n = 0$, we obtain

$$\|\vec{a}\|_{\nu} + 2\|\nabla S_j^{(0)}(x)\|_{\nu} + 2k^{-1/2-\beta}\sum_{s=1}^{n-1}\|\tilde{a}_j^{(s)}(x)\|_{\nu+1} \le ck^{1/2-\beta}\|\vec{A}\|_{\nu+1}.$$

Thus, $\|\tilde{a}_j^{(n)}(x)\|_{\nu} < c2^{\nu}k^{-2\beta}\|\vec{A}\|_{\nu+1}\|\tilde{a}_j^{(n-1)}(x)\|_{\nu+1}$. Using (22) for $\tilde{a}_j^{(n-1)}(x)$, we get (22) for $\tilde{a}_j^{(n)}(x)$. Estimate (19) follows from (22) and (31).

Let us prove the pair of estimates (20), (23). By elementary calculation we obtain from (13):

$$\Re S_j^{(0)} = -\sum_{r\in Z^2\setminus\{0\}} \frac{(2(\vec{p}_j(t),\vec{a}_r) + (\alpha a_0 - i\operatorname{div}\vec{a})_r)\, p_r^2(0)}{(p_{j+r}^2(t) - p_j^2(t))(p_{j-r}^2(t) - p_j^2(t))} e^{i(\vec{p}_r(0),x)}.$$

The estimate (20) for $n=0$ easily follows from the last formula and (29). Using this estimate and formula (15) for $\tilde{a}_j^{(1)}$, and taking into account (20) and (19) for $n=0$, it is easy to show that $\|\Im\tilde{a}_j^{(1)}\|_0 \le c2^\nu k^{1/2-3\beta}$, i.e., (23) holds for $n=1$. Formula (16) for $S_j(x)$ yields

$$\|\Re S_j^{(n)}\|_\nu < c\|\tilde{a}_j^{(n)}\|_{\nu+2} k^{-1-2\beta} + c\|\Im\tilde{a}_j^{(n)}\|_\nu k^{-1/2-\beta}.$$

Using the estimate (22), we obtain:

$$\|\Re S_j^{(n)}\|_\nu < c_{n,\nu}\|A\|_{\nu+n+2}^{n+1} k^{-2\beta(n+1)} + c\|\Im\tilde{a}_j^{(n)}\|_\nu k^{-1/2-\beta}. \tag{32}$$

Formulae (17) and (30) yield:

$$\|D^{\vec{\nu}}\Im\tilde{a}_j^{(n)}(x)\|_0 \le \|\Im\tilde{a}_j^{(n)}(x)\|_\nu$$

$$\le 2^\nu \left(\|\Re\nabla S_j^{(n-1)}(x)\|_\nu \|\vec{a}\|_\nu + 2\|\Re\nabla S_j^{(n-1)}(x)\|_\nu \sum_{s=0}^{n-1} \|\nabla S_j^{(s)}(x)\|_\nu \right. \tag{33}$$

$$\left. + 2\|\nabla S_j^{(n-1)}(x)\|_\nu \sum_{s=0}^{n-2} \|\Re\nabla S_j^{(s)}(x)\|_\nu \right).$$

Suppose (20) holds for $n = s-1$, and (23) for $n=s$, $s\ge 1$. Then, using (32), we conclude that (20) holds for $n=s$, and using (33), we get (23) for $n=s+1$.

Let us prove (21) and (24). Using formula (13) for $S_j^{(0)}$ it is easy to check (21) when $n=0$. Considering (15) we readily check (24) for $n=1$. It is easy to show that

$$\|S_j^{(n)} - S_m^{(n)}\|_\nu < c|j-m|\|\tilde{a}_j^{(n)}\|_\nu k^{-1-2\beta} + c\|\tilde{a}_j^{(n)} - \tilde{a}_j^{(m)}\|_\nu k^{-1/2-\beta}.$$

Using the estimate (22), we obtain:

$$\|S_j^{(n)} - S_m^{(n)}\|_\nu < c_{n,\nu}|j-m|\|A\|_{\nu+n}^{n+1} k^{-2\beta(n+1)} + c\|\tilde{a}_j^{(n)} - \tilde{a}_j^{(m)}\|_\nu k^{-1/2-\beta}. \tag{34}$$

Formulae (17) and (30) yield:

$$\|D^{\vec{\nu}}\left(\tilde{a}_j^{(n)}(x) - \tilde{a}_m^{(n)}(x)\right)\|_0 \le \|\tilde{a}_j^{(n)}(x) - \tilde{a}_m^{(n)}(x)\|_\nu$$

$$\le 2^\nu \left(\|\nabla S_j^{(n-1)}(x) - \nabla S_m^{(n-1)}(x)\|_\nu \|\vec{a}\|_\nu \right.$$

$$+2\|\nabla S_j^{(n-1)}(x) - \nabla S_m^{(n-1)}(x)\|_\nu \sum_{s-0}^{n-1} \|\nabla S_j^{(s)}(x)\|_\nu$$

$$\left. + 2\|\nabla S_j^{(n-1)}(x)\|_\nu \sum_{s=0}^{n-2} \|\nabla S_j^{(s)}(x) - \nabla S_m^{(n-1)}(x)\|_\nu \right). \tag{35}$$

Suppose (21) holds for $n = s - 1$, and (24) for $n = s$, $s \geq 1$. Then, using (34), we conclude that (21) holds for n=s, and using (35), we get (24) for $n = s + 1$. □

Lemma 2. *Under the conditions of Lemma* 1 *the function* $S_j(x)$ *defined by* (12)–(17) *satisfies the relations* (10), (11), *where*

$$C_{j,t,N} = \sum_{n=1}^{N+1} \tilde{a}_{j0}^{(n)}. \tag{36}$$

Corollary 2.

$$C_{j,t,N} < c\|\vec{A}\|_1^2 k^{1-2\beta}. \tag{37}$$

Estimate (37) follows from (36) and (22).

Corollary 3. *If* $j, m \in Z_0(k, \beta, \varepsilon)$, *then*

$$|C_{j,t,N} - C_{m,t,N}| < c\left\|\vec{A}\right\|_1^2 |m - j| k^{1/2-3\beta}. \tag{38}$$

This corollary follows from formulas (36) and (24).

Proof. Formula (13) and the straightforward calculation imply:

$$2i\left(\vec{p}_j(t), \nabla S_j^{(0)}\right) + \Delta S_j^{(0)} - 2(\vec{p}_j(t), \vec{a}) - \alpha a_0 + i\mathrm{div}\vec{a} = 0. \tag{39}$$

Using the definition of $S_j^{(1)}$ (see (14) and (15)) it is easy to check that

$$2i(\vec{p}_j(t), \nabla S_j^{(1)}) + \Delta S_j^{(1)} + 2i(\nabla S_j^{(0)}, \vec{a}) + \left(\nabla S_j^{(0)}, \nabla S_j^{(0)}\right) - (\vec{a}, \vec{a}) = \tilde{a}_{j0}^1,$$

we have got $\tilde{a}_{j0}^1$ because the term corresponding to $r = 0$ is omitted in formula (14). Formulae (16) and (17) easily lead to the equation for $S_j^{(n)}$:

$$\begin{gathered} 2i(\vec{p}_j(t), \nabla S_j^{(n)}) + \Delta S_j^{(n)} + 2i(\nabla S_j^{(n-1)}, \vec{a}) \\ + \left(\sum_{s=0}^{n-1} \nabla S_j^{(s)}, \sum_{s=0}^{n-1} \nabla S_j^{(s)}\right) - \left(\sum_{s=0}^{n-2} \nabla S_j^{(s)}, \sum_{s=0}^{n-2} \nabla S_j^{(s)}\right) = \tilde{a}_{j0}^{(n)}. \end{gathered} \tag{40}$$

Adding formulae (39)–(40) for $n = 0, 1, 2, \ldots, N{+}1$ and using (19) for $n = N, N{+}1$, we get (10), (11) and (36). □

Lemma 3. *The function* $\Psi_j(t, x)$ *satisfies the equation:*

$$H\Psi_j = \left(p_j^2(t) - C_{j,t,N}\right)\Psi_j + \delta\Psi_j, \tag{41}$$

$\delta\Psi_j$ *obeying the estimate*

$$\|\delta\Psi_j\|_0 \leq c(\vec{A}) k^{1-2(N+2)\beta}. \tag{42}$$

We prove this formula by direct substitution of (9) into equation and using (10). To get (42) from (11) we use (27). From the last lemma the next theorem follows immediately:

Theorem 1. *There is a set of eigenvalues λ_j of the operator $H(t)$, which can be parameterized by $j \in Z_0(k,\beta,\delta)$ and obey the asymptotic formula:*

$$\lambda_j(t) = p_j^2(t) - C_{j,t,N}, \tag{43}$$

$C_{j,t,N}$ being given by (36).

3. Properties of the Model Functions

Let G be the Gram matrix of the set of functions $\Psi_j(x)$, $j \in Z_0(k,\beta,\varepsilon)$ and I_0 be the identity matrix for $j \in Z_0(k,\beta,\varepsilon)$:

$$G_{lj} =< \Psi_j, \Psi_l >, \quad (I_0)_{jm} = \delta_{jm}, \ j,m \in Z_0(k,\beta,\varepsilon).$$

We estimate $\|G - I_0\|_*$, $\|\cdot\|_*$ being defined by (5).

Theorem 2. *Let $0 < \beta < 1/2$. Then,*

$$\|G - I_0\|_* =_{k\to\infty} O(k^{-2\beta}). \tag{44}$$

First, let us prove two lemmas.

Lemma 4. *Let $0 < \gamma_1 < \gamma_2$, $M \in \mathbf{N}$, $k > 1$. Suppose functions Ψ_j, Ψ_m, $j,m \in Z^d$, can be represented in the form $\Psi_j = f_j(x)\exp i(\vec{p}_j(t),x)$, $\Psi_m = f_m(x)\exp i(\vec{p}_m(t),x)$, where $|\vec{p}_j(t) - \vec{p}_l(t)| > k^{\gamma_2}$ and*

$$\int_Q |D^{\vec{\nu}} f_j(x)dx| < (ck^{\gamma_1})^{\nu}, \int_Q |D^{\vec{\nu}} f_m(x)|dx < (ck^{\gamma_1})^{\nu}$$

for any $\nu \le M$. Then,

$$\langle \Psi_j, \Psi_m \rangle = O\left(c^M k^{-M(\gamma_2-\gamma_1)}\right). \tag{45}$$

Proof. Let ν_{jm} be the unit vector in the direction of $\vec{p}_j(t) - \vec{p}_m(t)$. Noting that

$$\left(\nu_{jm}, \nabla e^{i(\vec{p}_j(t)-\vec{p}_m(t),x)}\right) = |\vec{p}_j(t) - \vec{p}_m(t)| e^{i(\vec{p}_j(t)-\vec{p}_m(t),x)},$$

and integrating M times by parts in the direction of ν_{jm}, we arrive at (45). □

Lemma 5. *Any model function Ψ_j, $j \in Z_0(k,\beta,\varepsilon)$, can be represented in the form:*

$$\Psi_j = f_j(x)\exp i(\vec{p}_j(t),x), \tag{46}$$

where

$$\int_Q |D^{\vec{\nu}} f_j(x)|dx < \left(C_{\|\vec{A}\|_0,\beta,\varepsilon} k^{1/2+\beta}\right)^{\nu}, \text{ if } \nu < k^{1/2+\beta-\varepsilon}. \tag{47}$$

Proof. If $j \in Z_0(k,\beta,\varepsilon)$, then the estimate (47) easily follows from formula (9) and estimates (25) and (27). □

Proof of Theorem 2. Suppose $|\vec{p}_j(t) - \vec{p}_m(t)| > k^{1/2+\beta+\delta}$, δ being a positive auxiliary parameter, $\delta \le \varepsilon$. Then using Lemmas 4 and 5, we get

$$\langle \Psi_m, \Psi_j \rangle = O\left(k^{(1/2+\beta)M} |\vec{p}_m(t) - \vec{p}_j(t)|^{-M}\right)$$
$$= O(k^{2+4\beta-\delta(M-4)} |\vec{p}_m(t) - \vec{p}_j(t)|^{-4})$$

for any $M < k^{1/2+\beta-\varepsilon}$. Choosing M large enough, say $M > 10d/\delta$, we obtain:

$$\sigma_1 = \sup_j \sum_{m:|\vec{p}_m(t)-\vec{p}_j(t)|>k^{1/2+\beta+\delta}} |\langle \Psi_m, \Psi_j \rangle| < C_{\|\vec{A}\|_0,\beta,\varepsilon,\delta} k^{-2}.$$

Let

$$|\vec{p}_m(t) - \vec{p}_j(t)| \le k^{1/2+\beta+\delta}.$$

Then,

$$\langle \Psi_m, \Psi_j \rangle = \frac{1}{|Q|} \int_Q \exp\left(i(\vec{p}_m(t) - \vec{p}_j(t), x) + \left(S_m + \bar{S}_j\right)(x)\right) dx. \tag{48}$$

For $m = j$, obviously,

$$\langle \Psi_j, \Psi_j \rangle = \frac{1}{|Q|} \int_Q e^{2\Re S_j(x)} dx = 1 + O(k^{-2\beta}),$$

the inequality (27) has been used to obtain the last estimate. Suppose $m \ne j$. Let us consider $S_m(x) + \bar{S}_j(x)$. Using formulae (26) and (28) we obtain:

$$\left\|D^{\vec{\nu}}\left(S_m(x) + \bar{S}_j(x)\right)\right\|_0 < C_{\nu,N} \|\vec{A}\|_{\nu+2} |m - j| k^{-2\beta}, \quad m \ne j.$$

Integrating (48) by parts in the direction $\vec{p}_m(t) - \vec{p}_j(t)$, we get:

$$\langle \Psi_m, \Psi_j \rangle = O(k^{-2\beta} |j - m|^{-M}) \text{ for any } M.$$

Therefore,

$$\sigma_2 = \sup_j \sum_{m \ne j} |\langle \Psi_m, \Psi_j \rangle| = O(k^{-2\beta}).$$

□

Let us denote by $\hat{Z}_0(k,\beta,\varepsilon)$ a smaller version of $Z_0(k,\beta,\varepsilon)$, namely, let $\hat{Z}_0(k,\beta,\varepsilon)$ be a set of $j \in Z^d$ satisfying the conditions:

(1) $$k/2 < p_j(t) < 3k/2.$$

(2) For any $q \in Z^d : 0 < |q| \le k^{\varepsilon}$:

$$|(\vec{p}_j(t), \vec{p}_q(0))| > 2k^{1/2+\beta}.$$

Let L be the linear span of $\{\Psi_j\}_{j \in Z_0(k,\beta,\varepsilon)}$, and L_0 be the linear span of $\{\Psi_j^0\}_{j \in \hat{Z}_0(k,\beta,\varepsilon)}$, $\Psi_j^0(x) = |Q|^{-1/2} e^{(\vec{p}_j(t),x)}$. Obviously, both L and L_0 are finite-dimensional subspaces of $L_2(Q)$. We denote by P and P_0 the projections on L and L_0 correspondingly. Let us consider the operator P_0P.

Theorem 3. *If* $\frac{3d+2}{2(3d+6)} < \beta < 1/2$, *then*

$$P_0 P = P_0 + \delta P,$$

where δP *is an operator satisfying the asymptotic estimate:*

$$\|\delta P\| =_{k\to\infty} O(k^{-k^{\mu}}),\ \ \mu < 2\beta. \tag{49}$$

Proof. It suffices to prove that each $\Psi_j^0(x)$, $j \in \hat{Z}_0(k,\beta,\varepsilon)$, can be represented as a linear combination of Ψ_m with a high accuracy:

$$\Psi_j^0(x) = \sum_{|m-j|<k^{1/2-\beta+\mu'}} c_{jm}\Psi_m(x) + \epsilon_j(x),\ \ \|\epsilon_j\|_0 = O(k^{-k^{\mu}}), \tag{50}$$

where $\mu < \mu' < 2\beta$.

Step 1. First we check that each $\Psi_j^0(x)$, $j \in \hat{Z}_0(k,\beta,\varepsilon)$, can be represented as a linear combination of Ψ_m with a good accuracy:

$$\Psi_j^0(x) = \sum_{m:|m-j|<k^{1/2-\beta+\delta}} c_{jm}\Psi_m(x) + \epsilon_j(x), \tag{51}$$

$$\|\epsilon_j\|_{L_2(Q)} = O(k^{-\gamma_0/2+\delta(d+2)/2}),$$

here $\gamma_0 = \min\{2\beta, \frac{\beta(d+6)}{2} - \frac{d+2}{4}\} > 0$ and $0 < \delta < \gamma_0/(d+2)$. To prove (51) it is enough to check that the projection of $\Psi_j^0(x)$ on L has the L_2-norm greater than $1 - k^{-\gamma_0+\delta(d+2)}$. This readily follows from the inequality

$$\sum_{m:|m-j|<k^{1/2-\beta+\delta}} |\langle\Psi_m, \Psi_j^0\rangle|^2 > 1 - k^{-\gamma_0+\delta(d+2)}, \tag{52}$$

since the system of functions Ψ_m is almost orthogonal (Theorem 2). Let us prove (52). If $j \in \hat{Z}_0(k,\beta,\varepsilon)$ and $|m-j| < k^{1/2-\beta+\delta}$, then, obviously, $m \in Z_q(k,\beta,\varepsilon)$ and

$$\langle\Psi_m(x), \Psi_j^0(x)\rangle = |Q|^{-1}\int_Q e^{i(\vec{p}_m(t)-\vec{p}_j(t),x)+S_m(x)}dx.$$

Using (26), we easily replace $S_m(x)$ by $-\bar{S}_j(x)$, with a small error following:

$$\langle\Psi_m(x), \Psi_j^0(x)\rangle = |Q|^{-1}\int_Q e^{-\bar{S}_j(x)+i(\vec{p}_m(t)-\vec{p}_j(t),x)}dx$$

$$+O\left(|m-j|k^{-2\beta} + k^{-2\beta}\right). \tag{53}$$

Considering that $|j-m| \le k^{1/2-\beta+\delta}$ we obtain

$$O\left(|m-j|k^{-2\beta} + k^{-2\beta}\right) = O(k^{1/2-3\beta+\delta}).$$

Note that the above integral is the $(j-m)$-th Fourier coefficient of the function $e^{-\bar{S}_j(x)}$. We will denote it by $(e^{-\bar{S}_j(x)})_{j-m}$. Thus,

$$\sum_{m:|m-j|<k^{1/2-\beta+\delta}} |\langle \Psi_m(x), \Psi_j^0(x)\rangle|^2$$
$$> \sum_{m:|m-j|<k^{1/2-\beta+\delta}} \left|(e^{-\bar{S}_j(x)})_{j-m} + O(k^{1/2-3\beta+\delta})\right|^2.$$

Using the Cauchy-Schwarz inequality and considering that the sum contains about $ck^{(1/2-\beta+\delta)d}$ terms, we obtain:

$$\sum_{m:|m-j|<k^{1/2-\beta+\delta}} |\langle \Psi_m(x), \Psi_j^0(x)\rangle|^2 > \sum_{m:|m-j|<k^{1/2-\beta+\delta}} |(e^{-\bar{S}_j(x)})_{j-m}|^2$$
$$+O(\|e^{-\bar{S}_j}\|_{L_2(Q)} k^{1/2-3\beta+\delta+(1/2-\beta+\delta)d/2}) + O(k^{1-6\beta+(1/2-\beta+\delta)d}). \tag{54}$$

Now we estimate an incomplete sum of the Fourier coefficients on the right-hand side of the last formula. We have the following obvious formula for the complete sum:

$$\sum_{m\in Z^d} |(e^{-\bar{S}_j(x)})_{j-m}|^2 = \|e^{-\bar{S}_j}\|^2_{L_2(Q)} = 1 + O(k^{-2\beta}). \tag{55}$$

Thus, it remains to estimate the tail of the series. Let us consider $(e^{-\bar{S}_j(x)})_{j-m}$ for $|j-m| \geq k^{1/2-\beta+\delta}$. Integrating $\tilde{M}$ times by parts the corresponding integral (see the right-hand side of (53)) in the direction of $\vec{p}_m(t) - \vec{p}_j(t)$ and using that $\|D^l S_j\| = O(k^{1/2-\beta})$ for all l, we get

$$|(e^{-\bar{S}_j(x)})_{j-m}| < C_{\tilde{M}} k^{(1/2-\beta)\tilde{M}} |m-j|^{-\tilde{M}}) < c_{\tilde{M}} k^{-\delta\tilde{M}}$$

for any positive $\tilde{M}$. Choosing $\tilde{M} > 6d/\delta$, we obtain:

$$\sum_{m:|j-m|\geq k^{1/2-\beta+\delta}} |\langle \Psi_m(x), \Psi_j^0(x)\rangle|^2 < k^{-2}.$$

Taking into account that β satisfies the estimate $\beta > \frac{d+2}{2(d+6)}$, we replace the $O(\cdot)$ in (54), (55) by $O(k^{-\gamma_0+\delta(d+2)/2})$.

Step 2. We prove that the function ε_j in (50) satisfies the estimate

$$\|\epsilon_j\|_0 = O(k^{-\gamma'}), \quad \gamma' = \gamma_0/2 - \delta(d+2)/2 - (1/2-\beta+\delta)d/2. \tag{56}$$

Let us consider the set W_1: $W_1 : \{m : |m-j| < 2k^{1/2-\beta+\delta}\}$ and let P_1^0 be the projection on the linear span of Ψ_m^0, $m \in W_1$. Estimate (56) easily follows from the two estimates:

$$\|P_1^0 \epsilon_j\|_0 = O(k^{-\gamma'}), \tag{57}$$

$$\|(I - P_1^0)\epsilon_j\|_0 < C_{\tilde{M}} k^{-\delta\tilde{M}}, \tilde{M} \in \mathbf{N}. \tag{58}$$

First we consider $P_1^0 \epsilon_j$. The estimate for ϵ_j in (51) yields

$$\|P_1^0 \epsilon_j\|_{L_2(Q)} < k^{-\gamma_0/2+\delta(d+2)/2}.$$

Considering that W_1 contains about $k^{(1/2-\beta+\delta)d}$ elements and using the Cauchy-Schwarz inequality, we get (57). To prove (58), we apply projection $I - P_1^0$ to both sides of the (51). Taking into account that $(I - P_1^0)\Psi_j^0 = 0$, $|(\Psi_m)_r| < C_M k^{1/2-\beta}|m - r|)^{-M}$ for any $r \notin W_1$ and $|c_{mj}| < 1 + o(1)$, we easily obtain (58). Thus, (56) holds.

Step 3. Let us consider the sets W_n: $W_n : \{m : |m - j| < (n+1)k^{1/2-\beta+\delta}\}$ and let P_n^0 be the corresponding projections on the linear spans of Ψ_m^0, $m \in W_n$. Clearly, $\cup_{n=1}^{N_1} W_n \in \hat{Z}_q$, $N_1 = [k^{\mu'}]$. Let us consider the reminder $\varepsilon_j(x)$ in (51). Using formula (51) for each $m \in W_1$, we get

$$P_1^0 \varepsilon_j = \sum_{m\in W_1} (\epsilon_j)_m \Psi_m^0(x) = \sum_{r\in W_2} c_{jr}\Psi_r(x) + \sum_{m\in W_1} (\epsilon_j)_m \epsilon_m(x),$$

here, obviously, $c_{jr} = \sum_m (\epsilon_j)_m c_{mr}$. Clearly,

$$\left\| \sum_{m\in W_1} (\epsilon_j)_m \epsilon_m(x) \right\|_0 \leq \|\epsilon_j\|_0 \sup_m \|\epsilon_m\|_0 \leq k^{-2\gamma'}.$$

Using (58) and the last estimate, we obtain:

$$\Psi_j^0(x) = \sum_{m\in W_2} c_{jr}\Psi_r(x) + \epsilon_j(x), \quad \|\epsilon_j\|_0 = O(k^{-2\gamma'}).$$

Repeating this procedure N_1 times, we get (50). □

References

[ReSi] Reed, M., Simon, B. *Methods of Modern Mathematical Physics.*, Vol IV, Academic Press, 3rd ed., New York – San Francisco – London (1987).

[BiSu] Birman, M.Sh., Suslina, T.A. *Two Dimensional Periodic Magnetic Hamiltonian is Absolutely Continuous.* Algebra i Analiz, **9** (1997), 1, 31–47; Engl. Transl.: St. Petersburg Math. J., **9**, 1.

[Sob] Sobolev A.V. *Absolute continuity of the periodic magnetic Schrödinger operator.* Invent. Math. **137** (1999), 1, 85–112.

[K] Karpeshina Yu.E. "Perturbation theory for the Schrödinger operator with a periodic potential", 352 pp., in series "Lecture Notes in Mathematics", # 1663, Springer-Verlag, 1997.

Yu.E. Karpeshina
Department of Mathematics
University of Alabama at Birmingham
AL 35294-1170, USA

Operator Theory:
Advances and Applications, Vol. 132, 233–244

The Laplace Operator, Null Set Perturbations and Boundary Conditions

W. Karwowski and S. Kondej

Abstract. We are interested in Schrödinger operator with the singular perturbations given by operators which act in space with measure supported by null set N. We insist on methods which are additive in the case of weakly — as well as strongly singular perturbations. Both cases are illustrated on the examples corresponding to the Schrödinger operator in $L^2(\mathbb{R}^2, dx)$ and $L^2(\mathbb{R}^3, dx)$ with the singular perturbations living on circle. We also solve the problem of the boundary conditions for two above mentioned models.

1. Introduction

This paper belongs to the line of research often called the singular perturbations of the Schrödinger operator. This sort of perturbation originated in quantum mechanics from the so-called point- or δ-interaction. We are pursuing the following idea. In the Hilbert space $L^2(\mathbb{R}^m, dx) = L^2$, $m=2,3$ the operator $A = -\Delta + \lambda$, $\lambda > 0$ defined on the Sobolev space $W^{2,2}(\mathbb{R}^m) \equiv W^{2,2}$ is strictly positive and self-adjoint. Let $N \subset \mathbb{R}^m$ be compact with vanishing Lebesgue measure and set

$$M_2 = C_0^\infty(\mathbb{R}^m \backslash N)^{cl,2},$$

where $cl, 2$ denotes closure in the norm of $W^{2,2}$. Then M_2 is a closed subspace of $W^{2,2}$, but it is dense in L^2. Consequently

$$\dot{A} := A \restriction M_2 \tag{1}$$

is a closed symmetric operator with the defect space $N_0 = AN_2 = \ker \dot{A}^*$ where $N_2 = W^{2,2} \ominus M_2$ and $\dot{A}^*$ is the adjoint operator to $\dot{A}$. In turn, the deficiency index of $\dot{A}$ is given by

$$n^\pm = \dim N_0 = \dim N_2.$$

If N consists of finite number of points then $n^\pm$ are finite; otherwise infinite.

Obviously, any extension of $\dot{A}$ can be characterized as $-\Delta_{bc} + \lambda$ where Δ_{bc} denotes the Laplacian in L^2 with apropriate boundary conditions on N. In particular, in [12] was described the set $D(\dot{A}^*)$ in the terms of boundary conditions on N, where N is a regular infinite or closed curve in $\mathbb{R}^3$.

In this paper we will be concerned with those self-adjoint extensions of $\dot{A}$ to which a physical interpretation can be attributed. Our reasoning is the following.

Suppose there is a Radon measure μ on $\mathbb{R}^m$ supported by N and consider the formal expression

$$A+\mu$$

which is the sort of generalization of δ-interaction. If $\widetilde{A}_\mu$ is a self-adjont extension of $\dot{A}$ related in some additive sense to μ we say that $\widetilde{A}_\mu$ is the perturbation of A *by the measure* μ.

Addressing to more general situation let us consider the formal sum

$$A+V$$

where V is a self-adjoint operator in $L^2(\mathbb{R}^m, \mu)$. In particular V may play a role of the Hamiltonian of a system living on N. Analogously, if $\tilde{A}_V$ is a self-adjont extension of $\dot{A}$ corresponding to $A+V$, then we interpret it as the Hamiltonian of composite system. We shall call $\tilde{A}_V$ a peturbation of A *by the dynamics of* V. It also seems natural to say that $\tilde{A}_V$ takes into account the internal structure of the system living on N, but this should be clearly distinguished from the "interaction with internal structure" as studied by B.S. Pavlov and collaborators (see [1]) and references therein. In their terminology the interaction with internal structure appears as generalized self-adjoint extension, while the models we are discussing belong to the class of standard extensions.

In this paper we are interested in the perturbations by the dynamics. There are different methods to link self-adjoint extensions of $\dot{A}$ with V [2, 3].

In this note we construct $\tilde{A}_V$ as the kind of the sum

$$A+V.$$

However we should say that operator $\tilde{A}_V$ only vagely recalls the usual meaning of addition.

Our main interest will be to construct $\tilde{A}_V$ and express it in terms of the boundary conditions on N. The general discussion of these problems can be found in [4]. Here we shall concentrate on two explicit models. In both cases we shall choose N to be a circle and V a function of the Laplace–Beltrami operator. This will be considered as perturbation of the Laplace operator in $\mathbb{R}^2$ and $\mathbb{R}^3$. A non-additive perturbation by such operator V has been constructed by one of us [5]. Let us mention also that the boundary condition on a curve for three-dimensional Laplace operator was discussed in [12]. However the main concept of this paper is different. Namely, in [12] it was selected some class of boundary conditions for which the appropriate extension of $\dot{A}$ is essentially self-adjoint. According to our construction of $\tilde{A}_V$ the boundary conditions considered in [12] correspond rather to perturbation by a measure than by dynamics.

The models discussed in this paper are interesting by themselves, but they can also be used as testing objects for further studies such as spectral analysis of the perturbed operators and the scattering theory.

For readers convenience we shall precede the explicite model presentation with a summary of general formulation.

1.1. General Definitions

As was mentioned above we consider the strictly positive operator $A = -\Delta + \lambda$ where $-\Delta$ stands for the self-adjoint Laplace operator in $L^2(\mathbb{R}^m, dx) = L^2$, m=*2,3* and $\lambda > 0$. The domain $D(A)$ of A coincides with the Sobolev space $W^{2,2}(\mathbb{R}^m) = W^{2,2}$. In further discussion we shall need the Sobolev spaces $W^{2,k}(\mathbb{R}^m) = W^{2,k}$, $-2 \leq k \leq 2$, equiped with the norms

$$\|f\|_k^2 = \int_{\mathbb{R}^m} \left|(-\Delta+\lambda)^{\frac{k}{2}} f(x)\right|^2 dx.$$

It is known that $W^{2,k}$ and $W^{2,-k}$ are mutually conjugate with respect to L^2. Put $\langle \cdot, \cdot \rangle$ for the duality between $W^{2,k}$ and $W^{2,-k}$.

Note, that A is unitary as the map from $W^{2,2}$ to L^2 and it acts isometrically from $W^{2,1}$ to $W^{2,-1}$ and from L^2 to $W^{2,-2}$. Thus the closure $\mathbb{A} \equiv A^{cl,2} : L^2 \to W^{2,-2}$ is unitary. We find it convenient to put $G := A^{-1}$ and $\mathbb{G} := \mathbb{A}^{-1}$.

Operator $\tilde{A}$ is called a singular perturbation of A if $\tilde{A}$ is invertible, self-adjoint in L^2 and the set

$$D = \{u \in D(\tilde{A}) \cap D(A) : \tilde{A}u = Au\}$$

is dense in L^2. The family of the singular perturbations of A will be denoted by $\mathcal{P}(A)$.

It is known [2, 3] that operators A and $\widetilde{A} \in \mathcal{P}(A)$ are connected by the following formula

$$\tilde{A}^{-1} = A^{-1} + \tilde{B}^{-1} \tag{2}$$

where $\tilde{B}^{-1} : L^2 \to N_0$ is self-adjoint. Note, that if $\tilde{B}^{-1}$ is bounded then zero belongs to the resolvent set of $\widetilde{A}$ and (2) presents the Krein formula for $\widetilde{A}$ at the point zero.

Given compact set $N \subset \mathbb{R}^m$ with vanishing Lebesque measure a self-adjoint, invertible extension of $\dot{A} = A \restriction C_0^\infty(\mathbb{R}^m \backslash N)$ belongs to $\mathcal{P}(A)$.

We shall say that the singular perturbation $\tilde{A}$ belongs to the weakly singular class, $\mathcal{P}_{ws}(A)$, if condition

$$D(\tilde{A}) \subseteq W^{2,1}$$

holds. The completion $\mathcal{P}_{ss}(A) \equiv \mathcal{P}(A) \backslash \mathcal{P}_{ws}(A)$ will be called the strongly singular class.

2. Additive method for weakly singular perturbation

We keep notations introduced in Sec. 1.1 *i.e.* $A = -\Delta + \lambda : D(A) \to L^2(\mathbb{R}^m, dx) = L^2$, $m = 2, 3$, $\lambda > 0$, $W^{2,k} = W^{2,k}(\mathbb{R}^m)$ and $\mathbb{A} \equiv A^{cl,2} : L^2 \to W^{2,-2}$.

Let $0 \neq \mathbb{V}: W^{2,1} \to W^{2,-1}$ be a closed, symmetric operator *i.e.* $\mathbb{V} \subset \mathbb{V}^*$. Note that adjoint operator, $\mathbb{V}^*$ is defined with respect to the dual inner product

$$\langle \mathbb{V}u, w \rangle = \langle u, \mathbb{V}w \rangle, \quad u, w \in D(\mathbb{V}) \subseteq W^{2,1}.$$

Further, we assume that

$$\ker \mathbb{V} \cap W^{2,2} \tag{3}$$

is dense in L^2.

Clearly $M_1 \equiv \ker \mathbb{V}$ is a proper subspace of $W^{2,1}$. Put N_1 for the orthogonal completion of M_1, $N_1 \equiv W^{2,1} \ominus M_1$.

The generalized sum of $\mathbb{A}$ and $\mathbb{V}$ is defined by [6], [7],[8]

$$D(\mathbb{A}\tilde{+}\mathbb{V}) = \{f \in D(\mathbb{V})\colon \mathbb{A}f + \mathbb{V}f \in L^2\}, \quad (\mathbb{A}\tilde{+}\mathbb{V})f = \mathbb{A}f + \mathbb{V}f. \tag{4}$$

The following theorem supplies condition for the self-adjointness of $\mathbb{A}\tilde{+}\mathbb{V}$.

Theorem 1 ([9, 10]). *If -1 is a regular point for $\mathbb{V}$, $-1 \in \rho(\mathbb{V})$, then operator $\mathbb{A}\tilde{+}\mathbb{V}$ is self-adjoint. Moreover we have $\mathbb{A}\tilde{+}\mathbb{V} \in \mathcal{P}_{ws}(A)$.*

3. Weakly singular perturbation of Laplace operator by object living on circle. Boundary conditions

Now, we shall be concerned with weakly singular perturbation of $A = -\Delta + \lambda :$ $D(A) \to L^2(\mathbb{R}^2, dx) \equiv L^2$, $\lambda > 0$. In this section we set $W^{2,2} = W^{2,2}(\mathbb{R}^2)$.

Then the operator $G = A^{-1}$ may be presented by the following integral kernel

$$G(x, x') = \frac{1}{(2\pi)^2} \int_{\mathbb{R}^2} dp \frac{e^{ip(x-x')}}{p^2 + \lambda}, \tag{5}$$

where $x = (x_1, x_2)$, $p = (p_1, p_2)$, $p^2 = p_1^2 + p_2^2$.

Let $N = \{(r = 1, \varphi \in \langle 0, 2\pi\rangle\}$ where r, φ are respectively, radial and angular coordinates.

Consider $-\frac{\partial^2}{\partial \varphi^2}$ as the self-adjoint operator in $L^2(\langle 0, 2\pi\rangle, d\varphi)$ with periodic boundary conditions. Let $V \in C(\mathbb{R})$ be real and define $V(-\frac{\partial^2}{\partial\varphi^2})$ in the sense of spectral theorem. As is well known the spectrum of $-\frac{\partial^2}{\partial\varphi^2}$ is $\{k^2\}_{k\in\mathbb{Z}}$. Set $v_k = V(k^2)$ for all $k \in \mathbb{Z}$.

Now our goal will be to construct a singular perturbation of A corresponding to formal sum

$$A + V(-\frac{\partial^2}{\partial\varphi^2})\delta(r-1).$$

Put for abbreviation $\mu_k \equiv e^{ik\varphi}\delta(r-1)$. The fact that $\mu_k \in W^{2,-1}$ allows to define

$$\mathbb{V} : D(\mathbb{V}) \to W^{2,-1},\ D(\mathbb{V}) = \{f \in W^{2,1}, \mathbb{V}f \in W^{2,-1}\}$$

by

$$\mathbb{V}f = \sum_{k\in\mathbb{Z}} v_k \langle f, \mu_k\rangle \mu_k, \tag{6}$$

where v_k are real numbers.

Clearly $\mathbb{V}$ given by (6) is symmetric. Besides, since $C_0^\infty(\mathbb{R}^2 \backslash N) \subset \ker \mathbb{V}$ condition (3) holds.

In accordance with the general definition (4) we construct operator $\mathbb{A}\tilde{+}\mathbb{V}$. Further, as follows from Theorem 1 $\mathbb{A}\tilde{+}\mathbb{V}$ is self-adjoint if $-1 \in \rho(\mathbb{V})$. Thus, we formulate conditions ensuring $-1 \in \rho(\mathbb{V})$.

Theorem 2. *-1 is a regular point for $\mathbb{V}$ if there exists $M > 0$ so that*

$$1 + v_k g_k > M, \quad \text{for all } k \in \mathbb{Z}, \tag{7}$$

where $g_k = \langle \mathbb{G}\mu_k, \mu_k \rangle$.

Proof. Note that $-1 \in \rho(\mathbb{V})$ iff operator

$$\mathbb{I} + \mathbb{G}\mathbb{V} : D(\mathbb{V}) \to W^{2,-1}$$

defined by

$$(\mathbb{I} + \mathbb{G}\mathbb{V})f = f + \sum_{k\in\mathbb{Z}} v_k \langle f, \mu_k \rangle \mathbb{G}\mu_k \tag{8}$$

is boundedly invertible. A technical calculation yields

$$\langle \mathbb{G}\mu_k, \mu_l \rangle = g_k \delta_{kl}. \tag{9}$$

Thus N_1 is spanned by $\{\mathbb{G}\mu_k\}_{k\in\mathbb{Z}'}$ where $\mathbb{Z}' = \{k \in \mathbb{Z}\colon v_k \neq 0\}$. Using (8), (9), and (7) we get

$$\begin{aligned} \|(\mathbb{I} + \mathbb{G}\mathbb{V})f\|_1^2 &= \left\| \sum_{k\in\mathbb{Z}} (g_k^{-1} + v_k) \langle f, \mu_k \rangle \mathbb{G}\mu_k \right\|_1^2 = \sum_{k\in\mathbb{Z}} (1 + v_k g_k)^2 g_k^{-1} |\langle f, \mu_k \rangle|^2 \\ &\geq M^2 \sum_{k\in\mathbb{Z}} g_k^{-1} |\langle f, \mu_k \rangle|^2 = M^2 \|f\|_1^2. \end{aligned}$$

Then by the fact that $(\mathbb{I} + \mathbb{G}\mathbb{V})v = v$ for $v \in M_1$ we get that $\mathbb{I} + \mathbb{G}\mathbb{V}$ is boundedly invertible. □

Note that if $v_k = k^2$ for all $k \in \mathbb{Z}$ then condition (7) is satisfied because $g_k > 0$. So, particularly we can take $\mathbb{V}$ which corresponds to formal expression $-\frac{\partial^2}{\partial\varphi^2}\delta(r-1)$.

Henceforth we assume that (7) is fulfilled. Then operator $\mathbb{A}\tilde{+}\mathbb{V}$ is self-adjoint.

Let us turn to formulating the boundary condition for $\mathbb{A}\tilde{+}\mathbb{V}$. For $w \in W^{2,1}$ we define

$$\sigma(w) = \lim_{r\to 1^+} \partial_r w(r,\varphi) - \lim_{r\to 1^-} \partial_r w(r,\varphi).$$

where $\partial_r = \frac{\partial}{\partial_r}$.

Setting ∂_i, $i = 1, 2$ for derivatives in the sense of distribution define the Laplace operator

$$\Delta_{bc} = \partial_1^2 + \partial_2^2 : D(\Delta_{bc}) \to L^2$$

with the following boundary condition

$$D(\Delta_{bc}) = \{w \in W_{loc}^{2,2}(\mathbb{R}^2 \backslash N) \cap D(\mathbb{V}) : \sigma(w) = \sum_{k\in\mathbb{Z}} v_k \langle w, \mu_k \rangle e^{ik\varphi}\}. \tag{10}$$

Using the generalized formula for integration by parts we get

$$\int_{\mathbb{R}^2}(-\Delta_{bc}+\lambda)f\bar{g}dx = \int_{\mathbb{R}^2}\nabla f\nabla\bar{g} + \lambda\int_{\mathbb{R}^2} f\bar{g}dx + \langle \mathbb{V}f, g\rangle \;\; f, g \in D(\Delta_{bc}),$$

where $\nabla = (\partial_1, \partial_2)$. Then one can show (for more details see [4])

$$D(\Delta_{bc}) = D(\mathbb{A}\tilde{+}\mathbb{V})$$

and

$$((-\Delta_{bc}+\lambda)f, g) = (\mathbb{A}\tilde{+}\mathbb{V}f, g) \;\; \textit{for } f, g \in D(\mathbb{A}\tilde{+}\mathbb{V}).$$

Thus the boundary conditions imposed in definition of $D(\Delta_{bc})$ uniquely determine the domain of $\mathbb{A}\tilde{+}\mathbb{V}$.

Remark. Let us discuss here how the construction of $\mathbb{A}\tilde{+}\mathbb{V}$ depends on the parameter λ. For this aim it is convenient to put $A_\lambda \equiv A = -\Delta + \lambda$ and $\mathbb{A}_\lambda \equiv \mathbb{A}$.

Firstly, it is easy to note that λ enters the construction of $\mathbb{A}_\lambda\tilde{+}\mathbb{V}$ additively i.e. given strictly positive constants λ, β we have

$$D(\mathbb{A}_\lambda\tilde{+}\mathbb{V}) = D(\mathbb{A}_\beta\tilde{+}\mathbb{V}) \quad \text{and} \quad \mathbb{A}_\lambda\tilde{+}\mathbb{V} = (\mathbb{A}_\beta\tilde{+}\mathbb{V}) + (\lambda - \beta). \tag{11}$$

In fact, (11) follows directly from the definition of $\mathbb{A}_\lambda\tilde{+}\mathbb{V}$.

The above construction of the generalized sum was provided for strictly positive operator A_λ. However we can extend this notation to the case of the "free" Laplacian. Namely, let us define operator $-\Delta\tilde{+}\mathbb{V}$ by

$$(-\Delta\tilde{+}\mathbb{V})f = -\overline{\Delta}f + \mathbb{V}f, \quad f \in D(-\Delta\tilde{+}\mathbb{V}) = \{g \in D(\mathbb{V}) : -\overline{\Delta}g + \mathbb{V}g \in L^2\}$$

where $-\overline{\Delta} = \mathbb{A}_\lambda - \lambda$.

Then, we have

$$D(-\Delta\tilde{+}\mathbb{V}) = D(\mathbb{A}_\lambda\tilde{+}\mathbb{V})$$

and

$$(-\Delta\tilde{+}\mathbb{V})f = ((\mathbb{A}_\lambda\tilde{+}\mathbb{V}) - \lambda)f \quad \text{for any } f \in D(-\Delta\tilde{+}\mathbb{V}).$$

So we see that operator $-\Delta\tilde{+}\mathbb{V}$ can be identified with $(\mathbb{A}_\lambda\tilde{+}\mathbb{V}) - \lambda$.

For $\mathbb{V}$ given by (6) operator $-\Delta\tilde{+}\mathbb{V}$ coincides with the Laplacian $-\Delta_{bc}$ with the boundary conditions described by (10).

4. Additive method for strongly singular perturbation

Now we shall discuss strongly singular perturbation of $A = -\Delta + \lambda : D(A) \to L^2(\mathbb{R}^3, dx) \equiv L^2$ where $\lambda > 0$ and $-\Delta$ denotes the Laplace operator in $L^2(\mathbb{R}^3, dx) \equiv L^2$. Operator $G^{-1} = A$, has the integral kernel

$$G(x, x') = \frac{1}{(2\pi)^3}\int_{\mathbb{R}^3} dp \frac{e^{ip(x-x')}}{p^2+\lambda} = \frac{1}{4\pi}\frac{e^{-\sqrt{\lambda}|x-x'|}}{|x-x'|}, \tag{12}$$

where $x = (x_1, x_2, x_3)$, $p = (p_1, p_2, p_3)$, $p^2 = p_1^2 + p_2^2 + p_3^2$. We also introduce convolution of $G(x, x')$ with itself

$$G^2(x, x') = \int_{\mathbb{R}^3} dx'' G(x, x'') G(x'', x') = \frac{1}{(2\pi)^3} \int_{\mathbb{R}^3} dp \frac{e^{ip(x-x')}}{(p^2 + \lambda)^2} \tag{13}$$

In that what follows we put $W^{2,k}(\mathbb{R}^3) = W^{2,k}$, $k = -2, 2$.

Set $\mathbb{V}_{-2} : D(\mathbb{V}) \subset C(\mathbb{R}^3) \to W^{2,-2}$ for operator satisfying the following conditions.

K) $ker \mathbb{V}_{-2} \cap W^{2,2}$ is dense in L^2.

R) $Ran \mathbb{V}_{-2} \cap W^{2,-1} = \{0\}$.

Similarly as in the previous sections we would like to construct self-adjoint realization for the sum $A + \mathbb{V}_{-2}$. The concept is based on an analogy with the generalized sum. Now, we shall summarize the main results presented in [4] and [11].

Put G_s, G_r for the integral operators with kernels

$$G_s(x, x\prime) = \frac{1}{4\pi} \frac{1}{|x - x\prime|} \text{ and } G_r(x, x\prime) = G(x, x\prime) - G_s(x, x\prime) \tag{14}$$

Next define

$$C_r := \mathbb{I} + G_r \mathbb{V}_{-2} : D(C_r) = \{g \in D(\mathbb{V}_{-2}) : C_r g \in W^{2,2}\} \to W^{2,2}. \tag{15}$$

and assume the invertibility for C_r. Then operator

$$C_s := C_r - G\mathbb{V}_{-2} = \mathbb{I} - G_s \mathbb{V}_{-2} : D(C_s) = D(C_r) \to L^2 \tag{16}$$

is invertible too [11].

Let $f \in D(C_s^{-1})$ and $f_r = C_s^{-1} f$. Define the set

$$D(A \hat{+} \mathbb{V}_{-2}) = \{f \in D(C_s^{-1}) : \mathbb{A} f + \mathbb{V}_{-2} f_r \in L^2\} \tag{17}$$

and operator $A \hat{+} \mathbb{V}_{-2}$ given by

$$(A \hat{+} \mathbb{V}_{-2}) f = \mathbb{A} f + \mathbb{V}_{-2} f_r, \quad f \in D(A \hat{+} \mathbb{V}_{-2}). \tag{18}$$

Let $f \in D(A \hat{+} \mathbb{V}_{-2})$. By a direct calculation one can show that f possesses representation in the following form [4]

$$f = Gg - \mathbb{G}\mathbb{V}_{-2} C_r^{-1} Gg, \quad g \in AD(C_r^{-1}). \tag{19}$$

Further, relying on the results of [4] we can formulate a condition ensuring the self-adjointness for $A \hat{+} \mathbb{V}_{-2}$.

Theorem 3. *Operator $A \hat{+} \mathbb{V}_{-2} \in \mathcal{P}_{ss}(A)$ if and only if $\mathbb{G}\mathbb{V}_{-2} C_r^{-1} G$ is self-adjoint and then its inverse is given by*

$$(A \hat{+} \mathbb{V}_{-2})^{-1} = A^{-1} - \mathbb{G}\mathbb{V}_{-2} C_r^{-1} G. \tag{20}$$

So, we see that $(A \hat{+} \mathbb{V}_{-2})^{-1}$ given by (20) is one of the realization of (2) and it is the Krein formula for $A \hat{+} \mathbb{V}_{-2}$ at the zero point provided operator $\mathbb{G}\mathbb{V}_{-2} C_r^{-1} Gg$ is bounded.

5. Strongly singular perturbation of Laplace operator by objects living on circle. Boundary conditions

Keep $L^2 = L^2(\mathbb{R}^3, dx)$, $A = -\Delta + \lambda : D(A) \to L^2$ and $W^{2,k} = W^{2,k}(\mathbb{R}^3)$, $k = -2, 2$.

In this section we put $N = \{r = 1, \theta = \frac{\pi}{2}, \varphi \in \langle 0, 2\pi\rangle\}$ where r, θ, φ, are spherical coordinates in $\mathbb{R}^3$. Similarly as before, given real function V from $C(\mathbb{R})$ define self-adjoint operator $V(-\frac{\partial^2}{\partial\varphi^2})$ in $L^2(\langle 0, 2\pi\rangle, d\varphi)$ where $\frac{\partial^2}{\partial\varphi^2}$ stands for the Laplacian with periodic boundary conditions. We also keep $v_k = V(k^2)$ for each $k \in \mathbb{Z}$.

Define a singular perturbation of A corresponding to the formal expression

$$A + V(-\frac{\partial^2}{\partial\varphi^2})\delta(r-1)\delta(\cos\theta). \tag{21}$$

Abbreviate $\eta_k = e^{ik\varphi}\delta(r-1)\delta(\cos\theta)$. Then one can show by technical calculation that

$$(\mathbb{G}\eta_k, \mathbb{G}\eta_l) = \|\mathbb{G}\eta_k\|^2 \delta_{kl}$$

i.e. $\mathbb{G}\eta_k$ are orthogonal in L^2.

Let

$$\mathbb{V}_{-2} : D(\mathbb{V}_{-2}) \to W^{2,-2} \tag{22}$$

be defined as follows

$$\mathbb{V}_{-2}f = \sum_{k\in\mathbb{Z}} v_k c_k(f)\eta_k, \; D(\mathbb{V}_{-2}) = \{f \in C(\mathbb{R}^3) : \mathbb{V}_{-2}f \in W^{2,-2}\} \tag{23}$$

where $c_k(f) = \int_0^{2\pi} d\varphi f(r = 1, \theta = \frac{\pi}{2}, \varphi)e^{-ik\varphi}$ and v_k are real numbers. One can easily show that $\mathbb{V}_{-2}$ satisfies K), R).

As in the general discussion we define operators C_r, C_s. In accord with (15) and (16) for $\mathbb{V}_{-2}$ given by (23) we get

$$C_s = \mathbb{I} - \sum_{k\in\mathbb{Z}} v_k c_k(\cdot)G_s\eta_k \tag{24}$$

and

$$C_r = \mathbb{I} + \sum_{k\in\mathbb{Z}} v_k c_k(\cdot)G_r\eta_k. \tag{25}$$

The following theorem provides the condition for the invertibility of C_r.

Theorem 4. *If for all* $k\in\mathbb{Z}$

$$1 + v_k q_k \neq 0, \quad q_k = c_k(G_r\eta_k) \tag{26}$$

then operator C_r *is invertible.*

Proof. Let (26) be satisfied. First we show that C_r is invertible. Similarly as (9) one can show by a technical manipulation

$$\int_0^{2\pi} d\varphi (G_r\eta_k)(r=1,\theta=\frac{\pi}{2},\varphi)e^{-il\varphi} = c_l(G_r\eta_k)\delta_{lk} = q_l\delta_{lk}. \tag{27}$$

Let $f \in \ker C_r$. Then, by (25) and (27) we obtain

$$c_k(C_rf) = c_k(f)(1+v_kq_k) = 0 \; for \; each \; k. \tag{28}$$

Since (26) is fulfilled (28) implies $c_k(f) = 0$ for all k. Then $C_rf = f$ *i.e* $f = 0$. Thus operator C_r is invertible. □

One can show that $q_k = \frac{1}{2}\ln(\lambda + k^2) - \gamma$, where $\gamma = 0,577\ldots$ is the Euler constant. Then it is easy to check by a direct calculation that for $v_k = k^2$ condition (26) is fulfilled.

Henceforth, we assume that (26) is satisfied. Then operator C_r is invertible. As follows from the general discussion C_s is invertible too and any function $f \in D(A\hat{+}\mathbb{V}_{-2})$ has the following representation

$$f = Gg - \mathbb{G}\mathbb{V}_{-2}C_r^{-1}Gg,$$

where

$$\mathbb{G}\mathbb{V}_{-2}C_r^{-1}Gg = \sum_{k\in\mathbb{Z}} v_k(1+q_kv_k)^{-1}(g,\mathbb{G}\eta_k)\mathbb{G}\eta_k$$

and

$$g \in AD(C_r^{-1}) = \{u \in L^2 : \sum_{k\in\mathbb{Z}} \|\mathbb{G}\eta_k\|^2 \left|(u,\mathbb{G}\eta_k)\right|^2 \left|v_k(1+v_kq_k)^{-1}\right|^2 < \infty\}.$$

Observing that $\mathbb{G}\mathbb{V}_{-2}C_r^{-1}G : AD(C_r^{-1}) \to L^2$ is self-adjoint we conclude from Theorem 3 that $A\hat{+}\mathbb{V}_{-2}$ is self-adjoint.

The problem at hand is to describe functions from $D(A\hat{+}\mathbb{V}_{-2})$ in the terms of boundary conditions on circle N. For this we shall need the following objects.

Put $r_N = ((x_1 - \cos\varphi)^2 + (x_2 - \sin\varphi)^2 + x_3^2)^{1/2}$.

Definition 5. *A function $f \in W_{loc}^{2,2}(\mathbb{R}^3\backslash N) \cap L^2$ belongs to Θ if the following expressions are finite*

$$\Xi(f)(\varphi) := \lim_{r_N\to 0} -\frac{1}{\ln r_N} f(x), \tag{29}$$

$$\Xi_k(f) := (2\pi)^{-1}\int_0^{2\pi} d\varphi \Xi(f)(\varphi)e^{ik\varphi} \quad for \; all \; k \in \mathbb{Z}, \tag{30}$$

$$\Omega(f)(\varphi) := \lim_{r_N\to 0}(f(x) - \sum_{k\in\mathbb{Z}} \Xi_k(f)G_s\eta_k(x), \tag{31}$$

$$\Omega_k(f) := (2\pi)^{-1}\int_0^{2\pi} d\varphi \Omega(f)(\varphi)e^{ik\varphi} \quad for \; all \; k \in \mathbb{Z}. \tag{32}$$

The following theorem supplies boundary conditions for functions from $D(A\hat{+}\mathbb{V}_{-2})$.

Theorem 6. *A function f belongs to $D(A\hat{+}\mathbb{V}_{-2})$ iff $f \in \Theta$ and satisfies the boundary condition given by*

$$\Xi_k(f) = v_k \Omega_k(f) \text{ for each } k \in \mathbb{Z}. \tag{33}$$

Proof. Let $f \in D(A\hat{+}\mathbb{V}_{-2})$. Then f is given by

$$f = Gg - \sum_{k\in\mathbb{Z}} v_k(1+q_k v_k)^{-1}(g, \mathbb{G}\eta_k)\mathbb{G}\eta_k, \tag{34}$$

where $g \in AD(C_r^{-1})$. Due to the definition of $D(A\hat{+}\mathbb{V}_{-2})$ (see (17) we have $f \in D(C_s^{-1})$. Then $f_r = C_s^{-1}f$ is given by

$$f_r = Gg - \sum_{k\in\mathbb{Z}} v_k(1+q_k v_k)^{-1}(g, \mathbb{G}\eta_k)G_r\eta_k.$$

Setting

$$f_s = \sum_{k\in\mathbb{Z}} v_k(1+q_k v_k)^{-1}(g, \mathbb{G}\eta_k)G_s\eta_k$$

we obtain decomposition of f

$$f = f_r + f_s$$

onto regular and singular components on N. To investigate the behaviour of f_s on N we shall need the following technical fact

$$\lim_{r_N\to 0} -\frac{1}{\ln r_N} G_s\eta_k = e^{ik\varphi}.$$

Then we get $\Xi_k(f) = \Xi_k(f_s) = -v_k(1+q_k v_k)^{-1}(g, \mathbb{G}\eta_k)$. In turn, it can be easily seen

$$\Omega_k(G_r\eta_k) = q_k \quad \text{and} \quad \Omega_k(G^2\eta_k) = \|\mathbb{G}\eta_k\|^2.$$

Thus $\Omega_k(f) = (1 - q_k v_k(1+q_k v_k)^{-1})(g, \mathbb{G}\eta_k)$ and condition (33) holds.

Proceeding analogously as in the proof of Theorem 4.1 [4] one can show that any function from Θ satisfying (33) has the form (34). □

Remark. Similarly as for the case of the generalized sum $A\tilde{+}\mathbb{V}$ the parameter λ enters additively the construction of $A_\lambda\hat{+}\mathbb{V}_{-2}$ where $A_\lambda \equiv A = -\Delta + \lambda$. i.e. for λ, $\beta > 0$ we have

$$D(A_\lambda\hat{+}\mathbb{V}_{-2}) = D(A_\beta\hat{+}\mathbb{V}_{-2}) \quad \text{and} \quad A_\lambda\hat{+}\mathbb{V}_{-2} = (A_\beta\hat{+}\mathbb{V}_{-2}) + (\lambda - \beta). \tag{35}$$

This fact follows from the definitions (17) and (18) which can be rewritten in the form

$$D(A_\lambda\hat{+}\mathbb{V}_{-2}) = \{f \in D(C_s^{-1}) : \mathbb{A}_\lambda f + \mathbb{V}_{-2}C_s^{-1}f \in L^2\}$$

and

$$(A_\lambda\hat{+}\mathbb{V}_{-2})f = \mathbb{A}_\lambda f + \mathbb{V}_{-2}C_s^{-1}f, \quad f \in D(A_\lambda\hat{+}\mathbb{V}_{-2}).$$

Since operator $\mathbb{V}_{-2}C_s^{-1}$ is independent of the parameter λ we can conclude that (35) is valid.

Analogously as in the remark of Section 3 we can extend the above construction of formal sum $A_\lambda\hat{+}\mathbb{V}_{-2}$ to the free Laplacian. For this aim let us define the set

$$D(-\Delta\hat{+}\mathbb{V}_{-2}) = \{f \in D(C_s^{-1}) : -\bar{\mathbf{\Delta}} f + \mathbb{V}_{-2} f_r \in L^2\}$$

where $-\bar{\mathbf{\Delta}} = \mathbb{A}_\lambda - \lambda$, $f_r = C_s^{-1} f$ and operator $A\hat{+}\mathbb{V}_{-2}$ which acts as

$$(-\Delta\hat{+}\mathbb{V}_{-2})f = -\mathbf{\Delta} f + \mathbb{V}_{-2} f_r, \quad f \in D(-\Delta\hat{+}\mathbb{V}_{-2}).$$

Obviously, we have

$$D(-\Delta\hat{+}\mathbb{V}_{-2}) = D(A_\lambda\hat{+}\mathbb{V}_{-2})$$

and

$$(-\Delta\hat{+}\mathbb{V}_{-2})f = ((A_\lambda\hat{+}\mathbb{V}_{-2}) - \lambda)f \quad \text{for any } f \in D(-\Delta\hat{+}\mathbb{V}_{-2}).$$

References

[1] S. Albeverio, P. Kurasov, *Singular Perturbation of Differential Operators,* London Math. Soc. Note Series, **271**, (2000).

[2] S. Albeverio, W. Karwowski, V. Koshmanenko, *Square Power of Singularly Perturbed Operators*, Math. Nachr. **173** (1995), 5–24.

[3] W. Karwowski, V. Koshmanenko, S. Ôta, *Schrödinger operator perturbed by operators related to null-sets*, Positivity **2**, no. 1 (1998), 77–99.

[4] S. Kondej, *Singular Pertrubation of Laplace Operator in the terms of Boundary Conditions*, (printed in Positivity).

[5] S. Kondej, *Singular Perturbation of Laplace Operator living on Null Sets,* master thesis (in Polish), (1996).

[6] Yu. Berezanskij, *The bilinear forms and Hilbert equipment, Spectral analysis of differential operators, Institute of Mathematics*, Kiev (1980).

[7] M.G. Krein and V.A. Yavrian, *Spectral shift functions arising in perturbations of a positive operator*, J. Operator Theory **6** (1981), 155–191.

[8] V. Koshmanenko, *Perturbation of self-adjoint operators by singular bilinear forms,* Ukrainian Math. J.43, no.**11** (1991), 1559–1566.

[9] V. Koshmanenko, *Singular Operator as a Parameter of Self-adjoint Extensions*, Op. Th. Adv. Appl., vol. **118**, (2000).

[10] S. Albeverio, W. Karwowski, V. Koshmanenko, *On Negative Eigenvalue problems of Generalized Laplace Operator*, to be published.

[11] S. Kondej, *On Eigenvalue Problems for Self-adjoint Operators with Singular Perturbation,* to be published (available at http://arXiv.org/abs/math.FA/0104028).

[12] Y.V. Kurylev, *Boundary conditions on a curve for a three-dimensional Laplace operator*, J.Sov.Math. **22**, 1072–1082, (1983).

W. Karwowski
Institute of Theoretical Physics
University of Wrocław
Pl. Maxa Borna 9
50-205 Wrocław, Poland

S. Kondej
Institute of Theoretical Physics
University of Wrocław
Pl. Maxa Borna 9
50-205 Wrocław, Poland

Operator Theory:
Advances and Applications, Vol. 132, 245–251

Ergodicity in the p-adic Framework

Andrei Khrennikov, Karl-Olof Lindahl, and Matthias Gundlach

Abstract. Monomial mappings, $x \mapsto x^n$, are topologically transitive and ergodic with respect to Haar measure on the unit circle in the complex plane. In this paper we obtain an analogous result for monomial dynamical systems over p-adic numbers. The process is, however, not straightforward. The result will depend on the natural number n. Moreover, in the p-adic case we never have ergodicity on the unit circle, but on the circles around the point 1.

1. Introduction

Investigations in p-adic quantum physics [1]–[5] (especially string theory [1], [2], [3]) stimulated an increasing interest in studying p-adic dynamical systems, see for example [5]–[11]. Some steps in this direction [5] demonstrated that even the simplest (monomial) discrete dynamical systems over the fields of p-adic numbers $\mathbb{Q}_p$ have quite complex behavior. This behavior depends crucially on the prime number p (which determines $\mathbb{Q}_p$). By varying p we can transform attractors into centers of Siegel discs and vice versa. The number of cycles and their lengths also depend crucially on p [5].

Some applications of discrete p-adic dynamical systems to cognitive sciences and neural networks were considered in [5], [6]. Some of these cognitive models are described by random dynamical systems in the fields of p-adic numbers, see [9]. In the present paper we study ergodicity of monomial p-adic dynamical systems on spheres. For a system $\psi_n(x) = x^n$, $n = 2, 3, \ldots$, the result depends crucially on the relation between n and p. Our proof is essentially based on p-adic analysis (analytic mappings), [12]. We remark that the corresponding fact for the field of complex numbers, $\psi_n : \mathbb{C} \to \mathbb{C}$, for the sphere $|z| = 1$, is rather trivial, see [13].

2. p-adic numbers

The field of p-adic numbers is denoted by the symbol $\mathbb{Q}_p$; $|\cdot|_p$ is the p-adic valuation. The p-adic valuation satisfies the strong triangle inequality

$$|x + y|_p \leq \max[|x|_p, |y|_p], \quad x, y \in \mathbb{Q}_p, \tag{1}$$

with equality in the case that $|x|_p \neq |y|_p$.

Write $B_r(a) = \{x \in \mathbb{Q}_p : |x-a|_p \leq r\}$ and $B_r^-(a) = \{x \in \mathbb{Q}_p : |x-a|_p < r\}$ where $r = p^n$ and $n = 0, \pm 1, \pm 2, \ldots$. These are the "closed" and "open" balls in $\mathbb{Q}_p$ while the sets $S_r(a) = \{x \in K : |x-a|_p = r\}$ are the spheres in $\mathbb{Q}_p$ of such radii r. Any p-adic ball $B_r(0)$ is an additive subgroup of $\mathbb{Q}_p$, while the ball $B_1(0)$ is also a ring, which is called the *ring of p-adic integers* and is denoted by $\mathbb{Z}_p$.

Let (m, n) denote the greatest common divisor of m and n. We say that n is a (multplicative) *unit* (with respect to the prime number p) iff $(n,p) = 1$. Let G_{p^l}, $l \geq 1$, be the multiplicative group of units in the residue field modulo p^l. Let us by $\langle n \rangle = \{n^N : \quad N \in \mathbb{N}\}$ denote the set *generated* by n.

Let ψ_n be a (monomial) mapping on $\mathbb{Z}_p$ taking x to x^n. Then all spheres $S_{p^{-l}}(1)$ are ψ_n-invariant iff n is a multplicative unit. This is a consequence of the following result in p-adic analysis, which can be obtained from Schikhof's book [12]:

Lemma 1. *Let $x, y \in S_1(0)$ and suppose $|x-y|_p < 1$. Then for all natural numbers n,*

$$|x^n - y^n|_p \leq |n|_p \, |x-y|_p \,, \tag{2}$$

with equality for $p > 2$. Moreover equality also holds for $p = 2$ if n is odd.

In particular ψ_n is an isometry on $S_{p^{-l}}(1)$ if and only if $(n,p) = 1$. Therefore we will henceforth assume that n is a unit. Also note that, as a consequence, $S_{p^{-l}}(1)$ is not a group under multiplication. Thus our investigations are not about the dynamics on a compact (abelian) group.

3. Minimality

Let us consider the dynamical system $x \mapsto x^n$ on spheres $S_{p^{-l}}(1)$. The result depends crucially on the following well-known result from group theory.

Lemma 2. *Let $p > 2$ and l be any natural number, then the natural number n is a generator of G_{p^l} if and only if n is a generator of G_{p^2}. G_{2^l} is noncyclic for $l \geq 3$.*

Recall that a dynamical system given by a continuous transformation ψ on a compact metric space X is called *topologically transitive* if there exists a dense orbit $\{\psi^n(x) : n \in \mathbb{N}\}$ in X, and (one-sided) *minimal*, if all orbits for ψ in X are dense. For the case of monomial systems $x \mapsto x^n$ on spheres $S_{p^{-l}}(1)$ topological transitivity means the existence of an $x \in S_{p^{-l}}(1)$ s.t. each $y \in S_{p^{-l}}(1)$ is a limit point in the orbit of x, i.e. can be represented as

$$y = \lim_{k \to \infty} x^{n^{N_k}}, \tag{3}$$

for some sequence $\{N_k\}$, while minimality means that such a property holds for any $x \in S_{p^{-l}}(1)$. Our investigations are based on the following theorem.

Theorem 3. *For $p \neq 2$ the set $\langle n \rangle$ is dense in $S_1(0)$ if and only if n is a generator of G_{p^2}.*

Proof. We have to show that for every $\epsilon > 0$ and every $x \in S_1(0)$ there is a $y \in \langle n \rangle$ such that $|x-y|_p < \epsilon$. Let $\epsilon > 0$ and $x \in S_1(0)$ be arbitrary. Because of the discreteness of the p-adic metric we can assume that $\epsilon = p^{-k}$ for some natural number k. But (according to Lemma 2) if n is a generator of G_{p^2}, then n is also a generator of G_{p^l} for every natural number l (and $p \neq 2$) and especially for $l = k$. Consequently there is an N such that $n^N = x \mod p^k$. From the definition of the p-adic metric we see that $|x-y|_p < p^{-k}$ if and only if $x \equiv y \mod p^k$. Hence we have that $\left|x - n^N\right|_p < p^{-k}$. □

Let us consider $p \neq 2$ and for $x \in B_{p^{-1}}(1)$ the p-adic exponential function $t \mapsto x^t$, see, for example [12]. This function is well defined and continuous as a map from $\mathbb{Z}_p$ to $\mathbb{Z}_p$. In particular, for each $a \in \mathbb{Z}_p$, we have

$$x^a = \lim_{k \to a} x^k, \quad k \in \mathbb{N}. \tag{4}$$

We shall also use properties of the p-adic logarithmic function, see, for example [12]. Let $z \in B_{p^{-1}}(1)$. Then $\log z$ is well defined. For $z = 1 + \lambda$ with $|\lambda|_p \leq 1/p$, we have:

$$\log z = \sum_{k=1}^{\infty} \frac{(-1)^{k+1}\lambda^k}{k} = \lambda(1 + \lambda\Delta_\lambda), \quad |\Delta_\lambda|_p \leq 1. \tag{5}$$

By using (5) we obtain that $\log : B_{p^{-1}}(1) \to B_{p^{-1}}(0)$ is an isometry:

$$|\log x_1 - \log x_2|_p = |x_1 - x_2|_p, \quad x_1, x_2 \in B_{1/p}(1) \, . \tag{6}$$

Lemma 4. *Let $x \in B_{p^{-1}}(1), x \neq 1, a \in \mathbb{Z}_p$ and let $\{m_k\}$ be a sequence of natural numbers. If $x^{m_k} \to x^a, k \to \infty$, then $m_k \to a$ as $k \to \infty$, in $\mathbb{Z}_p$.*

This is a consequence of the isometric property of log.

Theorem 5. *Let $p \neq 2$ and $l \geq 1$. Then the monomial dynamical system $x \mapsto x^n$ is minimal on the circle $S_{p^{-l}}(1)$ if and only if n is a generator of G_{p^2}.*

Proof. Let $x \in S_{p^{-l}}(1)$. Consider the equation $x^a = y$. What are the possible values of a for $y \in S_{p^{-l}}(1)$? We prove that a can take an arbitrary value from the sphere $S_1(0)$. We have that $a = \frac{\log x}{\log y}$. As $\log : B_{p^{-1}}(1) \to B_{p^{-1}}(0)$ is an isometry, we have $\log(S_{p^{-l}}(1)) = S_{p^{-l}}(1)$. Thus $a = \frac{\log x}{\log y} \in S_1(0)$ and moreover, each $a \in S_1(0)$ can be represented as $\frac{\log x}{\log y}$ for some $y \in S_{p^{-l}}(1)$.

Let y be an arbitrary element of $S_{p^{-l}}(1)$ and let $x^a = y$ for some $a \in S_1(0)$. By Theorem 3 if n is a generator of G_{p^2}, then each $a \in S_1(0)$ is a limit point of the sequence $\{n^N\}_{N=1}^{\infty}$. Thus $a = \lim_{k\to\infty} n^{N_k}$ for some subsequence $\{N_k\}$. By using the continuity of the exponential function we obtain (3).

Suppose now that, for some n, $x^{n^{N_k}} \to x^a$. By Lemma 4 we obtain that $n^{N_k} \to a$ as $k \to \infty$. If we have (3) for all $y \in S_{p^{-l}}(1)$, then each $a \in S_1(0)$ can be approximated by elements n^N. In particular, all elements $\{1, 2, \ldots, p-1, p+1, \ldots, p^2-1\}$ can be approximated with respect to mod p^2. Thus n is a generator of G_{p^2}. □

Example. In the case that $p = 3$ we have that ψ_n is minimal if $n = 2$, 2 is a generator of $U_{3^2} = \{1, 2, 4, 5, 7, 8\}$. But for $n = 4$ it is not; $\langle 4 \rangle \mod 3^2 = \{1, 4, 7\}$. We can also see this by noting that $S_{1/3}(1) = B_{1/3}(4) \cup B_{1/3}(7)$ and that $B_{1/3}(4)$ is invariant under ψ_4.

Corollary 6. *If a is a fixed point of the monomial dynamical system $x \mapsto x^n$, then this is minimal on $S_{p^{-1}}(a)$ if and only if n is a generator of G_{p^2}.*

Proof. The assertion follows immediately from Theorem 5 by topological conjugation via $x \mapsto x/a$, $S_{p^{-l}}(a) \mapsto S_{p^{-l}}(1)$. □

Example. Let $p = 17$ and $n = 3$. On $\mathbb{Q}_{17}$ there is a primitive 3rd root of unity. Moreover, 3 is also a generator of G_{17^2}. Therefore there exist nth roots of unity different from 1 around which the dynamics is minimal.

4. Unique ergodicity

In the following we will show that the minimality of the monomial dynamical system $\psi_n : x \mapsto x^n$ on the sphere $S_{p^{-l}}(1)$ is equivalent to its *unique ergodicity*. The latter property means that there exists a unique probability measure on $S_{p^{-l}}(1)$ and its Borel σ-algebra which is invariant under ψ_n. We will see that this measure is in fact the normalized restriction of the Haar measure on $\mathbb{Z}_p$. Moreover we will also see that the ergodicity with respect to Haar measure of ψ_n is also equivalent to its unique ergodicity. We should point out that — though many results are analogous to the case of the (irrational) rotation on the circle, our situation is quite different, in particular as we do not deal with dynamics on topological subgroups.

Lemma 7. *Assume that ψ_n is minimal. Then the Haar measure m is the unique ψ_n-invariant measure on $S_{p^{-l}}(1)$.*

Proof. First note that minimality of ψ_n implies that $(n, p) = 1$ and hence that ψ_n is an isometry on $S_{p^{-l}}(1)$, see equation (2). Then, as a consequence of Theorem 27.5 in [12], it follows that $\psi_n(B_r(a)) = B_r(\psi_n(a))$ for each ball $B_r(a) \subset S_{p^{-l}}(1)$. Consequently, for every open set $U \neq \emptyset$ we have $S_{p^{-l}}(1) = \cup_{N=0}^{\infty} \psi_n^N(U)$. Then it follows for a ψ_n-invariant measure μ that $\mu(U) > 0$.

Moreover we can split $S_{p^{-l}}(1)$ into disjoint balls of radii $p^{-(l+k)}$, $k \geq 1$, on which ψ_n acts as a permutation. In fact, for each $k \geq 1$, $S_{p^{-l}}(1)$ is the union,

$$S_{p^{-l}}(1) = \cup B_{p^{-(l+k)}}(1 + b_l p^l + \cdots + b_{l+k-1} p^{l+k-1}), \tag{7}$$

where $b_i \in \{0, 1, \ldots, p-1\}$ and $b_l \neq 0$.

We now show that ψ_n is a permutation on the partition (7). Recall that every element of a p-adic ball is the center of that ball (see Section 2), and as pointed out above $\psi_n(B_r(a)) = B_r(\psi_n(a))$. Consequently we have for all positve integers k, $\psi_n^k(a) \in B_r(a) \Rightarrow \psi_n^k(B_r(a)) = B_r(\psi_n^k(a)) = B_r(a)$ so that $\psi_n^{Nk}(a) \in B_r(a)$ for every natural number N. Hence, for a minimal ψ_n a point of a ball B of the partition (7) must move to another ball in the partition. Furthermore the minimality of ψ_n shows indeed that ψ_n acts as a permutation on balls. By invariance of μ all balls must have the same positive measure. As this holds for any k, μ must be the restriction of Haar measure m. □

The arguments of the proof of Lemma 7 also show that Haar measure is always ψ_n-invariant. Thus if ψ_n is uniquely ergodic, the unique invariant measure must be the Haar measure m. Under these circumstances it is known ([13], Theorem 6.17) that ψ_n must be minimal.

Theorem 8. *The monomial dynamical system $\psi_n : x \mapsto x^n$ on $S_{p^{-l}}(1)$ is minimal if and only if it is uniquely ergodic in which case the unique invariant measure is the Haar measure.*

Let us mention that unique ergodicity yields in particular the ergodicity of the unique invariant measure, i.e. the Haar measure m, which means that

$$\frac{1}{N}\sum_{i=0}^{N-1} f(x^{n^i}) \to \int f\,dm \text{ for all } x \in S_{p^{-l}}(1), \tag{8}$$

and all continuous functions $f : \quad S_{p^{-l}}(1) \to \mathbb{R}$.

On the other hand the arguments of the proof of Lemma 7, i.e. the fact that ψ_n acts as a permutation on each partition of $S_{p^{-l}}(1)$ into disjoint balls if and only if $\langle n\rangle = G_{p^2}$, proves that if n is not a generator of G_{p^2} then the system is not ergodic with respect to Haar measure. Consequently, if ψ_n is ergodic then $\langle n\rangle = G_{p^2}$ so that the system is minimal by Theorem 5, and hence even uniquely ergodic by Theorem 8. Since unique ergodicity implies ergodicity one has the following:

Theorem 9. *The monomial dynamical system $\psi_n : x \mapsto x^n$ on $S_{p^{-l}}(1)$ is ergodic with respect to Haar measure if and only if it is uniquely ergodic.*

Even if the monomial dynamical system $\psi_n : x \mapsto x^n$ on $S_{p^{-l}}(1)$ is ergodic, it never can be mixing, especially not weak-mixing. This can be seen from the fact that an abstract dynamical system is weak-mixing if and only if the product of such two systems is ergodic. If we choose a function f on $S_{p^{-l}}(1)$ and define a function F on $S_{p^{-l}}(1) \times S_{p^{-l}}(1)$ by $F(x,y) := f(\log x/\log y)$ (which is well defined as log does not vanish on $S_{p^{-l}}(1)$), we obtain a non-constant function satisfying $F(\psi_n(x), \psi_n(y)) = F(x,y)$. This shows (see [13], Theorem 1.6) that $\psi_n \times \psi_n$ is not ergodic, and hence ψ_n is not weak-mixing with respect to any invariant measure, in particular the restriction of Haar measure.

Let us consider the ergodicity of a perturbed system

$$\psi_q = x^n + q(x), \tag{9}$$

for some polynomial q such that $q(x) \equiv 0 \mod p^{l+1}$, $(|q(x)|_p < p^{-(l+1)})$. This condition is necessary in order to guarantee that the sphere $S_{p^{-l}}(1)$ is invariant. For such a system to be ergodic it is necessary that n is a generator of G_{p^2}. This follows from the fact that for each $x = 1 + a_l p^l + \dots$ on $S_{p^{-l}}(1)$ (so that $a_l \neq 0$) the condition on q gives

$$\psi_q^N(x) \equiv 1 + n^N a_l \mod p^{l+1}. \tag{10}$$

Now ψ_q acts as a permutation on the $p-1$ balls of radius $p^{-(l+1)}$ if and only if $\langle n \rangle = G_{p^2}$. Consequently, a perturbation (9) cannot make a nonergodic system ergodic.

References

[1] I. Volovich, p-adic string, *Class. Quant. Grav.*, **4**, 83–87, 1987.

[2] P. Freund, E. Witten, Adelic string amplitudes, *Phys. Lett.*, **B 199**, 191–195, 1987.

[3] V. Vladimirov, I. Volovich, and E. Zelenov, *p-adic Analysis and Mathematical Physics.* Singapore, World Scientific Publ., 1994.

[4] A. Khrennikov, *p-adic Valued Distributions in Mathematical Physics*, Dordrecht, Kluwer Academic Publ., 1994.

[5] A. Khrennikov, *Non-Archimedean Analysis: Quantum Paradoxes, Dynamical Systems and Biological Models*, Dordrecht, Kluwer Academic Publ., 1997.

[6] S. Albeverio, A. Khrennikov, P. Kloeden, Memory retrieval as a p-adic dynamical system, *Biosystems*, **49**, 105–115, 1999.

[7] S. Albeverio, A. Khrennikov, B. Tirozzi, p-adic neural networks, *Math. Models and Methods in Appl. Sc.*, **9**, 1417–1437, 1999.

[8] M. Baake, R. Moody, M. Schlottmann, Limit-(quasi) periodic point sets as quasicristals with p-adic internal spaces, *J. Phys. A: Math. Gen.*, **31**, 5755–5765, 1998.

[9] D. Dubischar, M. Gundlach, O. Steinkamp, A. Khrennikov, Attractors of random dynamical systems over p-adic numbers and a model of noisy cognitive processes, *Physica*, **D 130**, 1–12, 1999.

[10] A. Khrennikov, On the problem of small denominators in the field of complex p-adic numbers, *Reports of Vaxjo University*, **9**, 1–18, 2000.

[11] M. Gundlach, A. Khrennikov, *Ergodicity on p-adic sphere.* Abstracts of German Open Conference on Probability and Statistics, Hamburg, 2000, p. 61.

[12] W. Schikhof, *Ultrametric calculus*, Cambridge, University press, 1984.

[13] P. Walters, *An introduction to ergodic theory*, Berlin-Heidelberg-New York, Springer, 1982.

Andrei Khrennikov
School of Mathematics and Systems Engineering
Växjö University
Sweden
e-mail: Andrei.Khrennikov@msi.vxu.se

Karl-Olof Lindahl
School of Mathematics and Systems Engineering
Växjö University
Sweden
e-mail: Karl-Olof.Lindahl@msi.vxu.se

Matthias Gundlach
Institut für Dynamische Systeme
Universität Bremen
Germany

Operator Theory:
Advances and Applications, Vol. 132, 253–266

On the Resolvent Estimates for the Generators of Strongly Continuous Groups in the Hilbert Spaces

Alexander V. Kiselev

Abstract. C_0-groups of Hilbert space operators with polynomial and more general functional growth are considered. We investigate how the polynomial growth condition can be expressed in terms of integral estimates for the resolvent of the group generator. In the case of C_0-groups with functional growth such necessary and sufficient conditions are given when the generator admits a functional model representation and has a purely absolutely continuous spectrum.

1. Introduction

A group of bounded linear operators $T(t)$ acting in a Hilbert space H is called [1] a *strongly continuous*, or C_0*-group*, if

$$\lim_{t\to 0} T(t)x = x$$

for all $x \in H$. A C_0-group $T(t)$ is called a *group with polynomial growth* if, further, the following condition is satisfied:

$$\|T(t)\| \leq M(1+|t|^s)$$

for all $t \in \mathbb{R}$, a real $s > 0$ and $M \in (0, \infty)$.

The linear operator L, defined by

$$Lx = \lim_{t\to 0} \frac{1}{it}(T(t)x - x), \quad x \in D(L),$$

is called the *generator* of the corresponding C_0-group, the domain $D(L)$ of L being the set of elements $x \in H$, for which the limit exists in H.

The *uniform boundedness* (i.e., the condition $\|T(t)\| \leq M$ for all real t) of a C_0-group of operators in a Hilbert space is necessary and sufficient for the similarity of its generator to a self-adjoint operator [1].

The research was partially supported by the grants RFFI No. 00-01-00479, No. 01-01-06051.

In [10, 13] the condition

$$\sup_{\varepsilon>0} \varepsilon \int_{-\infty}^{\infty} \left(\|(L-k-i\varepsilon)^{-1}u\|^2 + \|(L^*-k-i\varepsilon)^{-1}u\|^2\right) dk \le C\|u\|^2 \tag{1}$$

for all $u \in H$ is proved to be equivalent to the uniform boundedness of the C_0-group $\exp(iLt)$, generated by the operator L.

In [12] this effective criterion, establishing the correspondence between the possible group growth and the resolvent properties of its generator, is generalized to the case of a C_0-group with polynomial growth. Namely, the following two theorems hold.

Theorem 1. *Let L be a densely defined closed operator in a Hilbert space H. Suppose that its spectrum $\sigma(L)$ is real. If there exist a $d > 0$ and a finite $M > 0$ such that the estimates*

$$\begin{aligned} \int_{\mathbb{R}} \|(L-k\pm i\varepsilon)^{-1}u\|^2 dk &\le M\frac{1}{\varepsilon}\left(1+\frac{1}{\varepsilon^d}\right)\|u\|^2 \\ \int_{\mathbb{R}} \|(L^*-k\pm i\varepsilon)^{-1}u\|^2 dk &\le M\frac{1}{\varepsilon}\left(1+\frac{1}{\varepsilon^d}\right)\|u\|^2 \end{aligned} \tag{2}$$

hold for all $\varepsilon > 0$ and all $u \in H$, then the operator L generates the C_0-group $\exp(iLt)$ in H and there exists a finite positive constant C such that

$$\|\exp(iLt)\| \le C(1+|t|^d) \tag{3}$$

for all $t \in \mathbb{R}$.

Theorem 2. *If a densely defined closed operator L generates the C_0-group $\exp(iLt)$ with polynomial growth in the Hilbert space H, i.e., there exist an $s > 0$ and a finite positive constant C such that*

$$\|\exp(itL)\| \le C(1+|t|^s)$$

for all $t \in \mathbb{R}$, then there exists a positive constant $M = M(C,s) < \infty$ such that the estimates (2) hold for all $\varepsilon > 0$, and $d = 2s$.

Note, that in Theorems 1, 2, contrary to the case of uniformly bounded C_0-groups in a Hilbert space H, there exists a gap between the necessary and sufficient conditions for polynomial growth.

In Section 2 we provide an explicit construction that proves the tightness of this gap in the power scale.

A further generalization of theorems 1 and 2 to the case of functionally bounded C_0-groups (see p. 262 for definition) is discussed in Section 3. Here we restrict ourselves to non-self-adjoint operators L which admit a functional model representation and have a purely absolutely continuous spectrum. The explicit model description of the operator $\exp(iLt)$ allows us to obtain results quite similar to those of Theorems 1 and 2, and these results coincide with the latter in the case of polynomially bounded C_0-groups.

The description of the functional model used is contained in Section 3.1. Theorems 4 and 5 of Section 3.2 present results, obtained by means of the functional model approach. These are then applied to the operator of the non-self-adjoint Friedrichs model in Section 3.3.

2. The gap between the necessary and sufficient conditions

In the present section we prove the following result showing that the conditions of Theorem 2 are exact at least in the power scale. The main result in this direction is given by Theorem 3, which with the help of the Cayley transformation is reduced to the explicit construction provided by Lemmas 1 and 2 below.

Theorem 3. *For any $\gamma > 0$ and any $\delta > 0$ there exist an operator L_δ in the Hilbert space H and a vector $u \in H$, such that:*

(1) $\|\exp(iL_\delta t)\| \leq 1 + |t|^\gamma$ *and* $\|\exp(iL_\delta t_i)\| = |t_i|^\gamma$ *for some subsequence t_i of real numbers, satisfying the condition* $\lim_{i\to\infty} |t_i| = \infty$;

(2) *there exists a positive constant $C = C(\delta, \gamma)$ such that the following estimate holds for all sufficiently small $\varepsilon > 0$:*

$$\int_{-\infty}^{\infty} \|(L_\delta - k + i\varepsilon)^{-1}u\|^2 dk \geq C\frac{1}{\varepsilon^{1+2\gamma-\delta}}.$$

In the following lemma we present an explicit construction of a bounded operator on a Hilbert space H which generates a discrete polynomially bounded C_0-group of operators on H (see [1]), and whose Cayley transformation has all the properties mentioned in Theorem 3 for the case $\gamma = 1$.

Lemma 1. *For any $\delta > 0$ there exist an operator T_δ acting in the Hilbert space H and a vector $u \in H$ such that:*

(1) $\|T_\delta^n\| \leqslant 1 + |n|$ *for all $n \in \mathbb{Z}$ and* $\|T_\delta^{n_j}\| = |n_j|$ *for some subsequence n_j of integers such that* $\lim_{j\to\infty} |n_j| = \infty$;

(2) *there exists a positive constant $C = C(\delta)$ such that the following estimate holds for all $r > 1$, r sufficiently close to 1*

$$\int_0^{2\pi} \|(T_\delta - re^{i\theta})^{-1}u\| d\theta \geqslant C\frac{1}{(r-1)^{3-\delta}}.$$

Proof. Without loss of generality we choose l^2 (the linear set of infinite sequences $u = \{u_i\}_{i=-\infty}^{\infty}$ closed w.r.t. the norm $\|u\| = (\sum_{i=-\infty}^{\infty} |u_i|^2)^{1/2}$) as the Hilbert space H and construct the weighted shift operator T_δ in it. Then we check that this operator satisfies the requirements of Lemma 1.

Let T_j be defined for each $j = 1, 2, \ldots$ by the following identity:

$$(T_j u)_{(k+1)} \equiv a_k u_k, \qquad u = \{u_k\}_{k=-\infty}^{\infty} \in l^2,$$

where for each $j = 1, 2, \ldots$ the sequence $a \equiv \{a_k\}_{k=-\infty}^{\infty}$ has the block structure

$$a \equiv \{\ldots, A_{k^j}, \ldots, A_{2^j}, A_1, A_0, A_1, A_{2^j}, \ldots, A_{k^j}, \ldots\}, \tag{4}$$

and the blocks A_n are chosen as follows:

$$A_n \equiv \left\{ \underbrace{n^{1/n}}_{\text{n times}}, \underbrace{\frac{1}{n^{1/n}}}_{\text{n times}} \right\}; \quad A_0 \equiv \{1\}.$$

Therefore, in the case of $j=1$ the sequence a is given by the following identity:

$$a \equiv \{ \underbrace{\cdots\cdots}_{\text{symmetrically defined}}, \underbrace{1}_{\text{0th cell}}, 1, \frac{1}{1}, \sqrt[2]{2}, \sqrt[2]{2}, \frac{1}{\sqrt[2]{2}}, \frac{1}{\sqrt[2]{2}}, \dots, \underbrace{\sqrt[n]{n}}_{n \text{ times}}, \underbrace{\frac{1}{\sqrt[n]{n}}}_{n \text{ times}}, \dots\} \quad (5)$$

Since T_j is a weighted shift operator, one easily obtains for the norms of its positive powers:

$$\|T_j^n\| = \max_{i_0} \prod_{i_0}^{i_0+n} a_i,$$

which, in turn, implies (by the analogous consideration of the inverse of T_j), that $\|T_j^n\| \le 1+|n|$ for all integers n and there exists a subsequence $\{n_k\}_{k=1}^{\infty}$ of integers such that $\lim_{k\to\infty} |n_k| = \infty$ and $\|T_j^{n_k}\| = |n_k|$.

If u is the vector

$$u \equiv (\dots, 0, \dots, 0, \underbrace{1}_{\text{1st cell}}, 0, \dots, 0, \dots)$$

then

$$(T_j - \lambda)^{-1} u = \left(\dots, 0, -\frac{1}{\lambda}, -\frac{a_1}{\lambda}, -\frac{a_1 a_2}{\lambda^2}, \dots, -\frac{a_1 \cdots a_n}{\lambda^{n+1}}, \dots \right)$$

and for arbitrary $\lambda = |\lambda| e^{i\theta}$ with $|\lambda| > 1$ we have

$$\int_0^{2\pi} \|(T_j - \lambda)^{-1} u\|^2 d\theta = 2\pi \|(T_j - \lambda)^{-1} u\|^2$$

$$= 2\pi \left(\frac{1}{|\lambda|^2} + \frac{a_1^2}{|\lambda|^4} + \dots + \frac{a_1^2 \cdots a_n^2}{|\lambda|^{2n+2}} + \dots \right) = 2\pi \sum_{k=0}^{\infty} \frac{\prod_{i=0}^k a_i^2}{|\lambda|^{2k+2}}. \quad (6)$$

To simplify the notation we denote $b_k \equiv \prod_{i=0}^k a_i$. The behaviour of $b_k \equiv b(k)$ is illustrated by Figure 1.

From the definition of b_k it is clear that in the case of $j=1$ the sequence of maxima of b_k for $k \geqslant 3$ is equal to the sequence $\{\|T_1^k\|\}_{k=3}^{\infty}$. Solving the equation $b_k \equiv b(k) = m$ with respect to k leads to $k(m) = m^2$, and thus $m = \sqrt{k}$, where m is the m-th member of the sequence of maxima of b_k (see Figure 1).

For $j=2$ we obtain the sequence of maxima of b_k equal to $\{1, 4, 9, 16, \dots\}$. In the general case (denoting the whole sequence of maxima of b_k by the same symbol m) we get $m = \{k^j\}_{k=1}^{\infty}$ and therefore $m = \{1, \{\|T_j^{k^j}\|\}_{k=2}^{\infty}\}$, $j = 2, 3, \dots$.

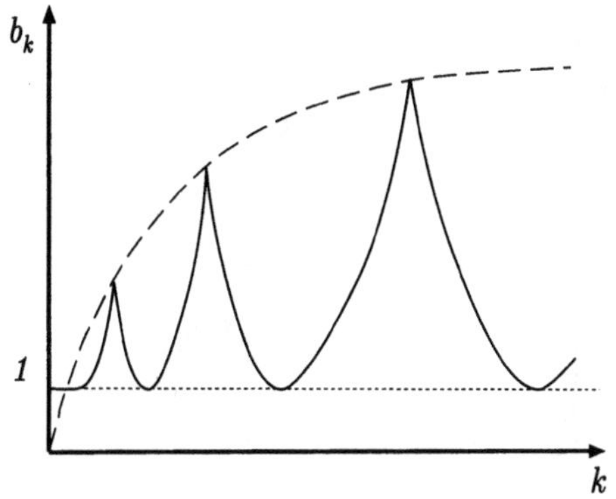

FIGURE 1. Illustration of the behaviour of $b_k \equiv b(k)$.

In the case of an arbitrary j the sequence $k_j(m)$ (the values k for which b_k is maximal) cannot be calculated in the same simple manner as for $j = 1$:

$$k_j(m) = 2\left(\sum_{l=1}^{m^{1/j}-1} l^j\right) + m$$
$$\equiv \frac{d_j}{2} m^{(j+1)/j} + d_j^{(1)} m + d_j^{(2)} m^{(j-1)/j} + \cdots + d_j^{(j)} m^{1/j},$$

but for large enough values of m we can use the asymptotic formula

$$k_j(m) \sim \frac{d_j}{2} m^{\frac{j+1}{j}}, \quad m \gg 1.$$

Leaving all the non-maximal values of b_k out of the sum in (6), we get the following estimate for it:

$$\int_0^{2\pi} \|(T_j - \lambda)^{-1} u\|^2 d\theta \geqslant \sum_{s=1}^{\infty} \frac{s^{2j}}{|\lambda|^{2k_j(s^j)+2}} \geqslant \sum_{s=s_0}^{\infty} \frac{s^{2j}}{|\lambda|^{2d_j s^{j+1}+2}},$$

where we have chosen s_0 depending on j such that

$$\frac{d_j}{2} s^{j+1} \geqslant d_j^{(1)} s^j + \cdots + d_j^{(j)} s \text{ for } s \geqslant s_0.$$

Taking into consideration that the terms of the latter sum are decaying with respect to s when

$$s > \left(\frac{j}{d_j(1+j)}\right)^{1/(1+j)} \left(\frac{1}{\log|\lambda|}\right)^{1/(1+j)} \equiv C_j (\log|\lambda|)^{-1/(1+j)},$$

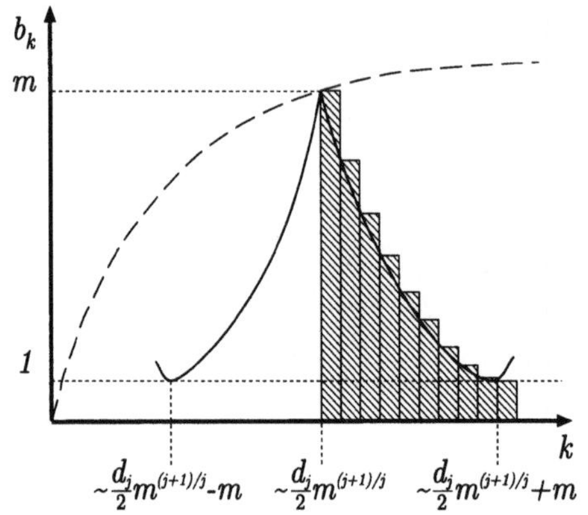

FIGURE 2. Detailed behaviour of $b_k \equiv b(k)$

we can easily obtain that for $|\lambda|$ sufficiently close to 1

$$\int_0^{2\pi} \|(T_j-\lambda)^{-1}u\|^2 d\theta \geqslant |\lambda|^{-2} \int_{C_j(\log|\lambda|)^{-1/(1+j)}}^{\infty} \frac{x^{2j}}{|\lambda|^{2d_j x^{j+1}}} dx$$

$$= \frac{1}{|\lambda|^2} \frac{1}{1+j} \left(\frac{1}{2d_j}\right)^{2-\frac{1}{1+j}} \left(\frac{1}{\log|\lambda|}\right)^{2-\frac{1}{1+j}} \int_{2d_j C_j^{j+1}}^{\infty} e^{-t} t^{\frac{j}{j+1}} dt$$

$$\geqslant \frac{1}{|\lambda|^2} \frac{1}{1+j} \left(\frac{1}{2d_j}\right)^{2-\frac{1}{1+j}} \frac{1}{(\log|\lambda|)^{2-\frac{1}{1+j}}} \int_2^{\infty} e^{-t} t^{\frac{j}{j+1}} dt. \quad (7)$$

Combined with (6) this yields

$$\int_0^{2\pi} \|(T_j-\lambda)^{-1}u\|^2 d\theta \geqslant C(j) \frac{1}{\epsilon^{2-1/(j+1)}}, \quad \text{where} \quad \epsilon \equiv |\lambda| - 1 \ll 1.$$

In the computation above we have left out all the non-maximal values of the sequence b_k. We now take them into consideration (see Figure 2, where a single "period" of the sequence b_k is shown) as follows. First we estimate the part of the sum (6), corresponding to the m-th "period" of this graph:

$$\sum_{k=k_j(m)-m}^{k_j(m)+m} \frac{b_k^2}{|\lambda|^{2k+2}} \geqslant \sum_{k=k_j(m)}^{k_j(m)+m} \frac{b_k^2}{|\lambda|^{2k+2}} \geqslant$$

$$\int_{k_j(m)}^{k_j(m)+m} \frac{(\sqrt[m]{m})^{2m+2k_j(m)-2x}}{|\lambda|^{2x+2}} dx = \int_0^m \frac{(\sqrt[m]{m})^{2y}}{|\lambda|^{2m+2k_j(m)-2y+2}} dy. \quad (8)$$

Here we have taken into account that the sequence $\frac{b_k^2}{|\lambda|^{2k+2}}$ is monotonously decaying within the interval $k_j(m) < k < k_j(m)+m$ (see Figure 2).

The integral in (8) can be calculated; thus

$$\sum_{k=k_j(m)-m}^{k_j(m)+m} \frac{b_k^2}{|\lambda|^{2k+2}} \geqslant \frac{1}{2} \frac{1}{\frac{\log m}{m} + \log|\lambda|} \left[\frac{m^2}{|\lambda|^{2k_j(m)+2}} - \frac{1}{|\lambda|^{2k_j(m)+2m+2}} \right]. \tag{9}$$

Combined with (6) this yields

$$\int_0^{2\pi} \|(T_j-\lambda)^{-1}u\|^2 d\theta \geqslant C \sum_{k=1}^{\infty} \frac{1}{\frac{j\log k}{k^j} + \log|\lambda|} \left[\frac{k^{2j}}{|\lambda|^{2k_j(k^j)+2}} - \frac{1}{|\lambda|^{2k_j(k^j)+2k^j+2}} \right]. \tag{10}$$

The second term of the sum (10) admits the following estimate:

$$\begin{aligned} \sum_{k=1}^{\infty} \frac{1}{\frac{j\log k}{k^j} + \log|\lambda|} \left[\frac{1}{|\lambda|^{2k_j(k^j)+2k^j+2}} \right] &\leqslant \sum_{k=1}^{\infty} \frac{1}{\log|\lambda|} \frac{1}{|\lambda|^{d_j k^{j+1}}} \\ &\leqslant C \frac{\Gamma(1+\frac{1}{1+j})}{d_j^{1/(1+j)}} \frac{1}{(\log|\lambda|)^{1+\frac{1}{1+j}}} \leqslant C(j) \frac{1}{\epsilon^{1+\frac{1}{1+j}}}, \quad \text{where } \epsilon \equiv |\lambda| - 1 \ll 1. \end{aligned} \tag{11}$$

Moreover, the first term of the sum (10) admits an estimate from below. Since for $k > C_j(\log|\lambda|)^{-1/(1+j)}$ (see (7)) we have

$$\begin{aligned} \frac{j\log k}{k^j} + \log|\lambda| \leqslant \frac{j}{C_j^j} (\log|\lambda|)^{\frac{j-1}{j+1}} + \log|\lambda| &\leqslant \log|\lambda| \left(1 + \frac{j}{C_j^j} (\log|\lambda|)^{-2/(1+j)}\right) \\ &\leqslant C_1(j) \epsilon^{1-\frac{2}{1+j}}, \quad \text{where } \epsilon \equiv |\lambda| - 1 \ll 1, \end{aligned}$$

quite analogously to (7) for all $|\lambda|$ sufficiently close to 1 we obtain the following estimate:

$$\begin{aligned} &\sum_{k=1}^{\infty} \frac{1}{\frac{j\log k}{k^j} + \log|\lambda|} \left[\frac{k^{2j}}{|\lambda|^{2k_j(k^j)+2}} \right] \\ &\quad \geqslant C_1(j) |\lambda|^{-2} \frac{1}{(\log|\lambda|)^{1-\frac{2}{1+j}}} \int_{C_j(\log|\lambda|)^{-1/(1+j)}}^{\infty} \frac{x^{2j}}{|\lambda|^{2d_j x^{j+1}}} dx \\ &\quad \geqslant C_1(j) \frac{1}{|\lambda|^2} \frac{1}{1+j} \left(\frac{1}{2d_j}\right)^{2-\frac{1}{1+j}} \frac{1}{(\log|\lambda|)^{3-\frac{3}{1+j}}} \int_2^{\infty} e^{-t} t^{\frac{j}{j+1}} dt. \end{aligned}$$

Combined with (11) the latter implies that for every j there exists a positive constant $\tilde{C}(j)$ such that for all $|\lambda|$ sufficiently close to 1

$$\int_0^{2\pi} \|(T_j-\lambda)^{-1}u\|^2 d\theta \geqslant \tilde{C}(j) \frac{1}{\epsilon^{3-\frac{3}{j+1}}}, \quad \text{where} \quad \epsilon = |\lambda| - 1 \ll 1.$$

Since $3 - \frac{3}{j+1}$ is arbitrarily close to 3 when $j \gg 1$, we finish the proof by choosing an appropriately large j for each $\delta > 0$. □

Lemma 1 together with the proof given can be easily generalized to the following one, dealing with the general case of an arbitrary γ, not necessarily equal to 1 as in Lemma 1.

Lemma 2. *For any $\gamma > 0$ and any $\delta > 0$ there exist an operator T_δ acting in the Hilbert space H and a vector $u \in H$ such that:*

(1) $\|T_\delta^n\| \leqslant 1 + |n|^\gamma$ *for all* $n \in \mathbb{Z}$ *and* $\|T_\delta^{n_j}\| = |n_j|^\gamma$ *for some subsequence* n_j *of integers such that* $\lim_{j\to\infty} |n_j| = \infty$*;*

(2) *there exists a positive constant* $C = C(\gamma, \delta)$ *such that the following estimate holds for all* $r > 1$*,* r *sufficiently close to 1,*

$$\int_0^{2\pi} \|(T_\delta - re^{i\theta})^{-1}u\| d\theta \geqslant C \frac{1}{(r-1)^{1+2\gamma-\delta}}.$$

In order to prove Lemma 2 one has to consider the sequence $\{a_i\}_{i=-\infty}^{\infty}$ constructed analogously to the one described by (4) from the blocks

$$A_n \equiv \left\{ \underbrace{(n^\gamma)^{1/n},}_{n \text{ times}} \underbrace{\frac{1}{(n^\gamma)^{1/n}}}_{n \text{ times}} \right\}; \quad A_0 \equiv \{1\}.$$

The rest of the proof of Lemma 1 remains intact, although some insignificant modifications to the sums and integrals in the estimates are to be made.

Theorem 3 now follows from an application of the Cayley transformation to the operator T_δ provided by Lemma 2 as in [10, 12].

3. C_0-groups with functional growth generated by operators with a purely absolutely continuous spectrum

In the present section we prove the results generalizing the ones from [12] (see also p. 254) to the case of an arbitrary C_0-group with functional growth acting in the Hilbert space H. We restrict ourselves to the situation where the generator of such C_0-group admits the functional model representation [1] and its spectrum is absolutely continuous. To achieve this goal we first provide the necessary background, following [1, 11, 9, 7, 8].

3.1. Functional model approach

We consider the class of operators of the form [9] $L = A + iV$, where A is a self-adjoint operator in H defined on the domain $D(A)$ and the perturbation V admits the factorization $V = \frac{\alpha J \alpha}{2}$, where α is a nonnegative self-adjoint operator in H and J is a unitary operator in $E \equiv \overline{R(\alpha)}$. This factorization corresponds to the polar decomposition of the operator V. In order that the expression $A + iV$ be meaningful, we impose the condition that V be (A)-bounded with the relative bound less then 1, i.e., $D(A) \subset D(V)$ and for some a and b $(a < 1)$ the condition

$\|Vu\| \le a\|Au\| + b\|u\|$, $u \in D(A)$ is satisfied, see [5]. Then the operator L is well defined on the domain $D(L) = D(A)$.

Alongside with the operator L we consider the maximal dissipative operator $L^{\|} = A + i\frac{\alpha^2}{2}$ and its adjoint, $L^{-\|} \equiv L^{\|*} = A - i\frac{\alpha^2}{2}$. Since the functional model for the dissipative operator $L^{\|}$ will be used below, we require that $L^{\|}$ be completely non-self-adjoint, i.e., that it has no reducing self-adjoint parts. This requirement is not restrictive in our case due to Proposition 1 in [9].

The remainder of the present section is devoted to a brief description (following [1, 11], see also [9]) of the construction of the self-adjoint dilatation of the completely non-self-adjoint dissipative operator $L^{\|}$.

The characteristic function $S(\lambda)$ of the operator $L^{\|}$ is a contractive, analytic operator-valued function acting in the Hilbert space E, defined for $Im\lambda > 0$ by

$$S(\lambda) = I + i\alpha(L^{-\|} - \lambda)^{-1}\alpha. \tag{12}$$

In the case that α is unbounded the characteristic function is first defined by the expression (12) on the manifold $E \cap D(\alpha)$ and then extended by continuity to the whole space E.

The formula (12) makes it possible to consider $S(\lambda)$ for $Im\lambda < 0$ with $S(\overline{\lambda}) = (S^*(\lambda))^{-1}$. Finally, $S(\lambda)$ possesses boundary values on the real axis in the strong sense: $S(k) \equiv S(k + i0)$, $k \in \mathbb{R}$ (see [1]).

Consider the model space $\mathcal{H} = L_2\left(\begin{smallmatrix} I & S^* \\ S & I \end{smallmatrix}\right)$, which is defined in [11] as the Hilbert space of two-component vector-functions $(\tilde{g}, g)$ on the real axis ($\tilde{g}(k), g(k) \in E, k \in \mathbb{R}$) with the metric

$$\left(\begin{pmatrix}\tilde{g}\\ g\end{pmatrix}, \begin{pmatrix}\tilde{g}\\ g\end{pmatrix}\right) = \int_{-\infty}^{\infty} \left(\begin{pmatrix} I & S^*(k) \\ S(k) & I \end{pmatrix}\begin{pmatrix}\tilde{g}(k)\\ g(k)\end{pmatrix}, \begin{pmatrix}\tilde{g}(k)\\ g(k)\end{pmatrix}\right)_{E\oplus E} dk.$$

It is assumed here that the set of two-component functions has been factored by the set of elements with norm equal to zero.

We define the following orthogonal subspaces in $\mathcal{H}$:

$$D_- \equiv \begin{pmatrix} 0 \\ H_2^-(E)\end{pmatrix}, \quad D_+ \equiv \begin{pmatrix} H_2^+(E) \\ 0\end{pmatrix}, \quad K \equiv \mathcal{H} \ominus (D_- \oplus D_+),$$

where $H_2^{+(-)}(E)$ denotes the Hardy class of analytic functions f in the upper (resp., lower) half plane with values in the Hilbert space E.

The subspace K can be described as $K = \{(\tilde{g}, g) \in \mathcal{H} : \tilde{g} + S^*g \in H_2^-(E), S\tilde{g} + g \in H_2^+(E)\}$. Let P_K be the orthogonal projection of $\mathcal{H}$ onto K:

$$P_K\begin{pmatrix}\tilde{g}\\ g\end{pmatrix} = \begin{pmatrix}\tilde{g} - P_+(\tilde{g} + S^*g) \\ g - P_-(S\tilde{g} + g)\end{pmatrix},$$

where $P_\pm$ are the orthogonal projections of $L_2(E)$ onto $H_2^\pm(E)$.

The following result holds [1, 11]: The operator $(L^{\|} - \lambda_0)^{-1}$ is unitarily equivalent to the operator $P_K(k - \lambda_0)^{-1}|_K$ for all λ_0 such that $Im\lambda_0 < 0$. In effect this means, that the operator of multiplication by k serves as the minimal ($clos_{Im\lambda\neq 0}(k - \lambda)^{-1}K = \mathcal{H}$) self-adjoint dilatation [1] of the operator $L^{\|}$.

The characteristic function of the operator L is defined by the following expression:

$$\Theta(\lambda) \equiv I + iJ\alpha(L^* - \lambda)^{-1}\alpha, \qquad Im\lambda \neq 0,$$

and is a meromorphic, J-contractive operator-function in the upper half-plane $(\Theta^*(\lambda)J\Theta(\lambda) \leq J, \quad Im\lambda > 0)$ [2]. The characteristic function $\Theta(\lambda)$ admits a factorization in the form of the ratio of two bounded analytic operator-functions (in the corresponding half-planes $Im\lambda < 0$, $Im\lambda > 0$) triangular with respect to the decomposition of the space E into the orthogonal sum

$$E = (\mathcal{X}_+ E) \oplus (\mathcal{X}_- E), \quad \mathcal{X}_\pm \equiv \frac{I \pm J}{2}.$$

Following [8], we define the subspaces $\hat{N}_\pm$ in $\mathcal{H}$ as follows:

$$\hat{N}_{+(-)} \equiv \left\{ \begin{pmatrix} \tilde{g} \\ g \end{pmatrix} : \quad \begin{pmatrix} \tilde{g} \\ g \end{pmatrix} \in \mathcal{H}, \quad \mathcal{X}_-(\tilde{g} + S^* g) = 0 \quad (\mathcal{X}_+(S\tilde{g} + g) = 0) \right\}$$

and introduce the following notation: $N_\pm = clos P_K \hat{N}_\pm$.

Then, as it is shown in [9], one has for $Im\lambda < 0$ ($Im\lambda > 0$) and $(\tilde{g}, g) \in \hat{N}_{-(+)}$, respectively:

$$(L - \lambda)^{-1} P_K \begin{pmatrix} \tilde{g} \\ g \end{pmatrix} = P_K \frac{1}{k - \lambda} \begin{pmatrix} \tilde{g} \\ g \end{pmatrix}.$$

The absolutely continuous and singular subspaces of the non-self-adjoint operator L are defined in [7]: Let $N \equiv \hat{N}_+ \cap \hat{N}_-$, $\tilde{N}_\pm \equiv P_K \hat{N}_\pm$, $\tilde{N}_e \equiv \tilde{N}_+ \cap \tilde{N}_-$, then[1]

$$\begin{aligned} N_e &\equiv clos\left(\tilde{N}_e\right) = clos P_K N \\ N_i &\equiv K \ominus N_e(L^*). \end{aligned} \tag{13}$$

We call the operator L an *operator with purely absolutely continuous spectrum* if $N_e = H$, i.e., the linear set $P_K N$ is dense in K.

3.2. A characterization of a C_0-group with functional growth in terms of its generator

Consider an arbitrary nonnegative continuous function on the right semiaxis, $f \in C(\mathbb{R}_+)$.

We call a C_0-group $T(t)$ in the Hilbert space H, satisfying the condition

$$\|T(t)\| \leq M f(|t|) \tag{14}$$

for all $t \in \mathbb{R}$ and some constant $M \in (0, \infty)$, a C_0-group *with functional growth* in the Hilbert space H.

Let the generator L of the group $T(t)$ with functional growth belong to the class described in the Section 3.1 (i.e., it admits the model representation described there). We further restrict ourselves to the case when the spectrum $\sigma(L)$ of the operator L is real.

[1]The linear set $\tilde{N}_e$ is called the set of "smooth" vectors of the operator L (see [9]).

We are primarily interested in obtaining necessary (respectively, sufficient) conditions for (14) to hold in the form of a pair of integral estimates for the resolvent of the operator L generating the group $T(t)$, considered for all $u \in H$ and all $\varepsilon > 0$:

$$\begin{aligned} \int_{\mathbb{R}} \|(L-k-i\varepsilon)^{-1}u\|^2 dk \leq C\frac{1}{\varepsilon}g(\varepsilon)\|u\|^2 \\ \int_{\mathbb{R}} \|(L^*-k-i\varepsilon)^{-1}u\|^2 dk \leq C\frac{1}{\varepsilon}g(\varepsilon)\|u\|^2, \end{aligned} \tag{15}$$

where C is an absolute constant and $g \in C(\mathbb{R}_+)$ is a continuous nonnegative function of $\varepsilon > 0$ (see also [3], where the necessary and sufficient conditions for functional growth are given in the framework of the possible generalizations of the Hille-Yosida theorem, i.e., in terms of estimates for the norms of the operators $(L-\lambda)^{-n}$, $n = 1, 2, \dots$).

One can prove the following results:

Theorem 4. *Let the spectrum of the operator L in the Hilbert space H be absolutely continuous. Then for every nonnegative function $g \in C(\mathbb{R}_+)$, satisfying the condition $g(\varepsilon) \leq \tilde{M} < \infty$ when $\varepsilon \gg 1$, the estimates* (15) *suffice for the operator L to generate the C_0-group* $\exp(iLt)$ *with*

$$\|\exp(iLT)\| \leq Mf(|t|) \text{ for all } t \in \mathbb{R},$$

where M is a finite positive constant, and the function $f(t)$ is defined by the identity $f(t) \equiv g(1/t)$ for $t > 0$.

Proof. Let $t > 0$. Then

$$e^{itk}e^{-\varepsilon t} = -\frac{1}{2\pi i}\int_{-\infty}^{\infty} e^{ixt}\left[(k-x+i\varepsilon)^{-1} - (k-x-i\varepsilon)^{-1}\right]dx$$

for any $\varepsilon > 0$.

Then from the conditions of the theorem we obtain for $u \in \tilde{N}_e$

$$e^{itL}e^{-\varepsilon t}u = -\frac{1}{2\pi i}\int_{-\infty}^{\infty} e^{ixt}\left[(L-x+i\varepsilon)^{-1} - (L-x-i\varepsilon)^{-1}\right]udx, \tag{16}$$

where $\tilde{N}_e$ is the set of "smooth" vectors of L [9] (provided, that the integral in (16) exists). Then

$$\begin{aligned} &\frac{\varepsilon}{\pi}\left|\int_{\mathbb{R}} e^{ixt}(\left[(L-x+i\varepsilon)^{-1}(L-x-i\varepsilon)^{-1}\right]u, v)dx\right| \\ &\qquad \leqslant \frac{\varepsilon}{\pi}\left|\int_{\mathbb{R}} \|(L-x-i\varepsilon)^{-1}u\| \cdot \|(L^*-x-i\varepsilon)^{-1}v\|dx\right| \leqslant \tilde{C}\|u\|\|v\|g(\varepsilon) \end{aligned}$$

for all $u, v \in \tilde{N}_e$, which justifies the formula (16) and yields the following estimate for all $u, v \in \tilde{N}_e$:

$$\left|(\exp(iLt)e^{-\varepsilon t}u, v)\right| \leq M\|u\|\|v\|g(\varepsilon).$$

We choose $\varepsilon \equiv 1/t$ and since $\tilde{N}_e$ is dense in H we obtain

$$|(\exp(iLt)u, v)| \leq M\|u\|\|v\|g(1/t)$$

for all $u, v \in H$ and all $t > 0$.

The boundedness of $\exp(iLt)$ when $t < 0$ can be shown in an analogous fashion. □

Theorem 5. *Let the spectrum of the operator L be absolutely continuous. Let the nonnegative function $f \in C(\mathbb{R}_+)$ be such that the Laplace-type transform*

$$g'(\varepsilon) \equiv \int_0^\infty e^{-2\varepsilon t}(f(t))^2 dt$$

is finite for all $\varepsilon \in (0, \infty)$ and $g'(\varepsilon) = O(1/\varepsilon)$ as $\varepsilon \to \infty$. Then the condition (15) *with $g(\varepsilon) = \varepsilon g'(\varepsilon)$ is necessary for the operator L to generate the functionally bounded C_0-group* $\exp(iLt)$,

$$\|\exp(iLt)\| \leq Mf(|t|) \quad \textit{for all } t \in \mathbb{R}.$$

Proof. For all $z \in \mathbb{C}_+$, $k \in \mathbb{R}$

$$\frac{1}{k - z} = i\int_0^\infty e^{izt}e^{-itk}dt.$$

Then from the conditions of the theorem we get for all $u \in \tilde{N}_e$ and $z \in \mathbb{C}_+$:

$$(L - z)^{-1}P_K\begin{pmatrix}\tilde{g}\\ g\end{pmatrix} = i\int_0^\infty dte^{izt}P_K e^{-itk}\begin{pmatrix}\tilde{g}\\ g\end{pmatrix} = i\int_0^\infty dte^{izt}\exp(-iLt)P_K\begin{pmatrix}\tilde{g}\\ g\end{pmatrix}, \tag{17}$$

provided that the latter integral exists. The Plancherel theorem applied to the right-hand side of the identity (17) yields

$$\begin{aligned}\int_{\mathbb{R}} \|(L - k - i\varepsilon)^{-1}u\|^2 dk &= 2\pi\int_0^\infty e^{-2\varepsilon t}\|\exp(-iLt)\|^2 dt \\ &\leqslant C\int_0^\infty e^{-2\varepsilon t}(f(t))^2\|u\|^2 dt = C\|u\|^2\frac{1}{\varepsilon}g(\varepsilon).\end{aligned}$$

for all $\varepsilon > 0$ and $u \in \tilde{N}_e$, i.e., for the linear manifold dense in H.

The corresponding estimate for $(L^* - k - i\varepsilon)^{-1}$ is obtained similarly based on the following identity:

$$\|\exp(iLt)\| = \|(\exp(iLt))^*\| = \|\exp(-iL^*t)\|.$$

□

3.3. The application to the Friedrichs model operator

Assume $\varphi, \psi \in L_2(\mathbb{R})$ and consider the operator L acting in the Hilbert space $L_2(\mathbb{R})$:

$$(Lu)(x) = xu(x) + (u, \varphi)\psi(x), \qquad u \in L_2(\mathbb{R}).$$

The determinant of perturbation $D(\lambda)$ in this case is given by the formula

$$D(\lambda) = 1 + \int \frac{\overline{\varphi(t)}\psi(t)}{t-\lambda}dt, \quad \lambda \in \mathbb{C} \setminus \mathbb{R},$$

and the resolvent $(L-\lambda)^{-1}$ is determined by the identity:

$$((L-\lambda)^{-1}u)(x) = \frac{u(x)}{x-\lambda} - \frac{1}{D(\lambda)}\left(\frac{u(t)}{t-\lambda}, \varphi(t)\right)\frac{\psi(x)}{x-\lambda}, \quad \lambda \in \mathbb{C} \setminus \mathbb{R}.$$

The necessary and sufficient conditions of the absolute continuity of the spectrum of the operator L are given in [6] in the special case of $(\varphi, \psi) = 0$.

In the simplest case that $\varphi(x)\psi(x) = 0$ for a. a. $x \in \mathbb{R}$ [12] the operator L generates the C_0-group $\exp(iLt)$ with polynomial growth and the following estimate holds:

$$\|\exp(iLt)\| \leq C(1+|t|), \quad t \in \mathbb{R}.$$

The following proposition generalizes this result:

Proposition 1. *Let L be an operator with purely absolutely continuous spectrum. Let the following conditions*

$$\|\frac{1}{D(k+i\varepsilon)}[\mathcal{P}(|\psi|^2)(k+i\varepsilon)]^{1/2} \cdot \mathbf{P}_+(\cdot\,\overline{\varphi})\| \leq Cg^{1/2}(\varepsilon)$$
$$\|\frac{1}{D(k-i\varepsilon)}[\mathcal{P}(|\varphi|^2)(k+i\varepsilon)]^{1/2} \cdot \mathbf{P}_+(\cdot\,\overline{\psi})\| \leq Cg^{1/2}(\varepsilon),$$

where

$$\mathcal{P}(u)(\lambda) \equiv \frac{1}{\pi}\int \frac{\varepsilon u(t)}{(k-t)^2+\varepsilon^2}dt$$
$$(\mathbf{P}_+u)(\lambda) \equiv \frac{1}{2\pi i}\int \frac{u(t)}{t-\lambda}dt$$

are the Poisson transformation of the function u and the Riesz projector from $L_2(\mathbb{R})$ onto the space H_2^+, respectively, be valid for some nonnegative function $g \in C(\mathbb{R}_+)$ such that $g(\varepsilon) \leq \tilde{M} < \infty$ when $\varepsilon \gg 1$. Then the operator L generates the C_0-group $\exp(iLt)$ in the space $L_2(\mathbb{R})$ and

$$\|\exp(iLt)\| \leq Mf(|t|)$$

for all $t \in \mathbb{R}$ and the function $f(t) \equiv g(1/t)$, $t > 0$.

To prove this proposition, one has to apply the results obtained in the previous section of the present paper (see also [10]) to the operator L described above.

Acknowledgements. The author is grateful to Prof. S.N. Naboko and Doc. M.M. Faddeev for their constant attention to his work and to Dr. M. Malejki for the possibility to get acquainted with the article [12] before it was published.

References

[1] Béla Sz.-Nagy and Ciprian Foiaş, *Analyse harmonique des opérateurs de l'espace de Hilbert*, Masson, Paris and Akad. Kiadó, Budapest, 1967.

[2] Brodskij M.S., *Triangular and Jordan representations of linear operators*, English transl. in Amer. Math. Soc., Providence, R.I., 1971.

[3] Chojnacki W., *Multiplier algebras, Banach bundles, and one-parameter semigroups*, Ann. Scuola Norm. Sup. Piza Cl. Sci. (4), Vol. XXVIII, 1999.

[4] K. Hoffman, *Banach spaces of analytic functions*, Prentice-Hall, Englewood Cliffs, N.J., 1962.

[5] Tosio Kato, *Perturbation theory for linear operators*, Springer-Verlag, 1966.

[6] Kiselev A.V., Faddeev M.M., *The similarity problem for non-self-adjoint operators with absolutely continuous spectrum*, Functional Analysis and its applications (2000), Vol. 34, pp. 140–142.

[7] Naboko S.N., *Absolutely continuous spectrum of a nondissipative operator and functional model. I.*, Zapiski Nauchnykh Seminarov LOMI AN SSSR, Vol. 65, pp. 90–102, 1976 (in Russian); English translation in J. Sov. Math.

[8] Naboko S.N., *Absolutely continuous spectrum of a nondissipative operator and functional model. II.*, Zapiski Nauchnykh Seminarov LOMI AN SSSR, Vol. 73, pp. 118–135, 1977 (in Russian); English translation in J. Sov. Math.

[9] Naboko S.N., *A functional model of the perturbation theory and its application to scattering theory*, Proceedings of the Steklov Institute of Mathematics (1981), Issue 2.

[10] Naboko S.N., *The conditions for similarity to unitary and self-adjoint operators*, Functional Analysis and its applications (1984), Vol. 18, pp. 16–27 (in Russian).

[11] Pavlov B.S., *On separation conditions for the spectral components of a dissipative operator*, English transl. in Math. USSR Izv. (1975), N 9.

[12] Maria Malejki, *C_0-groups with polynomial growth*, to be published.

[13] Van Casteren J., *Boundedness properties of resolvents and semigroups of operators*, Acta Sci. Math. Szeged. (1980), Vol. 48, N 1–2.

Alexander V. Kiselev
Dept. of Higher Math. and Mathematical Physics
Institute of Physics
Peterhoff, Ulianovskaya st., 1,
198904 St. Petersburg, Russia
e-mail: akiselev@mph.phys.spbu.ru

Operator Theory:
Advances and Applications, Vol. 132, 267–285

Supersymmetry of the Sturm–Liouville and Korteveg–de Vries Operators

Dimitry Leites

Abstract. In 70's A.A. Kirillov identified the (stationary) *Sturm–Liouville* operator $L_2 = \frac{d^2}{dx^2} + F$ with an element of the dual space $\hat{\mathfrak{g}}^*$ of the nontrivial central extension $\hat{\mathfrak{g}} = \mathfrak{vir}$, called the Virasoro algebra, of the Witt algebra $\mathfrak{witt} = \mathfrak{der}\,\mathbb{C}[x^{-1}, x]$. He interpreted the (stationary) *KdV operator* $L_3 = \frac{d^3}{dx^3} + \frac{d}{dx}F + F\frac{d}{dx}$ in terms of the stabilizer of L_2. He also found a supersymmetry that reduces solutions of $L_3(f) = 0$ to solutions of $L_2(g) = 0$ by studying the nontrivial central extension of a simplest super analog of the Virasoro algebra, the Neveu–Schwarz superalgebra. Kirillov also wrote the first superversion of KdV equation.

I extend Kirillov's results and show how to find all supersymmetric extension of the Sturm–Liouville and Korteveg–de Vries operators associated with the 10 *distinguished* stringy superalgebras, i.e., all the simple stringy superalgebras that possess a nontrivial central extension. There are 12 or 14 such extensions, depending on the point of view. Here I only consider scalar models.

Another approach to KdV is via Drinfeld–Sokolov's reduction. Khesin and Malikov extended Drinfeld–Sokolov's reduction to pseudodifferential operators and related the complex powers of Sturm–Liouville operators with the superized KdV-type hierarchies labelled by complex parameter. Similar approach to our Sturm–Liouville operators is also possible.

0. Introduction

A.A. Kirillov [Ki1] associated the (stationary) *KdV operator*

$$L_3 = \frac{d^3}{dx^3} + \frac{d}{dx}F + F\frac{d}{dx} \tag{KdV}$$

and the (stationary) *Sturm–Liouville* operator

$$L_2 = \frac{d^2}{dx^2} + F \tag{StL}$$

with the cocycle that determines the nontrivial central extension $\hat{\mathfrak{g}} = \mathfrak{vir}$ — the Virasoro algebra — of the Witt algebra $\mathfrak{g} = \mathfrak{witt} = \mathfrak{der}\mathbb{C}[x^{-1}, x]$. Moreover, Kirillov

Financial support of the NFR and help of Xuan Peiqi in 1995 are gratefully acknowledged.

found an explanation of the commonly known useful fact that the product of two solutions f_1, f_2 of the Sturm–Liouville equation satisfies $L_3(f_1f_2) = 0$. Kirillov's explanation is the existence of a supersymmetry [Ki2].

Kirillov's construction brings the KdV-type equations directly in the Lax form (analog of Euler's equation for a solid body) guaranteeing their complete integrability, cf. [Ku1], [OKh]. One should not forget here that the most profound dynamics, as Shander showed [Sh], is obtained with the help of $1|1$-dimensional time, rather than 1-dimensional.

Kirillov also classified the orbits of the coadjoint representation of $\mathfrak{vir}$ and showed its equivalence to the following important classification problems: the classification of symplectic leaves of the second Gelfand–Dickey structure on the second order differential operators, of projective structures on the circle, and of Hill equation, i.e., (StL) with periodic potential F. Kirillov's approach clarifies some earlier results by Poincaré, Kuiper, Lazutkin and Pankratova. The recent announcement of the classification of the simple stringy superalgebras and their central extensions [GLS] describes the scope of the problem: there are exactly 12 (or 14, depending on the interpretation) ways to superize the above results of Kirillov.

Kirillov himself partly considered one of these 12 or 14 possibilities to superize KdV, several more were considered by Kuperschmidt, Chaichian–Kulish, P. Mathieu, Khovanova, V. Ovsienko and O. Ovsienko, Khesin, Ivanov–Krivonos–Bellucci–Delduc–Toppan, and many others; from the sea of results we point out [Ku1]-[Ku3], [Kh], [KM], [BIK], [DI], [DIK], [IKT]. So far, the examples of N-extended superKdV equations are only connected with a part of the distinguished stringy superalgebras; for some reasons, of two contact series, the Neveu-Schwarz one is considered almost always whereas the Ramond one almost never.

In this paper I do not consider all of the 12 (14) cases either: only scalar ones. The four vector-valued cases are much more difficult technically and will be considered elsewhere.

0.1. Kirillov's interpretation of the Sturm–Liouville and Korteveg–de Vries operators

Let $\mathfrak{g} = \mathfrak{der}\mathbb{C}[x^{-1}, x]$. This algebra is often called Witt one (because Witt considered its version over finite fields) and denoted $\mathfrak{witt}$; let $\hat{\mathfrak{g}} = \mathfrak{vir}$ be the nontrivial central extension of $\mathfrak{g}$ given by the bracket

$$[f\frac{d}{dx} + az, g\frac{d}{dx} + bz] = (fg' - f'g)\frac{d}{dx} + c \cdot \mathrm{Res} fg''' \cdot z \text{ for } c \in \mathbb{C},$$

where z is the generator of the center of $\hat{\mathfrak{g}}$. Let $\mathcal{F} = \mathbb{C}[x^{-1}, x]$ be the algebra of functions; let $\mathcal{F}_\lambda$ for $\lambda \in \mathbb{C}$ be the rank 1 module over $\mathcal{F}$ spanned by dx^λ, where the λth power of dx is determined via analyticity of the formula for the $\mathfrak{g}$-action:

$$(f\frac{d}{dx})(dx^\lambda) = \lambda f' dx^\lambda.$$

In particular, $\mathfrak{g} \cong \mathcal{F}_{-1}$, as $\mathfrak{g}$-modules. Since the module *Vol* of volume forms is $\mathcal{F}_1$, the module dual to $\mathfrak{g}$ is $\mathfrak{g}^* = \mathcal{F}_2$: we use one dx to kill $\frac{d}{dx}$ and another dx

to integrate the product of functions. (We confine ourselves to *regular* generalized functions, i.e., we ignore the elements from the space of functionals on $\mathfrak{g}$ with 0-dimensional support, see [Ki1].) Explicitely,

$$\langle F(dx)^2, f\frac{d}{dx}\rangle = \text{Res } Ff.$$

0.1.1. THE KORTEVEG–DE VRIES OPERATOR

The Lie algebra of the stationary group of the element $\hat{F} = (F, c) \in \hat{\mathfrak{g}}^* = (\mathfrak{g}^*, \mathbb{C}\cdot z^*)$ is

$$\mathfrak{st}_{\hat{F}} = \{\hat{X} \in \hat{\mathfrak{g}} : \hat{F}([\hat{X}, \hat{Y}]) = 0 \text{ for any } \hat{Y} \in \hat{\mathfrak{g}}\}.$$

Let us describe $\mathfrak{st}_{\hat{F}}$ explicitely. Take $\hat{X} = g\frac{d}{dx} + az$, $\hat{Y} = f\frac{d}{dx} + bz$. Then

$$\hat{F}([\hat{X}, \hat{Y}]) = \hat{F}[(fg' - f'g)\tfrac{d}{dx} + \text{Res} fg''' \cdot z] =$$
$$\text{Res}[F(fg' - f'g) + cfg'''] \overset{\text{(partial integration)}}{=} \text{Res} f[Fg' + (Fg)' + cg'''].$$

Hence, $\hat{X} \in \mathfrak{st}_{\hat{F}}$ if and only if g is a solution of the equation $L_3 g = 0$, where L_3 is given by formula (KdV) above; it is the famous operator of the second Hamiltonian structure for the KdV. If $c \neq 0$ we can always rescale the equation and assume that

$$c = 1.$$

In what follows this is understood. Explicitely, the KdV operator is of the form

$$L_3 = (\text{the cocycle operator that determines } \hat{\mathfrak{g}}\ ) + \frac{d}{dx}F + F\frac{d}{dx}. \tag{0.1.1}$$

0.1.2. THE STURM–LIOUVILLE OPERATOR

The Sturm–Liouville operator $L_2 = \frac{d^2}{dx^2} + F$ is, clearly, selfadjoint. The factorization

$$F(dx)^2 + a \cdot z^* = (dx)^2(F + a\frac{d^2}{dx^2}z^*) \tag{0.1.2}$$

suggests to ignore z^* (though $\mathfrak{witt}$ nontrivially acts on it) and represent the elements of $\hat{\mathfrak{g}}^*$ as 2nd order selfadjoint differential operators: $\mathcal{F}_\lambda \longrightarrow \mathcal{F}_{\lambda+2}$. Since the operator is selfadjoint, $\mathcal{F}^*_{\lambda+2} = \mathcal{F}_{1-(\lambda+2)} = \mathcal{F}_\lambda$, hence, $\lambda = -\frac{1}{2}$. In what follows we consider the 10 scalar super analogs of this operator, 4 more (matrix ones) will be considered elsewhere.

0.1.3. KDV HIERARCHY

Assume that F depends on time, t. The *KdV hierarchy* is the series of evolution equations for $L = L_2$ or, equivalently, for F:

$$\dot{L} = [L, A_k], \text{ where } A_k = (\sqrt{L}^{2k-1})_+ \text{ for } k \in \mathbb{N}. \tag{0.1.3}$$

Here the subscript + singles out the differential part of the pseudodifferential operator. The case $k = 1$ is trivial and $k = 2$ corresponds to the original KdV equation.

• Drinfeld and Sokolov [DS] associated KdV-type equations with loop algebras and their twisted analogs; it turned out that earlier results by Gelfand and

Dikii are particular cases of their construction. Superization of Drinfeld-Sokolov's approach was started in [Kh] but, regrettably, aborted. We will reconsider this approach elsewhere.

• Khesin and Malikov [KM] observed that it is possible to extend the approach of Drinfeld and Sokolov from loops with values in $\mathfrak{sl}(n)$ to loops with values in $\mathfrak{sl}(\lambda)$, where $\lambda \in \mathbb{C}$, cf. [GL]. This leads to evolution equations for pseudodifferential operators — a continuous KdV hierarchy. Such an approach to evolution equations for L_2 is even more natural in the supersetting, when the Sturm–Liouville operator L_2 itself becomes a pseudodifferential one. This topic will be considered elsewhere.

0.2. Kirillov's interpretation of supersymmetry of the Sturm–Liouville and Korteveg–de Vries operators

In this subsection we need the technique of C-points, or *superfields*, see [D] or [L1], [L2]. In simple terms, without word "functor": replace each superspace V with the even space $(V \otimes C)_{\bar 0}$, called *the set of C-points of V*, for an arbitrary supercommutative superalgebra C and make sure that nothing depends on C in the final answer. Berezin called $(V \otimes C)_{\bar 0}$ the "Grassmann envelope" of V and considered just one C — the Grassmann algebra with infinitely many generators [B]; it is safer not to fix the number of generators but consider it "sufficiently large".

Kirillov suggested to replace in the above scheme (Section 0.1) $\mathfrak{g} = \mathfrak{witt}$ with the Lie superalgebra $\mathfrak{g} = \mathfrak{k}^L(1|1)$, see Section 1.1. The superalgebra $\mathfrak{g}$ has a nontrivial central extension, called the Neveu–Schwarz superalgebra $\mathfrak{ns} = \mathfrak{ns}(1)$ and the above scheme leads us to the $\mathfrak{ns}$-analog of the KdV operator

$$\mathcal{L}_5 = K_\theta K_1^2 + 2FK_1 + 2K_1F + (-1)^{p(F)}K_\theta F K_\theta \tag{0.2.1}$$

and the $\mathfrak{ns}$-analog of the Sturm–Liouville operator

$$\mathcal{L}_3 = K_\theta K_1 + F, \text{where } F \in (C[x^{-1}, x, \theta])_{\bar 1} \tag{0.2.2}$$

see Section 1.4. Let us calculate the stabilizer of an element of $\mathfrak{ns}(1; C)^*$.

In what follows we omit indicating C, but all the constructions are over C.

The straightforward calculations yield: $X = K_f \in \mathfrak{st}_{\hat F}$ if and only if f is a solution of the equation

$$\left(cK_\theta \frac{d^2}{dx^2} + 2\frac{d}{dx}F + 2F\frac{d}{dx} + (-1)^{p(F)}K_\theta F K_\theta\right) f = 0. \tag{0.2.3}$$

The operator

$$\begin{aligned}\mathcal{L}_5 = &\text{(the cocycle operator that determines } \hat{\mathfrak{g}})+\\ &2FK_1 + 2K_1F + (-1)^{p(F)}K_\theta K_1 K_\theta\end{aligned} \tag{0.2.4}$$

from the left-hand side of (0.2.3) will be called the $\mathfrak{ns}(1)$-*KdV operator.*

In components we have: $f = f_0 + f_1\theta$, $F = F_0 + F_1\theta$, where f_0 and F_1 are even functions (of x with values in C) while f_1 and F_0 are odd ones. Suppose $F_0 = 0$. Then (0.2.3) turns into a system

$$L_2 f_1 = 0, \qquad L_3 f_0 = 0.$$

Since formula (1.4.5) below implies that $\{g_1(x)\theta, g_2(x)\theta\}_{K.b.} = g_1g_2$, we see that the product of two solutions of the Sturm–Liouville equation $L_2g_i = 0$ is a solution of the KdV equation $L_3(g_1g_2) = 0$. **This is the supersymmetry Kirillov discovered.**

Remark. Kirillov only considered $\mathbb{C}$-points of $\mathfrak{g}$ and $\hat{\mathfrak{g}}$, that is why he missed all odd parameters ($F_0 \neq 0$) of the supersymmetry he found.

0.3. The result

I extend Kirillov's approach from $\mathfrak{witt}$ to all simple distinguished stringy Lie superalgebras — an elaboration of Remark from [L1], p. 167, where the importance of odd parameters in Kirillov's approach was first observed and the problem solved here was raised. To consider *all* superized à la Kirillov KdV and Sturm–Liouville operators was impossible before the list of distinguished stringy superalgebras ([GLS], [KvdL]) was completed.

0.4. On open problems

Passing to superization of the steps of Section 0.1, I interpret the elements of $\hat{\mathfrak{g}}^*$ for the distinguished stringy superalgebras $\mathfrak{g}$ as selfadjoint operators, perhaps, pseudodifferential, rather than differential. This, together with ideas applied by Khesin–Malikov to the usual Sturm–Liouville operator, requires generalizations of the Lie superalgebra of matrices of complex size associated and the analogs of superprincipal embeddings of $\mathfrak{osp}(N|2)$ for $N \leq 4$ (considered in [LSS] for $N = 1$). Such superizations were recently described (only for $N = 1$, see [GL]). It still remains to describe the corresponding W-superalgebras and Gelfand–Dickey superalgebras, and represent the results given below in components in order to compare with the results of physicists ([BIK], [DI], [DIK], [IKT] to name a few), describe superprincipal embeddings for $N > 1$ needed for a detailed superization of Drinfeld–Sokolov's and Khesin–Malikov's constructions.

There also remain: the invariants of superorbits and the four non-scalar cases: $\mathfrak{vect}^L(1|1)$, $\mathfrak{m}^L(1)$, $\mathfrak{vect}^L(1|2)$, and the most interesting $\mathfrak{svect}^L_\lambda(1|2)$.

Closely related to nontrivial central extensions of distinguished stringy superalgebras are superizations of the Schwarz derivative and Bott cocycle. When Radul gave his examples [Ra] several distinguished algebras were unknown; these cases should be considered.

There is also a possibility to superize Sturm-Liouville operators associated with the non-simple Lie superalgebras resembling $\mathfrak{witt}$ and admitting a nontrivial central extension, see [MOR] (where, as usual, Ramond superalgebra and its N-extended versions are neglected). Amazingly, these possibilities do not reduce to the extensions of contractions of simple stringy superalgebras, and there are not too many of them; not all possibilities are described yet.

Lastly, the simplest problem: in Section 1.3 there are described simplest modules a bit more general than $\mathcal{F}_\lambda$. Are there Sturm-Liouville operators acting in them?

1. Distinguished stringy superalgebras

We recall all the neccessary data. For the details of classification of simple vectorial Lie superalgebras see [LS] and [GLS]; for a review of the representation theory of simple Lie superalgebras including infinite dimensional ones see [L2], for basics on supermanifolds see [D], [M] or [L1, L2]. The ground field is $\mathbb{C}$.

1.1. Supercircle

A *supercircle* or (for a physicist) a *superstring* of dimension $1|n$ is the real supermanifold $S^{1|n}$ associated with the rank n trivial vector bundle over the circle. Let $x = e^{i\varphi}$, where φ is the angle parameter on the circle, be the even indeterminate of the Fourier transforms; let $\theta = (\theta_1, \dots, \theta_n)$, be the odd coordinates on the supercircle formed by a basis of the fiber of the trivial bundle over the circle. Then (x, θ) are the coordinates on $(\mathbb{C}^*)^{1|n}$, the complexification of $S^{1|n}$.

On $(\mathbb{C}^*)^{1|n}$, there are 5 series of simple "stringy" Lie superalgebras of vector fields and 4 exceptional such superalgebras. The 10 of these simple stringy superalgebras (or 12 if we consider algebras with distinct regradings as non-isomorphic) are distinguished: they admit nontrivial central extensions. Observe that as far as Sturm-Liouville operators are concerned the distinct regradings do produce distinct operators.

The **"main" 3 series** are:

- $\mathfrak{vect}^L(1|n) = \mathfrak{der}\mathbb{C}[x^{-1}, x, \theta]$ is the general vectorial Lie superalgebra;
- $\mathfrak{svect}^L_\lambda(1|n)$ the deformation of the divergence-free vectorial Lie superalgebra; the subalgebra of the vectorial Lie superalgebra, that preserve the volume form $x^\lambda vol$, where $vol = vol(x, \theta)$ is the volume element on $(\mathbb{C}^*)^{1|n}$. (Roughly speaking, vol "=" $dx \cdot \frac{\partial}{\partial\theta_1} \cdot \dots \cdot \frac{\partial}{\partial\theta_n}$. Recall that this is not equality: as shown in [BL], the change of variables acts differently on the left-hand side and right-hand side and only coincides for the simplest transformations.);
- $\mathfrak{k}^L(1|n)$ the subalgebra of $\mathfrak{vect}^L(1|n)$ that preserves the Pfaff equation $\alpha = 0$, where

$$\alpha = dx - \sum_{1\leq i\leq n} \theta_i d\theta_i$$

is the contact form. Usually, if $\left[\frac{n}{2}\right] = k$ we rename the first $2k$ indeterminates and express α for $n = 2k$ and $n = 2k+1$, respectively, as follows:

$$\alpha' = dx - \sum_{1\leq i\leq k} (\xi_i d\eta_i + \eta_i d\xi_i) \text{ or } \alpha' = dx - \sum_{1\leq i\leq k} (\xi_i d\eta_i + \eta_i d\xi_i) - \zeta d\zeta.$$

The superscript ^{L} indicates that we consider vector fields with Laurent coefficients, not polynomial ones. The Lie superalgebras of these 3 series are simple with the exception of $\mathfrak{svect}^L_\lambda(1|1)$ for any λ, $\mathfrak{svect}^L_\lambda(1|n)$ for $n > 1$ and $\lambda \in \mathbb{Z}$, and $\mathfrak{k}^L(1|4)$.

The **fourth series** is a simple ideal $\mathfrak{svect}^{\circ L}(1|n)$ of $\mathfrak{svect}^L_\lambda(1|n)$ for $n > 1$ and $\lambda \in \mathbb{Z}$, the quotient being spanned by $\theta_1 \ldots \theta_n \partial_x$ for any λ (observe that $\mathfrak{svect}^L_\lambda(1|n) \simeq \mathfrak{svect}^L_\mu(1|n)$ if $\lambda, \mu \in \mathbb{Z}$).

The *twisted supercircle* of dimension $1|n$ is the supermanifold that we denote $S^{1|n-1;M}$; it is associated with the Whitney sum of the trivial vector bundle of rank $n-1$ and the Möbius bundle. Since the Whitney sum of the two Möbius bundles is isomorphic to the trivial rank 2 bundle, we will only consider either $S^{1|n}$ or $S^{1|n-1;M}$.

Let $\theta = \sqrt{x}\theta_n$ be the corresponding to the Möbius bundle odd coordinate on $\mathbb{C}S^{1|n-1;M}$, the complexification of $S^{1|n-1;M}$. Set

$$\alpha^M = dx - \sum_{1<i\leq n-1} \theta_i d\theta_i - x\theta d\theta;$$

sometimes the following form is more convenient:

$$\alpha'^M = dx - \textstyle\sum_{1\leq i\leq k}(\xi_i d\eta_i + \eta_i d\xi_i) - x\theta d\theta;$$
$$\alpha'^M = dx - \textstyle\sum_{1\leq i\leq k}(\xi_i d\eta_i + \eta_i d\xi_i) - \zeta d\zeta - x\theta d\theta.$$

The **fifth series** is the Lie superalgebra $\mathfrak{k}^M(n)$ that preserves the Pfaff equation $\alpha^M = 0$.

One **exceptional superalgebra**, $\mathfrak{m}^L(1)$, is the Lie subsuperalgebra in $\mathfrak{vect}^L(1|2)$ that preserves the Pfaff equation given by the *even* contact form

$$\beta = d\tau + \pi dq - q d\pi$$

corresponding to the "odd mechanics" on $1|1$-dimensional supermanifold with $0|1$-dimensional time. Though the following regradings demonstrate the isomorphism of this superalgebra with the nonexceptional ones *considered as abstract* superalgebras, they are distinct as filtered superalgebras and to various realizations of these Lie superalgebras different Sturm–Liouville and KdV operators correspond.

Let t, θ be the indeterminates for $\mathfrak{vect}(1|1)$; let x, ξ, η be same for $\mathfrak{k}(1|2)$ (in the realization that preserves the Pfaff eq. $\alpha' = 0$); and let τ, q, π be the indeterminates for $\mathfrak{m}(1)$. Denote $\mathfrak{vect}(t, \theta)$ with the grading $\deg t = 2$, $\deg \theta = 1$ by $\mathfrak{vect}(t, \theta; 2, 1)$, etc. Then the following exceptional nonstandard degrees indicated after a semicolon provide us with the isomorphisms:

$$\mathfrak{vect}(t,\theta;2,1) \cong \mathfrak{k}(1|2); \qquad \mathfrak{k}(t,\xi,\eta;1,2,-1) \cong \mathfrak{m}(1);$$
$$\mathfrak{vect}(t,\theta;1,-1) \cong \mathfrak{m}(1); \qquad \mathfrak{m}(\tau,q,\pi;1,2,-1) \cong \mathfrak{k}(1|2).$$

Another, more serious, **exceptional Lie superalgebra** is $\mathfrak{k}^{L\circ}(1|4)$, the simple ideal of codimension 1 in $\mathfrak{k}^L(1|4)$, the quotient being generated by $\frac{\theta_1\theta_2\theta_3\theta_4}{x}$. The remaining exceptions, listed in [GLS], are not distinguished, so we ignore them in this paper.

1.2. The modules of tensor fields

To advance further, we have to recall the definition of the modules of tensor fields over the general vectorial Lie superalgebra $\mathfrak{vect}(m|n) = \mathfrak{der}\, \mathbb{C}[X]$, where $X = (x, \theta)$, and its subalgebras, see [BL]. Let $\mathfrak{g} = \mathfrak{vect}(m|n)$ and $\mathfrak{g}_{\geq} = \bigoplus\limits_{i\geq 0} \mathfrak{g}_i$, where $\deg X_i = 1$ for all i. Clearly, $\mathfrak{g}_0 \cong \mathfrak{gl}(m|n)$. Let V be the $\mathfrak{gl}(m|n)$-module with the *lowest* weight $\lambda = \mathrm{lwt}(V)$. Make V into a $\mathfrak{g}_{\geq}$-module setting $\mathfrak{g}_+ \cdot V = 0$ for $\mathfrak{g}_+ = \bigoplus\limits_{i>0} \mathfrak{g}_i$. Let us realize $\mathfrak{g}$ by vector fields on the $m|n$-dimensional linear supermanifold $\mathcal{C}^{m|n}$ with coordinates X. The superspace $T(V) = \mathrm{Hom}_{U(\mathfrak{g}_{\geq})}(U(\mathfrak{g}), V)$ is isomorphic, due to the Poincaré–Birkhoff–Witt theorem, to $\mathbb{C}[[X]] \otimes V$. Its elements have a natural interpretation as formal *tensor fields of type* V (or λ). When $\lambda = (a, \dots, a)$ we will simply write $T(\vec{a})$ instead of $T(\lambda)$. We usually consider irreducible $\mathfrak{g}_0$-modules.

For any other $\mathbb{Z}$-graded vectorial Lie superalgebra $\mathfrak{g} = \bigoplus\limits_{i\geq -d} \mathfrak{g}_i$ the construction of $T(V)$ as $\mathrm{Hom}_{U(\mathfrak{g}_{\geq})}(U(\mathfrak{g}), V)$ is identical.

Examples. $\mathit{Vol}(m|n) = T(1, \dots, 1; -1, \dots, -1)$ (the semicolon separates the first m coordinates of the weight with respect to the matrix units E_{ii} of $\mathfrak{gl}(m|n)$) is the superspace of *densities* or *volume forms*; clearly, $T(\vec{0})$ is the superspace of functions. We denote the generator of $\mathit{Vol}(m|n)$ corresponding to the ordered set of coordinates X by $\mathit{vol}(X)$. The space of λ-densities is $\mathit{Vol}^{\lambda}(m|n) = T(\lambda, \dots, \lambda; -\lambda, \dots, -\lambda)$. In particular, $\mathit{Vol}^{\lambda}(m|0) = T(\vec{\lambda})$ but $\mathit{Vol}^{\lambda}(0|n) = T(\overrightarrow{-\lambda})$.

1.3. Modules of tensor fields over stringy superalgebras

Denote by $T^L(V) = \mathbb{C}[t^{-1}, t] \otimes V$ the $\mathfrak{vect}(1|n)$-module that differs from $T(V)$ by allowing the Laurent polynomials as coefficients of its elements instead of just polynomials. Clearly, $T^L(V)$ is a $\mathfrak{vect}^L(1|n)$-module. Define the *twisted with weight* μ version of $T^L(V)$ by setting:

$$T^L_{\mu}(V) = \mathbb{C}[x^{-1}, x]x^{\mu} \otimes V. \tag{1.3.1}$$

• **The "simplest" modules — the analogues of the standard or identity representation of the matrix algebra**. The simplest modules over the Lie superalgebras of series $\mathfrak{vect}$ are, clearly, the modules of λ-densities, Vol^{λ}. These modules are characterized by the fact that they are of rank 1 over $\mathcal{F}$, the algebra of functions. Over stringy superalgebras, we can also twist these modules and consider $\mathit{Vol}^{\lambda}_{\mu}$. Observe that for $\mu \notin \mathbb{Z}$ this module has only one submodule, the image of the exterior differential d, see [BL], whereas for $\mu \in \mathbb{Z}$ there is, additionally, the kernel of the residue:

$$\begin{aligned} &\mathrm{Res} : \mathit{Vol}^L \longrightarrow \mathbb{C} \\ &f\mathit{vol}(x, \theta) \mapsto \text{ the coefficient of } \tfrac{\theta_1 \dots \theta_n}{x} \text{ in the expansion of } f. \end{aligned} \tag{1.3.2}$$

• Over $\mathfrak{svect}^L(1|n)$, all the spaces Vol^{λ} are, clearly, isomorphic, since $\mathit{vol}(x, \theta)$, hence their generators $\mathit{vol}(x, \theta)^{\lambda}$, are preserved. So all rank 1 modules over the algebra of functions are isomorphic to the module $\mathcal{F}_{0;\mu} = t^\mu \mathcal{F}$ of twisted functions for some μ.

Over $\mathfrak{svect}^L_\lambda(1|n)$, the simplest modules are generated (over the algebra of functions; twist is also possible) by $x^\lambda vol(x,\theta)$. The submodules of the simplest modules over $\mathfrak{svect}^L(1|n)$ and $\mathfrak{svect}^L_\lambda(1|n)$ are the same as those over $\mathfrak{vect}^L(1|n)$; but if the twist $\mu \in \mathbb{Z}$ there is, additionally, the trivial submodule generated by (a power of) $vol(x,\theta)$ or $x^\lambda vol(x,\theta)$, respectively.

• Over contact superalgebras, it is more natural to express the simplest modules not in terms of (twisted) λ-densities but via powers of the form α (or α') for the $\mathfrak{k}^L$ series, or α^M for the $\mathfrak{k}^M$ series, or β for $\mathfrak{m}^L(1)$. Set:

$$\mathcal{F}_\lambda = \begin{cases} \mathcal{F}\alpha^\lambda & \text{for } n=0 \\ \mathcal{F}\alpha^{\lambda/2} & \text{otherwise .} \end{cases} \tag{1.3.3}$$

Observe that $Vol^\lambda \cong \mathcal{F}_{\lambda(2-n)}$ as $\mathfrak{k}(1|n)$-modules. In particular, the Lie superalgebra of series $\mathfrak{k}$ does not distinguish between $\frac{\partial}{\partial x}$ and α^{-1}: their transformation rules are identical. Hence,

$$\mathfrak{k}(1|n) \cong \begin{cases} \mathcal{F}_{-1} & \text{if } n=0 \\ \mathcal{F}_{-2} & \text{otherwise .} \end{cases}$$

We denote the twisted versions by $\mathcal{F}_{\lambda;\mu}$.

For $n=2$ (and $\alpha = dx - \xi d\eta - \eta d\xi$) there are other rank 1 modules over $\mathcal{F} = \mathcal{F}_0$, the algebra of functions, namely:

$$T(\lambda,\nu;\mu) – \mathcal{F}_{\lambda;\mu} \cdot \left(\frac{d\xi}{d\eta}\right)^{\nu/2}. \tag{1.3.4}$$

• Over $\mathfrak{k}^M$, we should replace α with α^M and the definition of the $\mathfrak{k}^L(1|n)$-modules $\mathcal{F}_{\lambda;\mu}$ should be replaced with

$$\mathcal{F}^M_{\lambda;\mu} = \begin{cases} \mathcal{F}_{\lambda;\mu}(\alpha^M)^\lambda & \text{for } n=1 \\ \mathcal{F}_{\lambda;\mu}(\alpha^M)^{\lambda/2} & \text{for } n>1. \end{cases} \tag{1.3.5}$$

For $n=3$ and $\alpha^M = dx - \xi d\eta - \eta d\xi - t\theta d\theta$ there are other rank 1 modules over the algebra of functions $\mathcal{F}$, namely:

$$T^M(\lambda,\nu;\mu) = \mathcal{F}^M_{\lambda,\mu} \cdot \left(\frac{d\xi}{d\eta}\right)^{\nu/2}. \tag{1.3.6}$$

Examples.

1) The $\mathfrak{k}(2m+1|n)$-module of volume forms is $\mathcal{F}_{2m+2-n}$. In particular, $\mathfrak{k}(1|2) \subset \mathfrak{svect}(1|2)$.
2) As $\mathfrak{k}^L(1|n)$-module, $\mathfrak{k}^L(1|m)$ is isomorphic to $\mathcal{F}_{-1}$ for $m=0$ and $\mathcal{F}_{-2}$ otherwise. As $\mathfrak{k}^M(1|n)$-module, $\mathfrak{k}^M(1|n)$ is isomorphic to $\mathcal{F}_{-1}$ for $m=1$ and $\mathcal{F}_{-2}$ otherwise. In particular, $\mathfrak{k}^L(1|4) \simeq Vol$ and $\mathfrak{k}^M(1|5) \simeq \Pi(Vol)$; that is why both have ideals of codimension 1.

1.4. Convenient formulas

The four main series of stringy superalgebras are $\mathfrak{vect}^L(1|n)$, $\mathfrak{svect}^L_\lambda(1|n)$, $\mathfrak{k}^L(1|n)$ and $\mathfrak{k}^M(1|n)$. Obviously,

$$D = f\partial_x + \sum f_i\partial_i \in \mathfrak{svect}^L_\lambda(1|n) \quad \text{if and only if} \quad \lambda f = -x\mathrm{div}D. \qquad (1.4.1)$$

A laconic way to describe $\mathfrak{k}$, $\mathfrak{m}$ and their subalgebras is via *generating functions*.

• Odd form $\alpha = \alpha_1$. For $f \in \mathbb{C}[x,\theta]$ set :

$$K_f = (2-E)(f)\frac{\partial}{\partial x} - H_f + \frac{\partial f}{\partial x}E,$$

where $E = \sum\limits_i \theta_i\frac{\partial}{\partial\theta_i}$, and H_f is the hamiltonian field with Hamiltonian f that preserves $d\alpha_1$:

$$H_f = -(-1)^{p(f)}\left(\sum_{j\leq m}\frac{\partial f}{\partial\theta_j}\frac{\partial}{\partial\theta_j}\right).$$

The choice of α' instead of α only affects the form of H_f. We give it for $m = 2k+1$:

$$H_f = -(-1)^{p(f)}\sum_{j\leq k}\left(\frac{\partial f}{\partial\xi_j}\frac{\partial}{\partial\eta_j} + \frac{\partial f}{\partial\eta_j}\frac{\partial}{\partial\xi_j} + \frac{\partial f}{\partial\theta}\frac{\partial}{\partial\theta}\right).$$

• Even form $\beta = \alpha_0$. For $f \in \mathbb{C}[q,\pi,\tau]$ set:

$$M_f = (2-E)(f)\frac{\partial}{\partial\tau} - Le_f - (-1)^{p(f)}\frac{\partial f}{\partial\tau}E,$$

where $E = q\frac{\partial}{\partial q} + \pi\frac{\partial}{\partial\pi}$, and

$$Le_f = \frac{\partial f}{\partial q}\frac{\partial}{\partial\pi} + (-1)^{p(f)}\frac{\partial f}{\partial\pi}\frac{\partial}{\partial q}.$$

Since

$$L_{K_f}(\alpha_1) = K_1(f)\alpha_1, \qquad L_{M_f}(\alpha_0) = -(-1)^{p(f)}M_1(f)\alpha_0, \qquad (1.4.2)$$

it follows that $K_f \in \mathfrak{k}(1|m)$ and $M_f \in \mathfrak{m}(1)$. Observe that

$$p(Le_f) = p(M_f) = p(f) + \bar{1}.$$

• To the (super)commutators $[K_f, K_g]$ or $[M_f, M_g]$ there correspond *contact brackets* of the generating functions:

$$[K_f, K_g] = K_{\{f,g\}_{k.b.}}; \qquad [M_f, M_g] = M_{\{f,g\}_{m.b.}}$$

The explicit formulas for the contact brackets are as follows. Let us first define the brackets on functions that do not depend on x (resp. τ). The *Poisson bracket*

$\{\cdot,\cdot\}_{P.b.}$ is given by the formula

$$\begin{aligned}\{f,g\}_{P.b.} &= -(-1)^{p(f)}\sum_{j\leq m}\frac{\partial f}{\partial\theta_j}\frac{\partial g}{\partial\theta_j} \quad \text{or}\\ \{f,g\}_{P.b.} &= -(-1)^{p(f)}[\sum_{j\leq m}(\frac{\partial f}{\partial\xi_j}\frac{\partial g}{\partial\eta_j}+\frac{\partial f}{\partial\eta_j}\frac{\partial g}{\partial\xi_j})+\frac{\partial f}{\partial\theta}\frac{\partial g}{\partial\theta}].\end{aligned} \tag{1.4.3}$$

The *Buttin bracket* $\{\cdot,\cdot\}_{B.b.}$ is given by the formula (given here for $n = 1$)

$$\{f,g\}_{B.b.} = \sum_{i\leq n}(\frac{\partial f}{\partial q_i}\frac{\partial g}{\partial\xi_i}+(-1)^{p(f)}\frac{\partial f}{\partial\xi_i}\frac{\partial g}{\partial q_i}). \tag{1.4.4}$$

In terms of the Poisson and Buttin brackets the contact brackets take the form

$$\{f,g\}_{k.b.} = (2-E)(f)\frac{\partial g}{\partial x}-\frac{\partial f}{\partial x}(2-E)(g)-\{f,g\}_{P.b.}\ , \tag{1.4.5}$$

respectively,

$$\{f,g\}_{m.b.} = (2-E)(f)\frac{\partial g}{\partial\tau}+(-1)^{p(f)}\frac{\partial f}{\partial\tau}(2-E)(g)-\{f,g\}_{B.b.}. \tag{1.4.6}$$

It is not difficult to prove the following isomorphisms (as superspaces):

$\mathfrak{k}(1|n)\cong Span(K_f : f\in\mathbb{C}[x,\theta])$; $\mathfrak{m}(1)\cong Span(M_f : f\in\mathbb{C}[\tau,q,\pi])$.

- Define the *Möbius contact field* by the formula

$$\tilde{K}_f = (2-\tilde{E})(f)\mathcal{D}+\mathcal{D}(f)\tilde{E}+\tilde{H}_f, \tag{1.4.7}$$

where

$$\tilde{E} = \sum_{i<n}\theta_i\frac{\partial}{\partial\theta_i}+\theta\frac{\partial}{\partial\theta} \quad\text{and}\quad \mathcal{D}-\frac{\partial}{\partial x}-\frac{\theta}{2x}\frac{\partial}{\partial\theta}=\frac{1}{2}\tilde{K}_1 \tag{1.4.8}$$

and where

$$\tilde{H}_f = (-1)^{p(f)}(\sum\frac{\partial f}{\partial\theta_i}\frac{\partial}{\partial\theta_i}+\frac{1}{x}\frac{\partial f}{\partial\theta}\frac{\partial}{\partial\theta})$$

in the realization with form $\tilde{\alpha}^M$; in the realization with form $\tilde{\alpha}'^M$ for $n = 2k$ and $n = 2k+1$ we have, respectively:

$$\tilde{H}_f = (-1)^{p(f)}(\sum(\frac{\partial f}{\partial\xi_i}\frac{\partial}{\partial\eta_i}+\frac{\partial f}{\partial\eta_i}\frac{\partial}{\partial\xi_i})+\frac{1}{x}\frac{\partial f}{\partial\theta}\frac{\partial}{\partial\theta}),$$

$$\tilde{H}_f = (-1)^{p(f)}(\sum(\frac{\partial f}{\partial\xi_i}\frac{\partial}{\partial\eta_i}+\frac{\partial f}{\partial\eta_i}\frac{\partial}{\partial\xi_i})+\frac{\partial f}{\partial\zeta}\frac{\partial}{\partial\zeta}+\frac{1}{x}\frac{\partial f}{\partial\theta}\frac{\partial}{\partial\theta}).$$

The corresponding contact bracket of generating functions will be called the *Ramond bracket*; its explicit form is (see (1.4.8))

$$\{f,g\}_{R.b.} = (2-\tilde{E})(f)\mathcal{D}(g)-\mathcal{D}(f)(2-\tilde{E})(g)-\{f,g\}_{MP.b.}, \tag{1.4.9}$$

where the *Möbius-Poisson bracket* $\{\cdot,\cdot\}_{MP.b}$ is defined (in realization with the form $\tilde{\alpha}^M$) to be

$$\{f,g\}_{MP.b} = (-1)^{p(f)}\left(\sum\frac{\partial f}{\partial\theta_i}\frac{\partial g}{\partial\theta_i}+\frac{1}{x}\frac{\partial f}{\partial\theta}\frac{\partial g}{\partial\theta}\right). \tag{1.4.10}$$

Since (cf. (1.4.2))

$$L_{\tilde{K}_f}(\tilde{\alpha}) = \tilde{K}_1(f)\tilde{\alpha}. \tag{1.4.11}$$

It is easy to verify that $\mathfrak{k}(1|n) \cong Span(K_f : f \in \mathbb{C}[x^{-1}, t, \theta])$ whereas $\mathfrak{k}^M(1|n) = Span(\tilde{K}_f : f \in \mathbb{C}[x^{-1}, t, \theta])$. In other words, the spaces are identical but the brackets are nonisomorphic.

1.5. Distinguished stringy superalgebras

Stringy superalgebras considered above are introduced by a list. In the literature there are several definitions of these algebras, mostly self-contradicting ones. For example, in almost all physical papers stringy superalgebra are called "superconformal" ones (meaning conformal superalgebras). In reality, only $\mathfrak{witt}$, $\mathfrak{k}^L(1|1)$ and $\mathfrak{k}^M(1|1)$ are conformal in the original sense of the term ([GLS]); but even if other superalgebras were conformal, there central extensions are not, though the term "superconformal" is loosely applied to all. In [KvdL] and in several subsequent papers the "superconformal" superalgebras are defined as "simple ... such that", whereas everybody considers Virasoro, Neveu-Schwarz and Ramond superalgebras as "superconformal" ones, though they are not simple. Therefore, we suggest the term *stringy* for the general class of algebras inside of which some are conformal, some are simple, etc. An intrinsic definition of this class is given in [GLS] together with that for the loop *super*algebras (which, unlike simple loop algebras, may have no Cartan matrix, a usual key notion in their definition): both types of algebras are $\mathbb{Z}$-graded $\mathfrak{g} = \bigoplus_{i=-d}^{\infty} \mathfrak{g}_i$ of infinite depth d but in the adjoint representation they act differently:

for the loop algebras every root vector corresponding to any real root acts locally nilpotently in the adjoint representation,
for the stringy algebras this is not so.

In other words, for all stringy superalgebras there is an analog of the operator $\frac{d}{dx}$ which acts nontrivially on each homogeneous component, whereas for loop algebras there is no such operator. For the list of simple stringy superalgebras see [GLS] for a partial proof of its completeness see [KvdL].

Theorem. ([KvdL, GLS]) *The only nontrivial central extensions of the simple stringy Lie superalgebras are those given in the following table (for the scalar cases considered in this paper) and also for* $\mathfrak{m}^L(1)$, $\mathfrak{vect}^L(1|1)$, $\mathfrak{vect}^L(1|2)$, *and* $\mathfrak{svect}^L_\lambda(1|2)$.

The central extensions of the remaining four distinguished stringy superalgebras (see [GLS]) lead to matrix KdV and Sturm–Liouville operators to be considered elsewhere.

The operator ∇ introduced in the second column of the table by the formula $c: D_1, D_2 \mapsto \mathrm{Res}(D_1, \nabla(D_2))$ for an appropriate pairing $(\cdot,\cdot)$ will be referred to as the *cocycle operator*.

Let in this subsection and in Section 2.1 K_f be the common notation of both K_f and $\tilde{K}_f$, depending on whether we consider $\mathfrak{k}^L$ or $\mathfrak{k}^M$, respectively. Let further $\mathcal{K} = (2\theta\frac{\partial}{\partial\theta} - 1)\frac{\partial^2}{\partial x^2}$; to fit in width, the factor $(-1)^{p(f)}$ of cocycles c for $N = 2$ and 4 is not written.

algebra	$c: K_f, K_g \mapsto \mathrm{Res}(K_f, \nabla(K_g))$	The extended algebra
$\mathfrak{k}^L(1\|0)$	$\mathrm{Res} f K_1^3(g)$	Virasoro or $\mathfrak{vir}$
$\left.\begin{array}{l}\mathfrak{k}^L(1\|1)\\ \mathfrak{k}^M(1\|1)\end{array}\right\}$	$\mathrm{Res} f K_\theta K_1^2(g)$	Neveu-Schwarz or $\mathfrak{ns}$ Ramond or $\mathfrak{r}$
$\left.\begin{array}{l}\mathfrak{k}^L(1\|2)\\ \mathfrak{k}^M(1\|2)\end{array}\right\}$	$\mathrm{Res} f K_{\theta_1} K_{\theta_2} K_1(g)$	2-Neveu-Schwarz or $\mathfrak{ns}(2)$ 2-Ramond or $\mathfrak{r}(2)$
$\left.\begin{array}{l}\mathfrak{k}^L(1\|3)\\ \mathfrak{k}^M(1\|3)\end{array}\right\}$	$\mathrm{Res} f K_\xi K_\theta K_\eta(g)$	3-Neveu-Schwarz or $\mathfrak{ns}(3)$ 3-Ramond or $\mathfrak{r}(3)$
$\left.\begin{array}{l}\mathfrak{k}^{L\circ}(4)\\ \mathfrak{k}^M(1\|4)\end{array}\right\}$	(1) $\mathrm{Res} f K_{\theta_1} K_{\theta_2} K_{\theta_3} K_{\theta_4} K_1^{-1}(g)$ (2) $\mathrm{Res} f(x K_{x^{-1}}(g))$ (3) $\mathrm{Res} f K_1(g)$	(1) 4-Neveu-Schwarz = $\mathfrak{ns}(4)$; 4-Ramond = $\mathfrak{ns}(4)$ (2) 4′-Neveu-Schwarz = $\mathfrak{ns}(4')$; not defind for $\mathfrak{k}^M(1\|4)$ (3) 4^0-Neveu-Schwarz = $\mathfrak{ns}(4^0)$; not defind for $\mathfrak{k}^M(1\|4)$

2. Superized KdV and Sturm–Liouville operators

2.1. The KdV operators for the distinguished contact superalgebras

Clearly, the KdV operators corresponding to the supercircles associated with the cylinder and the Möbius bundle are absolutely different. To establish that similar is the situation with the Schrödinger operators, let us compair $\mathfrak{g}$ with $\mathfrak{g}^*$ for $\mathfrak{k}^L$ and $\mathfrak{k}^M$:

$\mathfrak{g} = \mathfrak{k}^L(1\|n)$	0	1	2	3	4	5	6	n
$\mathfrak{g}$	$\mathcal{F}_{-1}$	$\mathcal{F}_{-2}$	$\mathcal{F}_{-2}$	$\mathcal{F}_{-2}$	$\mathcal{F}_{-2}$	$\mathcal{F}_{-2}$	$\mathcal{F}_{-2}$	$\mathcal{F}_{-2}$
Vol	$\mathcal{F}_1$	$\Pi(\mathcal{F}_1)$	$\mathcal{F}_0$	$\Pi(\mathcal{F}_{-1})$	$\mathcal{F}_{-2}$	$\Pi(\mathcal{F}_{-3})$	$\mathcal{F}_{-4}$	$\Pi^n(\mathcal{F}_{2-n})$
$\mathfrak{g}^*$	$\mathcal{F}_2$	$\Pi(\mathcal{F}_3)$	$\mathcal{F}_2$	$\Pi(\mathcal{F}_1)$	$\mathcal{F}_0$	$\Pi(\mathcal{F}_{-1})$	$\mathcal{F}_{-2}$	$\Pi^n(\mathcal{F}_{4-n})$

$\mathfrak{g} = \mathfrak{k}^M(1\|n)$	1	2	3	4	5	6	7	n
$\mathfrak{g}$	$\mathcal{F}_{-1}$	$\mathcal{F}_{-2}$	$\mathcal{F}_{-2}$	$\mathcal{F}_{-2}$	$\mathcal{F}_{-2}$	$\mathcal{F}_{-2}$	$\mathcal{F}_{-2}$	$\mathcal{F}_{-2}$
Vol	$\Pi(\mathcal{F}_1)$	$\mathcal{F}_1$	$\Pi(\mathcal{F}_0)$	$\mathcal{F}_{-1}$	$\Pi(\mathcal{F}_{-2})$	$\mathcal{F}_{-3}$	$\Pi(\mathcal{F}_{-4})$	$\Pi^n(\mathcal{F}_{3-n})$
$\mathfrak{g}^*$	$\Pi(\mathcal{F}_2)$	$\mathcal{F}_3$	$\Pi(\mathcal{F}_2)$	$\mathcal{F}_1$	$\Pi(\mathcal{F}_0)$	$\mathcal{F}_{-1}$	$\Pi(\mathcal{F}_{-2})$	$\Pi^n(\mathcal{F}_{5-n})$

Remark. The comparison of $\mathfrak{g}$ with $\mathfrak{g}^*$ shows that there is a nondegenerate bilinear form on $\mathfrak{g} = \mathfrak{k}^L(1|6)$ and $\mathfrak{g} = \mathfrak{k}^M(1|7)$, even and odd, respectively. These forms are supersymmetric and given by the formula

$$(K_f, K_g) = \text{Res } fg.$$

Let in this subsection K_f be the common term for both K_f and $\tilde{K}_f$, as in Table 1.5, for $\mathfrak{g} = \mathfrak{k}^L$ or $\mathfrak{k}^{Lo}$ or $\mathfrak{k}^M$. The equation for $K_f \in \mathfrak{st}_{(F,1)}$, where $(F, 1) \in \hat{\mathfrak{g}}^*$, is of the form $\text{KdV}(f) = 0$, where the operators KdV are listed in the following table with the cocycle operators being the symmetrizations of the operators ∇ from Section 1.5; we write $\oplus$ instead of $+$ to grafically separate "the standard part". Explicitely:

n	KdV = the cocycle operator$\oplus$ "the standard part"
0	$K_1^3 \oplus FK_1 + K_1F$
1	$K_\theta(K_1)^2 \oplus (FK_1 + K_1F) + (-1)^{p(F)} K_\theta F K_\theta$
2	$(K_\xi K_\eta - K_\eta K_\xi)K_1 \oplus (FK_1 + K_1F) + (-1)^{p(F)}(K_\xi FK_\eta - K_\eta FK_\xi)$
3	$(K_\xi K_\eta - K_\eta K_\xi)K_\theta \oplus$ $(FK_1 + K_1F) + (-1)^{p(F)}(K_\xi FK_\eta - K_\eta FK_\xi + K_\theta FK_\theta)$
4_1	$(K_{\xi_1} K_{\eta_1} - K_{\eta_1} K_{\xi_1})(K_{\xi_2} K_{\eta_2} - K_{\eta_2} K_{\xi_2}) \int_x \oplus$ $(FK_1 + K_1F) + (-1)^{p(F)} \sum_{i=1,2}(K_{\xi_i} FK_{\eta_i} - K_{\eta_i} FK_{\xi_i})$
4_2	$xK_{x^{-1}} \oplus$ the standard part from 4_1
4_3	$K_1 \oplus$ the standard part from 4_1

2.2. The Sturm–Liouville operators as selfadjoint differential operators

• For the Neveu–Schwarz superalgebras we have the exact sequences

$$0 \longrightarrow \mathfrak{z} \longrightarrow \mathfrak{ns}(n) \longrightarrow \mathcal{F}_2 \longrightarrow 0. \tag{2.3.1}$$

Here $\mathfrak{z} = \mathbb{C} \cdot z$ if $n \neq 4$ and $\mathfrak{z} = \mathbb{C} \cdot z$, or $\mathbb{C} \cdot z_2$, or $\mathbb{C} \cdot z_3$ if $n = 4$, and centers correspond to the three cocycles.

Using the identification $Vol \cong \Pi^n(\mathcal{F}_{2-n})$ we dualize the above exact sequence and get:

$$\begin{aligned} &0 \longrightarrow \Pi^n(\mathcal{F}_{4-n}) \longrightarrow \mathfrak{ns}^*(n) \longrightarrow \mathfrak{z}^* \longrightarrow 0 \quad \text{for } n = 1, 2, 3, \\ &0 \longrightarrow \mathcal{F}_0/\mathbb{C} \longrightarrow \mathfrak{ns}^*(4) \longrightarrow \mathfrak{z}^* \longrightarrow 0, \text{ where } \mathbf{4} = 4 \text{ or } 4' \text{ or } 4^0 \end{aligned} \tag{2.3.2}$$

• For the Ramond superalgebras we similarly have the exact sequences

$$\begin{aligned} &0 \longrightarrow \mathfrak{z} \longrightarrow \mathfrak{r}(1) \longrightarrow \mathcal{F}_{-1} \longrightarrow 0 \\ &0 \longrightarrow \mathfrak{z} \longrightarrow \mathfrak{r}(n) \longrightarrow \mathcal{F}_{-2} \longrightarrow 0 \quad \text{for } n > 1. \end{aligned} \tag{2.3.3}$$

Using the identification

$$Vol \cong \begin{cases} \Pi(\mathcal{F}_1) & \text{for } n = 1 \\ \Pi^n(\mathcal{F}_{3-n}) & \text{for } n > 1 \end{cases}$$

we dualize the above exact sequence and get:

$$\begin{aligned} 0 \longrightarrow \Pi(\mathcal{F}_2) \longrightarrow \mathfrak{r}^*(1) \longrightarrow \mathfrak{z}^* \longrightarrow 0 \\ 0 \longrightarrow \Pi^n(\mathcal{F}_{5-n}) \longrightarrow \mathfrak{r}^*(n) \longrightarrow \mathfrak{z}^* \longrightarrow 0. \end{aligned} \tag{2.3.4}$$

Let us realize the elements of $\mathfrak{ns}^*(n)$ and $\mathfrak{r}^*(n)$ by selfadjoint (pseudo)differential operators $\hat{F} : \mathcal{F}_\lambda \longrightarrow \Pi^n(\mathcal{F}_\mu)$. We have already done this for $\mathfrak{vir}$ in the introduction.

If $\mathfrak{g}^*$ is of the form $\mathcal{F}_r$ of $\Pi(\mathcal{F}_r)$, then the order of the Sturm–Liouville operator determined by $\hat{\mathfrak{g}}^*$ is equal to r. In particular, the order of $\hat{F}$ is equal to $4 - n$ for $\mathfrak{ns}^*(n)$ and 0 for $n = 4'$ or 4^0 as well; it is equal to $5 - n$ for $\mathfrak{r}^*(n)$ if $n > 1$ and 2 for $\mathfrak{r}^*(1)$.

Now, let us solve the systems of two equations, of which the first equation counts the order of StL and the second one is the dualization condition:

$$\begin{aligned} &\mu = 2 + (2 - n) + \lambda, && \mu + \lambda = 2 - n && \text{for } \mathfrak{ns}(n) \\ &\mu = 1 + 1 + \lambda, && \mu + \lambda = 1 && \text{for } \mathfrak{r}(1) \\ &\mu = 2 + (3 - n) + \lambda, && \mu + \lambda = 3 - n && \text{for } \mathfrak{r}(n),\ n > 1. \end{aligned}$$

The solutions are:

$$\begin{aligned} &\mu = 3 - n, && \lambda = -1 && \text{for } \mathfrak{ns}(n) \\ &\mu = \tfrac{3}{2}, && \lambda = -\tfrac{1}{2} && \text{for } \mathfrak{r}(1) \\ &\mu = 4 - n, && \lambda = -1 && \text{for } \mathfrak{r}(n),\ n > 1. \end{aligned}$$

2.3. The list of Sturm-Liouville operators

For $\mathfrak{k}^L(1|n)$ and $n = 0, 1$ we can deduce the form of the Sturm-Liouville operators by factorization. For $n > 1$ and for $\mathfrak{k}^M(1|n)$ we define the Sturm-Liouville operators as self-adjoint operators equal to the sum of the operator given below with a potential F, where $p(F) \equiv n \pmod 2$ for $\mathfrak{k}^L(1|n)$ and $p(F) \equiv n + 1 \pmod 2$ for $\mathfrak{k}^M(1|n)$: the parities of the potential and the operator should be equal. Set $\Delta = \tilde{K}_\theta \tilde{K}_1^{-1} - \tilde{K}_1^{-1} \tilde{K}_\theta$. The operators are given with respect to forms α' and α'^M:

n	0	1	2	3
$\mathfrak{k}^L(1\|n)$	K_1^2	$K_\theta K_1$	$K_\xi K_\eta - K_\eta K_\xi$	$(K_\xi K_\theta K_\eta - K_\eta K_\theta K_\xi)(K_1)^{-1}$
$\mathfrak{k}^M(1\|n)$	–	$\Delta \tilde{K}_1^2$	$\Delta \tilde{K}_{\theta_1} \tilde{K}_1$	$\Delta(\ddot{K}_\xi \ddot{K}_\eta - \ddot{K}_\eta K_\xi)$

The answer for cocycle (3) supports physicists' belief — in spite of several independent proofs that there are three nontrivial cocycles on $\mathfrak{k}^{Lo}(1|4)$ — that "there are only two nontrivial cocycles":

$\mathfrak{k}^{Lo}(1\vert 4)$	(1) $(K_{\xi_1}K_{\eta_1} - K_{\eta_1}K_{\xi_1})(K_{\xi_2}K_{\eta_2} - K_{\eta_2}K_{\xi_2})(K_1)^{-2}$ (2) $xK_{x^{-1}}(K_1)^{-1} - (K_1)^{-1}xK_{x^{-1}}$ (3) any constant $c \neq 0$
$\mathfrak{k}^{M}(1\vert 4)$	(1) $\Delta\tilde{K}_{\theta_1}(\tilde{K}_\xi\tilde{K}_\eta - \tilde{K}_\eta\tilde{K}_\xi)(\tilde{K}_1)^{-1}$

For the Lie superalgebra $\mathfrak{vect}^L(1|1)$ the Sturm-Liouville operator is the same operator as for $\mathfrak{k}^L(1|2)$ but rewritten in the form of a matrix and with ∂_ξ instead of η.

We leave as an exercise to the reader the pleasure to write this matrix explicitely as well as to reexpress it in terms of the fields M_f for $\mathfrak{m}^L(1)$. For $\mathfrak{vect}^L(1|2)$ and $\mathfrak{svect}^L_\lambda(1|2)$ the Sturm-Liouville operators can be obtained from the Sturm-Liouville operator for $\mathfrak{k}^L(1|4)$ after restriction. All this will be done explicitely elsewhere.

2.4. The KdV hierarchies associated with the Sturm–Liouville operators

Let $L = L_r$ be the Sturm–Liouville operator of order r, see Section 2.2. As Shander taught us [Sh], the time parameter should run in super setting over a $1|1$-dimensional supermanifold, cf. also [MR]; so we define the KdV-type equations as the following Lax pairs:

$$D_T(L) = [L, A_k], \text{ where } A_k = (L^{k/r})_+ \text{ for } k \not\equiv r \pmod r \tag{2.5.1}$$

and where

$$D_T = \begin{cases} \frac{d}{dt} & \text{if } p(A_k) = \bar{0} \\ \frac{\partial}{\partial \tau} + \tau\frac{\partial}{\partial t} & \text{if } p(A_k) = \bar{1}. \end{cases}$$

Here the subscript $+$ singles out the differential part of the pseudodifferential operator. For complex k and for $\mathfrak{ns}(4)$ when L is a pseudodifferential operator, the differential part is not well defined and we shall proceed, *mutatis mutandis*, as Khesin–Malikov. The details will be given elsewhere.

References

[B] Berezin F., *Introduction to superanalysis.* Edited and with a foreword by A.A. Kirillov. With an appendix by V.I. Ogievetsky. Translated from the Russian by J. Niederle and R. Kotecký. Translation edited by Dimitry Leites. Mathematical Physics and Applied Mathematics, 9. D. Reidel Publishing Co., Dordrecht, 1987. xii+424 pp.

[BIK] Bellucci S., Ivanov E., Krivonos S. On $N = 3$ super Korteveg–de Vries Equation, J. Math. Phys., 34, 7, 1993, 3087–3097

[BL] Bernstein J., Leites D., Invariant differential operators and irreducible representations of Lie superalgebras of vector fields, Selecta Math. Sov., v. 1, 2, 1981,143–160

[CK] Chaichian M., Kulish P.P. Superconformal algebras and their relation to integrable nonlinear systems. Phys. Lett. B 183 (1987), no. 2, 169–174; Kulish, P.P. An analogue of the Korteweg–de Vries equation for the superconformal algebra. (Russian) Zap. Nauchn. Sem. Leningrad. Otdel. Mat. Inst. Steklov. (LOMI) 155 (1986), Differentsialnaya Geometriya, Gruppy Li i Mekh. VIII,142–149, 194 translation in J. Soviet Math. 41 (1988), no. 2, 970–975

[D] Deligne P. et al. (eds.) *Quantum fields and strings: a course for mathematicians.* Vol. 1, 2. Material from the Special Year on Quantum Field Theory held at the Institute for Advanced Study, Princeton, NJ, 1996–1997. AMS, Providence, RI; Institute for Advanced Study (IAS), Princeton, NJ, 1999. Vol. 1: xxii+723 pp.; Vol. 2: pp. i–xxiv and 727–1501

[DI] Delduc F., Ivanov E. $N = 4$ super KdV equation, Phys. Lett. B, v. 309, 1993, 312–319

[DIK] Delduc F., Ivanov E., Krivonos S., $N = 4$ super KdV hierarchy in $N = 4$ and $N = 2$ superspaces, J. Math. Phys., v. 37, 3, 1996, 1356–1381

[DS] Drinfeld, V.G.; Sokolov, V.V. Lie algebras and equations of Korteweg–de Vries type. Current problems in mathematics, Vol. 24, Itogi Nauki i Tekhniki, Akad. Nauk SSSR, Vsesoyuz. Inst. Nauchn. i Tekhn. Inform., Moscow, 1984, 81–180 (Russian, English translation: Soviet J. Math. v. 30, 1984, 1975—2036); id. Equations that are related to the Korteweg–de Vries equation. Dokl. Akad. Nauk SSSR 284 (1985), no. 1, 29–33 (Russian, English translation: Soviet Math. Dokl. 32 (1985), no. 2, 361–365.)

[GL] Grozman P., Leites D., Lie superalgebras of supermatrices of complex size. Their generalizations and related integrable systems. In: Vasilevsky N. et al. (eds.) *Proc. Intern. Symp. Complex Analysis and related topics*, Mexico, 1996. Operator Theory: Advances and Applications, 114. Birkhäuser Verlag, Basel, 2000, 73–105

[GLS] Grozman P., Leites D., Shchepochkina I., Lie superalgebras of string theories, hep-th 9702120; ; Acta Mathematica Vietnamica, v. 26, 2001, no. 1, 27–63

[IKT] Ivanov E., Krivonos S., Toppan F., $N = 4$ Sugawara construction on $\widehat{\mathfrak{sl}(2|1)}$, $\widehat{\mathfrak{sl}(3)}$ and mKdV-type superhierarchies. Modern Phys. Lett. A 14 (1999), no. 38, 2673–2686; id., $N = 4$ super NLS–mKdV hierarchies. Phys. Lett. B 405 (1997), no. 1-2, 85–94

[KvdL] Kac V., van de Leur, J.W., On classification of superconformal algebras. In: Gates S.J., Jr. et al. (eds.) *Strings '88. Proceedings of the workshop held at the University of Maryland, College Park, Maryland, May 24–28, 1988*, 77–106, World Sci. Publishing, Teaneck, NJ, 1989; Kac V., Superconformal algebras and transitive group actions on quadrics. Comm. Math. Phys. 186 (1997), no. 1, 233–252

[K2] Kac V., *Infinite dimensional Lie algebras*, Cambridge Univ. Press, Cambridge, 1991

[KM] Khesin B., Malikov F., Universal Drinfeld–Sokolov reduction and matrices of complex size, Commun. Math. Phys., **175** (1996), 113–134.

[Kh] Khovanova, T.G., Lie superalgebra $\mathfrak{osp}(1|2)$, Neveu–Schwarz superalgebra and superization of Korteweg–de Vries equation. In: Markov M. et al. (eds.) *Group theoretical methods in physics*, Vol. I (Yurmala, 1985), VNU Sci. Press, Utrecht, 1986, 307–314; id., The Korteweg–de Vries superequation connected with the Lie superalgebra of Neveu–Schwarz-2 string theory. (Russian) Teoret. Mat. Fiz. 72 (1987), no. 2, 306–312 (English translation: Theoret. and Math. Phys. 72 (1987), no. 2, 899–904)

[Ki1] Kirillov A., *Orbits of the group of diffeomorphisms of the circle and Lie superalgebras*, Sov. J. Funct. Analysis, v. 15, n.2, 1981, 75–76

[Ki2] Kirillov A., *The geometry of moments*, Lect. Notes in Math., Springer, Berlin ea., 1982, 101–123

[Ku1] Kupershmidt B., A super Korteweg–de Vries equation: an integrable system. Phys. Lett. A 102 (1984), no. 5–6, 213–215

[Ku2] Kupershmidt B. (ed.) *Integrable and superintegrable systems*, World Scientific Publishing Co., Inc., Teaneck, NJ, 1990. x+388 pp.; id. *Elements of superintegrable systems. Basic techniques and results.* Mathematics and its Applications, 34. D. Reidel Publishing Co., Dordrecht, 1987. xvi+187 pp.; id., A review of superintegrable systems. Nonlinear systems of partial differential equations in applied mathematics, Part 1 (Santa Fe, N.M., 1984), Lectures in Appl. Math., 23, Amer. Math. Soc., Providence, RI, 1986, 83–121

[Ku3] Kupershmidt B., ?⟶ Neveu–Schwarz–Ramond superalgebra→Virasoro algebra. Phys. Lett. A 113 (1985), no. 3, 117–120

[L1] Leites D. *Supermanifold Theory*, Karelia Branch of the USSR Acad. Sci., Petrozavodsk, 1983, 200 pp. (In Russian)

[L2] Leites D. (ed.), *Seminar on supermanifolds*, vv. 1–34. Reports of Stockholm University, 1987–1992, 2100 pp.

[LSS] Leites D., Saveliev M.V., Serganova V.V., Embeddings of $\mathfrak{osp}(N|2)$ and completely integrable systems. In: Markov M., Man'ko V. (eds.) *Proc. International Conf. Group-theoretical Methods in Physics*, Yurmala, May, 1985. Nauka, Moscow, 1986, 377–394 (English translation: VNU Sci Press, 1987, 255–297.

[LS] Leites D., Shchepochkina I., Classification of simple Lie superalgebras of vector fields with polynomial coefficients (to appear)

[M] Manin Y., *Gauge field theory and complex geometry.* Second edition. Grundlehren der Mathematischen Wissenschaften, 289. Springer-Verlag, Berlin, 1997. xii+346 pp.

[MR] Manin Yu.I., Radul A.O., A supersymmetric extension of the Kadomtsev–Petviashvili hierarchy. Comm. Math. Phys. 98 (1985), no. 1, 65–77

[MOR] Marcel P., Ovsienko V., Roger C., Extension of the Virasoro and Neveu–Schwarz algebras and generalized Sturm–Liouville operators. Lett. Math. Phys. 40 (1997), no. 1, 31–39

[OKh] Ovsienko V.Yu., Khesin B.A., The super Korteweg–de Vries equation as an Euler equation. Funktsional. Anal. i Prilozhen. 21 (1987), no. 4, 81–82. (Russian; English translation: Functional Anal. Appl. 21 (1988), no. 4, 329–331)

[Ra] Radul A. O., Superstring Schwarz derivative and the Bott cocycle. In: Yu. Manin et al. (eds.) *Numerical methods for solutions of boundary and initial value problems of differential equations*, Moscow Univ. Press, Moscow, 1986, 53–67 (Russian) and in: B. Kuperschmidt (ed.) *Integrable and superintegrable systems*, World Sci. Publishing, Teaneck, NJ, 1990, 336–351

[Sh] Shander V. N., Vector fields and differential equations on supermanifolds. (Russian) Funktsional. Anal. i Prilozhen. 14 (1980), no. 2, 91–92 (English translation: Functional Anal. Appl. 14 (1980), no. 2, 160–162)

Dimitry Leites
Department of Mathematics
University of Stockholm
Roslagsv. 101, Kräftriket hus 6
SE-106 91, Stockholm, Sweden
e-mail: mleites@matematik.su.se

Operator Theory:
Advances and Applications, Vol. 132, 287–322

Resonance Triadic Quantum Switch

A.B. Mikhaylova and B.S. Pavlov

Abstract. The mathematical design of a realistic three-position quantum switch controlled by the classical electric field is suggested in form of a circular quantum well — a unit disc on a plane — with four straight channels attached to it. This device implements a triple splitting of an input waveguide. The magnitude of the constant electric field directed parallel to the disc may be defined such that rotation of this field in the plane of the device permits manipulation of the electron current through the triple splitting. The problem of calculating of the current through the switch is reduced to the construction of scattered waves for the Schrödinger operator on the corresponding composite domain with the homogeneous Dirichlet conditions on the boundary. The Dirichlet boundary conditions are found to correspond most closely to real experimental conditions on the boundary of a deep quantum well. Explicit expression for transmission coefficient from one channel to another is obtained.

Technically the analysis of the corresponding infinitely-dimensional spectral problem is reduced to the analysis of a relevant finite-dimensional analytic matrix function. We estimate the errors that arise from replacement of infinite-dimensional operator by the finite matrix. Our main result is the calculation of the working point of the switch in the multi-dimensional space of the numerical parameters of the switch which permits the resonance manipulation of the current.

1. Introduction

It was noticed by R. Landauer in [36] that the problems of quantum conductance are dual to scattering problems on the corresponding networks, see also [8]. The problems of mathematical design of nano-devices from this point of view were considered first in [17], [1]. The problem of a quantum conductivity of a quasi-one-dimensional network in terms of relevant scattering problem was discussed first in [20]. Though the potentials on the arcs of the corresponding graph may be recovered from scattering data [21, 25], but the topology of the graph cannot [34] (see also [31, 30]). Use of periodic arrays of quantum dots for implementation of quantum transistors are discussed in [6, 52]. Interesting prospects for the development of this idea may be derived from [33], where the periodical structures with asymptotically growing lengths of gaps are considered.

In [6], [63] the theoretical design and the prospects of experimental implementation of a quantum relay based on simulated metal-isolator transition were discussed. The problem on design of a triadic quantum switch was formulated by doctor R. Compano in frames of our EU ESPRIT Project 28890 NTCONGS. It emerged in the course of work on the project that the idea of the resonance quantum switch may be based on observation which was done in the mathematical paper [18]: the scattering amplitude for resonator with a small opening depends on the value of the resonance eigenfunction of the resonator at the opening. In our previous papers [9], [45], [44], [51], [46] we developed this idea for simplest quantum networks. We produced some mathematical models of the resonance quantum switch designed to manipulate the quantum current through the splitting constructed of quantum wires and quantum domains (or rings) on the interface of an electrolyte and a narrow-gap semiconductor. Our efforts were inspired by the modern success of experimentalists which constructed quantum rings on the Silicon matrix using precise controlled diffusion of borons, see for instance the recent paper [7]. Nevertheless our previous papers were oriented rather to the qualitative mathematical modelling of the resonance processes in the switch. In particular in [46] and [45] we considered a solvable model of the Triadic Quantum Switch with Neumann boundary condition on the boundary of the domain and *sufficiently narrow wave-guides* modelled by zero-width lines. In [51] some prospects of using a model of the Quantum Switch based on a domain with Dirichlet boundary conditions are discussed "on a physical level of rigor".

In the present paper, which is independent from [51], we consider more realistic mathematical model of a Resonance Quantum Switch based on a deep circular quantum well Ω_0 designed as a unit disc [1] (in dimensionless units), and four straight channels Ω_s, $s = 1,2,3,4$ width δ attached radially to it centering at the points $a_1 \sim (\varphi = 0)$, $a_2 \sim (\varphi = \frac{\pi}{3})$, $a_3 \sim (\varphi = \pi), a_4 \sim (\varphi = \frac{5\pi}{3})$. The homogeneous Dirichlet boundary condition $u|_\Gamma = 0$ is applied on the boundary Γ of the joined domain $\Omega = \cup_{s=0}^4 \Omega_s$, and Meixner condition is required [42]. At the inner corners formed by the boundaries Γ_s, $s = 0, 1, 2, 3, 4$ of the channels and of the unit disc Ω_0. We consider a dimensionless Schrödinger operator in Ω

$$Lu = - \triangle u + V(x)u,$$

assuming that the potential is equal to zero on the wave-guides Ω_s, $s = 1, 2, 3, 4$ and linear $V(x) = \epsilon|x| \cos(x, \nu)$ inside the unit disc Ω on the complement $\hat{\Omega}_0$ of the channels Ω_s, $s = 1,2,3,4$. Our choice of the potential corresponds to the constant electric field $\nabla V = \epsilon\nu$ with the magnitude ϵ and the direction defined by the unit vector ν parallel to the plane of the domain Ω. The operator is defined as a Friedrichs extension of the restricted operator defined on smooth functions u with compact support. It is constructed via the closure of the corresponding quadratic

[1] If the domain is not specified another way, we assume in this paper, that it is a unit disc. But in the next section more general domain with a piece-wise smooth boundary is considered.

form

$$\int_{\Omega} [|\nabla u|^2 + (V - \min V)\, |u|^2] dm$$

and corresponds exactly to the Meixner condition at the inner corners.

The solution of the scattering problem for coupled wave-guides is rather a classical problem of mathematical physics. We mention here only some of a huge list of publications concerning relevant questions for homogeneous Dirichlet boundary conditions: [13], [5], [27], [26], [10], [11], [29], [4], [59], [14] which give examples of various technical approaches to the treating of the scattering problem for the Dirichlet Laplacian on composite domains with the techniques of the integral equations or asymptotic analysis.

Another approach to the problem may be based on the technique of matching asymptotic expansions suggested in [28] and developed in [19] for scattering by resonators with relatively small openings. We do not use this approach here since we do not assume now that the wave-guides λ_s, $s = 1, 2, 3, 4$ are narrow.

One more approach to the calculation of resonance characteristics of the corresponding scattering problem using a resonator with the small opening may adapted from analysis suggested in [62], [23], [50]. This approach is based on some integral equation with contracting operator. In fact we modify this approach to make it *independent* on the condition of narrowness of channels. As a result we obtain a rigorous version of the classical matching techniques successfully used by J. Schwinger during the period of the Second World War [54] for boundary problems in composite domains combined of radio-location wave-guides and resonators. In our case the corresponding integral equations will be written directly in terms of Green functions of the unperturbed problems on the "quantum well" Ω_0 and the channels Ω_s, $s = 1, 2, 3, 4$. The integral equations which appear in this way contain *finite-dimensional operator functions and small integral operators*, thus they can be reduced via couple of iterations to the relevant finite-dimensional problems and sometimes even solved in explicit form by iteration procedure.

The authors are grateful to Prof. B. Vainberg, Prof. Maz'ja and Dr. A. Schuchinskij for useful references.

2. Separation of singularities in the Green function, Poisson map and DN-map

In this section we prepare technical tools for analysis of the Schrödinger operator on a composite domain.We assume that $\Omega_0 \in R_2$ is a general compact domain with a piecewise twice differentiable boundary. The Green function $G_\lambda(x, y)$ of the operator L is a singular solution of the homogeneous equation

$$LG = -\Delta G + VG = \lambda G \tag{1}$$

with a special behaviour at the inner pole y:

$$G_\lambda(x,\, y) = \frac{1}{2\pi} \log \frac{1}{|x - y|} + O(1)$$

when $|x - y| \to 0$ and the homogeneous Dirichlet boundary conditions at the boundary $\partial\Omega := \Gamma_0$. For zero-potential on the whole plane the main singular solution is just the Hankel function of the first kind $\frac{i}{4}H_0^1(\sqrt{\lambda}|x-y|)$, and the Green function on the compact domain for zero-potential is represented in form of a sum of the above main singular solution and a regular solution of the corresponding homogeneous equation with proper boundary conditions:

$$\begin{aligned} G_\lambda^0(x, y) &= \frac{i}{4}H_0^1(\sqrt{\lambda}|x - y|) + g_\lambda(x, y), \\ -\Delta g &= \lambda g, \\ g|_{\partial\Omega} &+ \frac{i}{4}H_0^1(\sqrt{\lambda}|x - y|)|_{\partial\Omega} = 0. \end{aligned} \tag{2}$$

The Green function of the Schrödinger equation (1) on the domain may be found as a solution of the Lippmann-Schwinger equation

$$G_\lambda(x, y) = G_\lambda^0(x, y) + \int_\Omega G_\lambda^0(x, s)V(s)G_\lambda(s, y)ds^2. \tag{3}$$

One may deduce easily from this equation that it has the same logarithmic singularity at the pole as the Green function of the Laplace equation.

The non-perturbed Schrödinger operator L in $L_2(\Omega_0)$ for given potential with Dirichlet boundary condition is semi-bounded. Then the resolvent $[L + M]^{-1}$ of it with the spectral parameter $-M < \inf V$ is a bounded integral operator with the Green function $G_{-M}(x, y)$ as a kernel, and the norm of it in $L_2(\Omega_0) \times L_2(\omega_0)$ is small for large positive M. The following statement and relevant Corollary are valid for the Schrödinger operator on any compact domain Ω_0 with homogeneous Dirichlet boundary condition [2] on the piecewise twice differentiable boundary and Meixner condition at the inner corners:

Lemma 1. *For any regular point λ from the complement of the spectrum $\sigma(L)$ the resolvent of the operator L may be represented via the Hilbert Identity:*

$$G_\lambda(x, y) = G_{-M}(x, y) + (\lambda + M)\left\{G_{-M}G_\lambda\right\}(x, y)$$

where the second addend in the right side $(\lambda + M)\left\{G_{-M}G_\lambda\right\}(x, y) \equiv g_\lambda(x, y)$ is a continuous function of x, y and the spectral series of it on eigenfunctions φ_l of the operator L,

$$L\varphi_l = \lambda_l\varphi_l,$$

is absolutely and uniformly convergent in Ω_0. The separation of singularities at any finite set of eigenvalues λ_s is possible in any compact domain ω of the complex

[2]One may consider the operators with Dirichlet boundary condition only on some part of the boundary, and the Neumann condition on the rest, see the Corollary below.

plane of the spectral parameter:

$$g_\lambda(x,y) = (\lambda + M)\sum_{\lambda_l} \frac{\varphi_l(x)\varphi_l(y)}{(\lambda_l + M)(\lambda_l - \lambda)}$$
$$= (\lambda + M)\sum_{\lambda_l \leq K} \frac{\varphi_l(x)\varphi_l(y)}{(\lambda_l + M)(\lambda_l - \lambda)} + g^K_\lambda(x,y) \tag{4}$$

with uniformly and absolutely convergent series on ω for the correcting term $g^K_\lambda(x,y)$, and the uniform estimate for $g^K_\lambda(x,y)$, $|g^K_\lambda(x,y)| < \epsilon$ valid in $\Omega_0 \times \Omega_0$ for K large enough, $\lambda \in \omega$, uniformly with respect to the second argument y on any compact sub-domain $\Omega_0' \in \Omega$.

Proof. A proof of a similar statement with only one singular term separated is given in [50]. An extended discussion of the problem of separation of singularities is given in [45]. It is based on the analysis of the Lippmann-Schwinger equation (3) and shows that the Green function $G_{-M}(x,y)$ of the operator L admits a representation via the Hilbert Identity or the iterated Hilbert Identity. The proof suggested in [45] makes use of the fact of positivity of the integral operator $G_{-M} * G_{-M}$ which has a continuous kernel on the closed domain $\Omega_0 = \overline{\Omega_0}$. Then using smoothness of the normalized eigenfunctions

$$|\varphi_s|_{\mathrm{Lip}_{\frac{1}{2}}} \leq \mathbf{C}|\varphi_s|_{W_2^2(\Omega_0)}$$
$$\leq (\sup_\lambda |V(x)| + |\lambda_s|)$$

and the classical Mercer theorem we may check that the spectral series for it's kernel

$$G_{-M} * G_{-M}(x,y) = \sum_l \frac{\varphi_l(x)\varphi_l(y)}{(\lambda_l + M)^2}$$

is converging absolutely and uniformly in Ω_0. This implies the convergence of the series (4).

Note that practically for any compact domain ω on the complex plane λ one may optimize the above estimates proceeding in two steps : first choose the number M such that the norm of the resolvent $R_M = \int_{\Omega_0} G_{-M}(x,s) * dm(s)$ of the operator is less than $\epsilon/2$ and then choose the number K such that for $\lambda \in \omega$ the norm of the additional part g^K_λ of the resolvent is also small, $|g^K_\lambda| < \epsilon/2$. □

One may derive a similar statement for the Poisson map and the Dirichlet-to-Neumann map (DN-map) on the domain Ω_0 or the deformed domain, see below.

Consider the arc $[a,b] := \gamma_0 \in \partial\Omega_0$ on the boundary $\partial\Omega_0$ of the domain Ω_0. We call the arc *simple* if there exists a circle Σ_0 bordering the disc B_0 and passing through the points a,b such that the disc B_0 is divided by γ_0 into two simply connected subsets (homeomorphic to the ball)

$$B_0 = \{B_0 \cap \Omega_0\}\{B_0 \cap (R_2 \backslash \Omega_0)\}.$$

One can see that each sufficiently small arc of the piecewise C_2-smooth boundary of Ω_2 is a simple arc. But, in particular, for the unit disc each arc is simple.

Consider a crescent-like domain $\Omega_0' \subset \Omega$ bordered by two circular arcs Γ_1', $\Gamma_2' \in \Omega$ which are non-tangent both to the boundary $\partial\Omega_0$ and to the circle Σ_0 at the ends a, b of the simple arc γ_0, Γ_1' lies between γ_0 and Γ_2' and have no common points with the boundary except a, b.

The Dirichlet boundary-values problem

$$-\triangle u + Vu = \lambda u, \qquad u|_{\partial\Omega_0} = u_0,$$

may be solved by the Poisson map which is defined as an integral operator with the kernel calculated as a normal derivative of the Green function:

$$u(x) = -\int_{\partial\Omega_0} \frac{\partial G_\lambda}{\partial n_y}(x,y)u_0(y)d(\partial\Omega_0) := \mathcal{P}_\lambda u_0 \ (x).$$

Denote by P_λ the reduced Poisson map $\mathcal{P}_\lambda$ which acts from the space of continuous functions $C(\gamma_0)$ on the boundary supported by the arc γ_0 into the space of continuous functions $C(\Omega_0')$ on the crescent. It may be obtained as restriction onto Ω_0' of the solution $u = \mathcal{P}_\lambda u_0$ of the boundary problem

$$-\triangle u + Vu = \lambda u, \qquad u|_{\gamma_0} = u_0 \in C(\gamma_0). \tag{5}$$

We assume that the values of the spectral parameter λ lie outside of the spectrum of the corresponding operator L in $L_2(\Omega_0)$ which corresponds to the zero boundary conditions at the boundary $\partial\Omega_0 = \Gamma_0$.

Lemma 2. *The above restriction P_λ of the Poisson map $\mathcal{P}_\lambda$ admits the representation based on the Hilbert identity*

$$P_\lambda = P_{-M} + (\lambda + M)R_\lambda P_{-M}, \tag{6}$$

via the Poisson map with sufficiently large negative spectral parameter $-M <$ $\inf V$ outside of the spectrum σ of the Dirichlet operator L. The first term of the above decomposition P_{-M} is a **contraction operator** *from $C(\gamma_0)$ into $C(\Omega_0')$, and the second term $(\lambda + M)R_\lambda P_{-M}$ is a compact operator from $C(\gamma_0)$ into $C(\Omega_0')$. In any compact domain ω of the complex plane the restricted Poisson map P_λ is represented as a sum of a contraction from $C(\gamma_0)$ into $C(\Omega_0')$ and a finite-dimensional operator function of the spectral parameter.*

Proof. One may derive from the maximum principle for the Laplace equation that the solution v_0 of the Dirichlet problem outside the ball B_0:

$$-\triangle v_0 = 0, \qquad v_0|_{\Sigma_0\cap\Omega_0} = 1, \ v_0|_{\Sigma_0\cap(R_2\setminus\Omega_0)} = 0 \tag{7}$$

fulfills the condition $0 < v_0 < \sup_{\Omega_0'} v_0 < 1$ on the crescent. Then, comparing with v_0 the solution v of the Dirichlet problem

$$-\triangle v = 0, \qquad v|_{\gamma_0} = 1, \ v|_{\partial\Omega_0\setminus\gamma} = 0, \tag{8}$$

we obtain the estimate $0 < v < v_0 < \sup_{\Omega_0'} v_0 < 1$ on the crescent. Finally, using the maximum principle for proper values $-M < \inf V$ of the spectral parameter we obtain the estimation on the crescent Ω_0'

$$0 < v_{-M} < v < \sup_{\Omega_0'} v_0 < 1$$

for the solution $v_{_{-M}}$ of the equation

$$-\Delta v_{_{-M}} + V v_{_{-M}} = -M v_{_{-M}}, \qquad v_{_{-M}}|_{\gamma_0} = 1,\ v_{_{-M}}|_{\partial\Omega\setminus\gamma_0} = 0 \tag{9}$$

We may estimate now the norm of the solution u of the above equation with any continuous boundary values u_0:

$$-\Delta u + Vu = -Mu, \qquad u|_{\gamma_0} = u_0,\ u|_{\partial\Omega\setminus\gamma_0} = 0 \tag{10}$$

Really, due to the maximum principle the modulo of it on the whole domain Ω_0 does not exceed the solution v_{-M} of the above problem (9), hence on the crescent Ω_0' we have:

$$|u|_{C(\Omega_0')} \le \sup_{\Omega_0'} v_{_{-M}}\ |u_0|_{C(\Omega_0')}. \tag{11}$$

Thus the restriction of the Poisson map $\mathcal{P}_{_{-M}}$ from $C(\gamma_0)$ into $C(\Omega_0')$ is a contraction.

To prove compactness of the second addend in the right side of the formula (6) we notice first that the solution of the above boundary problem (9) is square integrable[3]. Hence, due to the maximum principle, the solution u of any boundary problem with continuous boundary values u_0 is square integrable as well, and $|u|_{L_2(\Omega_0)} \le |v_{_{-M}}|_{L_2(\Omega_0)}|u_0|_{C(\Omega_0')}$. Then $R_\lambda u \in W_2^2(\Omega_0)$ and, moreover, the image of the unit ball $B_1(C(\gamma_0))$ with the above operator $R_\lambda P_{-M}$ is compact in any Sobolev class compactly embedded into $W_2^2(\Omega_0)$, in particular it is compact in $C(\Omega_0')$.

To prove the last statement we may use the representation of the resolvent based on the iterated Hilbert Identity:

$$\begin{aligned} R_\lambda &= R_{-M} + (\lambda + M) R_\lambda R_{-M}, \\ R_\lambda P_{-M} &= R_{-M} P_{-M} + (\lambda + M) R_{-M} R_\lambda P_{-M}. \end{aligned} \tag{12}$$

The first addend in the right side of the last equation is a constant compact operator and hence may be represented as a sum of a constant finite-dimensional operator and a small operator. The second addend is a compact operator function from $C(\gamma_0)$ into $W_2^2(\Omega_0)$ and may be represented with use of the spectral decomposition of the resolvent $R_\lambda = R_\lambda^N \oplus R_\lambda'$ into the finite-dimensional part corresponding to the eigenvalues which lie inside the disc $|\lambda| < N$ and the complementary part which corresponds to the remaining eigenvalues. The term $(\lambda + M) R_{-M} R_\lambda' P_{-M}$ in the above representation of the Poisson map, which corresponds to the complementary component R_λ' of the resolvent, is uniformly small in ω as an operator from $C(\gamma_0)$ into $C(\Omega_0')$, if N is large enough. □

Corollary 3. *Similar representation of the reduced Poisson map is valid not only on the the crescent-like domain, but also on the semicircle which may be considered as a limit case of it.*

---

[3]Integration by parts gives even the estimate $\frac{d}{dM}\int_{\Omega_0} |v_{_{-M}}|^2 dm \le 0$, if $-M < \min V$.

Corollary 4. *In particular the statement (1) is true also for the Schrödinger operator* $\hat{L}$ *on the modified basic domain* $\hat{\Omega}_0 = \Omega_0 \setminus \cup_{s=1}^4 \Omega_s$ *— the part of the unit disc obtained by removing from it few segments enclosed by the bottom chords* δ_s *of the channels* Ω_s *and the corresponding arcs* γ_s *of the unit circle. The statement concerning separation of singularities of the Green functions remains also true for the Schrödinger operator* $\hat{\mathbf{L}}$ *defined in* $L_2(\hat{\Omega}_0)$ *with other boundary conditions, in particular with Dirichlet boundary condition on the chords* δ_s *replaced by the homogeneous Neumann condition.*

Together with Poisson maps $\mathcal{P}_\lambda$, $\hat{\mathcal{P}}_\lambda$, of operators L, $\hat{L}$ on basic domains with zero boundary conditions on $\partial\Omega_0$, $\partial\hat{\Omega}_0$ one may consider corresponding Dirichlet-to-Neumann maps (see [56]) Λ, $\hat{\Lambda}$ defined as limit values of the outer normal derivatives on the boundary

$$\Lambda u_0(x) = \frac{\partial}{\partial n_x} \mathcal{P}_\lambda u_0|_{\gamma_0}, \; \hat{\Lambda} u_0(x) = \frac{\partial}{\partial n_x} \hat{\mathcal{P}}_\lambda u_0|_{\delta_0}$$

of the solutions $\mathcal{P}_\lambda u_0$, $\hat{\mathcal{P}}_\lambda u_0$ of the above boundary-value problems for the corresponding differential equations $Lu = \lambda u$ and $\hat{L}u = \lambda u$ in Ω_0, $\hat{\Omega}_0$ respectively. Existence of normal derivatives of the double-layer potentials and actually existence of the DN-map on certain class of functions was established by A.M. Lyapunoff in his papers on potential theory, see [39], [40]. After the pioneering paper by A. Calderon, [12] DN-map is intensely used as a detail of solution of the inverse problems, see [58].

Being defined for the W_2^2-solutions of the Schrödinger equation outside the spectrum of the corresponding operator L on a domain with C_2-boundary, the DN-map acts from the space of the corresponding boundary data $W_2^{3/2}(\partial\Omega)$ onto $W_2^{1/2}(\partial\Omega)$, see [38], [35]. The inverse map Q_λ coincides with an integral operator on the boundary with a kernel equal to the restriction of the Green function $G_\lambda^N(x,y)$ of the Neumann problem onto the boundary:

$$u_0(x) = \int_{\partial\Omega} G_\lambda^N(x,y)\,(\Lambda u_0)\,(y)dy.$$

The DN-map is an analytic operator function with negative imaginary part in the upper half-plane of the spectral parameter λ and may serve as a multi-dimensional analog of the Weyl function, see [57]. The singularities of Λ_λ and the singularities of the inverse map Q_λ lie on the real axis $\Im\lambda = 0$ and coincide with the spectra of the corresponding operators defined by the above differential expression and Dirichlet and Neumann homogeneous boundary conditions on $\partial\Omega_0$ respectively.

Lemma 5. *The Dirichlet-to-Neumann map admits a representation based on Hilbert Identity or iterated Hilbert Identity, for instance:*

$$\Lambda_\lambda = \Lambda_{-M} - (\lambda + M)\mathcal{P}_{-M}^+ \mathcal{P}_{-M} - (\lambda + M)^2 \mathcal{P}_{-M}^+ R_\lambda \mathcal{P}_{-M}, \tag{13}$$

where R_λ *is the resolvent of the corresponding operator* L *outside of the spectrum. The third term in the right side of the above formula* (13) *is a compact operator in* $W_2^{3/2}(\partial\Omega_0)$. *It may be represented in each compact domain* ω *of the complex plane as a sum of a bounded operator in* $W_2^{3/2}$ *and a finite-dimensional operator function with a finite number of simple poles inside* ω *at the eigenvalues of the operator* L.

Proof. The proof may be obtained by differentiation with respect to the outer normal of the expression for the kernel of the Poisson map with iteration of the second term

$$R_\lambda \mathcal{P}_{-M} = R_{-M}\mathcal{P}_{-M} + (\lambda + M)R_{-M}R_\lambda\mathcal{P}_{-M}$$

and subsequent use of embedding theorems. The term $(\lambda + M)\mathcal{P}^+_{-M}\mathcal{P}_{-M}$ is a bounded operator in $W_2^{3/2}$ and the compact term $(\lambda + M)^2\mathcal{P}^+_{-M}R_\lambda\mathcal{P}_{-M}$ of the DN-map with the spectral parameter on any compact ω of the complex plane may be rewritten in spectral terms as an integral operator with the kernel:

$$(\lambda + M)^2\mathcal{P}^+_{-M}R_\lambda\mathcal{P}_{-M}(x,y)$$
$$= (\lambda + M)^2 \sum_{l=1}^{N} \frac{\frac{\partial\varphi_l(x)}{\partial n_x}\frac{\partial\varphi_l(y)}{\partial n_y}}{(\lambda_l + M)^2(\lambda_l - \lambda)} + \frac{\partial^2 g^K(x,y)}{\partial n_x \partial n_x}(x,y) \tag{14}$$

where the first term in the right side, for λ in the compact domain ω generates a finite-dimensional analytic operator function in $W_2^{3/2}(\partial\Omega_0)$ with simple poles at the eigenvalues of L and the last one corresponds to a small operator in $W^{3/2}(\partial\Omega_0)$ if the number N is large enough. □

In the next section we shall consider the boundary problems in composite domains with piecewise C_2-smooth boundary with inner corners $\alpha > \pi$. At the inner corners the classical statement [38], [35] on W_2^2-smoothness of the W_2^1-solution of the boundary problem with the boundary data from $W_2^{3/2}$ is not true any more,which may be seen from the simple example[4] of the harmonic function $u = \Im z^{\pi/\alpha}$ in the sector $0 < r < 1,\ 0 < \varphi < \alpha$. But it may be substituted by the weaker statement $u \in Lip_{\pi/\alpha}$ or $u \in W_p^2,\ p < \frac{2}{2-\pi/\alpha}$, see [61].

Representations similar to the above ones may be derived for the Green function, Poisson and Dirichlet to Neumann maps on the wave-guides Ω_s, s – 1, 2, 3, 4. But in our case all of them may be obtained more easily from the separation of variables. Denoting by $|\delta|$ the width of wave-guides Ω_s we may represent the solution of the Dirichlet problem with the boundary conditions $u|_{\delta_s} = u_s(y)$ at the bottom δ_s of the wave-guide Ω_s as

$$u(x,y) = \sum_l \sin\left(\frac{l\pi y}{|\delta|}\right)\frac{2}{|\delta|}e^{i\sqrt{\lambda - (\frac{l\pi}{|\delta|})^2}x}\int_0^{|\delta|} u_s(\eta)\sin\frac{l\pi\eta}{|\delta|}d\eta := \mathcal{P}_s u_s(x,y), \tag{15}$$

which gives an explicit formula for the Poisson map of the Dirichlet problem on the waveguide Ω_s, if the branches of square roots are defined on the spectral sheet

[4]This example was communicated to the authors by A. Hassel and M. Costabel. Similar example was published in PhD thesis of O.V. Guseva, [24].

of the variable λ by the condition

$$\Im\sqrt{\lambda-\left(\frac{\pi l}{|\delta|}\right)^2}\geq 0.$$

Note that for the spectral parameter on the upper shore of the first spectral band in the waveguide, $\lambda\in\left(\frac{\pi^2}{|\delta|^2},\frac{4\pi^2}{|\delta|^2}\right)$, all exponentials but the first are rapidly decreasing when $x\to\infty$.

The statements proven above actually permit to reduce the scattering problem in the composite domain Ω to solving of a finite-dimensional linear system with a matrix depending on λ, see below, Sections 3, 4.

3. Scattering problem

Our aim in this section is the derivation of an explicit formula connecting the scattering matrix in the composite domain $\Omega=\Omega_0\cup\Omega_1\cup\Omega_2\cup\Omega_3\cup\Omega_4$ with the Dirichlet-to-Neumann map of the basic compact domain. The spectrum of the operator L in $L_2(\Omega)$ consists (see for instance [55]) of a countable set of absolutely continuous branches $[(\frac{l\pi}{\delta})^2,\infty]$, $l=1,2,\dots$, each of them multiplicity 4, and possibly a finite number of eigenvalues λ_r of a finite multiplicity below the first threshold $\frac{\pi^2}{\delta^2}$. There may be some embedded eigenvalues, as a simple example of the Schrödinger operator in a single channel $\Omega_1=(0,\delta)\times(0,\infty)$ attached to the rectangle $\Omega_0=(0,\delta)\times(a,0)$ with a piecewise-constant compactly-supported potential V shows:

$$V=\begin{cases}0 & \text{if } (x,y)\in\Omega_1,\\ V<-\frac{4\pi^2}{|\delta|^2}+\lambda & \text{if } (x,y)\in\Omega_0,\end{cases}$$

$$a=\frac{-1}{\sqrt{-V-\frac{4\pi^2}{\delta^2}+\lambda}}\tan^{-1}\frac{\sqrt{-V-\frac{4\pi^2}{|\delta|^2}+\lambda}}{\sqrt{\frac{4\pi^2}{|\delta|^2}-\lambda}}<0.$$

The eigenfunction corresponding to the eigenvalue λ embedded into the first spectral band $(\frac{\pi}{|\delta|})^2<\lambda<\frac{4\pi^2}{|\delta|^2}$ is

$$\psi=\begin{cases}\frac{|\delta|\sin\frac{2\pi y}{|\delta|}}{2\pi}\left[\cos\sqrt{-V-\frac{4\pi^2}{|\delta|^2}+\lambda}\,x-\sqrt{\frac{4\pi^2}{\delta^2}-\lambda}\,\frac{\sin\sqrt{-V-\frac{4\pi^2}{|\delta|^2}+\lambda}\,x}{\sqrt{-V-\frac{4\pi^2}{|\delta|^2}+\lambda}}\right], & \text{if } (x,y)\in\Omega_1,\\ \frac{|\delta|\sin\frac{2\pi y}{|\delta|}}{2\pi}e^{-\sqrt{\frac{4\pi^2}{|\delta|^2}-\lambda}\,x}, & \text{if } (x,y)\in\Omega_0\end{cases}$$

The eigenvalues below the first threshold are not involved into the process of quantum conductivity, but the embedded eigenvalues, see for instance [15] and [16], may be transformed to resonances by small perturbations. We shall consider this transformation in the next section in connection with resonance eigenvalue on the basic domain.

The total multiplicity of the absolutely-continuous spectrum is (countable) infinite. The eigenfunctions of the absolutely-continuous spectrum are so-called scattered waves, see [54]. These are solutions Ψ of the homogeneous equation $L\Psi = \lambda\Psi$, which fulfill some (asymptotic) conditions. For instance, the scattered wave *initiated by the plane wave incoming from the channel* Ω_t has the components $\Psi_{st},\ s = 1,2,3,4$, are represented on the first spectral band as $(\frac{\pi y}{|\delta|})^2 < \lambda < (\frac{2\pi y}{|\delta|})^2$

$$\Psi_{tt}(x,y) = \sqrt{\frac{2}{|\delta|}} \sin\frac{\pi y}{|\delta|} e^{-i\sqrt{\lambda-(\frac{\pi}{|\delta|})^2}\, x} + S_{tt}\sqrt{\frac{2}{|\delta|}} \sin\frac{\pi y}{|\delta|} e^{i\sqrt{\lambda-(\frac{\pi}{|\delta|})^2}\, x}$$
$$+\sum_{l>1} s_{tt}^{l}\sqrt{\frac{2}{|\delta|}} \sin\frac{l\pi y}{|\delta|} e^{-\sqrt{(\frac{l\pi}{|\delta|})^2-\lambda}\, x},$$
$$\Psi_{st}(x,y) = S_{st}^{1}\sqrt{\frac{2}{|\delta|}} \sin\frac{\pi y}{|\delta|} e^{i\sqrt{\lambda-(\frac{\pi}{|\delta|})^2}\, x} + \sum_{l>1} s_{st}^{l}\sqrt{\frac{2}{|\delta_s|}} \sin\frac{l\pi y}{|\delta|} e^{-\sqrt{(\frac{l\pi}{|\delta_s|})^2-\lambda}\, x},$$
$$s = 1,2,3,4. \tag{16}$$

The coefficients S_{st}^{1}, s = 1, 2, 3, 4 in front of the oscillating exponentials are elements of the *scattering matrix* on the first spectral band below threshold, and s_{st}^{l} are just the amplitudes of the exponentially decreasing modes. If λ sits on the upper spectral band, then higher oscillating modes are present.

To construct the scattered wave on the first spectral band we must find the solution of the Schrödinger equation on the basic domain $\hat{\Omega}_0$ or Ω_0 which satisfy the matching conditions on the bottom chords δ_s of the wave-guides Ω_t (in case of $\hat{\Omega}_0$) or on the arcs γ_s of the unit circle spanned by the chords δ_s, in case of Ω_0

$$\begin{aligned} \Psi_0 - \Psi_s|_{\delta_s} &= 0, \quad \frac{\partial}{\partial n}(\Psi_0 - \Psi_s)|_{\delta_s} = 0. \\ \Psi_0 - \Psi_s|_{\gamma_s} &= 0, \quad \frac{\partial}{\partial n}(\Psi_0 - \Psi_s)|_{\gamma_s} = 0. \end{aligned} \tag{17}$$

We shall suggest in the next section a convenient integral equation which is equivalent to this boundary problem. Now we assume that all DN-maps on $\Omega_0, \hat{\Omega}_0$ and the channels $\Omega_s,\ s = 1,2,3,4$ are already constructed. Then we derive an important formula for the 4×4 scattering matrix on the first spectral band (below the second threshold $\lambda = \frac{4\pi^2}{|\delta|^2}$) and similar formulae for higher spectral bands, connecting is with the DN-map $\hat{\Lambda}_0$ of the Schrödinger operator $\hat{L}$ in $L_2(\hat{\Omega}_0)$ on the modified basic domain $\hat{\Omega}_0$. Note that the restriction onto δ_s of the DN-map $\hat{\Lambda}_0$ is connected to the restriction $P_{\delta_s\ \gamma_s}$ of the Poisson map $\mathcal{P}$ in Ω_0 onto $\delta_s \times \gamma_s$:

$$\hat{\Lambda}_0|_{\delta_s} P_{\delta_s,\gamma_s} = \frac{\partial}{\partial n_{\delta_s}} P_{\delta_s,\gamma_s},$$

and hence is uniquely defined for all values of the spectral parameter outside of the discrete spectrum of operators $L,\ \hat{L}$.

In fact the scattering matrix is a *sub-matrix* of an infinite matrix which is defined by the matrix elements $S_{st},\ s_{st}$ defined above. Consider the normalized eigenfunctions $e_s^l = \sqrt{\frac{2}{\delta}} \sin \frac{\pi l y}{\delta}$ of the part of the Laplacian on the cross-sections of the wave-guides Ω_s. Then the infinite matrix may on the first spectral band is a matrix of the operator in the channel space $H = \oplus \sum_{s=1}^{4} L_2(\delta_s)$ defined as

$$\mathbf{S}(\lambda) = \sum_{s,t=1}^{4} S_{st}^1 e_s^1 >< e_t^1 + \sum_{s,t=1}^{4} \sum_{l=2}^{\infty} s_{st}^l e_s^l >< e_t^1 := \mathcal{S}^1(\lambda) + \mathbf{s}^1(\lambda). \tag{18}$$

We shall call it the *parent S-matrix.* Then the part $\mathcal{S}^1(\lambda)$ of the parent S-matrix $\mathbf{S}(\lambda)$ in the *open* first channel is exactly the scattering matrix on the first spectral band (below the second threshold): for any *initiating* linear combination of incoming modes from the first spectral band $\sum_s \psi_s e_s^1 e^{-i\sqrt{\lambda - (\frac{\pi}{|\delta|})^2}\, x} := \Psi_{in} e^{-i\sqrt{\lambda-(\frac{\pi}{|\delta|})^2}\, x}$ the *initiated* scattered wave is constructed as

$$\Psi = \Psi_{in} e^{-i\sqrt{\lambda-(\frac{\pi}{|\delta|})^2}\, x} + \mathcal{S}^1(\lambda)\Psi_{in} e^{i\sqrt{\lambda-(\frac{\pi}{|\delta|})^2}\, x} + \mathbf{s}^1(\lambda)\Psi_{in},$$

where the last term is exponentially decreasing below the threshold: $\mathbf{s}^1(\lambda) = \sum_{l=2}^{\infty} e^{i\sqrt{\lambda-(\frac{l\pi}{|\delta|})^2}\, x} s_{st}^l e_s^l >< e_t^l$ and both operators $\mathcal{S}^1(\lambda),\ \mathbf{s}^1(\lambda)$ act in $L_2(\delta) = \oplus \sum_{s=1}^{4} L_2(\delta_s)$ as matrices with elements $S_{st},\ s_{st}$ uniquely defined from the condition of matching of the initiated scattered wave Ψ in the channels Ω_s with the solution of the Schrödinger equation on the basic domain $\hat{\Omega}_0$.

To describe the matching conditions in convenient form in general case when several channels are open[5], we consider the decomposition of the space $L_2(\delta) = H$ of the square-integrable functions on the cross-sections into the orthogonal sum of the channel-spaces $H = H_+ \oplus H_-$ which correspond to the cross-sections of *open* $(l \leq m)$ and *closed* $(l \geq m+1)$ channels:

$$H_{+,s} = \bigvee_{l=1}^{l=m} e_s^l,\ \sum_{s=1}^{4} H_{+,s} := H_+,$$

$$H_{-,s} = \bigvee_{l=m+1}^{l=\infty} e_s^l,\ \sum_{s=1}^{4} H_{-,s} := H_-$$

There are two types of oscillating exponential modes $e^{\pm\, i\sqrt{\lambda-(\frac{l\pi}{|\delta|})^2}\, x} e_s^l$ in the open channels and only one type of a bounded exponential mode

$$e^{i\,\sqrt{\lambda-(\frac{l\pi}{|\delta|})^2}\, x} e_s^l = e^{-\sqrt{(\frac{l\pi}{|\delta|})^2-\lambda}\, x} e_s^l$$

[5]We deliberately use the term *channel* both for the domains Ω_s, $s = 1,2,3,4$, and for the invariant subspaces of the Laplacian on them. Still we hope that any confusion is eliminated, since the term is used for subspaces with adjectives "open" or "closed".

in the closed channels. These modes fulfill inside the channels the Helmholtz equation with the spectral parameter λ inside the channels and the Dirichlet boundary conditions on the walls $y = 0,\ \delta$. The diagonal operators

$$\mathcal{K}_s = \oplus \sum_{l=1}^{\infty} i\sqrt{\lambda - \frac{l^2\pi^2}{|\delta|^2}} e_s^l\rangle\langle e_s^l \quad \text{and} \quad \mathcal{K} = \sum_{s=1}^{4} \mathcal{K}_s .$$

The operator $\mathcal{K}$ for complex values of the spectral parameter,

$$\Im\sqrt{\lambda - \frac{\pi^2 l^2}{|\delta|^2}} > 0$$

plays the role of the Dirichlet-to-Neumann map on the channel space H. The limit values of it on the real axis of the spectral parameter exist in each channel, and may be decomposed into the orthogonal sum of two parts

$$\mathcal{K} = \mathcal{K}^+ \oplus \mathcal{K}^-, \qquad \mathcal{K}_s = \mathcal{K}_s^+ \oplus \mathcal{K}_s^-$$

which correspond to the open and closed channels respectively:

$$\mathcal{K}_s^+ = \oplus \sum_{l=1}^{m} i\sqrt{\lambda - \frac{l^2\pi^2}{|\delta|^2}}\ e_s^l\rangle\langle e_s^l\ ,\ \mathcal{K}^+ = \sum_{l=1}^{m} \mathcal{K}_s^+ ,$$

$$\mathcal{K}_s^- = \oplus \sum_{l=m+1}^{\infty} i\sqrt{\lambda - \frac{l^2\pi^2}{|\delta|^2}}\ e_s^l\rangle\langle e_s^l\ ,\ \mathcal{K}^- = \sum_{l=m+1}^{\infty} \mathcal{K}_s^- .$$

One may notice that both operators $\mathcal{K}^{\pm}$ have bounded inverse in H, if λ does not coincide with thresholds. The operator $[\mathcal{K}^-]\,(\lambda')^{-1}$ is small for given $\lambda' \ll \frac{\pi^2 m^2}{|\delta|^2}$ if $m \gg 1$ and acts as an operator from $W_2^{1/2}(\delta)$ into $W_2^{3/2}(\delta)$. It is self-adjoint and negative on the real axis of the spectral parameter λ below the threshold $\frac{\pi^2(m+1)^2}{|\delta|^2}$. The operator $\mathcal{K}^+$ is anti-hermitian (the operator $i\mathcal{K}^+$ is self-adjoint).

One may consider a similar decomposition for the DN-map $\hat{\Lambda}$ of the modified domain $\hat{\Omega}_0$, restricted onto the cross-sections of the channels $\hat{\Lambda}|_H = \hat{\Lambda}|_{H_+} \oplus \hat{\Lambda}|_{H_-} := \hat{\Lambda}_+ \oplus \hat{\Lambda}_-$. The restrictions of the eigenfunctions e_s^l on the bottom cross-sections δ_s of the channels Ω_s may be continued from $\delta = \cup_{s=1}^4 \delta_s$ onto the complement $\partial\hat{\Omega}_0 \backslash \delta$ by zero. We denote the resulting functions on $\partial\hat{\Omega}_0$ we denote by e_s^l as well. They belong to the domain of $W_2^{3/2}\left(\partial\hat{\Omega}_0\right)$ of the operator $\mathcal{K}$. This fact permits to consider the restriction of the DN-map $\hat{\Lambda}$ onto the channel space H as densely-defined unbounded operator in H. In particular for real values of the spectral variable λ one may consider the corresponding symmetric operator enclosed by the orthogonal projections[6] P_H acting from $L_2(\partial\hat{\Omega}_0)$ onto H:

$$P_H \hat{\Lambda}|_H .$$

[6]The orthogonal projections act in $L_2(\partial\hat{\Omega}_0)$ as multiplications by the indicators ξ_s of $\delta_s \subset \partial\hat{\Omega}_0$.

It is defined in H on the linear variety of all elements from $W_2^{3/2}(\delta)$ and takes values in $W_2^{1/2}(\delta)$ on the section. We may consider the representation of it by a Hermitian matrix in the above orthogonal decomposition $H = H_+ \oplus H_-$, $P_H = P_+ \oplus P_-$:

$$\begin{aligned} P_H \hat{\Lambda}|_H &= P_+\hat{\Lambda}P_+ + P_+\hat{\Lambda}P_- + P_-\hat{\Lambda}P_+ + P_-\hat{\Lambda}P_- \\ &:= \hat{\Lambda}_{++} + \hat{\Lambda}_{+-} + \hat{\Lambda}_{-+} + \hat{\Lambda}_{--} \\ &= \begin{pmatrix} \hat{\Lambda}_{++} & \hat{\Lambda}_{+-} \\ \hat{\Lambda}_{-+} & \hat{\Lambda}_{--} \end{pmatrix} \end{aligned}$$

Theorem 6. *The scattering matrix $\mathcal{S}$ may be represented via the restricted DN-map $P_H\hat{\Lambda}P_H$ as*

$$\mathcal{S} = \left\{\hat{\Lambda}_{++} - \hat{\Lambda}_{+-}\left[\hat{\Lambda}_{--} - \mathcal{K}^-\right]^{-1}\hat{\Lambda}_{-+} + \mathcal{K}^+\right\}^{-1} \left\{\hat{\Lambda}_{++} - \hat{\Lambda}_{+-}\left[\hat{\Lambda}_{--} - \mathcal{K}^-\right]^{-1}\hat{\Lambda}_{-+} - \mathcal{K}_+\right\}.$$

Here the operator $\hat{\Lambda}_{++} - \hat{\Lambda}_{+-}\left[\mathcal{K}^- - \hat{\Lambda}_{--}\right]^{-1}\hat{\Lambda}_{+-} := \mathcal{D}(\lambda)$ is a hermitian matrix-function in H^+ on the real axis of the spectral parameter λ, outside of the spectrum of $\hat{L}$, $\mathcal{K}^+$ is anti-hermitian, $\mathcal{K}^-$ is real negative and the whole matrix $\mathcal{S}$ is unitary in H_+.

Proof. Consider the *background* scattering problem for the Schrödinger operator on the channels Ω_t with zero boundary conditions on the boundary $\partial\Omega_t$, $t = 1, 2, 3, 4$. Components of the corresponding scattered waves on the open channel H_l may be presented as

$$\Psi_t^0 = \psi_{0l}^t e_t^l \left[e^{-i\sqrt{\lambda - \frac{\pi^2 l^2}{|\delta|^2}}} - e^{i\sqrt{\lambda - \frac{\pi^2 l^2}{|\delta|^2}}}\right], \qquad l \le m,$$

and the whole background scattered wave is

$$\psi_0(x) = \left[e^{-\mathcal{K}x} - e^{\mathcal{K}x}\right]\vec{\psi}_0,$$

where $\vec{\psi}_0 = \sum_{t,l}\psi_{0l}^t e_t^l$. It will play a role of an eigenfunction of the absolutely continuous spectrum of the Laplacian in $\oplus\sum_{s=1}^4 L_2(\Omega_s)$. Then the components of the *perturbed* eigenfunctions Ψ on the channels may be presented as linear combinations of the non-perturbed wave ψ_0 and *outgoing*[7] exponential solution combined of both oscillating $l \le m$ and exponentially decreasing components, in

[7] Following [37] we call the solution of the homogeneous equation *outgoing* if it may be analytically continued as a bounded function onto the spectral plane λ.

all channels:

$$\begin{aligned}\psi = \left[e^{-\mathcal{K}x} - e^{\mathcal{K}x}\right]\vec{\psi}_0 + \sum_{l,l'\leq m;t,t'} e_t^l e^{i\sqrt{\lambda - \frac{\pi^2 l^2}{|\delta|^2}}\,x}\, \mathcal{T}_{ll'}^{tt'}\psi_{0l'}^{t'} \\ + \sum_{l\geq m+1,l'\leq m;t,t'} e_t^l e^{i\sqrt{\lambda - \frac{\pi^2 l^2}{|\delta|^2}}\,x}\, s_{ll'}^{tt'}\psi_{0l'}^{t'} \\ := \left[e^{-\mathcal{K}x} - e^{\mathcal{K}x}\right]\vec{\psi}_0 + \mathcal{K}\left[\mathcal{T} + \mathbf{s}\right]\vec{\psi}_0.\end{aligned} \tag{19}$$

The operator $\mathcal{T}$ — T-matrix — in H introduced by (19) is connected to the parent scattering matrix $\mathbf{S}$ defined above (18) by the formula:

$$\mathbf{S} = -I + \mathcal{T} + \mathbf{s}$$

The matching condition of the scattered wave on the bottom cross-sections δ_t of the channels $\Omega_t,\ t = 1,2,3,4$ with the solution of the Schrödinger equation inside the modified domain $\hat{\Omega}_0$ gives, after cancelling the initiating channel-vector $\vec{\psi}_0$ the following

$$\hat{\Lambda}\left(\mathcal{T} + \mathbf{s}\right)|_\delta\vec{\psi}_0 = \frac{\partial}{\partial n_\delta}\psi = \left[-2\mathcal{K} + \mathcal{K}\left(\mathcal{T} + \mathbf{s}\right)\right]\vec{\psi}_0, \tag{20}$$

which may be transformed into an operator equation by cancelling the arbitrary initiating vector $\vec{\psi}_0 \in H_+$. One can rewrite this equation in form:

$$\hat{\Lambda}\left(\mathcal{T} + \mathbf{s}\right) = -\left[\hat{\Lambda} - \mathcal{K}\right]^{-1} 2\mathcal{K}. \tag{21}$$

After framing the result by the projections P_+ onto the open channels we obtain

$$\mathcal{T} = -P_+\frac{1}{\hat{\Lambda} - \mathcal{K}}P_+ 2\mathcal{K}^+ \qquad \text{or} \qquad \mathbf{S} = -\left[\hat{\Lambda} - \mathcal{K}\right]^{-1}\left[\hat{\Lambda} + \mathcal{K}\right].$$

Now the scattering matrix may be obtained as $\mathcal{S} - P_+\mathbf{S}P_+$ with use of the techniques for operator matrices, developed in [2], [43]. Really, the operator $\mathcal{K} - \hat{\Lambda}$ has a non-negative finite-dimensional imaginary part $\mathcal{K}^+$, hence is invertible, if the operator operator $\Delta = \hat{\Lambda}_{--} - \mathcal{K}^-$ is invertible. Due to the Hilbert identity

$$\Delta = \mathcal{K}^{--}\left[\hat{\Lambda}_{--}^{-1} - \left(\mathcal{K}^-\right)^{-1}\right]\hat{\Lambda}_{--}$$

the operator Δ is invertible if and only if neither of eigenvalues of the compact in H_- operator $\hat{\Lambda}_{--}^{-1} - \left(\mathcal{K}^-\right)^{-1}$ coincides with zero. Vice versa, if it has a zero eigenvalue at the real spectral point λ_0 with the eigenvector $\vec{\psi}_0 \in W_2^{1/2}(\delta) \in H^-$, then the operator $\hat{L}$ has an embedded eigenvalue λ_0 with the eigenvector

$$\Psi_0 - \begin{cases} e^{\mathcal{K}^-\,x}\vec{\psi}_0, & \text{on the channels } \Omega_s,\ s = 1,2,3,4 \\ \hat{P}_0\vec{\psi}_0 & \text{inside the domain } \hat{\Omega}. \end{cases}$$

If the operator $\hat{\Lambda}_{--} - \mathcal{K}^-$ is invertible, then the restriction of the parent S-matrix $\mathbf{S}$ onto the open channels H^+ may be calculated in explicit form. Really, the equation

(21) may be reduced to the pair of equations obtained by multiplication of (21) from the left side by $P_\pm$ respectively:

$$\hat{\Lambda}_{++}\mathcal{T} + \hat{\Lambda}_{+-}\mathbf{s}_{-+} = -2\mathcal{K}^+ + \mathcal{K}^+\mathcal{T}, \quad \hat{\Lambda}_{-+}\mathcal{T} + \hat{\Lambda}_{--}\mathbf{s}_{-+}. \tag{22}$$

Subject to invertibility of the operator $\hat{\Lambda}_{--} - \mathcal{K}^-$ we obtain the formula for T-matrix $\mathcal{T}$

$$\mathcal{T} = -\frac{I}{\hat{\Lambda}_{++} - \hat{\Lambda}_{+-}\frac{I}{\hat{\Lambda}_{--}-\mathcal{K}^-}\hat{\Lambda}_{-+} - \mathcal{K}^+} 2\mathcal{K}^+,$$

and the explicit expression for the scattering matrix

$$\begin{aligned} \mathcal{S} = -&\left\{\hat{\Lambda}_{++} - \hat{\Lambda}_{+-}\frac{I}{\hat{\Lambda}_{--} - \mathcal{K}^-}\hat{\Lambda}_{-+} - \mathcal{K}^{++}\right\}^{-1} \\ &\left\{\hat{\Lambda}_{++} - \hat{\Lambda}_{+-}\frac{I}{\hat{\Lambda}_{--} - \mathcal{K}^-}\hat{\Lambda}_{-+} + \mathcal{K}^{+}\right\}. \end{aligned} \tag{23}$$

Comparing this expression with the previous expression for the matrix $\mathcal{T}$

$$\mathcal{T} = -P_+\frac{1}{\hat{\Lambda} - \mathcal{K}}P_+ 2\mathcal{K}^+$$

we may observe that

$$\frac{I}{\hat{\Lambda}_{++} - \hat{\Lambda}_{+-}\frac{I}{\hat{\Lambda}_{--}-\mathcal{K}^-}\hat{\Lambda}_{-+} - \mathcal{K}^+} = P_+\frac{1}{\hat{\Lambda} - \mathcal{K}}P_+,$$

which means that the left side is an operator function with a positive imaginary part, since both $\hat{\Lambda}$ and $-\mathcal{K}$ have negative imaginary parts in upper half-plane. The positive imaginary part of the expression in the left side is caused by presence of the anti-hermitian term $\mathcal{K}^+$, $\Im\mathcal{K}^+ > 0$. The matrix function $\hat{\Lambda}_{++} - \hat{\Lambda}_{+-}\frac{I}{\hat{\Lambda}_{--}-\mathcal{K}^-}\hat{\Lambda}_{-+}$ is a self-adjoint matrix in the channel space H_+, and the S-matrix is unitary in H^+. □

The general representation for the scattering matrix obtained in the preceding theorem contains the matrix function

$$P_+\left(\hat{\Lambda} - \mathcal{K}^-\right)P_+ = \hat{\Lambda}_{++} - \hat{\Lambda}_{+-}\frac{I}{\hat{\Lambda}_{--} - \mathcal{K}^-}\hat{\Lambda}_{-+} := \mathcal{D}(\lambda) \tag{24}$$

which is hermitian on the real axis λ and has a negative imaginary part in the upper half-plane. It may have simple poles at the eigenvalues of the operator $\hat{L}_0$, in particular when $\hat{\Lambda}_{+-} = 0$, but in the generic case, when the perturbation is non-degenerated the "old" poles appearing from the eigenvalues of the basic operator $\hat{L}_0$ are "compensated" and new poles appear from the zeroes of the denominator $\hat{\Lambda}_{--} - \mathcal{K}^-$. The corresponding residues *do not coincide* generally with any spectral characteristics of the operator $\hat{L}$. The simplest statement concerning compensation of singularities at the eigenvalues of the non-perturbed operator $\hat{L}_0$ may be found in [9], [44], where the residues are still expressed in terms of values of the resonance eigenfunction of the Schrödinger operator on the basic domain (ring). Generally these residues play an essential role in scattering processes, see the last theorem (8) in this section. We begin the proof of this result with a general observation concerning the role of poles of the function $\mathcal{D}$ in scattering processes.

Let us consider a pole λ_α of the matrix function $\mathcal{D}$ on the first spectral band, where $\mathcal{K}^+ = i\sqrt{\lambda - \frac{\pi^2}{|\delta|^2}}\sum_{s=1}^4 e_s\rangle\langle e_s = i\sqrt{\lambda - \frac{\pi^2}{|\delta|^2}}P_+$. Then the following representation is valid for $\mathcal{D}$ near the pole:

$$\mathcal{D}(\lambda) = \alpha\frac{P_\alpha}{\lambda - \lambda_\alpha} + \mathcal{Q}(\lambda), \tag{25}$$

where α is a positive constant, $\mathcal{Q}(\lambda)$ is a hermitian analytic operator function on a small neighborhood of λ_α, with a negative imaginary part in the upper half-plane and P_α is an orthogonal projection in H_+. We assume that the projection is one-dimensional in H_+, $P_\alpha = e_\alpha\rangle\langle e_\alpha$, $|e_\alpha| = 1$. We shall use an orthogonal decomposition of H_+ with respect to the basis of subspaces $P_\alpha H_+ \oplus (I - P_\alpha)\, H_+ := H_+^\alpha \oplus H_+^\perp$. Denoting by $P_\perp$ the orthogonal projection $I - P_\alpha$ we may present the term $\mathcal{Q}(\lambda)$ by the matrix

$$\mathcal{Q} = \left\{ \begin{array}{cc} P_\alpha \mathcal{Q} P_\alpha & P_\alpha \mathcal{Q} P_\perp \\ P_\perp \mathcal{Q} P_\alpha & P_\perp \mathcal{Q} P_\perp \end{array} \right\} = \left\{ \begin{array}{cc} \mathcal{Q}_{\alpha\alpha} & \mathcal{Q}_{\alpha\perp} \\ \mathcal{Q}_{\perp\alpha} & \mathcal{Q}_{\perp\perp} \end{array} \right\}.$$

The operator $\mathcal{K}^+ = i\sqrt{\lambda - \frac{\pi^2}{|\delta|^2}}I_{H_+}$ is anti-hermitian, hence the sum $\Delta_\perp = \mathcal{Q}_{\perp\perp} - \mathcal{K}^+ P_\perp$ is invertible in $H_\perp$, and the sum

$$\frac{\alpha P_\alpha}{\lambda - \lambda_\alpha} + \mathcal{Q}_{\alpha\alpha} - \mathcal{K}^+ P_\alpha - \mathcal{Q}_{\alpha\perp}\frac{I}{\Delta_\perp}\mathcal{Q}_{\perp\alpha} := d(\lambda) \tag{26}$$

is an invertible operator in the one-dimensional subspace H_α. The following statement is an analog of similar statements in [9], [45], [44] which describe the resonance structure of factors of scattering matrices:

Lemma 7. *The inverse of the operator function*

$$\mathcal{D}(\lambda) - \mathcal{K}^+ = \alpha\frac{P_\alpha}{\lambda - \lambda_\alpha} + \mathcal{Q}(\lambda) - \mathcal{K}^+$$

may be calculated in terms of an orthogonal decomposition $H_+ := H_+^\alpha \oplus H_+^\perp$ *as*

$$\left(\mathcal{D}(\lambda) - \mathcal{K}^+\right)^{-1} = \left\{ \begin{array}{cc} d^{-1} & -d^{-1}\mathcal{Q}_{\alpha\perp}\frac{I}{\Delta_\perp} \\ -\frac{I}{\Delta_\perp}\mathcal{Q}_{\perp\alpha}d^{-1} & \frac{I}{\Delta_\perp}\mathcal{Q}_{\perp\alpha}d^{-1}\mathcal{Q}_{\alpha\perp}\frac{I}{\Delta_\perp} + \frac{I}{\Delta_\perp} \end{array} \right\}. \tag{27}$$

Proof. The proof may be obtained as a solution of the relevant equation presented in terms of the orthogonal decomposition $H = H_+ \oplus H_-$, see, for instance [2]. □

To accomplish the practical calculation of the scattering matrix based on the preceding formula (27) we may proceed in two steps.

On the first step we may interpret the preceding formula (27) in terms of an intermediate operator constructed on the orthogonal sum of spaces $L_2(\hat\Omega_0) \oplus H_-$ as a common extension of the operators $\hat L_{0-}$ on the basic domain and $-\Delta\big|_{H_- \times L_2(0,\infty)}$ in the channel space. We reduce preliminary the above sum to the sub-domain of functions vanishing near the bottom sections δ_s of the channels and then extended with the *partial matching conditions* for the functions $\hat{u}$ on the basic domain $\hat{\Omega}$ and $u^- = (u_1^-, u_2^-, u_3^-, u_4^-)$ on the channels:

$$\hat{u} - u^- \big|_\delta = 0,\; P_- \frac{\partial}{\partial n}\left(\hat{u} - u^-\right)\big|_\delta = 0. \tag{28}$$

The total operator $\hat{L}_{0-}$ combined of the above differential expression L of Schrödinger type on the basic domain and the Laplacian on the channels is symmetric. We consider below the *generic case* when it is self-adjoint, assuming that any solution $\hat{u}$ of the equation $L\hat{u} = \lambda\hat{u}$ supplied with the boundary conditions at the bottom chords $\delta = \cup_{s=1}^4 \delta_s$

$$\hat{u}|_\delta = 0,\; P_- \frac{\partial \hat{u}}{\partial n} = 0,$$

and with homogeneous Dirichlet boundary condition on the complement of δ in $\partial\hat{\Omega}_0$ is equal to zero identically inside the basic domain. This means, in particular, that there are no eigenfunctions of the operator $\hat{L}_{0-}$ which are equal to identical zero on the channels[8]. The absolutely-continuous spectrum of the operator coincides with $[\frac{4\pi^2}{|\delta|^2}, \infty]$ and has the multiplicity $m_- = m - 4$, where m is the multiplicity of the spectrum of the operator L on the whole composite domain Ω. The poles λ_α of the function $\mathcal{D}$ on the first spectral band coincide with eigenvalues of the operator $\hat{L}_{0-}$ below the threshold $\frac{4\pi^2}{|\delta|^2}$.

On the second step one may construct the perturbation of the operator $\hat{L}_{0-}$ via attaching to it the restriction $-\triangle\,\big|_{H_+ \times L_2(0,\infty)}$ of the Laplacian onto the open channel. This attachment may be described via imposing partial boundary conditions on the corresponding functions $\hat{u}$ on the basic domain and $u^+ = (u^1, u^2, u^3, u^4)$ on the open channels:

$$P_+\hat{u} - u^+ \big|_\delta = 0,\; \frac{\partial}{\partial n}\left(P_+\hat{u} - u^+\right)\big|_\delta = 0. \tag{29}$$

This two-step procedure permits to eliminate calculation of the parent scattering matrix. We describe below most important technical details of this approach, assuming that the spectral parameter belongs to the first spectral band. We impose a physically-sensible condition on the width of channels which guarantee the efficiency of the construction. Actually we assume that the part of the intermediate operator $\hat{\mathcal{L}}_{0-}$ below the threshold may be obtained by "continuous deformation" of the corresponding part of the basic operator: the eigenvalues $\hat{\lambda}_{s,-}$ of the intermediate operator $\hat{L}_{0-}$ below the threshold do not deviate too much from the corresponding eigenvalues $\hat{\lambda}_{s,0}$ of the non-perturbed operator $\hat{L}_0$ and may be obtained by the standard perturbation procedure connecting the perturbed and the

[8]The above assumption is natural, but it is non-trivial, because obvious examples may be constructed for which a non-zero solution exists which fulfills the above condition with *some* orthogonal projection P_-.

basic operator $\hat{L}_0$:

$$|\hat{\lambda}_{s,-} - \hat{\lambda}_{s,0}| < \frac{1}{2}\min_{t\neq s} |\hat{\lambda}_{s,0} - \hat{\lambda}_{t,0}| = \epsilon_s. \tag{30}$$

Practically this condition appears below in rather technical form, see (31), involving an operator which appears in perturbation procedure.

We may advance with straightforward perturbation techniques assuming that λ sits inside the first spectral band "near" to the simple *resonance* eigenvalue λ_0 of the basic operator $\hat{L}_0$, $\lambda = \mu\frac{\pi^2}{|\delta|^2} + (1-\mu)\frac{4\pi^2}{|\delta|^2}$, $0 < \mu < 1$. The corresponding normalized eigenfunction will be denoted by φ_0. From [61] follows that this function is continuously differentiable on the bottom sections δ_s of the channels with respect to the exterior normal, except the corner points at the extremities of the chords δ_s, and the derivative of it is square integrable on δ_s. Then separating the singular term

$$\frac{P_-\frac{\partial\varphi_0}{\partial n}\rangle\langle P_-\frac{\partial\varphi_0}{\partial n}}{\lambda - \lambda_0} := \frac{h_0\rangle\langle h_0}{\lambda - \lambda_0}$$

in the above representation we may present the homogeneous equation as

$$Be_0 + \frac{P_-\frac{\partial\varphi_0}{\partial n}\rangle\langle\frac{\partial\varphi_0}{\partial n}}{\lambda - \lambda_0} e_0 = 0,$$

where B is a hermitian operator function appearing from the separation of singularities in DN-map with use of the formula (13):

$$B(\lambda) = P_-\hat{\Lambda}_{-M}P_- - (\lambda + M)\, P_-\hat{\mathcal{P}}^+_{-M}\hat{\mathcal{P}}_{-M}\, P_- - P_-\mathcal{K}^- P_-$$
$$- (\lambda + M)^2 \sum_{s\neq 0} \frac{P_-\frac{\partial\varphi_s}{\partial n}\rangle\langle P_-\frac{\partial\varphi_s}{\partial n}}{(\lambda_s + M)(\lambda_s - \lambda} - (\lambda + \lambda_0 + 2M)(P_-\frac{\partial\varphi_0}{\partial n}\rangle\langle P_-\frac{\partial\varphi_0}{\partial n}).$$

This operator function depends smoothly on λ near the resonance eigenvalue $\hat{\lambda}_{0,0} := \lambda_0$ and may be invertible, at least if the channels Ω_s are "not too wide". Really, we may estimate the positive operator function $-\mathcal{K}^-$ from below as

$$-\mathcal{K}^- \geq \frac{\pi}{\delta}\sqrt{3\mu},$$

and use the facts that the DN-map $\hat{\Lambda}_{-M}$ is positive for large negative $-M$, and the sum of the operator$P_-\hat{\mathcal{P}}^+_{-M}\hat{\mathcal{P}}_{-M}\, P_-$ together with the remaining compact part $(\lambda+M)^2 P_-\hat{\mathcal{P}}^+_{-M} R_\lambda \hat{\mathcal{P}}_{-M} P_- - \mathcal{K}^-$ are bounded in the channel space H_- if λ sits in a small neighborhood of λ_0 and hence is separated from others eigenvalues of the basic operator $\hat{L}_0$. We impose on the operator B the following condition, which may be interpreted as a technical version of the above condition of the existence of a "continuous deformation" of the part of the basic operator below the second threshold $\frac{4\pi^2}{|\delta|^2}$ into the corresponding part of the intermediate operator:

The operator B is invertible near the eigenvalue λ_0 on the operator $\hat{L}_0$ and the equation $\lambda_0 - \lambda - \langle h_0|B^{-1}|h_0\rangle = 0$ has only one solution $\lambda = \lambda_\alpha$ in a neighborhood ω_α

$$\omega_\alpha = \{\lambda : |\lambda - \lambda_0| < \epsilon_0\}\,. \tag{31}$$

defined above by the condition (30).

This condition is obviously fulfilled for narrow channels $\delta << \pi\sqrt{3\mu}$. Optimization of the choice of parameters implying the invertibility of the operator B is an important technical problem involving extended computations. It will be discussed separately in detail in the subsequent publications.

A straightforward calculation of the resolvent of the hermitian operator

$$\Delta_- = B(\lambda) + \frac{P_-\frac{\partial\varphi_0}{\partial n}\rangle\langle P_-\frac{\partial\varphi_0}{\partial n}}{\lambda - \lambda_0} \quad \text{gives} \quad \Delta_-^{-1} = B^{-1} + \frac{B^{-1}h_\alpha\rangle\langle B^{-1}h_\alpha}{\lambda_0 - \lambda - \langle h_\alpha|B^{-1}|h_\alpha\rangle},$$

where $h_\alpha = P_-\frac{\partial\varphi_0}{\partial n}|_\delta$. This implies an equation for the calculation of the pole λ_α of the resolvent $\lambda_0 - \lambda - \langle h_0|B^{-1}|h_0\rangle = 0$ and the expression for the corresponding eigenvector $e_\alpha = B^{-1}h_\alpha$.

Now we may take into account the fact that the operator $\hat{\Lambda}_{-+}$ is bounded from the four-dimensional channel space H_+ into H_-, and the adjoint operator $\hat{\Lambda}_{+-}$ is bounded from H_- into H_+. Then denoting by $P_\alpha := e_\alpha\rangle\langle e_\alpha$ the one-dimensional projection operator onto the subspace in H_+ spanned by $\hat{\Lambda}_{+-}h_\alpha = \hat{\Lambda}_{+-}B^{-1}P_-\frac{\partial\varphi_0}{\partial n}$ and by $\alpha = \left|\hat{\Lambda}_{+-}B^{-1}P_-\frac{\partial\varphi_0}{\partial n}\right|^2$ the corresponding coefficient, we may, generically, present the function $\mathcal{D}$ near the pole λ_α in asymptotic form (25), and calculate the asymptotic of the function d:

$$d^{-1}(\lambda) \approx \frac{\lambda - \lambda_\alpha}{\alpha}P_\alpha.$$

Based on this asymptotic we may find the asymptotic of the expression (27) for $\left(\mathcal{D}(\lambda) - \mathcal{K}^+\right)^{-1}$ near the pole λ_α as:

$$\left\{\begin{array}{cc} \frac{\lambda-\lambda_\alpha}{\alpha}P_\alpha & -\frac{\lambda-\lambda_\alpha}{\alpha}P_\alpha\mathcal{Q}_{\alpha\perp}\frac{I}{\Delta_\perp} \\ -\frac{I}{\Delta_\perp}\mathcal{Q}_{\perp\alpha}\frac{\lambda-\lambda_\alpha}{\alpha}P_\alpha & -\frac{I}{\Delta_\perp}\mathcal{Q}_{\perp\alpha}\frac{\lambda-\lambda_\alpha}{\alpha}\mathcal{Q}_{\alpha\perp}\frac{I}{\Delta_\perp} + \frac{I}{\Delta_\perp} \end{array}\right\}, \tag{32}$$

where the spectral parameter λ in all non-zero terms at the pole λ_α should be substituted by λ_α. In particular straight at the pole λ_α we have

$$\left(\mathcal{D}(\lambda) - \mathcal{K}^+\right)^{-1}\Big|_{\lambda_\alpha} = \frac{I}{\mathcal{Q}_{\perp\perp} - \mathcal{K}^+P_\perp},$$

and the scattering matrix at the pole λ_0 is equal to

$$\mathcal{S}(\lambda_0) = -I - \left[\mathcal{Q}_{\perp\perp} - \mathcal{K}^+P_\perp\right]^{-1}2\mathcal{K}^+ = I - 2P_\alpha + \frac{2i\delta}{\pi\sqrt{3(1-\mu)}}\mathcal{Q}_{\perp\perp}, \tag{33}$$

in agreement with similar results for the solvable models of the switch (quantum relay) [9], [45], [44].

Summarizing the above results we may formulate the following tentative statement:

Theorem 8. *If the channels are "non-too-wide" (i.e. fulfill the above "continuous-deformation condition" (31)), then for each resonance eigenvalue λ_0 of the basic operator $\hat{L}_0$ inside the first spectral band there exist a corresponding eigenvalue λ_α of the intermediate operator $\hat{L}_{0-}$ which may be found as a solution of the equation $\lambda_\alpha = \lambda_0 - \langle h_\alpha \big| B^{-1}(\lambda) \big| h_\alpha \rangle$. This eigenvalue serves as a pole of the function $\mathcal{D} = \alpha \frac{P_\alpha}{\lambda - \lambda_\alpha} + \mathcal{Q}(\lambda)$ with $\alpha = \big| \hat{\Lambda}_{+-} B^{-1} P_- \frac{\partial \varphi_0}{\partial n} \big|^2$ and P_α is an orthogonal projection onto one-dimensional subspace spanned by the vector $\hat{\Lambda}_{+-} B^{-1} P_- \frac{\partial \varphi_0}{\partial n}$. The scattering matrix at λ_α has the following asymptotic for "narrow" channels:*

$$\mathcal{S}(\lambda) = I - 2P_\alpha + \frac{2i\delta}{\pi\sqrt{3(1-\mu)}} Q_{\perp\perp} \tag{34}$$

The asymptotic formula (34) shows that the transmission coefficient from one channel to another is essentially defined by the products of components of the channel-vectors $\hat{\Lambda}_{+-} B^{-1} P_- \frac{\partial \varphi_0}{\partial n}$ in the channel-spaces $H_{+,s}$. If the channels are narrow, then the operator B contains a dominating term $\mathcal{K}^-$ and hence the operator B^{-1} is essentially reduced to multiplication by the constant $\frac{\delta}{\pi\sqrt{3\mu}}$. But the factor $\hat{\Lambda}_{+-}$ and adjoint factor bear essential spectral information on $\hat{L}_0$ which is encoded in components of the normal derivatives of the resonance eigenfunction in the channel spaces $H_\pm$ on the bottom sections of channels:

$$P_+ \frac{\frac{\partial \varphi_0}{\partial n} \rangle \langle \frac{\partial \varphi_0}{\partial n}}{\lambda - \lambda_0} P_- .$$

The resonances-zeroes of the scattering matrix in upper halfplane- also may be calculated asymptotically for narrow channels with use of the operator version of Rouche theorem, see [22]. It will be done elsewhere.

4. Integral equation

To construct the scattered waves iniciated by the plane waves incoming from the first channel we proceed in the following way:

1. We take the ansatz for the initiating wave as a reflected wave inside the closed first channel Ω_1 with homogeneous Dirichlet boundary condition on the boundary $\partial\Omega_1$:

$$\psi_1 = e_1 \left[e^{-i\sqrt{\lambda - (\frac{\pi}{|\delta|})^2}\, x} - e^{i\sqrt{\lambda - (\frac{\pi}{|\delta|})^2}\, x} \right], \quad e_1 = \sqrt{\frac{2}{\delta}} \sin \frac{\pi y}{\delta}$$

2. We assume that the scattered waves on the chords δ_s are equal to the functions Ψ_s, which are to be defined eventually from the ongoing procedure. Note that the restriction of the scattered wave onto the channels Ω_s may be calculated with use of the corresponding outgoing Poisson map $\mathcal{P}_s$. In particular on the corresponding arcs γ_s of the unit circle, $\gamma_s \in \Omega_s$ we obtain:

$$\Psi\big|_{\gamma_s} = \mathcal{P}_s \Psi_s \big|_{\gamma_s} + \delta_{s1} \psi_1 \big|_{\gamma_s}.$$

Denote the total function reduced onto the sum of arcs $\gamma = \cup_{s=1}^4$ by $\Psi_\gamma + \psi_\gamma 1$.

Note that the Poisson map in each channel may be presented by the formula

$$\begin{aligned} \mathcal{P}_s(\lambda) &= \mathcal{P}_s(-M) + (\lambda+M) R^s_{-M} \mathcal{P}_s(\lambda) \\ &= \mathcal{P}_s(-M) + (\lambda+M) \sum_{l=1}^{N} e^{i\sqrt{\lambda - \frac{\pi^2 l^2}{|\delta|^2}}\, y} \mathbf{e}_l\rangle\langle \mathbf{e}_l \\ &\quad + (\lambda+M) \sum_{l=N+1}^{\infty} e^{i\sqrt{\lambda - \frac{\pi^2 l^2}{|\delta|^2}}\, y} \mathbf{e}_l\rangle\langle \mathbf{e}_l \\ &:= \mathcal{P}_s(-M) + \mathcal{P}_s^N(\lambda) + \mathbf{p}_{N,s}(\lambda). \end{aligned} \tag{35}$$

Consider the circles Σ_s radius $\frac{\delta}{\sqrt{2}}$ passing through the ends of the chords δ_s and centered at the points c_s inside Ω_0 $\operatorname{dist}(c_s, \delta_s) = \delta/2$. Parts of the circles inside the channels Ω_s we denote by Σ_{ss} and parts of them inside the domain $\Omega_0 \backslash \cup_{s=1}^4 \Omega_s$ we denote by Σ_{0s}. Note that, due to results of Section 2, that the restriction of the first term in the right side of the above map (35) onto any crescent-like sub-domain or an arc $\Sigma_{ss} \in \Omega_s$ is a contracting operator from $C(\delta_s)$ into $C(\Sigma_s)$, and the sum of remaining terms is a compact operator, $|\mathbf{p}_{N,s}(\lambda)|_C < \epsilon$, if $N > N(\epsilon)$. Hence the sum $\mathcal{P}_s(-M) + \mathbf{p}_{N,s}(\lambda)$ is a contraction, if N is large enough. The operator function $\mathcal{P}_s^N$ is finite-dimensional.

3. We calculate the component of the scattered wave Ψ inside the unit disc Ω_0 as a solution of the Dirichlet problem for the corresponding Schrödinger equation with the boundary data at the unit circle:

$$\Psi|_{\gamma_s} = \mathcal{P}_s \Psi_s|_{\gamma_s} + \delta_{s1}\psi_1|_{\gamma_s}, \text{ or } \Psi|_\gamma + \psi_\gamma, \text{ on } \gamma = \cup_{s=1}^4 \gamma_s.$$

$$\Psi\big|_{\partial\Omega \backslash \cup_{t=1}^4 \gamma_t} = 0.$$

This solution may be obtained via the *Poisson map* $\mathcal{P}_0$ *for the unit disc* Ω_0:

$$\begin{aligned} \Psi|_{\Omega_0} &= \mathcal{P}_0 \Psi_\gamma + \mathcal{P}_0 \psi_\gamma \\ &= \mathcal{P}_{0,-M}\Psi_\gamma + (\lambda+M) \sum_{l \le N} \frac{< \frac{\partial \varphi_l(x)}{\partial n_x}, \Psi_\gamma > \varphi_l(y)}{(\lambda_l + M)(\lambda_l - \lambda)} \\ &\quad + (\lambda+M) \sum_{l \ge N} \frac{< \frac{\partial \varphi_l(x)}{\partial n_x}, \Psi_\gamma > \varphi_l(y)}{(\lambda_l + M)(\lambda_l - \lambda)} + \mathcal{P}_0 \psi_1 \\ &:= \mathcal{P}_{0,-M}\Psi_\gamma + \mathcal{P}_0^N \Psi_\gamma + p_0^N \Psi_\gamma + \mathcal{P}_0 \psi_1. \end{aligned}$$

Here, according to results of the second section, the map $\mathcal{P}_{0,-M}$ is contracting from $C\gamma$ into $C(\cup_{s=1}^{s=4} \Sigma_{0s})$, the sum

$$(\lambda+M) \sum_{l=1}^{\infty} \frac{\langle \frac{\partial \varphi_l(x)}{\partial n_x}, \Psi_\gamma \rangle \varphi_l(y)}{(\lambda_l + M)(\lambda_l - \lambda)}$$

is compact and

$$|(\lambda+M)\sum_{l\geq N}\frac{\langle\frac{\partial\varphi_l(x)}{\partial n_x},*\rangle\varphi_l(y)}{(\lambda_l+M)(\lambda_l-\lambda)}|_C\leq\epsilon$$

if $N\geq N_\epsilon$.

4. We consider the functions on the circles defined as a direct sum obtained in previous paragraphs:

$$\mathcal{P}_0\left(\Psi_\gamma|_{\Sigma_{0s}}+\psi_\gamma\right)\big|\big|_{\Sigma_{0s}}\oplus\left(\mathcal{P}_s\Psi_s+\delta_s1\psi_1\right)\big|\big|_{\Sigma_{ss}}:=\Psi_{\Sigma_s}.$$

5. We represent the scattered wave inside the disc D_s bordered by the circle Σ_s as a solution of the Dirichlet problem in D_s for the corresponding equation. Reduction of this solution $\Psi|_{D_s}$ onto the bottom sections δ_s of the channels coincides with $\Psi|_{\delta_s}$:

$$\Psi|_{\delta_s}=\mathcal{P}_{D_s}\Psi_{\Sigma_s}|_{\delta_s},\, s=1,\,2,\,3,\,4.$$

The last equation is actually the required integral equation for Ψ_s, $s=1,\,2,\,3,\,4$. It may be rewritten in detail as

$$\begin{aligned}\Psi_s=&\ \mathcal{P}_{D_s}(\lambda+M)\sum_{l\leq N}\frac{<\frac{\partial\varphi_l(x)}{\partial n_x},\sum_t\mathcal{P}_t\Psi_t>\varphi_l}{(\lambda_l+M)(\lambda_l-\lambda)}\\&+\mathcal{P}_{D_s}(\lambda+M)\sum_{l\leq N}\frac{<\frac{\partial\varphi_l(x)}{\partial n_x},\psi_1>\varphi_l}{(\lambda_l+M)(\lambda_l-\lambda)}\\&+\mathcal{P}_{D_s}\left(\mathcal{P}_{-M}+\mathbf{p}_N\right)\left(\sum_t^4\mathcal{P}_t\Psi_t+\psi_1\right)|_{\gamma_s}\\:=&\ \left\{\mathcal{K}_N[\Psi+\psi_1]+\mathcal{K}_\epsilon[\Psi+\psi_1]\right\}_s,\quad s=1,2,3,4,\end{aligned}\tag{36}$$

or, in vector form

$$\Psi-\mathcal{K}_\epsilon\Psi=\left[\mathcal{K}_\epsilon+\mathcal{K}_N\right]\psi_1+\mathcal{K}_N\Psi.\tag{37}$$

The integral operator in this equation is represented as a sum of the finite-dimensional operator function $\mathcal{K}_N$

$$\sum_t\mathcal{K}_{st}\Psi_t=\mathcal{P}_{D_s}(\lambda+M)\sum_{l\leq N}\frac{<\frac{\partial\varphi_l(x)}{\partial n_x},\sum_t\mathcal{P}_t\Psi_{t}>\varphi_l}{(\lambda_l+M)(\lambda_l-\lambda)}$$

acting in $\oplus\sum_{s=1}^4L_2(\delta_s)$ and the infinitely-dimensional part

$$\left\{\mathcal{P}_{D_s}\left(\mathcal{P}_{-M}+\mathbf{p}\right)\mathcal{P}_t\right\}_{st}:=\left\{\mathcal{K}_\epsilon\right\}_{st}\sim\mathcal{K}_\epsilon,$$

which is a contracting operator if N is large enough.

7. To accomplish the procedure we may rewrite the equation in finite-dimensional form, using the invertibility of the operator $I-\mathcal{K}_\epsilon$:

$$\Psi=\left[I-\mathcal{K}_\epsilon\right]^{-1}\left[\mathcal{K}_\epsilon+\mathcal{K}_N\right]\psi_1+\left[I-\mathcal{K}_\epsilon\right]^{-1}\mathcal{K}_N\Psi.\tag{38}$$

One may construct the scattered waves basing on solutions of this equation. We may use the corresponding homogeneous equation to find resonances or estimate the life-time of them. This job requires extended computational work.

5. Approximate scattering matrix, solvable model and design of a resonance triadic quantum switch

Neither of above expressions (23) and (38) for the scattering matrix gives any simple formula for calculation of the transmission coefficients in explicit form. Nevertheless, under assumption that the widths of the channels are "not too large" we may transform the formula (23) to more effective formula (34) which gives a qualitative description of the transmission coefficient from one channel to another. In fact first two essential terms of this formula were obtained by substituting $\mathcal{D}$ by the polar expression and the last term serves just to estimate the error. The calculation of the second term in formula (34) gives an approximate expression for the resonance transmission coefficient

$$\mathcal{S}_{st}(\lambda_\alpha) \approx 2\frac{\varphi_s^1\varphi_t^1}{\sum_{s=1}^{s=4}} e_s^1 \rangle\langle e_t^1,$$

where $\varphi_s^1 = \int_{\delta_s} \frac{\partial\varphi_0}{\partial n} e_s^1 dy$ is the first Fourier coefficient of the restriction of normal derivative $\frac{\partial\varphi_0}{\partial n}|_{\delta_s}$ of the resonance eigenfunction on the bottom section δ_s. Calculating of the correcting terms in $\mathcal{Q}$ involves the higher Fourier coefficients appearing from the projection $P_{(-)}$.

The first order result is in agreement with our approximate calculation in ([51]). If the channel Ω_s is centered at the zero $\hat{a}_s$ of restriction of the normal derivative of the resonance eigenfunction on the bottom section $\frac{\partial\varphi_0}{\partial n}|_{\delta_s}$, then the above dot-product may be expanded as

$$\begin{aligned}\varphi_s^1 = \int_{\delta_s} \frac{\partial\varphi_0}{\partial n} e_s^1 dy &\approx \frac{1}{2}\sqrt{\frac{2}{|\delta|}} \int_{-|\delta|/2}^{|\delta|/2} \frac{\partial^2\varphi_0}{\partial n\partial y}(\hat{a}_s)\, y^2 \cos\frac{\pi y}{|\delta|} dy \\ &\approx \sqrt{\frac{2}{|\delta|}\frac{\partial^2\varphi_0}{\partial n\partial y}(\hat{a}_s)\frac{|\delta|^3}{\pi^3}}\end{aligned}$$

But if the value of the normal derivative at the center of the channel a_s is not equal to zero, then

$$\varphi_s^1 = \int_{\delta_s} \frac{\partial\varphi_0}{\partial n} e_s^1 dy \approx 2\sqrt{\frac{2}{|\delta|}\frac{\partial\varphi_0}{\partial n}(\hat{a}_s)\frac{\delta}{\pi}}.$$

These estimates imply approximate expressions for the transmission coefficients from one channel Ω_1 to another Ω_s at the resonance frequency $\sqrt{\lambda_0}$, assuming that other channels Ω_t, $t \neq 1, s$ are centered at the zeroes $\hat{a}_t$ of the normal derivative of the resonance eigenfunction:

$$
\begin{aligned}
\mathcal{S}_{s1} &\approx \frac{\frac{\partial\varphi_0}{\partial n}(\hat{a}_s)\frac{\partial\varphi_0}{\partial n}(\hat{a}_1)}{|\frac{\partial\varphi_0}{\partial n}(\hat{a}_s)|^2 + |\frac{\partial\varphi_0}{\partial n}(\hat{a}_1)|^2 + \sum_{r\neq 1,s}\left|\frac{\partial^2\varphi_{\hat{a}_r}}{\partial n\partial y}(a_r)\right|^2 \frac{|\delta|^4}{4\pi^4}}, \\
\mathcal{S}_{t1} &\approx \frac{\frac{|\delta|^2}{\pi^2}\, \frac{\partial^2\varphi_0}{\partial n\partial y}(\hat{a}_s)\, \frac{\partial\varphi_0}{\partial n}(\hat{a}_1)}{2|\frac{\partial\varphi_0}{\partial n}(\hat{a}_s)|^2 + 2|\frac{\partial\varphi_0}{\partial n}(a_1)|^2 + \frac{1}{2}\sum_{r\neq 1,s}|\frac{\partial^2\varphi_0}{\partial n\partial y}(a_r)|^2\, \frac{|\delta|^4}{\pi^4}}.
\end{aligned}
\tag{39}
$$

In fact these expressions for the transmission coefficients are corresponding to the solvable model of the switch where the Laplacian on the channels Ω_s is substituted by the part of the Laplacian $-\frac{d^2u_s}{dx^2}P_+ + \frac{\pi^2}{\delta^2}P_+$ on the open channel with the matching conditions on the bottom chord

$$
P_+u\big|_{\delta_s} - u_s = 0,\ \frac{\partial}{\partial n}P_+u\big|_{\delta_s} - \frac{du_s}{dx} = 0.
$$

We do not discuss now the general connections between the mathematical design of nano-electronic devices and the modern system of solvable models in Quantum Mechanics, see [3], [49]. In fact the whole philosophy of solvable models is based on the prominent Krein Formula, see [32], [47]. The simplest solvable model of the triadic quantum switch mentioned above may be used when choosing the contact points $\{a_s\}$ on the unit circle — on the boundary of the circular quantum well — and selection of the form of the potential inside of it such that switching of the electrons current from one channel to another may be manipulated by some simple change of the potential. A similar problem for the solvable model of the quantum switch based on a circular quantum well Ω_0 with Neumann boundary conditions was solved in [45]. The transmission coefficient in [45] appeared to be proportional to the values of the resonance eigenfunction at the point-contacts of the disc Ω_0 with the one-dimensional wave-guides (quantum wires). The switching effect in [45] was observed when the zeroes of the resonance eigenfunction were localized at the contact points of two wave-guides meanwhile the resonance eigenfunction has non-zero values at the contact points with another two wave-guides. The geometrical idea of the device suggested in [45] was based on the observation that for a constant electric field oriented along the vector ν all eigenfunctions of the non-perturbed operator L_0 in the circular domain may be represented as functions of the polar coordinates r, θ where θ *may be chosen as the azimuth with respect to the vector* ν, so that for the vector ν rotated by the angle $\delta\theta$ the corresponding potential[9] $V = \epsilon\langle x, \nu\rangle$ is rotated by the angle $\delta\theta$ and each eigenfunction is rotated by the same angle $\delta\theta$ together with the vector ν. This observation suggests that the problem of switching the electron's current will be solved if we may choose the intensity ε such that some zeroes b_s of the resonance eigenfunction on the boundary of the unit disc Ω_0 satisfy the condition $\frac{|b_s - b_t|}{|b_s - b_r|} = 2$. In the simplest case there may be just two zeroes on the unit circle b_1 and b_2 dividing it in the ratio $2 \div 1$, for instance $|b_1 - b_2| = \frac{2\pi}{3}, |b_2 - b_1| = \frac{4\pi}{3}$. Then attaching the wires to the unit circle at the

[9] We assume that the electron's charge is included in the coefficient ϵ.

points $a_1 = 0$, $a_2 = \frac{\pi}{3}$, $a_3 = \pi$, $a_4 = -\frac{\pi}{3}$ such that $|a_2 - a_3| = |a_3 - a_4| = \frac{2\pi}{3}$ we may choose the direction of the vector ν such that zeroes of the corresponding resonance eigenfunction coincide either with a_2, a_3 or with a_3, a_4, or with a_2, a_4 and at the same time $\varphi_0(a_1)$, $\varphi_0(a_4) \neq 0$ in the first case and $\varphi_0(a_1)$, $\varphi_0(a_2) \neq 0$ in the second case, and $\varphi_0(a_1)$, $\varphi_0(a_3) \neq 0$ in the third case. Then the only essential transmission coefficient for resonance energy will be S_{14} in the first case, or S_{13} in the second case, or S_{13} in the last case.

The above approximate formulae (39) show hat we may use the same geometrical idea for the triadic quantum switch based on a disc with zero boundary condition. According to the approximate formulae for the transmission coefficients we may expect the switching phenomenon to be connected with the movement of zeroes of the normal derivatives of the resonance eigenfunction on the boundary of a circular quantum well. Hence we may use the potential defined by the constant electric field, as before, and find the intensity ϵ of the field such that for some eigenfunction the zeroes of its normal derivative are distributed such that they divide the circle in ratio $2 \div 1$.

We shall do all spectral calculations for the non-dimensional equation on the quantum well modeled as a unit disc, and then we may select the physical parameters of the model using proper change of variables, see the remark below at the end of this section.

Using Mathematica we succeeded choosing the intensity $\varepsilon = 18.86$ and the eigenfunction of the corresponding Schrödinger operator in the unit disc with Dirichlet boundary conditions such that the normal derivative of the resonance eigenfunction has only two zeroes on the unit circle which divide the circle in ratio $2 \div 1$. Then attaching the wires at the points with azimuth $a_1 = 0$, $a_2, a_4 = \pm\frac{\pi}{3}$, $a_3\pi$ we obtain the pattern of contact points which satisfies the required properties so that we may administer the flow of electrons from the channel Ω_1 centered at a_1 to only one of the waveguides Ω_2, Ω_3 or Ω_4 centered at the points a_2, a_3, a_4,with two complementary channels blocked, by adjusting the direction of the field V in such a way that both zeroes of the normal derivative of the resonance eigenfunction φ_0 of the operator L_0 would coincide with a_2 and a_3, or a_2 and a_4, or a_3 and a_4 — so that only one outgoing channel (respectively Ω_4, or Ω_3, or Ω_3) would be opened. The control over the flow is carried out by the change of the direction ν of the gradient of the electrostatic potential $V = \varepsilon < x, \nu >$. We assume below that this direction defines the reference point for θ so that $\theta = 0$ corresponds to the point of the lowest potential. The formulae for the transmission coefficient (39) shows that we really may manipulate the quantum current across the domain this way, if we choose the direction of the vector $\mathbf{e}$ such that the transmission coefficients are equal

$$S_{1k} \approx 0 \quad \begin{cases} \text{for } k = 2,3\,; & \text{and } S_{14} \neq 0 \\ \text{or for } k = 2,4\,; & \text{and } S_{13} \neq 0. \\ \text{or for } k = 3,4\,; & \text{and } S_{12} \neq 0 \end{cases}$$

Now we add a few remarks concerning the change of variables which may reduce the real problem to the problem for non-dimensional equation. In reality we should consider dynamical processes described by the two-dimensional Schrödinger equation written in polar coordinates with real electrostatic potential $V = e\mathcal{E}r\cos(\theta) + V_0$:

$$(-\frac{\hbar^2}{2m_e}\Delta\Psi + (e\mathcal{E}r\cos(\theta) + V_0))\Psi = E_g\Psi \tag{40}$$

and with the Dirichlet boundary condition $\Psi|_R = 0$ on a circular domain with radius R on the surface of a narrow-gap semiconductor with the effective electron-mass m_e, the Fermi level E_g and the potential shift V_0, and the electron's charge e. The radius R should be chosen such that $2R$ is less or equal to the free path of the electron, and V_0 should be chosen such that E_g is exactly equal to the resonance energy. The typical values of these parameters may be found in any reference book on solid-state physics, see also [6].

In fact we may reduce the spectral analysis of the real equation (40) to the spectral analysis of the corresponding non-dimensional equation on the disc of radius $r = 1$

$$-\Delta\Psi + \varepsilon r\cos(\theta)\Psi = \lambda\Psi \tag{41}$$

with Dirichlet boundary condition at the boundary: $\Psi|_{r=1} = 0$ and obtain the required values of the parameters from the corresponding change of variables.

Really, by the first step we transform the real equation just by division through $\frac{\hbar^2}{2m_e}$, followed by the change of variable $r \to \xi = \frac{r}{R}$, and then multiplication by R^2. This results in a non-dimensional equation for the properly redefined eigenfunction $\varphi_0(\xi,\theta) = \Psi(x,\theta)$

$$-\Delta_\xi\varphi_0 + \mathcal{E}e\frac{2m_eR^3}{\hbar^2}|\xi|\cos\theta\varphi_0 = (E_g - V_0)\frac{2m_eR^2}{\hbar^2}\varphi_0. \tag{42}$$

in the unit disc with the Dirichlet boundary condition at the boundary. Hence we obtain the non-dimensional equation with parameters $\epsilon = e\mathcal{E}\frac{2m_eR^3}{\hbar^2}$ and $\lambda_0 = (E_g - V_0)\frac{2m_eR^2}{\hbar^2}$

6. Perturbation procedure

Selection of proper values of the parameter ϵ is actually a special problem of Spectral Analysis which may be reduced to some problem of linear algebra. In this section we reduce the spectral problem for the Schrödinger operator to the spectral problem for some specially arranged finite matrix and obtain explicit expressions for its eigenfunctions and numerical values of its eigenvalues. We also estimate the error arising from this simplification. In order to obtain a solution of (41) we develop the perturbation procedure for the non-dimensional operator

$$L\varphi = -\Delta_\xi\varphi + \varepsilon\xi\varphi$$

in the unit disc with Dirichlet boundary condition using the representation of it in form of an infinite matrix with respect to the basis of normalized eigenvectors Φ_{ns} of the *non-perturbed operator*

$$-\Delta_\xi \Phi^{c,s}_{ns} = (k^s_n)^2 \Phi^{c,s}_{ns},$$

which may be constructed in explicit form:

$$\Phi^c_{0s} = \frac{J_0(k^s_0 r)}{2\pi(\int_0^1 |J_n(k^s_n r)|^2 r dr)^{1/2}}, \quad n = 0,$$

$$\Phi^c_{ns} = \frac{J_n(k^s_n r)\cos(n\phi)}{\pi(\int_0^1 |J_n(k^s_n r)|^2 r dr)^{1/2}}, \quad n = 1, 2, \dots$$

$$\Phi^s_{ns} = \frac{J_n(k^s_n r)\sin(n\phi)}{\pi(\int_0^1 |J_n(k^s_n r)|^2 r dr)^{1/2}}, \quad n = 1, 2, \dots$$

Here J_n is n-th Bessel function, and k^s_n is s-th root of J'_n. Using this basis we may present the perturbed operator as an orthogonal sum of two block-symmetric matrices $\mathcal{A}^c \oplus \mathcal{A}^s$, the first addend corresponding to the cosine-part of the basis (including $n = 0$), the second one corresponding to the sine-part of it ($n > 0$). The second addend $\mathcal{A}^s$ coincides in fact with the sub-matrix of the first addend for $n \geq 1$, hence we may calculate the first addend $\mathcal{A}^c$ only. It may be represented as a sum of the diagonal matrix $\mathcal{A}^{diag} = \text{diag}(k^s_n)^2$ which corresponds to the non-perturbed operator and the perturbation V caused by the homogeneous field $\epsilon\xi$:

$$\mathcal{A} = \mathcal{A}^{diag} + V$$

$$\begin{pmatrix} A_{00} + V_{00} & V_{01} & 0 & 0 & \dots \\ V_{10} & A_{11} + V_{11} & V_{12} & 0 & \dots \\ 0 & V_{21} & A_{22} + V_{22} & V_{23} & \dots \\ . & . & . & . & \dots \end{pmatrix}.$$

Here $V_{ik} = V^+_{ki}$ and A_{ii} are infinite diagonal blocks with elements

$$A^{jj}_{ii} = (k^j_i)^2$$

$$V^{st}_{01} = \varepsilon\pi \frac{\int_0^1 J_1(k^s_1 r) J_0(k^t_0 r) r^2 dr}{\left(\int_0^1 [J_0(k^t_0 r)]^2 r dr\right)^{1/2} \left(\int_0^1 [J_1(k^t_1 r)]^2 r dr\right)^{1/2}}$$

$$V^{st}_{nm} = \frac{\varepsilon\pi}{2} \frac{\int_0^1 J_n(k^s_n r) J_m(k^t_m r) r^2 dr}{\left(\int_0^1 [J_n(k^t_n r)]^2 r dr\right)^{1/2} \left(\int_0^1 [J_m(k^t_m r)]^2 r dr\right)^{1/2}}, \qquad m = n \pm 1.$$

We know that all the eigenvalues of the non-perturbed operator have multiplicity 2 except the ones which correspond to the constant angular factor because for any $\lambda = (k^s_n)^2$ there are 2 eigenfunctions:

$$\frac{J_n(k^s_n r)\cos(n\phi)}{(\int_0^1 |J_n(k^s_n r)|^2 r dr{+}\pi)^{1/2}} \quad \text{and} \quad \frac{J_n(k^s_n r)\sin(n\phi)}{(\int_0^1 |J_n(k^s_n r)|^2 r dr{+}\pi)^{1/2}}.$$

Having in mind the future estimates of errors of the perturbation procedure it is convenient to rearrange the matrix A in the following way. We consider the matrix A as a set of infinite rows, each one containing one "diagonal" element $(k_n^s)^2$. One may deduce easily from the properties of Bessel functions that the roots k_n^s depend monotonically on each index: $k_n^s < k_n^t$, if $s < t$, and $k_n^s < k_m^s$, if $n < m$. It means that for any fixed positive number M we may find a curve $\{n, s\}_M$ which divides the lattice Z_2 into two parts: the first one is I where $(k_m^t)^2 \leq M$ and the second one II where for any m, t, holds: $(k_m^t)^2 > M$. Now we rearrange the elements of basis of eigenfunctions φ_{mt} of the non-perturbed operator in increasing (non decreasing) order of the eigenvalues $(k_n^s)^2$ of the non-perturbed operator $-\Delta$, and form an orthogonal decomposition of the Hilbert space $L_2(\Omega_0)$ into sum of two subspaces $H_I = \bigvee_{mt\in I} \varphi_{mt}$ and $H_{II} = \bigvee_{mt\in II} \varphi_{mt}$, $H = H_I \oplus H_{II}$. Denote the rearranged matrix of the perturbed operator $-\Delta + V$ by $\mathbf{A}$ and its "diagonal" sub-matrix by $\mathbf{A}^{diag} := \operatorname{diag}(k_n^s)^2$. According the ordering of the eigenvalues of the non-perturbed problem we have $\mathbf{A}_{ll}^{diag} < \mathbf{A}_{l+1\ l+1}^{diag}$. Hence for any fixed number M we may find some finite-dimensional subspace $H_I = E_{_N}$ and the corresponding block $\mathbf{A}_{_{NN}}$, whose "diagonal elements" are not greater than M, and the complementary block $\mathbf{A}_{_{N^\perp N^\perp}}$ in$H_{II} = E_{_N}^\perp$ so that the rearranged matrix looks like

$$A = \begin{pmatrix} \mathbf{A}_{_{NN}} & \begin{matrix} 0 & 0 \\ V_{_{NN^\perp}} & 0 \end{matrix} \\ \begin{matrix} 0 & V_{_{N^\perp N}} \\ 0 & 0 \end{matrix} & \mathbf{A}_{_{N^\perp N^\perp}} \end{pmatrix}$$

with all "diagonal elements" of $\mathbf{A}_{_{N^\perp N^\perp}}$ greater then $M_{_N}$.

We may represent $\mathbf{A}_{_{N^\perp N^\perp}}$ as a sum of the diagonal matrix $\mathbf{A}^{diag}_{_{N^\perp N^\perp}}$ and a bounded matrix $V_{_{N^\perp N^\perp}} = P_{_{N^\perp}} V P_{_{N^\perp}}$:

$$\mathbf{A}_{_{N^\perp N^\perp}} = \mathbf{A}^{diag}_{_{N^\perp N^\perp}} + V_{_{N^\perp N^\perp}}$$

where

$$V_{_{N^\perp N^\perp}} - P_{_{N^\perp}} V P_{_{N^\perp}}.$$

Our goal is to find Ψ^s eigenvectors of $\mathbf{A}$:

$$\mathbf{A}\Psi^s = \mu^s \Psi^s. \tag{43}$$

In fact, we may easily find the eigenvalues $\mu^s_{_N}$ and eigenvectors $\psi^s_{_N}$ of the finite matrix $\mathbf{A}_{_{NN}}$

$$\mathbf{A}_{_{NN}} \psi^s_{_N} = \mu^s_{_N} \psi^s_{_N}. \tag{44}$$

Then we may estimate between the eigenvalues of the full matrix $\mathbf{A}$ and the corresponding[10] eigenvalues of the finite matrix $\mathbf{A}_{_{NN}}$ and the difference $\Psi^s - \psi^s_{_N}$ between the corresponding eigenvectors. It appears that *both are asymptotically*

[10] We arrange all eigenvalues μ_s of $\mathbf{A}$ and $\mu^s_{_N}$ of $A_{_{NN}}$ in an increasing order.

small for any fixed s when $N \to \infty$, or, equivalent, $M \to \infty$. The following correspondence rule is true for large values of N.

Theorem 9. *Assume that the s-th eigenvalue μ^s of $\mathbf{A}$ is simple*

$$\mathbf{A}\Psi^s = \mu^s \Psi^s, \ \min_t (\mu^s - \mu^t) = \delta_s > 0.$$

Then for any M large enough there exists a subspace $E_{_N}$, $N = N(M)$, $\dim E_{_N} = N$, such that

1. *The corresponding block $\mathbf{A}_{NN}$ of $\mathbf{A}$ in $E_{_N}$ has all "diagonal" elements not greater than M.*
2. *All "diagonal" elements of the complementary diagonal block $\mathbf{A}^{diag}_{N_\perp N_\perp}$ are greater than M.*
3. *If $\mu^s_{_N}$ is the eigenvalue number s of the block $\mathbf{A}_{NN}$, then for large N it is also isolated and simple and*

$$\lim_{N\to\infty} \mu^s_{_N} = \mu^s$$

and the eigenvector $\psi^s_{_N}$ of the block-matrix $\mathbf{A}_{NN}$ tends to the eigenvector of $\mathbf{A}$

$$\lim_{N\to\infty} \psi^s_{_N} = \Psi^s.$$

Proof. The proof of the theorem may be found in [45]. □

In course of proof the explicit expressions for the errors $\delta\mu^s$ and $\delta\Psi_s$ of finite-dimensional approximations of the eigenvalue and the eigenvector are obtained in dependence of N (which determines the order of the matrix $A_{_{NN}}$) and estimates the number s of the eigenvalue μ_s). For large enough M

$$\delta\mu^s \approx \langle P_s R(\mu_s)\psi_{_N}, \psi_{_N}\rangle$$

and similar statement holds for $\delta\Psi$:

$$\delta\Psi \approx [\hat{\mathbf{A}}_{_{NN}} - \mu_s I]^{-1} P_s^{\perp} R(\mu_s)\psi_{_N}.$$

Here P_s is an orthogonal projection of $\mathbf{A}_{_{NN}}$

$$\mathbf{A}_{_{NN}} = \sum_t \mu^t_{_N} P_t$$

corresponding to $\mu^s_{_N}$,

$$R_{_N}(\lambda) = V_{_{NN\perp}} \times (\mathbf{A}_{_{N\perp N\perp}} - \lambda I_{_{N\perp}} + V_{_{N\perp N\perp}})^{-1} \times V_{_{N\perp N}}$$

and

$$\hat{\mathbf{A}}_{_{NN}} = \sum_{\mu^t_{_N} \neq \mu^s_{_N}} \mu^t_{_N} P_t$$

where $\mu^s_{_N}$ is the eigenvalue of $\mathbf{A}_{_{NN}}$, which is the nearest to μ^s for large N:

$$\mathbf{A}_{_{NN}} \psi^s_{_N} = \mu^s_{_N} \psi^s_{_N}.$$

Using Mathematica we have found eigenvalues and eigenfunctions of $\mathbf{A}_{NN}$. We chose $M = 150$, in this case $N = 18$ — that is, there are 18 diagonal elements k_n^s that satisfy the requirement $(k_n^s)^2 \leq 150$.

7. Numerical results for dimensionless equation

After obtaining the explicit expressions for the eigenfunctions and numerical values of the eigenvalues we have two criteria for choosing one of these 18 eigenfunctions. Firstly, the normal derivative of it should have only two zeroes on the unit circle, and these zeroes should divide the circle in ratio $2 \div 1$. Actually positions of the zeroes can be adjusted by proper choice of ε (see below), but not for all functions with two zeroes. Secondly, the corresponding eigenvalue should not be too close to its neighbors for the reason of temperature stability of the device and better convergence of the perturbation procedure, see the estimates of convergence in [45]. It can be easily seen that a few eigenfunctions satisfy both requirements, namely $\Psi_{18}^2, \Psi_{18}^4, \Psi_{18}^6$. But it occurred that functions Ψ_{18}^4 and Ψ_{18}^6 do not allow the mentioned adjustment — that is, they do not change in desirable way with the change of ε. The function Ψ_{18}^{13} has the required design, but its eigenvalue $\mu_{18}^{13} = 103.60$ is too close its neighbor $\mu_{18}^{12} = 98.83$. Eigenvalues and eigenfunctions were found for different values of ε. Then by interpolation the value of ε was obtained which gives the desired positions of zeroes of the normal derivative of the chosen eigenfunction with a precision of 4 significant digits. So, it was found that for $\varepsilon = 18.86$ (in dimensionless case) the eigenfunction that corresponds to the second smallest eigenvalue $\mu_{18}^2 = 14.62$ has 2 zeroes of the normal derivative on the unit circle which divide it in the ratio $2 \div 1$. This permits to choose the contact points a_s on the circle such that the switching of electron current from one ray to another may be controlled and manipulated by the change of the direction of the electric field, as required. Then the selected resonance eigenvalue $\mu_{18}^2 = 14.62$ is not really too close to the neighboring eigenvalues $\mu_{18}^1 = 2.09$ and $\mu_{18}^3 = 25.82$ in comparison with other eigenvalues, for proper choice of ϵ, which also fulfill the above conditions concerning the zeroes of eigenfunctions.

It should be noted that the error arising from using the finite matrix $\mathbf{A}_{NN}$ can actually be done as small as desired by increasing M and, correspondingly, N, that is, by increasing the order of $\mathbf{A}_{NN}$. For the case of $M = 150$ it is still 12.5%.

8. Conclusion

In the actual paper we developed the analytic techniques of estimation of the transmission coefficients of a quantum switch and suggested a geometrical idea of the resonance manipulation of the electron's current through the triple splitting of waveguides. Using the first-order approximation of the perturbation theory we obtained the explicit expression for the transmission coefficient for the triadic switch and numerical value for the magnitude of the govering constant electric field such that the manipulation of the quantum current is possible via the change of the direction of the field in a plane parallel to the plane of the switch. Using Mathematica we checked that for a special value $\epsilon = 18.86$ of the dimensionless magnitude of the constant electric field in the domain Ω_0 zeroes of the normal derivative of some eigenfunction φ_0 divide the unit circle in ratio $1 \div 2$, which permits to apply our geometrical idea to fabricate our device, at least "in vitro".

Acknowledgments. We are grateful to the Commission of the European Communities for financial support in the frame of EC-Russia Exploratory Collaborative Activity under EU ESPRIT Project 28890 NTCONGS and partial support from the Russian Academy of Sciences (Grant RFFI 97-01-01149) and the Staff Rerearch Grant from the University of Auckland.

References

[1] V. Adamjan, *Scattering matrices for microschemes*, *Operator Theory: Adv. and Appl.*, **59** (1992), 1–10.

[2] V. Adamjan and H. Langer, *Spectral properties of a class of rational operator-valued functions*, *J. Operator Th.*, **33**(1995), 259–277.

[3] S. Albeverio and P. Kurasov, *Singular Perturbations of Differential Operators*, London Math. Society Lecture Notes Series 271, Cambridge Univ. Press, 2000, 429 p.

[4] I.E. Anar and A.O.Celebi, *A Scattering Problem in R_n*, *Bull. of the Inst. of Math. Academia Sinica*, **22**(1994), 323–340.

[5] C. Anne, *Fonctions propres sur des variétés avec des anses fines, application à la multiplicité*, *Comm. Partial Diff. Eq.*, **15**(1990), 1617–1630 (in French).

[6] I.Antoniou, B. Pavlov and Y. Yafyasov, *Quantum electronic Devices, based on Metal-Dielectric Transition in Low-Dimensional Quantum Structures*, In: D.S. Bridges, C. Calude, J. Gibbons, S. Reeves, I. Witten (eds.). *Combinatorics, Complexity, Logic, Proceedings of DMTCS'96*, Springer-Verlag, Singapore, 1996.

[7] N.T. Bagraev, V.K. Ivanov, L.E. Kljachkin, A.M. Maljarenko, S.A. Rykov, and I.A. Shelyh, *Interferention of charge carriers in one-dimensional semiconductor rings*, *Fizika i tehnika poluprovodnikov*, **34**(2000), 89–99 (in Russian).

[8] C.W.J. Beenakker and H. van-Houten, *Quantum Transport in semiconductor nanostructures*, in the series *Solid State Physics. Advances and Applications*, ed. by H. Ehrenreich and D. Turnbull, Vol. 44, Acad. Press, San Diego, 1991, pp. 1–228.

[9] V. Bogevolnov, A. Mikhailova, B. Pavlov, and A. Yafyasov, *About Scattering on the ring*, *Operator Theory: Advances and Application*, **124**(2001), 155–187.

[10] M. Brown, P.D. Hislop, and A. Martinez, *Lower bounds on the interaction between cavities connected by a thin tube*, *Duke Math. J.*, **73**(1994), 163–176.

[11] M. Brown, P.D. Hislop, and A. Martinez, *Eigenvalues and resonances for domains with tubes: Neumann boundary conditions*, *J. Diff. Eq.*, **115**(1995), 458–476.

[12] A.P. Calderón, *On an inverse boundary value problem*, Seminar on Numerical Analysis and its Applications to Continuum Physics, publ. by Soc. Brasiliera de Mathematica, Rio de Janeiro, (1980), pp. 65–73.

[13] Y. Colin de Verdiere, *Sur la multiplicité de la première valeur propre non nulle du Laplacien*, *Comment. Math. Helv.*, **63**(1988), 184–193 (in French).

[14] D. Colton and R. Kress, *Integral Equation methods in Scattering Theory*, John Wiley and Sons, 1983.

[15] P. Duclos and P. Exner, *Curvature-induced bound states in quantum waveguides in two and three dimensions*, *Rev. Math. Phys.*, **7**(1995), 73–102.

[16] P. Exner, E.M. Harrell and M. Loss, *Optimal eigenvalues for some Laplacians and Schrödinger operators depending on curvature*, (Proc. of the Conf. Mathematical results in quantum mechanics, Prague 1998), *Operator Theory Adv. Appl.*, **108**(1999), 47–58.

[17] P. Exner and P. Seba, *A new type of quantum interference transistor*, *Phys. Lett. A*, **129**(1988), 477–480.

[18] M.Faddeev and B.Pavlov, *Scattering by resonator with the small opening*, *Proc. LOMI*, **126**(1983) (*J. of Sov. Math.*, **27**(1984), 2527–2533).

[19] R. Gadylshin, *Scattering by bodies with narrow channels*, *Matematicheskij Sbornik*, **185**(1994), 39–62 (English Translation in *Russian Acad. Sb. Math.*, **82**(1995), 293–313).

[20] N.I. Gerasimenko and B.S. Pavlov *Scattering problems on noncompact graphs*, *Teoret. Mat. Fiz.*, **74** (1988), 345–359; translation in *Theoret. and Math. Phys.*, **74** (1988), 230–240.

[21] N.I. Gerasimenko *Inverse scattering problem on a noncompact graph*, *Teoret. Mat. Fiz.*, **75** (1988), 187–200; translation in *Theoret. and Math. Phys.*, **75**(1988), 460–470.

[22] I.S. Gohberg and E.I. Sigal, *Operator extension of the theorem about logarithmic residue and Rouchet theorem*, *Mat. sbornik*, **84**(1971), 607.

[23] V. Gotlib, *DAN USSR*, **287**(1986), 1109–1113 (in Russian).

[24] O.V. Guseva, О краевых задачах для сильно еллиптических систем (On boundary problems for strongly-elliptic systems), (Russian) *DAN SSSR*, **102**(1955), 1069–1072.

[25] M. Harmer, Hermitian Symplectic Geometry and Extension Theory, *J. Phys. A*, **33** (2000), 9193–9203.

[26] B. Helfer, *Semi-classical analysis for the Schrödinger operator and applications*, Lecture Notes in Mathematics., 1336, Springer, Berlin 1988.

[27] P.D.Hislop and A.Martinez, *Scattering resonances of a Helmholtz resonator*, *Indiana Math. J.*, **40**(1991), 767–788.

[28] A.M. Iljin, Согласование асимптотических разложений краевых задач, Nauka, Moscow, 1989, 334 p. (in Russian).

[29] S. Jimbo, *Remarks on the behaviour of certain eigenvalues on a singularly perturbed domain with several thin channels*, *Comm. Partial Diff. Eq.*, **17**(1992), 523–552.

[30] A. Kiselev, *Some examples in one-dimensional scattering on manifolds*, *J. Math. Anal. Appl.*, **212** (1997), 263–280.

[31] V. Kostrykin and R. Schrader, *Kirchhoff's rule for quantum wires*, *J. Phys. A: Math. Gen.*, **32**(1999), 595–630.

[32] M. Krein, *Concerning the resolvents of an Hermitian operator with the deficiency-index (m, m)*, *Comptes Rendue (Doklady) Acad. Sci. URSS (N.S.)*, **52**(1946), 651–654.

[33] P. Kurasov and J. Larson, **Spectral asymptotics for Schrödinger operators with periodic point interactions**, Dept. of Mathematics, Stockholm Univ., Research report in Mathematics N3, 2001 (to be published in *J. Math. Anal. Appl.*).

[34] P. Kurasov and F. Stenberg, **On the inverse scattering problem on branching graphs**, Dept. of Mathematics, Stockholm Univ., Research report in Mathematics N7, 2001 (to be published in *J. Phys. A: Math. Gen.*).

[35] O. Ladyzhenskaya and N. Uraltseva, Линейные и квазилинейные уравнения эллиптического типа (Linear and quasilinear equations of elliptic type), Second edition, Nauka, Moskow, 1973, 576 p.

[36] R. Landauer, *IBM Journal for Research and Development*, **1**(1957), 223.

[37] P. Lax, R. Phillips, *Scattering theory*, Academic Press, New York, 1967.

[38] J.-L. Lions and E. Magenes, *Problèmes aux limites non homogènes et applications* Vol.1 (French) Travaux et Recherches mathématiques, No.17, Dunod, Paris (19680) 372 p.

[39] A.M. Lyapunoff, *Sur quelques questions attachées au problème de Dirichlet*, *Journal de Mathématiques pure et appliquées*, 1898, 5-ème séries, Paris, IV, 241–311.

[40] A.M. Lyapunoff, *On fundamental principle of Neumann's method in Dirichlet problem.*, *Transactions of Kharkov Math. Soc. Ser II*, **VII**(1902), 229–252.

[41] V.M. Maz'ja. *Sobolev Spaces,* Translated from the Russian by T.O. Shaposhnikova, Springer Series in Soviet Mathematics. Springer-Verlag, Berlin, 1985, 486 p.

[42] J. Meixner, *The behaviour of electromagnetic waves at edges*, *IEEE Trans.*, AP-20, 442–446.

[43] R. Mennicken and A. Shkalikov, *Spectral decomposition of symmetric operator matrices*, *Math. Nachrichten*, **179**(1996), 259–273.

[44] A.Mikhailova, B.Pavlov, I.Popov, T.Rudakova, and A.M.Yafyasov, *Scattering on a compact domain with few semiinfinite wires attached: resonance case*, Report Series of the department of Mathematics, the University of Auckland, April 2000, No 420, 17p. (Accepted for publication in "Mathematishe Nachrichten".)

[45] A. Mikhaylova and B. Pavlov, *Quantum domain as a triadic relay*, in: Unconventional Models of Computations UMC'2K (Proceedings of the UMC'2K Conference, Brussels, Dec 2000) eds.I. Antoniou, C. Calude, M.J. Dinneen, Springer Verlag Series for Discrete Mathematics and Theoretical Computer Science (2001), pp. 167–186.

[46] A. Mikhailova and B. Pavlov, *Quantum Domain as a triadic relay*, April 2000, Department of Mathematics Report series No.420, ISSN 1173-0889, the University of Auckland, Auckland, NZ, 16p. New technologies for narrow-gap semiconductors. Esprit project N 28890 ESPRIT NTCONGS Progress Reports (July 1, 1999–December 31,1999).

[47] M. Naimark, *Self-adjoint extensions of the second kind of a symmetric operators*, *Bull. AN USSR, Ser. Math.*, **4**(1940), 53–104.

[48] S.P. Novikov, *Schrödinger operators on graphs and symplectic geometry*, in: The Arnol'd fest (Proceedings of the Fields Institute Conference in Honour of the 60th

Birthday of Vladimir I. Arnol'd), eds. E. Bierstone, B. Khesin, A. Khovanskii, and J. Marsden, to appear in the Fields Institute Communications Series.

[49] B. Pavlov, *The theory of extensions and explicitly solvable models*, *Uspekhi Mat. Nauk*, **42**(1987), 99–131.

[50] B. Pavlov, *Splitting of acoustic resonances in domains connected by a thin channel*, *New Zealand J. of Mathematics*, **25**(1996), 199–216.

[51] B. Pavlov, I. Popov, V. Geyler, and O. Pershenko, *Possible construction of a quantum multiplexer*, *Europhyics Letters*, **52**(2000), 196–202.

[52] B. Pavlov, G. Roach, and A. Yafyasov, *Resonance scattering and design of quantum gates*, Unconventional models of computation (Auckland, 1998), 336–351, Springer Ser. Discrete Math. Theor. Comput. Sci., Springer, Singapore, 1998.

[53] M. Reed and B. Simon, *Methods of modern mathematical physics*, Academic Press, New York-London, 1972.

[54] J. Schwinger and D.S. Saxon, *Discontinuities in Waveguides*, Gordon and Breach Science Publishers, New York, London, Paris, 1968.

[55] A. Silbergleit and Yu. Kopilevich, *Spectral theory of guided waves*, Institute of Physics publishing, Bristol, 1996, 310 p.

[56] J. Sylvester and G. Uhlmann, *The Dirichlet to Neumann map and applications*. In: Proceedings of the Conference "Inverse problems in partial differential equations (Arcata,1989)", SIAM, Philadelphia, 1990, pp. 101–139.

[57] E.C. Titchmarsh, *Eigenfunctions expansions associated with second order differential equations*, Clarendon Press, Oxford (1962), Vol. 1, 203 p.

[58] G. Uhlmann, *Development in inverse problems since Calderon's foundational paper.* In the book: "Harmonic Analysis and partial differential Equations. Essay in honor of Alberto P. Calderón", publ. by the Univ. of Chicago Press, 1999, pp. 295–345.

[59] F. Ursell, *On exterior problems of acoustics*, *Proc. Cambridge Phil. Society*, **74**(1973), 117–125.

[60] B. Vainberg, *Asymptotic methods in equations of Mathematical Physics*, Gordon & Breach Science Publishers, New York, 1989, 498 p.

[61] G. Verzbinski and V. Maz'ya, *The closure in $[L_p]$ of the operator of the Dirichlet problem in a domain with conical points* (Russian), *Izv. Vys. Ucebn. Zaved. Matematika*, 1974, no. 6(145), 8–19.

[62] I.I. Vorovich and V.A. Babeshko, *Dynamical problems on elasticity for non-classical domains*, Nauka, Moscow, 1979 (in Russian).

[63] A.M. Yafyasov, V.B. Bogevolnov, and T.V. Rudakova, *Quantum interferentional electronic transistor (QIET). Theoretical analysis of electronic properties for low-dimensional systems*, (Progress report,1995), Preprint IPRT N 99-95.

A.B. Mikhaylova
Laboratory of Complex Systems Theory
Institute for Physics
St.-Petersburg State University
Ulyanovskaya 1
St.-Petersburg 198904, Russia

B.S. Pavlov
Laboratory of Complex Systems Theory
Institute for Physics
St.-Petersburg State University
Ulyanovskaya 1
St.-Petersburg 198904, Russia

Department of Mathematics
University of Auckland
Private Bag 92019
Auckland, New Zealand

Operator Theory:
Advances and Applications, Vol. 132, 323–332

The Hamilton Operator and Quantum Vacuum for Nonconformal Scalar Fields in the Homogeneous and Isotropic Space

Yu.V. Pavlov

Abstract. The diagonalization of the metrical and canonical Hamilton operators of a scalar field with an arbitrary coupling, with a curvature in N-dimensional homogeneous isotropic space is considered in this paper. The energy spectrum of the corresponding quasiparticles is obtained and then the modified energy-momentum tensor is constructed; the latter coincides with the metrical energy-momentum tensor for conformal scalar field. Under the diagonalization of corresponding Hamilton operator the energies of relevant particles of a nonconformal field are equal to the oscillator frequencies, and the density of such particles created in a nonstationary metric is finite. It is shown that the modified Hamilton operator can be constructed as a canonical Hamilton operator under the special choice of variables.

1. Introduction

Our aim in this paper is the investigation of the Hamiltonian diagonalization method and the definition of the Hamilton operator and quantum vacuum of nonconformal scalar field in nonstationary homogeneous isotropic space. Quantum field theory in curved space-time (see monographs [1, 2]) has important applications to cosmology and astrophysics. However there are several problems that have not been finally solved until the present time. One of them is the definition of vacuum state and the notion of elementary particle in curved space-time; this is due to the absence of the group of symmetries such as the Poincaré group in the Minkowsky space. This problem for nonconformal scalar field is under active discussion even in the case of homogeneous isotropic space [3, 4, 5]. As a consequence of various definitions of vacuum states we have a variety of calculated quantum characteristics of nonconformal scalar fields in curved space.

In [6, 7] it was shown that in the case of arbitrary coupling of scalar fields with curvature additional, nonconformal contributions are dominant in vacuum averages of the energy-momentum tensor. It should be also mentioned that the investigation of nonconformal scalar fields is not only of independent interest; this investigation is caused by impossibility of preservation of conformal invariation of effective action and the usual action in the case of interacting quantized field [2].

In the definition of vacuum and in the formulation of the problem of particles creation in curved space-time two known approaches are widely used: the Hamiltonian diagonalization procedure [1] offered in [8, 9], and the so-called "adiabatic" procedure [2] offered in [10]. Supposing that a quantum of energy corresponds to a particle, then observation of particles at some moment (according to quantum mechanics) means to find the Hamilton operator's eigenstate; this is taken into account automatically in the diagonalization approach. In nonstationary metrics the diagonalization procedure is realized by the time-dependent Bogolyubov transformations (see below). If the operators of these transformations are Hilbert-Schmidt operators then the representations of commutation relations are unitarily equivalent for both the old and the new creation and annihilation operators [11]. However, the use of the Hamilton operator constructed from the metrical energy-momentum tensor, successful in the conformal case [1], leads to the difficulties related to an infinite density of created particles in the nonconformal case [12]. At the same time essential problems and ambiguities take place in adiabatic approach also [5].

In this paper we will consider the complex scalar field with arbitrary coupling, with curvature in N-dimensional homogeneous isotropic space. In Section 2 all necessary information is given and nonconformal scalar field quantizing in N-dimensional homogeneous isotropic space is defined. In Section 3 the metrical and canonical Hamilton operators diagonalization is carried out, and the energies of corresponding quasiparticles are calculated, and conditions connected with demand of the Hamilton operators diagonality are investigated. In Section 4 the modified energy-momentum tensor is defined so that the quasiparticles from diagonalization of corresponding Hamilton operator have energies coinciding with the oscillator frequency of the wave equation. It is shown that such Hamiltonian can be defined as canonical under a certain choice of canonical variables. It is proved that the density of particles being created in a nonstationary metric is finite and the results of given investigations are summarized.

The system of units in which the Planck constant ($\hbar$) and light velocity are equal 1 is used in the paper.

2. Quantizing of scalar field in homogeneous isotropic space

We consider the complex scalar field $\phi(x)$ of mass m satisfying the equation

$$(\nabla_i \nabla^i + \xi R + m^2)\, \phi(x) = 0\,, \tag{1}$$

where ∇_i is covariant differentiation, R is the scalar curvature, $x = (t, \mathbf{x})$, ξ is the coupling constant. The value $\xi = \xi_c = (N-2)/\,[\,4\,(N-1)]$ corresponds to conformal coupling in space-time of dimension N ($\xi_c = 1/6$ if $N = 4$). The equation (1) is conformally invariant if $\xi = \xi_c$ and $m = 0$; the value $\xi = 0$ reduces to the case of minimal coupling.

The metric of N-dimensional homogeneous isotropic space-time is

$$ds^2 = g_{ik}dx^i\,dx^k = dt^2 - a^2(t)\,dl^2 = a^2(\eta)\,(d\eta^2 - dl^2)\,, \tag{2}$$

where $dl^2 = \gamma_{\alpha\beta}\,dx^\alpha\,dx^\beta$ is the metric of $(N-1)$-dimensional space with the constant curvature $K = 0, \pm 1$.

The equation (1) can be obtained by varying the action with Lagrangian density

$$L(x) = \sqrt{|g|}\,\left[g^{ik}\partial_i\phi^*\partial_k\phi - (m^2 + \xi R)\phi^*\phi\right], \tag{3}$$

where $g = \det(g_{ik})$.

The canonical energy-momentum tensor of the scalar field is

$$T_{ik}^{can} = \partial_i\phi^*\partial_k\phi + \partial_k\phi^*\partial_i\phi - g_{ik}|g|^{-1/2}L(x)\,. \tag{4}$$

The metrical energy-momentum tensor which can be obtained by varying the action of g_{ik} has a form [13]:

$$T_{ik} = T_{ik}^{can} - 2\,\xi\,[R_{ik} + \nabla_i\nabla_k - g_{ik}\nabla_j\nabla^j]\,\phi^*\phi\,, \tag{5}$$

where R_{ik} is Ricci tensor.

In the metric (2) the equation (1) takes the form

$$\phi'' + (N-2)\left(\frac{a'}{a}\right)\phi' - \Delta_{N-1}\,\phi + (m^2 + \xi R)\,a^2\phi = 0\,, \tag{6}$$

where Δ_{N-1} is the Laplace-Beltrami operator in $(N-1)$-dimensional space, and the prime denotes the derivative with conformal time η.

For the function $\tilde\phi = a^{(N-2)/2}\phi$ the equation (6) takes the form without the first derivative in time

$$\tilde\phi'' - \Delta_{N-1}\,\tilde\phi + \left(m^2a^2 - \Delta\xi a^2 R + ((N-2)/2)^2\,K\right)\tilde\phi = 0\,, \tag{7}$$

where $\Delta\xi = \xi_c - \xi$. The variables in the equations (6), (7) can be separated; namely, for $\tilde\phi = g_\lambda(\eta)\Phi_J(\mathbf{x})$ we have

$$g_\lambda''(\eta) + \Omega^2(\eta)\,g_\lambda(\eta) = 0\,, \tag{8}$$

and

$$\Delta_{N-1}\,\Phi_J = -(\lambda^2 - ((N-2)/2)^2\,K)\,\Phi_J\,; \tag{9}$$

$\Omega(\eta)$ is the oscillator frequency

$$\Omega^2(\eta) = m^2a^2 + \lambda^2 - \Delta\xi\,a^2 R\,, \tag{10}$$

J is a set of indices (quantum numbers) numbering the eigenfunctions of the Laplace-Beltrami operator. It should be noted that the eigenvalues of the operator $-\Delta_{N-1}$ are not negative and we have the inequality

$$\lambda^2 - ((N-2)/2)^2\,K \geq 0\,.$$

For quantization we decompose the field $\tilde\phi(x)$ by the complete set of the solutions of (7), i.e.

$$\tilde\phi(x) = \int d\mu(J)\left[\tilde\phi_{\bar J}^{(-)}\,a_{\bar J}^{(-)} + \tilde\phi_J^{(+)}\,a_J^{(+)}\right]; \tag{11}$$

here $d\mu(J)$ is the measure in the space of the Laplace-Beltrami Δ_{N-1} eigenvalues

$$\tilde{\phi}_J^{(+)}(x) = \frac{1}{\sqrt{2}}\, g_\lambda(\eta)\, \Phi_J^*(\mathbf{x}) \,, \quad \tilde{\phi}_{\bar{J}}^{(-)}(x) = \frac{1}{\sqrt{2}}\, g_\lambda^*(\eta)\, \Phi_{\bar{J}}(\mathbf{x}) \,, \tag{12}$$

$\Phi_J(\mathbf{x})$ is orthonormal eigenfunctions of Δ_{N-1} operator, and $\bar{J}$ is a set of quantum numbers of the function complex conjugated to the function Φ_J. In $(N-1)$-dimensional spherical coordinates for $J=\{\lambda,l,\dots,m\}$ we have $\bar{J}=\{\lambda,l,\dots,-m\}$.

Substituting expansion (11) in the expression for conserved charge

$$Q = i \int\limits_{\Sigma} \left(\tilde{\phi}^* \partial_0 \tilde{\phi} - (\partial_0 \tilde{\phi}^*)\, \tilde{\phi} \right) \sqrt{\gamma}\, d^{N-1}x \,, \tag{13}$$

where $\gamma = \det(\gamma_{\alpha\beta})$, Σ is a space-like hypersurface $\eta = \mathrm{const}$, and imposing the normalization condition

$$g_\lambda\, g_\lambda^{*\prime} - g_\lambda^\prime\, g_\lambda^* = -2i \,, \tag{14}$$

we obtain

$$Q = \int d\mu(J) \left(\overset{*}{a}{}_J^{(+)} a_J^{(-)} - \overset{*}{a}{}_{\bar{J}}^{(-)} a_{\bar{J}}^{(+)} \right) . \tag{15}$$

The metrical Hamiltonian is expressed in terms of the metrical energy-momentum tensor (5) by [1]:

$$H(\eta) = \int\limits_{\Sigma} \zeta^i\, T_{ik}(x)\, d\sigma^k = \int\limits_{\eta=\mathrm{const}} \zeta^0\, T_{00}(x)\, g^{00} \sqrt{|g|}\, d^{N-1}x = $$
$$= a^{N-2}(\eta) \int\limits_{\eta=\mathrm{const}} T_{00}(x) \sqrt{\gamma}\, d^{N-1}x \,, \tag{16}$$

where $(\zeta^i) = (1, 0, \dots, 0)$ is the time-like conformal Killing vector.

The quantization is realized by the commutation relations

$$\left[a_J^{(-)},\, \overset{*}{a}{}_{J'}^{(+)} \right] = \left[\overset{*}{a}{}_J^{(-)},\, a_{J'}^{(+)} \right] = \delta_{JJ'} \,, \quad \left[a_J^{(\pm)},\, a_{J'}^{(\pm)} \right] = \left[\overset{*}{a}{}_J^{(\pm)},\, \overset{*}{a}{}_{J'}^{(\pm)} \right] = 0 \,. \tag{17}$$

The Hamilton operator (16) can be written, through the $a_J^{(\pm)}$, $\overset{*}{a}{}_J^{(\pm)}$ operators in the form

$$\begin{aligned} H(\eta) \;&=\; \int d\mu(J) \Big\{ E_J(\eta) \left(\overset{*}{a}{}_J^{(+)} a_J^{(-)} + \overset{*}{a}{}_{\bar{J}}^{(-)} a_{\bar{J}}^{(+)} \right) \\ &+\; F_J(\eta)\, \overset{*}{a}{}_J^{(+)} a_{\bar{J}}^{(+)} + F_J^*(\eta)\, \overset{*}{a}{}_{\bar{J}}^{(-)} a_J^{(-)} \Big\} , \end{aligned} \tag{18}$$

where

$$E_J(\eta) = \frac{1}{2} \left\{ |g_\lambda^\prime|^2 + D_\lambda(\eta)\, |g_\lambda|^2 - Q(\eta)\, (|g_\lambda|^2)^\prime \right\} , \tag{19}$$

$$F_J(\eta) = \frac{(-1)^m}{2} \left\{ g_\lambda^{\prime\,2} + D_\lambda(\eta)\, g_\lambda^2 - Q(\eta)\, (g_\lambda^2)^\prime \right\} , \tag{20}$$

$$D_\lambda(\eta) = m^2 a^2 + \lambda^2 + \Delta\xi\, (N-1)\, (N-2)\, (c^2 - K) \,, \tag{21}$$

$$Q(\eta) = \Delta\xi\, 2\, (N-1)\, c \,, \tag{22}$$

and $c = a'(\eta)/a(\eta)$. The Hamilton operator corresponding to canonical energy-momentum tensor (4) has the form (18) with (21) and (22) replaced by

$$D_\lambda(\eta) = m^2a^2 + \lambda^2 + (N-1)(N-2)\left((\xi+\xi_c)c^2 - \Delta\xi K\right) + 2\xi(N-1)c', \tag{23}$$

$$Q(\eta) = c(N-2)/2. \tag{24}$$

3. The diagonalization of the metrical and canonical Hamilton operators

The Hamilton operator (18) is diagonal at time moment η_0 in the operators $\overset{*}{a}{}^{(\pm)}_J$, $a^{(\pm)}_J$, which in this case are the creation and annihilation operators of particles and antiparticles under $F_J(\eta_0) = 0$. Utilizing (20)–(24) it may be shown that this condition is consistent with normalization (14) only when $p^2_\lambda(\eta_0) > 0$, where for the metrical Hamilton operator

$$p_\lambda(\eta) = \sqrt{m^2a^2(\eta) + \lambda^2 + 4\Delta\xi\,(N-1)^2\,(\xi\,c^2 - \xi_c K)} \tag{25}$$

and for the canonical Hamilton operator

$$p_\lambda(\eta) = p_{can,\lambda}(\eta) = \sqrt{(m^2 + \xi R)\,a^2(\eta) + \lambda^2 - ((N-2)/2)^2 K}. \tag{26}$$

The requirement for the metrical Hamilton operator to be diagonal at the instant η_0, i.e., $F_J(\eta_0) = 0$, and the normalization condition lead to the initial conditions on the functions $g_\lambda(\eta_0)$:

$$g'_\lambda(\eta_0) = (2\Delta\xi(N-1)\,c + i p_\lambda(\eta_0))\,g_\lambda(\eta_0), \quad |g_\lambda(\eta_0)| = 1/\sqrt{p_\lambda(\eta_0)}. \tag{27}$$

For the canonical Hamilton operator the initial conditions are

$$g'_\lambda(\eta_0) = ((N-2)\,c/2 + i\,p_{can,\lambda}(\eta_0))\,g_\lambda(\eta_0), \quad |g_\lambda(\eta_0)| = 1/\sqrt{p_{can,\lambda}(\eta_0)}. \tag{28}$$

The state of vacuum $|\,0>$ corresponding to (27), (28) is defined in the standard form

$$a^{(-)}_J|\,0> = \overset{*}{a}{}^{(-)}_J|\,0> = 0. \tag{29}$$

For arbitrary time moment η we diagonalize the Hamilton operator in terms of $\overset{*}{b}{}^{(\pm)}_J$, $b^{(\pm)}_J$ operators which connected with $\overset{*}{a}{}^{(\pm)}_J$, $a^{(\pm)}_J$, by the time-dependent Bogolyubov transformations:

$$\begin{cases} a^{(-)}_J = \alpha^*_J(\eta)\,b^{(-)}_J(\eta) - (-1)^m\beta_J(\eta)\,b^{(+)}_{\bar J}(\eta), \\ \overset{*}{a}{}^{(-)}_J = \alpha^*_J(\eta)\;\overset{*}{b}{}^{(-)}_J(\eta) - (-1)^m\beta_J(\eta)\;\overset{*}{b}{}^{(+)}_{\bar J}(\eta), \end{cases} \tag{30}$$

where $\alpha_J(\eta)$, $\beta_J(\eta)$ are the functions satisfying the initial conditions $|\alpha_J(\eta_0)| = 1$, $\beta_J(\eta_0) = 0$ and the identity

$$|\alpha_J(\eta)|^2 - |\beta_J(\eta)|^2 = 1. \tag{31}$$

(In the homogeneous and isotropic space $\alpha_J = \alpha_\lambda$, $\beta_J = \beta_\lambda$ [1].)

The substitution of the decomposition (30) in (18) gives, if we demand coefficients before nondiagonal terms $\overset{*}{b}{}_J^{(\pm)} b_J^{(\pm)}$ equal to 0, the equation

$$2(-1)^{m+1}\alpha_J\beta_J E_J + F_J\alpha_J^2 + F_J^*\beta_J^2 = 0\,. \tag{32}$$

It can be shown that the condition (32) is consistent with the normalization (14), only if $p_\lambda^2(\eta) > 0$. In that case

$$|\beta_J|^2 = E_J/\,(2p_\lambda) - 1/2 = |F_J|^2/\,(2p_\lambda\,(E_J + p_\lambda))\,. \tag{33}$$

In obtaining (33) we take into account, the result that can be checked,

$$E_J^2 - |F_J|^2 = p_\lambda^2(\eta)\cdot\left[-(g_\lambda\, g_\lambda^{*\prime} - g_\lambda^\prime\, g_\lambda^*)^2/\,4\right]\,. \tag{34}$$

(The multiplier in square brackets equals 1 under the normalization condition (14).)

In the case of (32) and $p_\lambda^2(\eta) > 0$, the Hamilton operator (18) takes the form

$$H(\eta) = \int d\mu(J)\, p_\lambda(\eta)\, \left(\overset{*}{b}{}_J^{(+)} b_J^{(-)} + \overset{*}{b}{}_{\bar J}^{(-)} b_{\bar J}^{(+)}\right)\,. \tag{35}$$

So $p_\lambda(\eta)$ has the meaning of energy of quasiparticles corresponding to the diagonal form of the metrical Hamilton operator (and $p_{can,\lambda}(\eta)$ for the canonical Hamilton operator). For the 4-dimensional space-time the equation (25) corresponds to energy values obtained in [3] and [14].

The quasiparticle energy $p_\lambda(\eta)$ differs from the oscillator frequency $\Omega(\eta)$ of the wave equation for nonconformal field, and this leads to a series of difficulties. Thus the conditions $p_\lambda^2(\eta) > 0$ and $\Omega^2(\eta) > 0$ may be in contradiction for a nonconformal field in some cases. For example, in the case of quasi-Euclidean space ($K = 0$) and zero-mass field the condition $p_\lambda^2(\eta) > 0$ (with arbitrary λ) for the metrical case reduces to $\xi \in [0, \xi_c]$; but if $\xi < \xi_c$, $m = 0$ and $R > 0$ for low λ then we have $\Omega^2(\eta) < 0$.

It should be noted that for $p_\lambda^2(\eta) < 0$ the condition of diagonalization reduces to the vanishing of norm, energy and charge of the state with $\phi(x) \neq 0$, and this situation does not have any physical foundation.

The vacuum state defined by the equations

$$b_J^{(-)}\,|\,0_\eta\!> = \overset{*}{b}{}_J^{(-)}\,|\,0_\eta\!> = 0\,, \tag{36}$$

depends on time in the nonstationary metric. Under the initial conditions (27), (28) we have $b_J^{(\pm)}(\eta_0) = a_J^{(\pm)}$ and $|\,0_{\eta_0}\!> = |\,0\!>$. In the Heisenberg representation, the state $|\,0\!>$, which is vacuum at the instant η_0, is no longer a vacuum for $\eta \neq \eta_0$. It contains $|\beta_J(\eta)|^2$ quasiparticle pairs corresponding to the operators $\overset{*}{b}{}_J^{(\pm)}, b_J^{(\pm)}$ in every mode [1]. The number of the created pairs of quasiparticles in the unit of space volume (for $N = 4$) is [1]

$$n(\eta) = \frac{1}{2\pi^2 a^3(\eta)} \int d\mu(J)\, |\beta_\lambda(\eta)|^2\,. \tag{37}$$

For asymptotic solutions of equation (8) (see [15]), normalized according to (14), we can obtain from (19)–(24) that $E_J \sim \lambda$ and for nonstationary metrics $|F_J(\eta)| \sim$

$|Q(\eta)|$ in $\lambda \to \infty$. Therefore, according to (33), this is corrected with the substitution of $p_\lambda \to p_{can,\lambda}$, and we have $|\beta_\lambda|^2 \sim \lambda^{-2}$. Consequently, the density of created quasiparticles, proportional to the integral in (37), is infinite.

So, in the diagonalization procedure, both for the metrical and the canonical Hamilton operators in nonconformal scalar fields, there is a problem of infinite density of quasiparticles created in the nonstationary metrics. In both cases the energies of corresponding quasiparticles differ from the oscillator frequency of the wave equation. It is shown below that these difficulties are absent in the case of the Hamilton operator corresponding to the modified energy-momentum tensor.

4. Modified energy-momentum tensor and modified Hamilton operator

Let us consider the modified energy-momentum tensor

$$T_{ik}^{mod} = T_{ik}^{can} - 2\xi_c \left[R_{ik} + \nabla_i \nabla_k - g_{ik} \nabla_j \nabla^j\right] \phi^* \phi \,. \tag{38}$$

From the definition (38) it is clear that for conformal scalar fields (i.e. if $\xi = \xi_c$) T_{ik}^{mod} coincides with the metrical energy-momentum tensor (5). The structure of the Hamiltonian constructed by T_{ik}^{mod} similarly to (16) is

$$\begin{aligned} H^{mod}(\eta) &= \int h(x)\, d^{N-1}x = \int d^{N-1}x \sqrt{\gamma} \left\{ \tilde{\phi}^{*\prime} \tilde{\phi}' + \gamma^{\alpha\beta} \partial_\alpha \tilde{\phi}^* \partial_\beta \tilde{\phi} \right. \\ &\qquad \left. + \left[m^2 a^2 - \Delta\xi\, a^2 R + \Big((N-2)/2\Big)^2 K \right] \tilde{\phi}^* \tilde{\phi} \right\}. \end{aligned} \tag{39}$$

We show that the modified Hamiltonian (39) can be obtained in homogeneous isotropic space as canonical under the certain choice of variables describing scalar field. If we add N-divergence $(\partial J^i/\partial x^i)$, to the Lagrangian density (3), where in the $(\eta, \mathbf{x})$ system of coordinates the N-vector $(J^i) = (\sqrt{\gamma}\, c \tilde{\phi}^* \tilde{\phi} (N-2)/2,\, 0, \dots, 0)$, the movement equations (1) are invariant under this addition. Choosing $\tilde{\phi}(x) = a^{(N-2)/2}(\eta)\phi(x)$ and $\tilde{\phi}^*(x)$, i.e., the variables in terms of which the equation (1) has the form (7), for the field's coordinates and using the Lagrangian density $L^\Delta(x) = L(x) + (\partial J^i/\partial x^i)$, we obtain that the Hamiltonian density $\tilde{\phi}' (\partial L^\Delta)/(\partial \tilde{\phi}') + \tilde{\phi}^{*\prime} (\partial L^\Delta)/(\partial \tilde{\phi}^{*\prime}) - L^\Delta(x)$ is equal to $h(x)$, from (39). This is why the Hamiltonian (39) is a canonical one for the scalar field, if $\tilde{\phi}(x)$ and $\tilde{\phi}^*(x)$ are chosen as the field's variables.

The modified Hamilton operator can be written in form (18), but in that case $Q(\eta) = 0$ and $D_\lambda(\eta) = \Omega^2(\eta)$; under its diagonalization by $\overset{*}{b}{}_J^{(\pm)}, b_J^{(\pm)}$, operators we obtain (35) with the change $p_\lambda \to \Omega$. The oscillator frequencies $\Omega(\eta)$ then coincide with the energy of corresponding particles. The initial conditions for $g_\lambda(\eta)$, corresponding to the diagonal form in the time moment η_0 with operators $\overset{*}{a}{}_J^{(\pm)}, a_J^{(\pm)}$ (17), are

$$g_\lambda'(\eta_0) = i\,\Omega(\eta_0)\, g_\lambda(\eta_0)\,, \quad |g_\lambda(\eta_0)| = 1/\sqrt{\Omega(\eta_0)}\,. \tag{40}$$

They coincide with the initial conditions used in [7] if $\arg g_\lambda(\eta_0) = 0$ is fixed. In the case of radiation dominated background ($R = 0$) they coincides with conditions used in [6, 16].

We show that the density of the particles corresponding to the diagonal form of H^{mod} and created in the nonstationary metric is finite. For this, we find the asymptotic behavior of the functions $|\beta_\lambda(\eta)|^2$ as $\lambda \to \infty$. The functions $\beta_\lambda(\eta)$ and $\alpha_\lambda(\eta)$ that are the solutions of (32) and satisfy identity (31) can be represented as

$$\beta_\lambda(\eta) = \frac{i}{2} \frac{e^{i\,\Theta(\eta_0,\eta)}}{\sqrt{\Omega}} \left(g'(\eta) - i\,\Omega\, g(\eta) \right), \tag{41}$$

$$\alpha_\lambda(\eta) = \frac{i}{2} \frac{e^{i\,\Theta(\eta_0,\eta)}}{\sqrt{\Omega}} \left(g^{*\prime}(\eta) - i\,\Omega\, g^*(\eta) \right), \tag{42}$$

where $\Theta(\eta_1,\eta_2) = \int\limits_{\eta_1}^{\eta_2} \Omega(\eta)\, d\eta$. In consequence of (41), (42) and equation (8) the functions $s_\lambda(\eta) = |\beta_\lambda(\eta)|^2$ and $f_\lambda(\eta) = 2\,\alpha_\lambda(\eta)\,\beta_\lambda(\eta) \exp[-2i\,\Theta(\eta_0,\eta)]$ satisfy the system of equations:

$$\begin{cases} s'_\lambda(\eta) = \dfrac{\Omega'}{2\Omega}\, \mathrm{Re} f_\lambda(\eta)\,, \\[2ex] f'_\lambda(\eta) + 2\,i\,\Omega\, f_\lambda(\eta) = \dfrac{\Omega'}{\Omega}\,(1 + 2 s_\lambda(\eta))\,. \end{cases} \tag{43}$$

Taking into account the initial condition $s_\lambda(\eta_0) = f_\lambda(\eta_0) = 0$ (as $\beta_\lambda(\eta_0) = 0$) we write the system of differential equations (43) in the equivalent form of the system of Volterra integral equations

$$f_\lambda(\eta) = \int_{\eta_0}^{\eta} w(\eta_1)\,(1 + 2 s_\lambda(\eta_1)) \exp[-2\,i\,\Theta(\eta_1,\eta)]\, d\eta_1\,, \tag{44}$$

$$s_\lambda(\eta) = \frac{1}{2} \int_{\eta_0}^{\eta} d\eta_1\, w(\eta_1) \int_{\eta_0}^{\eta_1} d\eta_2\, w(\eta_2)\,(1 + 2 s_\lambda(\eta_2)) \cos[2\,\Theta(\eta_2,\eta_1)]\,, \tag{45}$$

where $w(\eta) = \Omega'(\eta)/\,\Omega(\eta)$. To find the asymptotic behavior of $s_\lambda(\eta)$, we restrict our consideration to the first iteration of integral equation (45) and, taking into account that $\Theta(\eta_2,\eta_1) \to \lambda(\eta_1 - \eta_2)$ as $\lambda \to \infty$, represent (45) as

$$s_\lambda(\eta) \approx \frac{1}{4} \left| \int_{\eta_0}^{\eta} w(\eta_1) \exp(2\,i\,\lambda\,\eta_1)\, d\eta_1 \right|^2 . \tag{46}$$

Consequently, we have $s_\lambda \sim \lambda^{-6}$, and the integral in (37) is therefore convergent. Thus in this case the density of created particles is finite for 4-dimensional space-time. In the case of finite volume space ($K = +1$) the total number of created particles is finite also, the Bogolyubov transformations realized by Hilbert-Schmidt operators, and the representations of commutation relations for operators $\overset{*}{b}{}_J^{(\pm)}(\eta)\,, b_J^{(\pm)}(\eta)$ are unitarily equivalent for all time.

In the presented work the metrical, canonical and introduced modified Hamilton operators are investigated. It is shown that the density of particles created in nonstationary homogeneous isotropic space metric is finite only in the case of modified Hamiltonian (39) and the energies of such particles are equal to the oscillator frequency.

The modified energy-momentum tensor (38), introduced above, coincides with the metrical one for a conformal scalar field. In homogeneous isotropic space T_{ik}^{mod} results in the modified Hamiltonian (39) which can be obtained as well as canonical under the special choice of field's variables.

It can be seen that considering a line combination of metrical (5) and canonical (4) tensors we can certainly obtain the modified tensor (38) if the quasiparticles' energy coincides with oscillator frequency. It should be stressed that the metrical energy-momentum tensor cannot be changed to T_{ik}^{mod} in the right-hand sides of Einstein's equations because T_{ik}^{mod} is not covariant conservation. However under the corpuscular interpretation of the nonconformal scalar field and when the diagonalization procedure is used, the modified Hamilton operator constructed by T_{ik}^{mod} is preferable in comparison with the metrical Hamilton operator.

Acknowledgments. The author is grateful to Prof. A.A. Grib for helpful discussions.

References

[1] A.A. Grib, S.G. Mamayev and V.M. Mostepanenko, Vacuum quantum effects in strong fields, Friedmann Laboratory Publishing, St. Petersburg, 1994.

[2] N.D. Birrell and P.C.W. Davies, Quantum fields in curved space, Cambridge University Press, 1982.

[3] V.B. Bezerra, V.M. Mostepanenko and C. Romero, Hamiltonian diagonalization for a nonconformal scalar field in an isotropic gravitational background, *Int. J. Mod. Phys. D* **7** (1998) 249.

[4] I.H. Redmount, Natural vacua in hyperbolic Friedmann-Robertson-Walker spacetimes, *Phys. Rev. D* **60** (1999) 104004.

[5] J. Lindig, Not all adiabatic vacua are physical states, *Phys. Rev. D* **59** (1999) 064011.

[6] M. Bordag, J. Lindig, V.M. Mostepanenko and Yu.V. Pavlov, Vacuum stress-energy tensor of nonconformal scalar field in quasi Euclidean gravitational background, *Int. J. Mod. Phys. D* **6** (1997) 449.

[7] M. Bordag, J. Lindig and V.M. Mostepanenko, Particle creation and vacuum polarization of a non-conformal scalar field near the isotropic cosmological singularity, *Class. Quantum Grav.* **15** (1998) 581.

[8] A.A. Grib and S.G. Mamayev, On field theory in the Friedmann space, *Yad. Fiz.* **10** (1969) 1276. (Engl. trans. in *Sov. J. Nucl. Phys. (USA)* **10** (1970) 722).

[9] A.A. Grib and S.G. Mamayev, Creation of matter in the Friedmann model of the Universe, *Yad. Fiz.* **14** (1971) 800. (Engl. trans. in *Sov. J. Nucl. Phys. (USA)* **14** (1972) 450).

[10] L. Parker, Quantized fields and particle creation in Expanding Universe, I, *Phys. Rev.* **183** (1969) 1057.

[11] F.A. Berezin, The method of second quantization, Academic Press, New York, 1966.

[12] S.A. Fulling, Remarks on positive frequency and Hamiltonians in expanding universes, *Gen. Relativ. Gravit.* **10** (1979) 807.

[13] N.A. Chernikov and E.A. Tagirov, Quantum theory of scalar field in de Sitter spacetime, *Ann. Inst. H. Poincaré A* **9** (1968) 109.

[14] M. Castagnino and R. Ferraro, Observer-dependent quantum vacua in curved space, *Phys. Rev. D* **34** (1986) 497.

[15] M.V. Fedoryuk, Asymptotic methods for linear ordinary differential equations, Nauka, Moscow, 1983.

[16] S.G. Mamayev, V.M. Mostepanenko and V.A. Shelyuto, Dimensional regularization method for quantized fields in non-stationary isotropic spaces, *Theor. Math. Phys.* **63** (1985) 366.

Yu.V. Pavlov
Institute of Mechanical Engineering
Russian Academy of Sciences
61 Bolshoy, V.O.
St. Petersburg, 199178, Russia
e-mail: `pavlov@ipme.ru`

Operator Theory:
Advances and Applications, Vol. 132, 333–346

Boundary Conditions for Singular Perturbations of Self-adjoint Operators

Andrea Posilicano

Abstract. Let $A : D(A) \subseteq \mathcal{H} \to \mathcal{H}$ be an injective self-adjoint operator and let $\tau : D(A) \to \mathcal{X}$, $\mathcal{X}$ a Banach space, be a surjective linear map such that $\|\tau\phi\|_{\mathcal{X}} \le c\,\|A\phi\|_{\mathcal{H}}$. Supposing that Kernel τ is dense in $\mathcal{H}$, we define a family A^{τ}_{Θ} of self-adjoint operators which are extensions of the symmetric operator $A_{|\{\tau=0\}}$. Any ϕ in the operator domain $D(A^{\tau}_{\Theta})$ is characterized by a sort of boundary conditions on its univocally defined regular component ϕ_{reg}, which belongs to the completion of $D(A)$ w.r.t. the norm $\|A\phi\|_{\mathcal{H}}$. These boundary conditions are written in terms of the map τ, playing the role of a trace (restriction) operator, as $\tau\phi_{\text{reg}} = \Theta\, Q_\phi$, the extension parameter Θ being a self-adjoint operator from $\mathcal{X}'$ to $\mathcal{X}$. The self-adjoint extension is then simply defined by $A^{\tau}_{\Theta}\phi := A\,\phi_{\text{reg}}$. The case in which $A\phi = \Psi * \phi$ is a convolution operator on $L^2(\mathbb{R}^n)$, Ψ a distribution with compact support, is studied in detail.

1. Introduction

Let

$$A : D(A) \subseteq \mathcal{H} \to \mathcal{H}$$

be a self-adjoint operator on the complex Hilbert space $\mathcal{H}$ (to prevent any misunderstanding we remark here that all over the paper we will avoid to identify a Hilbert space with its strong dual). As usual $D(A)$ inherits a Hilbert space structure by introducing the scalar product leading to the graph norm

$$\|\phi\|_A^2 := \langle\phi,\phi\rangle_{\mathcal{H}} + \langle A\phi, A\phi\rangle_{\mathcal{H}}\,.$$

Considering then a linear bounded operator

$$\tau : D(A) \to \mathcal{X}\,, \qquad \tau \in \mathsf{B}(D(A),\mathcal{X})\,,$$

$\mathcal{X}$ a complex Banach space, we are interested in describing the self-adjoint extensions of the symmetric operator $A_{|\{\tau=0\}}$. In typical situations A is a (pseudo-)differential operator on $L^2(\mathbb{R}^n)$ and τ is a trace (restriction) operator along some null subset $F \subset \mathbb{R}^n$ (see e.g. [1]–[4], [6]–[8], [16]–[19], [21], [22] and references therein).

Denoting the resolvent set of A by $\rho(A)$, we define $R(z) \in \mathsf{B}(\mathcal{H}, D(A))$, $z \in \rho(A)$, by

$$R(z) := (-A + z)^{-1}$$

and we then introduce, for any $z \in \rho(A)$, the operators $\breve{G}(z) \in \mathsf{B}(\mathcal{H}, \mathcal{X})$ and $G(z) \in \widetilde{\mathsf{B}}(\mathcal{X}', \mathcal{H})$ by

$$\breve{G}(z) := \tau \cdot R(z)\,, \qquad G(z) := C_{\mathcal{H}}^{-1} \cdot \breve{G}(z^*)'\,. \tag{1}$$

Here the prime $'$ denotes both the strong dual space and the (Banach) adjoint map, and $C_{\mathcal{H}}$ indicates the canonical conjugate-linear isomorphism on $\mathcal{H}$ to $\mathcal{H}'$ (the reader is refered to Section 2 below for a list of definitions and notations). As an immediate consequence of the first resolvent identity for $R(z)$ we have (see [**19**, Lemma 2.1])

$$(z - w)\, R(w) \cdot G(z) = G(w) - G(z) \tag{2}$$

and so

$$\forall\, w, z \in \rho(A), \qquad \text{Range}(G(w) - G(z)) \subseteq D(A)\,. \tag{3}$$

In [**19**, Thm. 2.1], by means of a Kreĭn-like formula, and under the hypotheses

$$\tau \text{ is surjective} \tag{h1}$$

$$\text{Range}\, \tau' \cap \mathcal{H}' = \{0\}\,, \tag{h2}$$

we constructed a family A_{Θ}^{τ} of self-adjoint extension of $A_{|\{\tau=0\}}$ by giving its resolvent family. The hypothesis (h1) could be weakened, see [19], but here we prefer to use a simpler framework. In formulating (h2) we used the embedding of $\mathcal{H}'$ into $D(A)' \supseteq \text{Range}\, \tau'$ given by $\varphi \mapsto \langle C_{\mathcal{H}}^{-1}\varphi, \cdot\rangle_{\mathcal{H}}$. Such a hypothesis is then equivalent to the denseness, in $\mathcal{H}$, of the set $\{\tau = 0\}$. Indeed there exists $\ell \in \mathcal{X}'$ such that $\tau'\ell \in \mathcal{H}'$ if and only if there exists $\psi \in \mathcal{H}$ (necessarily orthogonal to Kernel τ) such that for any $\phi \in D(A)$ one has $\langle \psi, \phi \rangle_{\mathcal{H}} = \ell(\tau\phi)$.

The advantage of the formula given in [19] over other approaches (see e.g. [20], [9], [10], [12] and references therein) is its relative simplicity, being expressed directly in terms of the map τ; moreover the domain of definition of A_{Θ}^{τ} can be described, interpreting the map τ as a trace (restriction) operator, in terms of a sort of boundary conditions (see [**19**, Remark 2.10]). In the case $0 \notin \sigma(A)$, $\sigma(A)$ denoting the spectrum of A, this description becomes particularly expressive since $A_{\Theta}^{\tau}\phi$ can be simply defined by the original operator applied to the regular component of ϕ. Such a regular component $\phi_0 \in D(A)$ is univocally determined by the natural decomposition which enter in the definition of $D(A_{\Theta}^{\tau})$ and it has to satisfy the boundary condition

$$\tau\phi_0 = \Theta\, Q_{\phi}\,.$$

More precisely, by (h1), (h2) and by [**19**, Lemma 2.2, Thm. 2.1, Prop. 2.1, Remarks 2.10, 2.12], we have the following

Theorem 1. *Let $A : D(A) \subseteq \mathcal{H} \to \mathcal{H}$ be self-adjoint with $0 \notin \sigma(A)$, let $\tau : D(A) \to \mathcal{X}$ be continuous and satisfy* (h1) *and* (h2)*. If $\Theta \in \widetilde{\mathsf{L}}(\mathcal{X}', \mathcal{X})$ is self-adjoint, $G := G(0)$ and*

$$D(A^\tau_\Theta) := \{\phi \in \mathcal{H} : \phi = \phi_0 + GQ_\phi,\ \phi_0 \in D(A),\ Q_\phi \in D(\Theta),\ \tau\phi_0 = \Theta Q_\phi\}\,,$$

then the linear operator

$$A^\tau_\Theta : D(A^\tau_\Theta) \subseteq \mathcal{H} \to \mathcal{H}\,, \qquad A^\tau_\Theta \phi := A\phi_0$$

is self-adjoint and coincides with A on the kernel of τ; the decomposition entering in the definition of its domain is unique. Its resolvent is given by

$$R^\tau_\Theta(z) := R(z) + G(z) \cdot (\Theta + \Gamma(z))^{-1} \cdot \breve{G}(z)\,, \qquad z \in W^-_\Theta \cup W^+_\Theta \cup \mathbb{C}\backslash\mathbb{R}\,,$$

where

$$\Gamma(z) := \tau \cdot (G - G(z))$$

and

$$W^\pm_\Theta := \{\, \lambda \in \mathbb{R} \cap \rho(A) \ :\ \gamma(\pm\Gamma(\lambda)) > -\gamma(\pm\Theta) \,\}\ .$$

Remark 2. *By (h1) one has $\mathcal{X} \simeq D(A)/\mathrm{Kernel}\,\tau \simeq (\mathrm{Kernel}\,\tau)^\perp$ and so*

$$D(A) \simeq \mathrm{Kernel}\,\tau \oplus \mathcal{X}\,.$$

*This implies that $\mathcal{X}$ inherits a Hilbert space structure and we could then identify $\mathcal{X}'$ with $\mathcal{X}$. Even if this gives some advantage (see [**19**, Remarks 2.13–2.16, Lemma 2.4]) here we prefer to use only the Banach space structure of $\mathcal{X}$.*

The purpose of the present paper is to extend the above theorem to the case in which A is merely injective. Thus, denoting the pure point spectrum of A by $\sigma_{pp}(A)$, we require $0 \notin \sigma_{pp}(A)$ but we do not exclude the case $0 \in \sigma(A)\backslash\sigma_{pp}(A)$; this is a typical situation when A is a differential operator on $L^2(\mathbb{R}^n)$. In order to carry out this program we will suppose that the map τ has a continuos extension to $\widehat{D}(A)$, the completion of $D(A)$ with respect to the norm $\|A\phi\|_\mathcal{H}$ (note that $\widehat{D}(A) = D(A)$ when $0 \notin \sigma(A)$). This further hypothesis allows then to perform the limit $\lim_{\epsilon\to 0} G(i\epsilon) - G(z)$ (see Lemma 3); thus an analogue on the above Theorem 1 is obtained (see Theorem 5). Such an abstract construction is successively specialized to the case in which $A\phi = \Psi * \phi$ is a convolution operator on $L^2(\mathbb{R}^n)$, where Ψ is a distribution with compact support (so that this comprises the case of differential-difference operators). In this situation the results obtained in Theorem 5 can be made more appealing (see Theorem 11). The case in which $A = \Delta : H^2(\mathbb{R}^n) \to L^2(\mathbb{R}^n)$, $n > 4$, and τ is the trace (restriction) operator along a d-set with a compact closure of zero Lebesgue measure, $0 < n - d < 4$, is explicitly studied (see Example 14). Of course, since $-\Delta$ is not negative, in this case one could apply Theorem 1 to $-\Delta + \lambda$, $\lambda > 0$, and then define $-\Delta^\tau_\Theta := (-\Delta + \lambda)^\tau_\Theta - \lambda$. However this alternative definition looks a bit artificial and has the drawback of giving rise to boundary conditions which depend on the arbitrary parameter λ. The starting motivation of this work was indeed the desire to get rid of such a dependence.

2. Definitions and notations

- Given a Banach space $\mathcal{X}$ we denote by $\mathcal{X}'$ its strong dual;
- $\mathsf{L}(\mathcal{X},\mathcal{Y})$, resp. $\widetilde{\mathsf{L}}(\mathcal{X},\mathcal{Y})$, denotes the space of linear, resp. conjugate linear, operators from the Banach space $\mathcal{X}$ to the Banach space $\mathcal{Y}$; $\mathsf{L}(\mathcal{X}) := \mathsf{L}(\mathcal{X},\mathcal{X})$, $\widetilde{\mathsf{L}}(\mathcal{X}) := \widetilde{\mathsf{L}}(\mathcal{X},\mathcal{X})$.
- $\mathsf{B}(\mathcal{X},\mathcal{Y})$, resp. $\widetilde{\mathsf{B}}(\mathcal{X},\mathcal{Y})$, denotes the (Banach) space of bounded, everywhere defined, linear, resp. conjugate linear, operators on the Banach space $\mathcal{X}$ to the Banach space $\mathcal{Y}$.
- Given $A \in \mathsf{L}(\mathcal{X},\mathcal{Y})$ and $\widetilde{A} \in \widetilde{\mathsf{L}}(\mathcal{X},\mathcal{Y})$ densely defined, the closed operators $A' \in \mathsf{L}(\mathcal{Y}',\mathcal{X}')$ and $\widetilde{A}' \in \widetilde{\mathsf{L}}(\mathcal{Y}',\mathcal{X}')$ the are the adjoints of A and $\widetilde{A}$ respectively, i.e.
$$\forall x \in D(A) \subseteq \mathcal{X}, \quad \forall \ell \in D(A') \subseteq \mathcal{Y}', \qquad (A'\ell)(x) = \ell(Ax)\,,$$
$$\forall x \in D(\widetilde{A}) \subseteq \mathcal{X}, \quad \forall \ell \in D(\widetilde{A}') \subseteq \mathcal{Y}', \qquad (\widetilde{A}'\ell)(x) = (\,\ell(\widetilde{A}x)\,)^*$$
where * denotes complex conjugation.
- $J_\mathcal{X} \in \mathsf{B}(\mathcal{X},\mathcal{X}'')$ indicates the injective map (an isomorphism when $\mathcal{X}$ is reflexive) defined by $(J_\mathcal{X}\, x)(\ell) := \ell(x)$.
- A closed, densely defined operator $A \in \mathsf{L}(\mathcal{X}',\mathcal{X}) \cup \widetilde{\mathsf{L}}(\mathcal{X}',\mathcal{X})$ is said to be self-adjoint if $J_\mathcal{X} \cdot A = A'$.
- For any self-adjoint $A \in \mathsf{L}(\mathcal{X}',\mathcal{X}) \cup \widetilde{\mathsf{L}}(\mathcal{X}',\mathcal{X})$ we define
$$\gamma(A) := \inf\{\,\ell(A\ell),\ \ell \in D(A),\ \|\ell\|_{\mathcal{X}'} = 1\,\}\,.$$
- If $\mathcal{H}$ is a complex Hilbert space with scalar product (conjugate linear w.r.t. the first variable) $\langle\cdot,\cdot\rangle$, then $C_\mathcal{H} \in \widetilde{\mathsf{B}}(\mathcal{H},\mathcal{H}')$ denotes the isomorphism defined by $(C_\mathcal{H}\, y)(x) := \langle y, x\rangle$. The Hilbert adjoint of the densely defined linear operator A is then given by $A^* = C_\mathcal{H}^{-1} \cdot A' \cdot C_\mathcal{H}$.
- $\mathcal{F}$ and $*$ denote Fourier transform and convolution respectively.
- $\mathcal{D}'(\mathbb{R}^n)$ denotes the space of distributions and $\mathcal{E}'(\mathbb{R}^n)$ is the subspace of distributions with compact support.
- $H^s(\mathbb{R}^n)$, $s \in \mathbb{R}$, is the usual scale of Sobolev-Hilbert spaces, i.e. $H^s(\mathbb{R}^n)$ is the space of tempered distributions with a Fourier transform which is square integrable w.r.t. the measure with density $(1+|x|^2)^s$. As usual the strong dual of $H^s(\mathbb{R}^n)$ will be represented by $H^{-s}(\mathbb{R}^n)$.
- c denotes a generic strictly positive constant which can change from line to line.

3. Singular perturbations and boundary conditions

Given the injective self-adjoint operator $A : D(A) \subseteq \mathcal{H} \to \mathcal{H}$, we denote by $\widehat{D}(A)$ the Banach space given by the completion of $D(A)$ with respect to the norm
$$\|\phi\|_{(A)} := \|A\phi\|_\mathcal{H}\,.$$

As usual $D(A)$ will be treated as a (dense) subset of $\widehat{D}(A)$ by means of the canonical embedding $\mathcal{J} : D(A) \to \widehat{D}(A)$ which associates to ϕ the set of all Cauchy sequences converging to ϕ.

As in the introduction we consider then a continuous linear map

$$\tau : D(A) \to \mathcal{X},$$

$\mathcal{X}$ is a Banach space, and we will suppose that it satisfies, besides (h1) and (h2), the further hypothesis

$$\|\tau\phi\|_{\mathcal{X}} \le c\,\|A\phi\|_{\mathcal{H}}\,. \tag{h3}$$

By (h3) τ admits an extension belonging to $\mathsf{B}(\widehat{D}(A), \mathcal{X})$; analogously A admits an extension belonging to $\mathsf{B}(\widehat{D}(A), \mathcal{H})$. By abuse of notation we will use the same symbols τ and A to denote these extensions.

Let us now take a sequence $\{\epsilon_n\}_1^\infty \subset \mathbb{R}$ converging to zero. By functional calculus one has

$$\|(-A\cdot R(i\epsilon_n) - I)\phi\|_{\mathcal{H}}^2 = \int_{\sigma(A)} d\mu_\phi(\lambda)\, \frac{\epsilon_n^2}{\lambda^2+\epsilon_n^2}$$

with $\mu_\phi(\{0\}) = 0$ since $0 \notin \sigma_{pp}(A)$. Thus

$$1 \ge \frac{\epsilon_n^2}{\lambda^2+\epsilon_n^2} \longrightarrow 0\,, \qquad \mu_\phi\text{-a.e.}$$

and, by dominated convergence theorem,

$$\mathcal{H}\text{-}\lim_{n\uparrow\infty} -A\cdot R(i\epsilon_n)\phi = \phi\,.$$

So $\{R(i\epsilon_n)\phi\}_1^\infty$ is a Cauchy sequence in $D(A)$ with respect to the norm $\|\cdot\|_{(A)}$. We can therefore define $R \in \mathsf{B}(\mathcal{H}, \widehat{D}(A))$ by

$$R\phi := \widehat{D}\text{-}\lim_{n\uparrow\infty} R(i\epsilon_n)\phi\,,$$

and then $K(z) \in \widetilde{\mathsf{B}}(\mathcal{X}', \widehat{D}(A))$ by

$$K(z) := zR\cdot C(z)\,.$$

Alternatively, using (2), $K(z)$ can be defined by

$$K(z)\phi := \widehat{D}\text{-}\lim_{n\uparrow\infty} (G(i\epsilon_n) - G(z))\,\phi\,.$$

This immediately implies, using (3),

$$\forall\, w, z \in \rho(A), \qquad \text{Range}(K(w) - K(z)) \subseteq D(A)$$

and

$$\forall\, w, z \in \rho(A), \qquad K(w) - K(z) = G(z) - G(w)\,. \tag{4}$$

Also note that

$$-A\cdot K(z) = zG(z)\,. \tag{5}$$

Lemma 3. *The map*

$$\Gamma : \rho(A) \to \widetilde{\mathsf{B}}(\mathfrak{X}', \mathfrak{X}) , \qquad \Gamma(z) := \tau \cdot K(z)$$

satisfies the relations

$$\Gamma(z) - \Gamma(w) = (z - w)\, \breve{G}(w) \cdot G(z) \tag{6}$$

and

$$J_{\mathfrak{X}} \cdot \Gamma(z^*) = \Gamma(z)' , \tag{7}$$

Proof. Since $K(z)$ is the strong limit of $G(\pm i\epsilon_n) - G(z)$, one has

$$\forall \ell \in \mathfrak{X}' , \qquad \Gamma(z)\ell = \lim_{n \uparrow \infty} \hat{\Gamma}_n(z)\ell ,$$

where

$$\hat{\Gamma}_n(z) : \mathfrak{X}' \to \mathfrak{X} , \qquad \hat{\Gamma}_n(z) := \tau \cdot \left(\frac{G(i\epsilon_n) + G(-i\epsilon_n)}{2} - G(z) \right) .$$

Thus $\Gamma(z)$ satisfies (6) and

$$\forall \ell_1, \ell_2 \in \mathfrak{X}' , \qquad \ell_1(\Gamma(z^*)\ell_2) = (\ell_2(\Gamma(z)\ell_1))^*$$

(which is equivalent to (7)) since $\hat{\Gamma}_n(z)$ does (see [**19**, Lemma 2.2]). □

Before stating the next lemma we introduce the following definition: Given $\phi \in \mathcal{H}$ and $\psi \in \widehat{D}(A)$, the writing $\phi = \psi$ will mean that ϕ is in $D(A)$ and $\mathcal{J}\phi = \psi$.

Lemma 4. *Given $\phi \in \mathcal{H}$, $z \in \rho(A)$, suppose there exist $\psi \in \widehat{D}(A)$ and $Q \in \mathfrak{X}'$ such that*

$$\phi - G(z)Q = \psi + K(z)Q . \tag{8}$$

Then the couple (ψ, Q) is unique and z-independent.

Proof. Let (ψ_1, Q_1), (ψ_2, Q_2) both satisfy (8). Then

$$G(z)\,(Q_2 - Q_1) = (\psi_1 - \psi_2) + K(z)\,(Q_1 - Q_2) .$$

By (h2) and the definition of $G(z)$ one has $\operatorname{Range} G(z) \cap D(A) = \{0\}$ and so $Q_1 - Q_2 \in \operatorname{Kernel} G(z)$. But (h1) implies the injectivity of $G(z)$ (see [**19**, Remark 2.1]). Therefore $(\psi_1, Q_1) = (\psi_2, Q_2)$. The proof is then concluded observing that z-independence follows by (4). □

We now can extend Theorem 1 to the case in which A is injective:

Theorem 5. *Let $A : D(A) \subseteq \mathcal{H} \to \mathcal{H}$ be self-adjoint and injective, let $\tau : D(A) \to \mathfrak{X}$ satisfy* (h1)–(h3). *Given $\Theta \in \widetilde{\mathsf{L}}(\mathfrak{X}', \mathfrak{X})$ self-adjoint, let $D(A^\tau_\Theta)$ be the set of $\phi \in \mathcal{H}$ for which there exist $\phi_{\text{reg}} \in \widehat{D}(A)$, $Q_\phi \in D(\Theta)$ such that*

$$\phi - G(z)Q_\phi = \phi_{\text{reg}} + K(z)Q_\phi \qquad \textit{and} \qquad \tau\phi_{\text{reg}} = \Theta\, Q_\phi .$$

Then

$$A^\tau_\Theta : D(A^\tau_\Theta) \subseteq \mathcal{H} \to \mathcal{H} , \qquad A^\tau_\Theta\, \phi := A\, \phi_{\text{reg}} .$$

is a self-adjoint operator which coincides with A on the kernel of τ and its resolvent is given by

$$R^{\tau}_{\Theta}(z) := R(z) + G(z) \cdot (\Theta + \Gamma(z))^{-1} \cdot \breve{G}(z), \qquad z \in W^-_{\Theta} \cup W^+_{\Theta} \cup \mathbb{C}\backslash\mathbb{R}.$$

where

$$\Gamma(z) := \tau \cdot K(z).$$

Proof. For brevity we define

$$\phi_z := \phi - G(z)Q_{\phi}, \qquad \phi \in D(A^{\tau}_{\Theta})$$

and

$$\Gamma_{\Theta}(z) := \Theta + \Gamma(z).$$

Then one has

$$\Gamma_{\Theta}(z)Q_{\phi} = \tau\phi_{\text{reg}} + \tau \cdot K(z)Q_{\phi} = \tau\phi_z.$$

Since $\Gamma(z)$ is a bounded operator satisfying (6) and (7), and Θ is self-adjoint, by (h1) and [**19**, Prop. 2.1, Remark 2.12], $\Gamma_{\Theta}(z)$ has a bounded inverse for any $z \in W^-_{\Theta} \cup W^+_{\Theta} \cup \mathbb{C}\backslash\mathbb{R}$. Therefore

$$Q_{\phi} = \Gamma_{\Theta}(z)^{-1} \cdot \tau\phi_z$$

and

$$D(A^{\tau}_{\Theta}) \subseteq \left\{\phi \in \mathcal{H} \,:\, \phi = \phi_z + G(z) \cdot \Gamma_{\Theta}(z)^{-1} \cdot \tau\,\phi_z,\, \phi_z \in D(A)\right\}.$$

Let us now prove the reverse inclusion.

Given $\phi \in \mathcal{H}$,

$$\phi = \phi_z + G(z) \cdot \Gamma_{\Theta}(z)^{-1} \cdot \tau\,\phi_z, \qquad \phi_z \in D(A),$$

we define $\phi_{\text{reg}} \in \widehat{D}(A)$ by

$$\phi_{\text{reg}} := \phi_z - K(z) \cdot \Gamma_{\Theta}(z)^{-1} \cdot \tau\,\phi_z.$$

Thus one has

$$\begin{aligned}\tau\phi_{\text{reg}} &= \tau\phi_z - \tau \cdot K(z) \cdot \Gamma_{\Theta}(z)^{-1} \cdot \tau\,\phi_z \\ &= \tau\phi_z - \tau\phi_z + \Theta \cdot \Gamma_{\Theta}(z)^{-1} \cdot \tau\,\phi_z = \Theta\, Q_{\phi},\end{aligned}$$

with

$$Q_{\phi} := \Gamma_{\Theta}(z)^{-1} \cdot \tau\,\phi_z.$$

In conclusion

$$D(A^{\tau}_{\Theta}) = \left\{\phi \in \mathcal{H} \,:\, \phi = \phi_z + G(z) \cdot \Gamma_{\Theta}(z)^{-1} \cdot \tau\,\phi_z,\, \phi_z \in D(A)\right\}.$$

Since, by (5),

$$A\phi_{\text{reg}} = A\phi_z - A \cdot K(z)Q_{\phi} = A\phi_z + zG(z)Q_{\phi}$$

we have

$$(-A^{\tau}_{\Theta} + z)\phi = (-A + z)\phi_z.$$

Thus A^{τ}_{Θ} coincides with the operator constructed in [**19**, Thm. 2.1]; therefore, by (h2), this operator is self-adjoint, has resolvent given by $R^{\tau}_{\Theta}(z)$ and is equal to A on the kernel of τ. □

4. Singular perturbations of convolution operators

Let $\Psi \in \mathcal{E}'(\mathbb{R}^n)$. By the Paley-Wiener theorem we know that $\mathcal{F}\Psi$ is a smooth function which is, together with its derivatives of any order, polynomially bounded. Then we define the continuous convolution operator

$$\Psi* : \mathcal{D}'(\mathbb{R}^n) \to \mathcal{D}'(\mathbb{R}^n), \qquad \phi \mapsto \Psi * \phi,$$

and, supposing $\mathcal{F}\Psi$ real-valued, the restriction of this operator to the dense subspace

$$D(\widetilde{\Psi}) := \{\phi \in L^2(\mathbb{R}^n) \ : \ \Psi * \phi \in L^2(\mathbb{R}^n)\}$$

provide us with the self-adjoint convolution operator

$$\widetilde{\Psi} : D(\widetilde{\Psi}) \subseteq L^2(\mathbb{R}^n) \to L^2(\mathbb{R}^n), \qquad \widetilde{\Psi}\phi := \Psi * \phi .$$

Evidently $\widetilde{\Psi}$ is injective if and only if the set of real zeroes of $\mathcal{F}\Psi$ is a null set. From now on we will therefore suppose that $\mathcal{F}\Psi$ is real and has a null set of real zeroes. This implies, since $(\text{Range}\,\widetilde{\Psi})^\perp = \text{Kernel}\,\widetilde{\Psi}^*$, that $\widetilde{\Psi}$ has a dense range and, using (h3), the following lemma becomes then obvious:

Lemma 6. *The map $\tau : D(\widetilde{\Psi}) \to \mathcal{X}$ can be extended to*

$$\mathcal{D}(\widetilde{\Psi}) := \{\varphi \in \mathcal{D}'(\mathbb{R}^n) \ : \ \Psi * \varphi \in L^2(\mathbb{R}^n)\} ,$$

by defining

$$\tau\varphi := \lim_{n\uparrow\infty} \tau\phi_n ,$$

where $\{\phi_n\}_1^\infty \subset D(\widetilde{\Psi})$ is any sequence such that

$$L^2\text{-}\lim_{n\uparrow\infty} \Psi * \phi_n = \Psi * \varphi .$$

Now we will moreover suppose that $\mathcal{F}\Psi$ is slowly decreasing, i.e. (see [11]) we will suppose that there exist $k > 0$ such that for any $\zeta \in \mathbb{R}^n$ we can find a point $\xi \in \mathbb{R}^n$ such that

$$|\zeta - \xi| \le k \log(1 + |\zeta|),$$

$$|\mathcal{F}\Psi(\xi)| \ge (k + |\xi|)^{-k} .$$

By [**11**, Thm. 1] we know that $\Psi* : \mathcal{D}'(\mathbb{R}^n) \to \mathcal{D}'(\mathbb{R}^n)$ is surjective if and only if $\mathcal{F}\Psi$ is slowly decreasing. This is certainly true when $\widetilde{\Psi}$ is a differential operator, i.e. when $\mathcal{F}\Psi$ is a polynomial. The hypotheses we made on $\mathcal{F}\Psi$ permit us to state the following

Lemma 7. *Given Ψ as above, one has the identification*

$$\widehat{D}(\widetilde{\Psi}) \simeq \mathcal{D}(\widetilde{\Psi})/\sim,$$

where

$$\varphi_1 \sim \varphi_2 \quad \iff \quad \Psi * \varphi_1 = \Psi * \varphi_2 .$$

This identification is given by the isometric maps which to the equivalence class of Cauchy sequences $[\{\phi_n\}_1^\infty] \in \widehat{D}(\widetilde{\Psi})$ *associates the equivalence class of distributions* $[\varphi] \in \mathcal{D}(\widetilde{\Psi})/\sim$ *such that*

$$L^2\text{-}\lim_{n\uparrow\infty} \Psi * \phi_n = \Psi * \varphi .$$

Proof. Given $[\{\phi_n\}_1^\infty] \in \widehat{D}(\widetilde{\Psi})$, the sequence $\{\Psi * \phi_n\}_1^\infty$ is a Cauchy one in $L^2(\mathbb{R}^n)$ and so it converges to some $f \in L^2(\mathbb{R}^n)$. Then, by [**11**, Thm. 1], there exists $\varphi \in \mathcal{D}(\widetilde{\Psi})$ such that $\Psi * \varphi = f$. Conversely let $\varphi \in \mathcal{D}(\widetilde{\Psi})$; since $\widetilde{\Psi}$ has a dense range there exists a (unique in $\widehat{D}(\widetilde{\Psi})$) sequence $\{\phi_n\}_1^\infty \subset D(\widetilde{\Psi})$ such that $\Psi * \phi_n$ converges in $L^2(\mathbb{R}^n)$ to $\Psi * \varphi$. □

Defining

$$\widetilde{\mathcal{D}}'(\mathbb{R}^n) := \mathcal{D}'(\mathbb{R}^n)/\sim$$

and then the sum of $\phi \in L^2(\mathbb{R}^n) \subset \mathcal{D}'(\mathbb{R}^n)$ plus $\psi = [\varphi] \in \widehat{D}(\widetilde{\Psi}) \simeq \mathcal{D}(\widetilde{\Psi})/\sim\subseteq \widetilde{\mathcal{D}}'(\mathbb{R}^n)$ by

$$\phi + \psi := [\phi + \varphi] \in \widetilde{\mathcal{D}}'(\mathbb{R}^n),$$

we can introduce the linear operator

$$G : \mathcal{X}' \to \widetilde{\mathcal{D}}'(\mathbb{R}^n), \qquad G := G(z) + K(z) .$$

According to Lemma 4, Theorem 5 and the definition of G, for any $\phi \in D(\widetilde{\Psi}^\tau_\Theta)$ we can give the unique decomposition

$$\phi = \phi_{\text{reg}} + GQ_\phi .$$

Thus we can define $D(\widetilde{\Psi}^\tau_\Theta)$ as the set of $\phi \in L^2(\mathbb{R}^n)$ for which there exists $Q_\phi \in D(\Theta)$ such that

$$\phi - GQ_\phi =: \phi_{\text{reg}} \in \widehat{D}(\widetilde{\Psi})$$

and

$$\tau\phi_{\text{reg}} = \Theta\, Q_\phi .$$

Lemma 8. *The definition of* G *is* z*-independent and*

$$\forall \ell \in \mathcal{X}', \qquad G\ell = [G_*\ell] ,$$

where

$$G_* : \mathcal{X}' \to \mathcal{D}'(\mathbb{R}^n),$$

is any conjugate linear operator such that

$$-\Psi * G_*\ell = \tau^*\ell . \tag{9}$$

Here $\tau^* : \mathcal{X}' \to \mathcal{D}'(\mathbb{R}^n)$ *is defined by*

$$\tau^*\ell(\varphi) := (\,\ell(\tau\varphi^*)\,)^* , \qquad \varphi \in C_0^\infty(\mathbb{R}^d) .$$

Proof. z-independence is an immediate consequence of (4). By the definition of G and by (5) there follows

$$-\Psi * G_* \ell = (-\Psi * + z) G(z) \ell$$

and the proof is concluded by the relation

$$(-\Psi * + z) \cdot G(z) = \tau^* \tag{10}$$

which can be obtained proceeding as in [**19**, Remark 2.4]. □

Remark 9. *By [11, Thm. 1], as $\mathcal{F}\Psi$ is slowly decreasing, the equation (9) is always resoluble; in particular, denoting the fundamental solution of $-\Psi *$ by $\mathcal{G}$, when the convolution $\mathcal{G} * \tau^* \ell$ is well defined (e.g. when $\tau^*\ell \in \mathcal{E}'(\mathbb{R}^n)$), one has*

$$G_* : \mathcal{X}' \to \mathcal{D}'(\mathbb{R}^n)\,, \qquad G_* \ell = \mathcal{G} * \tau^* \ell\,.$$

Analogously, denoting the fundamental solution of $-\Psi * + z$ by $\mathcal{G}_z$, one has

$$G(z) : \mathcal{X}' \to L^2(\mathbb{R}^n)\,, \qquad G(z)\ell = \mathcal{G}_z * \tau^* \ell\,.$$

Remark 10. *Note that $\widetilde{\Psi}\phi_{\text{reg}} = \Psi * \varphi$ and $\tau\phi_{\text{reg}} = \tau\varphi$ for any $\varphi \in \mathcal{D}(\widetilde{\Psi})$ such that $\phi_{\text{reg}} = [\varphi]$. Here we implicitly used the extension given in Lemma 6 and the identification given in Lemma 7. This also implies that $\Gamma(z)$ in Lemma 3 can be rewritten as*

$$\Gamma(z) = \tau \cdot (G_* - G(z))\,.$$

By Remark 9, when the convolution is well defined, one can also write

$$\Gamma(z)\ell = \tau\left((\mathcal{G} - \mathcal{G}_z) * \tau^* \ell\right)\,.$$

In conclusion, by making use of the previous lemmata and remarks, we can restate Theorem 5 in the following way:

Theorem 11. *Let $\Psi \in \mathcal{E}'(\mathbb{R}^n)$ with $\mathcal{F}\Psi$ real-valued, slowly decreasing and having a null set of real zeroes, let $\widetilde{\Psi} : D(\widetilde{\Psi}) \subseteq L^2(\mathbb{R}^n) \to L^2(\mathbb{R}^n)$, $\widetilde{\Psi}\phi := \Psi * \phi$, let $\tau : D(\widetilde{\Psi}) \to \mathcal{X}$ satisfy* (h1)–(h3). *Given $\Theta \in \widetilde{\mathsf{L}}(\mathcal{X}', \mathcal{X})$ self-adjoint, let $D(\widetilde{\Psi}^\tau_\Theta)$ be the set of $\phi \in L^2(\mathbb{R}^n)$ for which there exists $Q_\phi \in D(\Theta)$ such that*

$$\phi - G_* Q_\phi =: \varphi_{\text{reg}} \in \mathcal{D}(\widetilde{\Psi})$$

and

$$\tau\varphi_{\text{reg}} = \Theta\, Q_\phi\,.$$

Then

$$\widetilde{\Psi}^\tau_\Theta : D(\widetilde{\Psi}^\tau_\Theta) \subseteq L^2(\mathbb{R}^n) \to L^2(\mathbb{R}^n)\,, \qquad \widetilde{\Psi}^\tau_\Theta\, \phi := \Psi * \varphi_{\text{reg}}\,,$$

is a self-adjoint operator which coincides with $\widetilde{\Psi}$ on the kernel of τ and its resolvent is given by

$$R^\tau_\Theta(z) := R(z) + G(z) \cdot (\,\Theta + \Gamma(z)\,)^{-1} \cdot \breve{G}(z)\,, \qquad z \in W^-_\Theta \cup W^+_\Theta \cup \mathbb{C}\backslash\mathbb{R}\,,$$

where

$$\Gamma(z) := \tau \cdot (G_* - G(z))\,.$$

Remark 12. *The boundary conditions and the operators* $\Gamma(z)$ *and* $\widetilde{\Psi}^{\tau}_{\Theta}$ *appearing in the previous theorem are independent of the choice of the representative (see Lemma 8)* G_* *entering in the definition of* φ_{reg}. *Indeed any different choice will not change the equivalence class to which* φ_{reg} *belongs, and both* τ *and* $\widetilde{\Psi}$ *do not depend on the representative in such a class (see Remark* 10*).*

Remark 13. *Proceeding as in [**19**, Remark 2.4] one can give the following alternative definition of* $\widetilde{\Psi}^{\tau}_{\Theta}$ *where only* $Q_\phi \in D(\Theta)$ *appears:*

$$\widetilde{\Psi}^{\tau}_{\Theta}\phi := \Psi * \phi + \tau^* Q_\phi \,.$$

This is an immediate consequence of identity (10).

Example 14. *Let us consider the case* $A = \Delta : H^2(\mathbb{R}^n) \to L^2(\mathbb{R}^n)$. *Obviously* A *is an injective convolution operator, thus we can apply to it the previous theorem.*

A Borel set $F \subset \mathbb{R}^n$ *is called a d-set,* $d \in (0, n]$, *if*

$$\exists c_1,\, c_2 > 0 \;:\; \forall x \in F,\ \forall r \in (0,1), \qquad c_1 r^d \le \mu_d(B_r(x) \cap F) \le c_2 r^d \,,$$

where μ_d *is the d-dimensional Hausdorff measure and* $B_r(x)$ *is the closed n-dimensional ball of radius r centered at the point x (see [**14**, §1.1, Chap. VIII]). Examples of d-sets are d-dimensional Lipschitz submanifolds and (when d is not an integer) self-similar fractals of Hausdorff dimension d (see [**14**, Chap. II, Example 2]). Moreover a finite union of d-sets which intersect on a set of zero d-dimensional Hausdorff measure is a d-set.*

In the case $0 < n - d < 4$ *we take as the linear operator* τ *the unique continuous surjective (thus (h1) holds true) map*

$$\tau_F : H^2(\mathbb{R}^n) \to B^{2,2}_{\alpha}(F) \,, \qquad \alpha = 2 - \frac{n-d}{2}$$

such that, for μ_d*-a.e.* $x \in F$,

$$\tau_F\phi(x) \equiv \left\{\phi^{(j)}_F(x)\right\}_{|j|<\alpha} = \left\{\lim_{r\downarrow 0} \frac{1}{\lambda_n(r)} \int_{B_r(x)} dy\, D^j\phi(y)\right\}_{|j|<\alpha} \,, \tag{11}$$

where $j \in \mathbb{Z}^n_+$, $|j| := j_1 + \cdots + j_n$, $D^j := \partial_{j_1} \cdots \partial_{j_n}$ *and* $\lambda_n(r)$ *denotes the n dimensional Lebesgue measure of* $B_r(x)$. *We refer to [**14**, Thms. 1 and 3, Chap. VII] for the existence of the map* τ_F; *obviously it coincides with the usual evaluation along F when restricted to smooth functions. The definition of the Besov-like (actually Hilbert, see Remark* 1*) space* $B^{2,2}_{\alpha}(F)$ *is quite involved and we will not reproduce it here (see [**14**, §2.1, Chap. V]). In the case* $0 < \alpha < 1$ *(i.e.* $2 < n - d < 4$*) things simplify and* $B^{2,2}_{\alpha}(F)$ *can be defined (see [**14**, §1.1, Chap. V]) as the Hilbert space of* $f \in L^2(F; \mu_F)$ *having finite norm*

$$\|f\|^2_{B^{2,2}_{\alpha}(F)} := \|f\|^2_{L^2(F)} + \int_{|x-y|<1} d\mu_F(x)\, d\mu_F(y)\, \frac{|f(x) - f(y)|^2}{|x-y|^{d+2\alpha}} \,,$$

where μ_F *denotes the restriction of the d-dimensional Hausdorff measure* μ_d *to the set F. When* $\alpha > 1$ *and F is a generic d-set the functions* $\phi^{(j)}_F \in L^2(F; \mu_F)$

are not uniquely determined by $\phi_F^{(0)}$; *contrarily we may then identify* $\{\phi_F^{(j)}\}_{|j|<\alpha}$ *with the single function* $\phi_F^{(0)}$. *This is possible when F preserves Markov's inequality (see [***14***, §2, Chap. II]). Sets with such a property are closed d-sets with* $d > n-1$ *(see [***14***, Thm. 3, §2.2, Chap. II]), a concrete example being e.g. the boundary of von Koch's snowflake domain in* $\mathbb{R}^2$ *(a d-set with* $d = \log 4/\log 3$*, see [24]). If F has some additional differential structure then* $B_\alpha^{2,2}(F) \simeq H^\alpha(F)$*, where* $H^\alpha(F)$ *denotes the usual (fractional) Sobolev-Slobodeckiĭ space. Some known cases where* $B_\alpha^{2,2}(F) \simeq H^\alpha(F)$ *(for any value of* $\alpha > 0$*) are the following:*
- F is the graph of a Lipschitz function $f : \mathbb{R}^d \to \mathbb{R}^{n-d}$ *(see [***5***, §20]);*
- F is a bounded manifold of class C_γ, $\gamma > \min(3, \max(1,\alpha))$*, i.e. F has an atlas where the transition maps are of class* C^k, $k < \gamma \le k+1$*, and have derivatives of order less or equal to k which satisfy Lipschitz conditions of order* $\gamma - k$ *(see [13] for the case* $\gamma > \max(1,\alpha)$ *and see [***5***, §24] for the case* $\gamma > 3$ *);*
*- F is a connected complete Riemannian manifold with positive injectivity radius and bounded geometry, in particular a connected Lie group (see [***23***, §7.4.5, §7.6.1]).*

Supposing now that F has a compact closure, let χ *be a smooth function with a compact support B such that* $\chi = 1$ *on F. Then by Sobolev's inequality one has (from now on* $n > 4$*),*

$$\begin{aligned}\|\tau_F\phi\|^2_{B_\alpha^{2,2}(F)} = \|\tau_F\chi\phi\|^2_{B_\alpha^{2,2}(F)} &\le c\,\|\chi\phi\|^2_{H^2(\mathbb{R}^n)}\\ &\le c\,(\,\|\phi\|^2_{L^2(B)} + \|\nabla\phi\|^2_{L^2(B)} + \|\Delta\phi\|^2_{L^2(B)}\,)\\ &\le c\,(\,\|\phi\|^2_{L^{\frac{2n}{n-4}}(B)} + \|\nabla\phi\|^2_{L^{\frac{2n}{n-2}}(B)} + \|\Delta\phi\|^2_{L^2(\mathbb{R}^n)}\,)\\ &\le c\,\|\Delta\phi\|_{L^2(\mathbb{R}^n)}\,,\end{aligned}$$

and so τ_F *satisfies (h3). Moreover, since*

$$\mathcal{D}(\Delta) = \{\varphi \in \mathcal{D}'(\mathbb{R}^n) \,:\, \Delta\phi \in L^2(\mathbb{R}^n)\} \subseteq H^2_{\text{loc}}(\mathbb{R}^n)$$

and $\bar F$ *is supposed to be compact, the extension of* τ_F *to* $\mathcal{D}(\Delta)$ *is again defined by* (11). *Denoting the dual of* $B_\alpha^{2,2}(F)$ *by* $B_{-\alpha}^{2,2}(F)$ *(the space* $B_{-\alpha}^{2,2}(F)$ *can be explicitly characterized in the case* $0 < \alpha < 1$ *or when F preserves Markov's inequality, see* [15]*), hypothesis (h2) is equivalent to* $\tau_F'\ell \notin L^2(\mathbb{R}^n)$ *for any* $\ell \in B_{-\alpha}^{2,2}(F)\setminus\{0\}$*, where* $\tau_F'\ell \in H^{-2}(\mathbb{R}^n)$ *is defined by*

$$\tau_F'\ell(\phi) := \ell(\tau_F\phi)\,.$$

Therefore, as the support of $\tau_F'\ell$ *is given by* $\bar F$*, (h2) is certainly verified when* $\bar F$ *has zero Lebesgue measure. Considering the fundamental solution of* $-\Delta$*, given by*

$$\mathcal{G}(x) = \frac{1}{(n-2)\sigma_n}\,\frac{1}{|x|^{n-2}}\,,$$

σ_n *the measure of the unitary sphere in* $\mathbb{R}^n$*, the convolution* $\mathcal{G}*\tau_F'\ell$ *is a well-defined distribution as* τ_F' *is in* $\mathcal{E}'(\mathbb{R}^n)$. *Therefore, by Lemma 8 we can choose* G_* *to be the map*

$$G_* : B_{-\alpha}^{2,2}(F) \to \mathcal{D}'(\mathbb{R}^n)\,, \qquad G_*\ell := \mathcal{G} * \tau_F^*\ell\,.$$

Thus, by the previous theorem (and Remark 13*), supposing that the d-set F has a compact closure of zero Lebesgue measure, given any self-adjoint operator* $\Theta \in \widetilde{\mathsf{L}}(B^{2,2}_{-\alpha}(F), B^{2,2}_{\alpha}(F))$, $\Theta \in \mathsf{L}(B^{2,2}_{\alpha}(F))$ *if one uses the identification* $B^{2,2}_{-\alpha}(F) \simeq B^{2,2}_{\alpha}(F)$*, we have then the self-adjoint operator*

$$\Delta^F_\Theta\, \phi := \Delta\varphi_{\text{reg}} \equiv \Delta\phi + \tau^*_F Q_\phi\,,$$

where

$$\phi = \varphi_{\text{reg}} + \mathcal{G} * \tau^*_F Q_\phi\,, \qquad \varphi_{\text{reg}} \in \mathcal{D}(\Delta)\,, \quad Q_\phi \in D(\Theta)\,,$$

and

$$\left\{\lim_{r\downarrow 0} \frac{1}{\lambda_n(r)} \int_{B_r(x)} dy\, D^j\,(\phi - \mathcal{G} * \tau^*_F Q_\phi)\,(y)\right\}_{|j|<\alpha} = \Theta\, Q_\phi(x)\,,$$

$|j| = 0$ *if F preserves Markov's inequality or* $B^{2,2}_{\alpha}(F) \simeq H^\alpha(F)$*. When* $F = M$*, M a compact Riemannian manifold, a natural choice for Θ is given by* $\Theta = (-\Delta_{LB})^{-\alpha}$*, where Δ_{LB} denotes the Laplace-Beltrami operator. The case in which* $A = (\Delta - \lambda) : H^2(\mathbb{R}^3) \to L^2(\mathbb{R}^3)$, $\lambda > 0$ *(note that here* $0 \notin \sigma(A)$*, thus Theorem* 1 *directly applies), and F is a plane circle, is treated, without giving boundary conditions, in* [16] *(also see [***19***, Example* 3.2*] for connections with Birman-Kreĭn-Vishik theory).*

References

[1] S. Albeverio, J.R. Fenstad, R. Høegh-Krohn, W. Karwowski, T. Lindstrøm: Schrödinger Operators with Potentials Supported by Null Sets. Published in *Ideas and Methods in Mathematical Analysis, Vol. II.* New-York, Cambridge: Cambridge Univ. Press 1991

[2] S. Albeverio, F. Gesztesy, R. Høegh-Krohn, H. Holden: *Solvable Models in Quantum Mechanics.* Berlin, Heidelberg, New York: Springer-Verlag 1988

[3] S. Albeverio, V. Koshmanenko: On Schrödinger Operators Perturbed by Fractal Potentials. *Rep. Math. Phys.* **45** (2000), 307–326

[4] S. Albeverio, P. Kurasov: *Singular Perturbations of Differential Operators.* Cambridge: Cambridge Univ. Press 2000

[5] O.V. Besov, V.P. Il'in, S.M. Nikol'skiĭ: *Integral Representations of Functions and Embedding theorems. Vol. II.* Washington: Winston & Sons 1979

[6] J.F. Brasche: Generalized Schrödinger Operators, an Inverse Problem in Spectral Analysis and the Efimov Effect. Published in *Stochastic Processes, Physics and Geometry.* Teaneck, New Jersey: World Scientific Publishing 1990

[7] J.F. Brasche, A. Teta: Spectral Analysis and Scattering Theory for Schrödinger Operators with an Interaction Supported by a Regular Curve. Published in *Ideas and Methods in Mathematical Analysis, Vol. II.* New-York, Cambridge: Cambridge Univ. Press 1991

[8] S. E. Cheremshantsev: Hamiltonians with Zero-Range Interactions Supported by a Brownian Path. *Ann. Inst. Henri Poincaré,* **56** (1992), 1–25

[9] V.A. Derkach, M.M. Malamud: Generalized Resolvents and the Boundary Value Problems for Hermitian Operators with Gaps. *J. Funct. Anal.*, **95** (1991), 1–95

[10] V.A. Derkach, M.M. Malamud: The Extension Theory of Hermitian Operators and the Moment Problem. *J. Math. Sciences*, **73** (1995), 141–242

[11] L. Ehrenpreis: Solution of Some Problems of Division IV. *Am. J. Math.* **82** (1960), 522–588

[12] F. Gesztesy, K.A. Makarov, E. Tsekanovskii: An Addendum to Krein's Formula. *J. Math. Anal. Appl.* **222** (1998), 594–606

[13] A. Jonsson: Besov Spaces on Submanifolds of $\mathbb{R}^n$. *Analysis* **8** (1988), 225–269

[14] A. Jonsson, H. Wallin: Function Spaces on Subsets of $\mathbb{R}^n$. *Math. Reports* **2** (1984), 1–221

[15] A. Jonsson, H. Wallin: The Dual of Besov Spaces on Fractals. *Studia Math.* **112** (1995), 285–298

[16] W. Karwowski, V. Koshmanenko, S. Ôta: Schrödinger Operators Perturbed by Operators Related to Null Sets. *Positivity* **2** (1998), 77–99

[17] A.N. Kochubei: Elliptic Operators with Boundary Conditions on a Subset of Measure Zero. *Funct. Anal. Appl.* **16** (1978), 137–139

[18] Y.V. Kurylev: Boundary Condition on a Curve for a Three-Dimensional Laplace Operator. *J. Sov. Math.* **22** (1983), 1072–1082

[19] A. Posilicano: A Kreĭn-like Formula for Singular Perturbations of Self-Adjoint Operators and Applications. *J. Funct. Anal.* **183** (2001), 109–147

[20] Sh.N. Saakjan: On the Theory of Resolvents of a Symmetric Operator with Infinite Deficiency Indices. *Dokl. Akad. Nauk Arm. SSR* **44** (1965), 193–198 [in Russian]

[21] Yu.G. Shondin: On the Semiboundedness of δ-Perturbations of the Laplacian Supported by Curves with Angle Points. *Theor. Math. Phys.* **105** (1995), 1189–1200

[22] A. Teta: Quadratic Forms for Singular Perturbations of the Laplacian. *Publ. RIMS Kyoto Univ.* **26** (1990), 803–817

[23] H. Triebel: *Theory of Function Spaces II.* Basel, Boston, Berlin: Birkhäuser 1992

[24] H. Wallin: The Trace to the Boundary of Sobolev Spaces on a Snowflake. *Manuscripta Math.* **73** (1991), 117–125

Andrea Posilicano
Dipartimento di Scienze
Università dell'Insubria
I-22100 Como, Italy
e-mail: posilicano@mat.unimi.it

Operator Theory:
Advances and Applications, Vol. 132, 347–359

Asymptotical and Topological Constructions in Hydrodynamics

Andrei I. Shafarevich

Abstract. For the Navier–Stokes equations, we study asymptotic solutions (solitary vortices and vortical threads) localized in a small neighborhood of a point or a curve in the three-dimensional space. A relationship is established between such solutions and topological invariants of vector fields on the plane and of Liouville foliations of three-dimensional space into two-dimensional tori. Equations describing vortical threads and solitary vortices are also obtained. One of the variables in these equations ranges in the corresponding topological invariant (namely, the Reeb graph or the Fomenko invariant). For the vortical thread equations, we obtain an infinite series of integral identities that become conservation laws in the vanishing viscosity limit.

1. Introduction

Asymptotic theory of equations of hydrodynamics is an old and highly developed branch of mathematical physics and theory of nonlinear PDE's. Among first and most famous results is the classical Prandtl theory of boundary layers; in the simplest version this theory describes the behavior of a two-dimensional incompressible fluid with small viscosity near horizontal rectilinear boundary.

The main term of the velocity field of the fluid (more precisely, the main term of the formal asymptotic solution to the steady Navier–Stokes equations) satisfies the Prandtl equations

$$w\frac{\partial w}{\partial s}+a\frac{\partial w}{\partial I}+\frac{\partial \pi}{\partial s}=\nu\frac{\partial^2 w}{\partial I^2},\quad \frac{\partial w}{\partial s}+\frac{\partial a}{\partial I}=0. \tag{1}$$

where s and $y=\varepsilon I$ are horizontal and vertical coordinates on the plane, w and εa denote horizontal and vertical components of the velocity field of the fluid near the boundary, $\pi(s)$ is a given function (pressure of the external flow), $\nu = const.$

Topological hydrodynamics is a modern and rapidly developing area of mathematics (see, e.g. [1]); one of the most famous results of this theory describe geometrical properties of Euler equations for ideal fluids. Namely, consider the Euler

The author is grateful to S. Albeverio, A.V. Bolsinov, O. Chkhetiani, V.G. Danilov, S.Yu. Dobrokhotov, A.T. Fomenko, A.G. Kulikovskii, V.P. Maslov, G.A. Omel'yanov, and A.I. Shnirel'man for useful discussions.

equations for an ideal incompressible fluid

$$\frac{\partial u}{\partial t} + (u, \nabla)u + \nabla p = 0, \quad (\nabla, u) = 0, \tag{2}$$

where $u(x,t)$ is a vector field in a certain domain in $\mathbf{R}^3$ or $\mathbf{R}^2$ (velocity field of the fluid), $p(x,t)$ is a scalar function (pressure). These equations can be treated as Euler equations on the Lie algebra of divergence-free vector fields [1]. Remind the definition of this type of equations. Let G be a Lie group with left-invariant (or right-invariant) Riemannian metric; equations of geodesics on G can be rewritten as a system of equations on the Lie algebra. These equations admit remarkable properties; particularly, orbits of the coadjoint representation of G form invariant manifolds for Euler equations and the restriction of the corresponding vector field to the orbit is Hamiltonian with respect to the Kirillov form. These general properties can be used to study equations (2); for example, in 2D-case functions of the form

$$\int Q(\operatorname{curl} u)dx$$

are constant on the orbits of coadjoint representations for any smooth function $Q : \mathbf{R} \to \mathbf{R}$ (recall that the curl of a 2D-vector field is a scalar function). This leads to the infinite number of first integrals for 2D Euler equations of an ideal incompressible fluid

$$\frac{\partial}{\partial t} \int Q(\operatorname{curl} u)dx = 0. \tag{3}$$

Another famous result of topological hydrodynamics describes the behavior of trajectories of steady flows [1]. Namely, consider equations (2) in 3D-case and let $u(x)$ be the steady (i.e. time-independent) smooth solution of these equations. Denoting by Ω the curl of the vector field u, one can rewrite (2) in the form

$$u \times \Omega = \nabla B, \quad B = \frac{u^2}{2} + p,$$

and applying curl to these equations, one obtains

$$[u, \Omega] = 0,$$

where $[,]$ denotes the commutator of vector fields. Latter formulae describe the topology of trajectories of u. Consider a smooth compact nonsingular level surface of the function B (Bernoulli surface); a pair of commuting fields u and Ω are tangent to this surface, so, according to the Liouville theorem, each connected component of the surface is diffeomorphic to the 2-torus and the motion on this torus, defined by the vector field u, is conditionally periodic.

The main aim of this paper is to describe certain connections between asymptotical and geometrical parts of mathematical hydrodynamics as well as between the first one and topological theory of integrable Hamiltonian systems. We illustrate these connections using two concrete examples of asymptotic problems — the problems of description of stretched and solitary vortices in an incompressible fluid. A more detailed description of these problems as well as the proofs of the propositions of this paper, can be found in [2]–[5].

2. Description of stretched vortices via equations on Reeb graphs

The stationary velocity field $u(x)$ of an incompressible fluid (a vector field in R^3) satisfies the Navier–Stokes equations

$$(u, \nabla)u + \nabla p = \mu \Delta u, \quad (\nabla, u) = 0. \tag{4}$$

Here $p(x)$ is a scalar function (pressure) and $\mu > 0$ is the viscosity coefficient. We consider the solution of these equations, consisting of a smooth vector field $V(x)$ (external flow) and a perturbation, localized in a small neighborhood of some (*a priori* unknown) curve γ in the three-dimensional space. The characteristic width ε of this neighborhood (compared with the characteristic scale of the external field V) is the small parameter in our problem. A perturbation of this kind can naturally be treated as a narrow jet (or vortical thread) flowing in the fluid. One can show (e.g., see [6]) that such rapidly varying solutions can exist only if the viscosity coefficient is sufficiently small, more precisely, if $\mu = O(\varepsilon^2)$; otherwise, viscosity destroys the jet. Hence in what follows we set $\mu = \varepsilon^2\nu, \quad \nu = O(1)$.

Let $x = R(s)$ be the natural parameterization of γ (that is, s is the arc length). Consider some neighborhood $\mathcal{G}'$ of γ. We assume that $\mathcal{G}'$ is independent of ε (but sufficiently small). In this neighborhood, we introduce coordinates $\{s, y_1, y_2\}$, where y are euclidean coordinates in the normal planes to γ.

We seek an asymptotic solution u, p of the three-dimensional Navier–Stokes equations (4) in a smaller neighborhood $\mathcal{G} \subset \mathcal{G}'$ in the form

$$u = V(x) + U\Big(\frac{y}{\varepsilon}, s\Big) + U_1\Big(\frac{y}{\varepsilon}, x\Big) + \dots, \quad P = P(x) + \pi\Big(\frac{y}{\varepsilon}, s\Big) + \pi_1\Big(\frac{y}{\varepsilon}, x\Big) + \dots, \tag{5}$$

where the functions $U(z,s), \pi(z,s)$ smoothly depend on all arguments for $z \in \mathbf{R}^2$, $s \in \gamma$, and moreover, $U(z,s), \pi(z,s) \to 0$ as $|z| \to \infty$ no slower than $O(|z|^{-2})$. Here and in the following, by z we denoted the two-dimensional vector of "dilated" coordinates in the normal plane to γ: $z = y/\varepsilon$. Outside $\mathcal{G}$, we multiply the functions U and π by a smooth cutoff function vanishing outside $\mathcal{G}'$. Below we obtain and study equations governing the vector field U. Denote by $u(z,s)$ the projection of the field $U(z,s) + V|_\gamma$ onto the normal planes (to γ) and by $w(z,s)$ the projection of the same field to the corresponding tangent line. We look at v as a vertical field (section) of the two-dimensional vector bundle ρ on γ (its fibers are z-planes — "stretched" normal planes to γ), and on w as a function on this bundle.

Assertion 1. *Let the vector field* (5) *satisfy the Navier–Stokes equations* (4) mod $O(1)$ *as* $\varepsilon \to 0$. *Then*

a) *The two-dimensional vector-field* v *satisfies steady Euler equations.*

b) *The function* w *is constant along the trajectories of the field* v:

$$(v, \nabla_z)v + \nabla_z \pi = 0, \quad (\nabla_z, v) = 0, \quad (v, \nabla_z)w = 0. \tag{6}$$

Throughout the following, we consider solutions v of the Euler equations on the plane such that v satisfies the following conditions.

1) The field v has only finitely many singular points, they all are nondegenerate (since v is divergence-free, it follows that these points are either saddles or centers), and each separatrix contains exactly one singular point.
2) Almost all trajectories of v are closed.

Evolution of the vortex along the curve (i.e. dependence of v, w on the "slow" coordinate s) can be described in terms of topological invariants of the two-dimensional divergence-free vector field v. The most principle describes a phase portrait of v, i.e. the number and positions of rest points and separatrices of this field. In order to represent this information it is convenient to consider the set formed by the trajectories of v (i.e. the quotient space of the z-plane by the trajectories). This topological space is a graph Γ; vertices of this graph correspond to the rest points of v and edges correspond to domains on the z-plane, smoothly foliated by closed trajectories. Two vertices are connected with an edge if the corresponding rest points can be connected by a continuous curve which does not intersect separatrices. So each vertex of G is either of degree 1 (corresponding rest point is of center type) or of degree 3 (corresponding rest point is of saddle type). Moreover, Γ does not contain cycles (this is due to the fact that the Euclidean plane is simply connected) and therefore forms a tree.

Consider the parameterization of this graph by the area; namely, to each point of the arbitrary edge of Γ (i.e. closed trajectory l of the vector field v) we associate a number $I = (2\pi)^{-1} \int_l z_2 dz_1 = (2\pi)^{-1} S(l)$, where S is the area of the domain bounded by l.

Remark. Each smooth divergence-free vector field on the plane is Hamiltonian (the Hamiltonian is traditionally referred to as the *stream function* in hydrodynamics). The trajectories of v are level curves of the stream function, and the singular points of v are critical points of the Hamiltonian. The graph Γ (the set of level curves of the stream function) is called the *Reeb graph* (e.g., see [7]); this is an object well known in Morse theory (and generalizations of Morse theory). The parameter I is the action variable of the corresponding Hamiltonian system.

The Bernoulli integral

$$B_0(z) = \frac{v^2}{2} + \pi_0$$

is an important characteristic of a vector field v satisfying the Euler equations (6). The function B_0 is constant on the trajectories of v, that is, is well defined and continuous on Γ. If (v, π_0) is a smooth solution of the Euler equations (we consider only such solutions), then B_0 is a smooth function of I on each edge of Γ. Thus, to each smooth solution of the Euler equations (6) we can assign a pair consisting of a parameterized graph Γ and a continuous function B_0 on Γ that is smooth on the edges of Γ.

Conjecture 1. *Let Γ be a parameterized binary tree (i.e. a tree with vertices of degrees 1 and 3 only) and B_0 be a continuous function defined on Γ and smooth on the edges. There exists an open (in an appropriate sense) subset of the set of pairs (Γ, B_0), such that for each pair in this subset there exists a smooth solution (v, π_0) of the Euler equations for which Γ is the quotient of the z-plane with respect to the trajectories of v, and B_0 is the Bernoulli integral.*

Numerous motivations for this hypothesis were suggested by different authors (see, e.g. [1], [5], [8]). Here we mention only one of them.

A variational principle for the Euler equations [1]. Let $\mathcal{K}$ be a space of divergence-free vector fields v that can be obtained from each other (or from some given field v_0) with the use of area-preserving diffeomorphisms of the plane. (The fields are assumed to be sufficiently rapidly decaying at infinity, and the diffeomorphisms are assumed to stabilize at infinity.) We consider the energy integral $J = \int_{R^2} v^2(z)dz$ on $\mathcal{K}$. The following assertion holds. *The Euler equations* (6) *are the equations of extremals for the energy functional J on the space $\mathcal{K}$.*

Clearly, the parameterized graph Γ is invariant with respect to area-preserving diffeomorphisms. Moreover, there is another natural invariant on the edges of Γ, namely, the frequency (or the period) of a closed trajectory. Therefore, the set of spaces $\mathcal{K}$ of vector fields obtained from each other by canonical diffeomorphisms of the plane is parameterized (at least) by a graph and a function on the graph. Hence one can expect that these "parameters" are inherited by the solutions of the corresponding variational problems. In other words, if the minimum of the functional J is attained on a space $\mathcal{K}$ corresponding to some values of the invariants, then one can naturally expect that the minimum also exists on spaces corresponding to close values of the invariants.

Recall that w is constant on the trajectories of v and hence is also well defined and continuous on Γ and smooth on the edges. Thus, we eventually obtain the parameterization of solutions of system (6) by two functions B_0 and w on the graph Γ. In what follows, we obtain equations that determine the dependence of w and B_0 on the slow variable s.

Let us write down the equations obtained by comparing the coefficient in front of ε^0 in the expression obtained by the substitution of the functions (5) into the Navier–Stokes equations (4) with zero. We have

$$\begin{aligned} -(v, \nabla_z)w_1 - (v^1, \nabla_z)w &= H(z, s), \\ -(v, \nabla_z)v^1 - (v^1, \nabla_z)v - \nabla_z \pi_1 &= F(z, s), \\ -(\nabla_z, v^1) &= G(z, s). \end{aligned} \tag{7}$$

Here the functions H, F, G can be expressed explicitly in terms of $U(z, s)$ and its derivatives.

The above system consists of the linearized equations (6) with nonzero right-hand sides. The dependence of B_0 and w on the coordinate s is determined by the compatibility conditions for this system. Prior to writing out these conditions, let us give the corresponding assertion about the cokernel of the linearized Euler operator. Consider the equations

$$(v, \nabla_z)\xi - \frac{\partial v^*}{\partial z}\xi = \nabla_z \chi, \quad (\nabla_z, \xi) = 0, \tag{8}$$

which are formally adjoint to the linearized Euler equations in the space of divergence-free vector fields. Here ξ is a two-dimensional vector field and χ is a scalar function.

Theorem 1. *([2]) Let $v(z)$ be a smooth solution of (6) with properties 1) and 2). Then every divergence-free vector field ξ commuting with $v(z)$ satisfies (8).*

Remark. Divergence-free vector fields, commuting with v obviously form an infinite-dimensional vector space. This subspace in the cokernel of the linearized Euler operator is generated by the variation of the arbitrary function $B_0(z)$ and can be interpreted as a space of functions on Γ. Indeed, if a divergence-free field ξ commutes with v, then it has the same trajectories as v, since the Hamiltonians of ξ and v are in involution. Hence the Hamiltonian b of ξ is a continuous function on the Reeb graph Γ and is smooth on the edges. Thus, the set of divergence-free fields commuting with v can be interpreted as the set of functions b of one variable I ranging in Γ. This is an argument in favor of our conjecture about the parameterization of the set of solutions to the Euler equations.

Let us now give the compatibility conditions for system (7).

Theorem 2. *([2]) Suppose that there exists a smooth solution (v^1, w_1, π_1) of system (7). Then the functions B_0, w satisfy the equations*

$$\begin{aligned} &w\tfrac{\partial w}{\partial s} + a\tfrac{\partial w}{\partial I} + \langle \pi_s \rangle = \nu \tfrac{\partial}{\partial I} D^2 \tfrac{\partial}{\partial I} w, \\ &\tfrac{\partial w}{\partial s} + \tfrac{\partial a}{\partial I} = 0, \\ &w\tfrac{\partial B}{\partial s} + a\tfrac{\partial B}{\partial I} = \nu\Big(\tfrac{\partial}{\partial I} D^2 \tfrac{\partial}{\partial I} B - \Lambda^2\Big), \end{aligned} \tag{9}$$

where

$$B = B_0 + \frac{w^2}{2}, \quad \langle \pi_s \rangle = \frac{1}{2\pi} \oint_l \frac{\partial}{\partial s}\Big(\pi_0 - \frac{1}{2} V^2|_g\Big) d\varphi,$$

the functions D^2, Λ^2 can be expressed in terms of v, w, and l is a point of Γ (i.e. a closed trajectory of v).

Remark. The above conditions arise from the conditions of orthogonality of the right-hand sides of (7) to the subspace, indicated in Theorem 2.1, of the cokernel of the linearized Euler operator. Note that to write out the conditions of orthogonality to the cited infinite-dimensional space, we have chosen a "basis in the rigging of that subspace" in a special way. Roughly speaking, the basis consists of delta-like fields supported on trajectories of v. It is this choice of a basis that permits one to reduce finding the orthogonality conditions to averaging along the trajectories of v.

The parameter I in (9) ranges on the edges of Γ. It appears that solvability conditions for (7) imply additional relations on the behavior of the functions w, B in the edges of the graph Γ. Namely, the following assertion holds

Theorem 3. *Let a smooth solution of* (7) *exist. Then the functions w and B are continuous on Γ, the function a vanishes at all 1-vertices and the Kirchhoff conditions*

$$\begin{gathered} a_1 + a_2 = a_3, \quad (D^2 \tfrac{\partial B}{\partial I})_1 + (D^2 \tfrac{\partial B}{\partial I})_2 = \left(D^2 \tfrac{\partial B}{\partial I}\right)_3, \\ \left(D^2 \tfrac{\partial w}{\partial I}\right)_1 + \left(D^2 \tfrac{\partial w}{\partial I}\right)_2 = \left(D^2 \tfrac{\partial w}{\partial I}\right)_3, \end{gathered} \tag{10}$$

of the theory of electric circuits are satisfied at each 3-vertex, where the subscripts 1 *and* 2 *indicate the limits of the corresponding functions at the vertex along incoming edges and the subscript* 3 *indicates the limit along the outgoing edge* (*the graph is directed in such a way that the parameter I increases*).

Remark. From the viewpoint of differential equations on graphs (e.g., see [9]), the Kirchhoff conditions (10) are a special case of the so-called α-smoothness conditions that are specified at each vertex of a graph and relate the derivatives of the unknown function on the edges incident to the vertex.

Remark. The first pair of equations in (9) is similar to the Prandtl equations (1). However, the term $\langle \pi_s \rangle$ in the first equation in (9) cannot be treated as a given function in general; it is related to the Bernoulli function B of transverse circulations in the vortex. This makes equations (9) similar to the so-called Prandtl equations with self-induced pressure (e.g., see [10]).

Remark. The right-hand sides of (9) are due to the nonzero viscosity of the fluid. Apart from the coefficient ν, they contain the function D^2, which in general depends on the vortex velocity field itself (that is, on the Bernoulli function B). Similar functions occur in equations describing turbulent flows (the so-called turbulent viscosity coefficients; e.g., see [11]). We note, however, that in contrast to the above-mentioned equations, equations (9) (and, in particular, the expression for the coefficient D^2) are obtained as a mathematical consequence of the fact that the Navier–Stokes equations possess a solution describing a vortical thread rather than from physical considerations.

Despite the fairly complicated form of the equations (9), we have been able to establish a number of interesting properties of these equations. Namely, the solutions satisfy infinitely many integral identities (similar to the energy balance equation), which become conservation laws for $\nu = 0$ (that is, in an ideal fluid). Furthermore, equations (9) have two additional conservation laws that are valid for arbitrary values of ν.

Theorem 4. ([4]) *Suppose that the functions v, π_0, and w satisfy equations* (6), *and the corresponding functions w and B, together with some function a that is smooth on the edges of Γ, satisfy equations* (9) *on Γ. Next, suppose that a, B, and*

w satisfy the Kirchhoff conditions (10) *and moreover, the functions w, v and π_0 decay as $I \to \infty$ at the rate of $O(I^{-1-\delta})$, $\delta > 0$. Then the following identities hold:*

$$\frac{\partial}{\partial s}\int_\Gamma (w - |V||_\gamma)dI = 0, \quad \frac{\partial}{\partial s}\int_\Gamma \Big(w^2 - V^2|_\gamma - \Big\langle\frac{v^2}{2}\Big\rangle\Big)dI = 0, \tag{11}$$

$$\frac{\partial}{\partial s}\int_\Gamma wQ(B)dI = -\nu\int_\Gamma \Big[Q''(B)D^2\Big(\frac{\partial B}{\partial I}\Big)^2 + \Lambda^2 Q'(B)\Big]dI. \tag{12}$$

Here $Q(t)$ is an arbitrary smooth function vanishing at zero and

$$\Big\langle\frac{v^2}{2}\Big\rangle = \frac{1}{2\pi}\oint_l \frac{v^2}{2}d\varphi.$$

Remark. Relations (11) are conservation laws for (9) (the longitudinal momentum of the vortex and the *difference* between the energy of the longitudinal flow and the energy of transverse circulations are conserved). Relations (12) resemble energy balance equations: the right-hand side describes viscous dissipation. For $\nu = 0$ (that is, if the fluid is ideal or the viscosity coefficient is much less than ε^2), relations (12) also become the conservation laws

$$\frac{\partial}{\partial s}\int_\Gamma wQ(B)dI = 0.$$

These equations resemble the well-known series of conservation laws (3) for the two-dimensional nonstationary Euler equations. It is possible that the existence of the integrals (12) is related to certain algebraic properties of the vortical thread equations (9) close to those of the Euler equations (see Sect.1).

3. An asymptotic description of point (solitary) vortices. Vortex equations and Fomenko invariants

In this section, we consider solutions to the Navier–Stokes equations localized in a small neighborhood of a moving point, representing solitary vortices in an incompressible fluid. The motion of such a vortex is described by the Cauchy problem for the nonstationary Navier–Stokes equations

$$\begin{aligned} &\tfrac{\partial u}{\partial t} + (u,\nabla)u + \nabla p = \varepsilon^2\nu\Delta u, \quad (\nabla, u) = 0,\\ &u|_{t=0} = V_0(x) + v_0\Big(\tfrac{x-y}{\varepsilon}\Big), \quad (\nabla, V_0) = 0, \quad (\nabla, v_0) = 0, \end{aligned} \tag{13}$$

where $V_0(x)$ is a smooth vector field bounded together with all derivatives (the external flow) and the vector function $v_0(z)$ (the solitary vortex) belongs to the Schwartz space. We consider initial vortices of a special form (the conditions on v_0 are given below). We seek an asymptotic solution of this problem in the form

$$\begin{aligned} u &= V(x,t) + \varepsilon V_1(x,t) + \cdots + v(\tfrac{x-R(t)}{\varepsilon}, t) + \varepsilon v^1(\tfrac{x-R(t)}{\varepsilon}, x, t, \varepsilon) + \cdots,\\ P &= P(x,t) + \varepsilon P_1(x,t) + \cdots + \pi_0(\tfrac{x-R(t)}{\varepsilon}, t) + \varepsilon\pi_1(\tfrac{x-R(t)}{\varepsilon}, x, t, \varepsilon) + \cdots \end{aligned} \tag{14}$$

where $V(x,t)$, $V_1(x,t)$, $P(x,t)$, $P_1(x,t)$, $v(z,t)$ and $\pi_0(z,t)$ are smooth vector fields and functions, and moreover, v, v^1, π, and π_1 decay sufficiently rapidly as $|z| \to \infty$.

We assume that $R(t)$ is a three-dimensional vector function such that the curve $x = R(t)$ is a nonsingular trajectory of the vector field $V(x,t)$. Let us substitute (14) into (13) and consider the resulting equations outside some neighborhood (independent of ε) of the point $R(t)$. By equating the coefficient of ε^0 with zero, we obtain the Euler equations

$$\frac{\partial V}{\partial t} + (V,\nabla)V + \nabla P = 0, \quad (\nabla, V) = 0, \quad V|_{t=0} = V_0(x) \tag{15}$$

for the field $V(x,t)$. Throughout the following, we assume that $V(x,t)$ is a smooth solution of this problem for $t \in [0,T]$.

Now consider equations (13) in a neighborhood of the point $R(t)$. By killing terms of the order of ε^{-1} and by taking into account the fact that $R(t)$ is a trajectory of V, we obtain the three-dimensional stationary Euler equations

$$(v,\nabla_z)v + \nabla_z \pi_0 = 0, \quad (\nabla_z, v) = 0 \tag{16}$$

for the field $v(z,t)$, just as in Section 2.

By analogy with Section 2, we arrive at the question as to how the solutions of these equations can be parameterized. Note that the variational principle is valid both in the two- and in the three-dimensional case, and hence it is natural to expect that the solutions of (16) are parameterized by topological invariants of three-dimensional divergence-free vector fields with respect to volume-preserving diffeomorphisms of the three-dimensional space. The problem of constructing such invariants is apparently beyond human possibilities for arbitrary divergence-free fields, but the topological properties of solutions of the Euler equations in general position are quite special, which permits one to indicate some of the desired topological invariants.

Recall that the topology of stationary Euler fields is similar to the topology of completely integrable Hamiltonian systems with two degrees of freedom: every field of this kind specifies a foliation of the three-dimensional flow region into two dimensional tori, similar to the Liouville foliation of a constant energy surface (see Section 1). This permits one to introduce an invariant of such fields by analogy with the Fomenko invariant (molecule [7]) arising in the theory of Hamiltonian systems. Namely, let Γ be the set of compact level surfaces of the Bernoulli integral $B = v^2/2 + \pi_0$. We consider fields v such that Γ is a graph, and moreover, the interior points of edges of Γ correspond to nonsingular level surfaces (that is, two-dimensional tori).

Clearly, the graph Γ corresponding to a field v remains the same if we transform the field by a volume-preserving diffeomorphism of the three-dimensional space. Moreover, just as in Section 2, the graph bears a natural parameterization invariant with respect to such diffeomorphisms. Specifically, to each point of an arbitrary edge of the graph (that is, a torus invariant with respect to the flows generated by v and $\Omega = \operatorname{curl}_z v$), we assign the number I equal to the volume of the corresponding solid torus divided by $4\pi^2$. The parameterized graph Γ is a topological invariant of v. By analogy with Section 2, there are additional invariants on this graph, namely, the frequency functions of v. Specifically, consider an arbitrary edge of Γ; it defines a region foliated into two-dimensional tori in the three-dimensional space. In this region, we choose a basis of cycles on each invariant torus such that the basis smoothly depends on the torus and denote the corresponding angular coordinates on the tori by $\varphi = \{\varphi_1, \varphi_2\}$, $\varphi_j \in [0, 2\pi]$. In the coordinates (I, φ) (which, by analogy with the theory of integrable Hamiltonian systems, will be called the action-angle variables; see [1]), the field v has the form

$$v = \omega_1(I)\frac{\partial}{\partial\varphi_1} + \omega_2(I)\frac{\partial}{\partial\varphi_2};$$

the functions $\omega_j(I)$ (the frequencies) are invariant with respect to diffeomorphisms. Thus we have defined a set of invariants of v, consisting of a parameterized graph Γ and a pair of smooth functions (ω_1, ω_2) on the edges of Γ.

Conjecture 2. *There exists an open (in an appropriate sense) subset of the set of triples* $(\Gamma, \omega_1, \omega_2)$, *where* Γ *is a parameterized graph and the* ω_j *are smooth functions on the edges of* Γ, *such that for each triple in this open subset there exists a smooth solution* (v, π_0) *of the Euler equations* (16) *for which* Γ *is the set of compact level surfaces of the Bernoulli integral and the functions* ω_j *are the frequencies of* v *on the corresponding tori.*

Now let us kill the terms of the order of ε^0 occurring after the substitution of (14) into (13). By analogy with Section 2, we obtain

$$-(v, \nabla_z)v^1 - (v^1, \nabla_z)v - \nabla_z\pi_1 = \Phi(z,t), \qquad (\nabla_z, v^1) = 0, \tag{17}$$

where

$$\Phi = \frac{\partial v}{\partial t} + \frac{\partial V}{\partial x}v + (\frac{\partial V}{\partial x}z, \nabla_z)v - \nu\Delta_z v$$

and the matrix $\partial V/\partial x$ of derivatives of the field V is computed at $x = R(t)$. System (17) consists of the nonhomogeneous linearized Euler equations. Just as in the vortical thread problem, the solvability conditions for these equations result in equations on Γ that must be satisfied by the frequencies $\omega_j(I,t)$. These conditions are related to the structure of the cokernel of the linearized Euler operator, that is, the set of solutions ξ of the equations

$$(v, \nabla_z)\xi - \frac{\partial v^*}{\partial z}\xi = \nabla_z\chi, \qquad (\nabla_z, \xi) = 0. \tag{18}$$

Here ξ is a three-dimensional vector field and χ is a scalar function.

Theorem 5. *([5]) Let* $v(z)$ *be a smooth solution of* (16). *Then every divergence-free field* ξ *commuting with both* $v(z)$ *and* $\Omega(z) = \operatorname{curl}_z v$ *satisfies* (18).

Remark. Divergence-free vector fields indicated in the theorem obviously form an infinite-dimensional linear space. This subspace of the cokernel of the linearized Euler operator is generated by the variation of arbitrary functions ω_1 and ω_2 and can be interpreted as a space of pairs of functions on Γ. Indeed, let us introduce the above-described action-angle variables (I, φ) in an arbitrary domain of the three-dimensional z-space such that the domain is smoothly foliated into tori $B = c$. In these coordinates, the fields v and Ω have the form $\omega_1(I)\partial/\partial\varphi_1 + \omega_2(I)\partial/\partial\varphi_2$ and $\lambda_1(I)\partial/\partial\varphi_1 + \lambda_2(I)\partial/\partial\varphi_2$, respectively, where $\omega_j(I)$ and $\lambda_j(I)$ are the frequencies of these fields on the invariant torus corresponding to the parameter I. On the other hand, every divergence-free vector field ξ commuting with v and Ω has the form $b_1(I)\partial/\partial\varphi_1 + b_2(I)\partial/\partial\varphi_2$ in these coordinates (here the $b_j(I)$ are arbitrary smooth functions). In other words, the Liouville tori of v are also invariant manifolds of ξ, and so ξ is determined by its frequency vector (b_1, b_2). Thus, the set of divergence-free fields commuting with v and Ω can be interpreted as the set of pairs (b_1, b_2) of functions of one variable I ranging on Γ. This is an argument in favor of the conjecture made about the parameterization of the set of solutions of the Euler equations in the preceding subsection.

The above described structure of the cokernel of the linearized Euler equations results in the following solvability conditions for system (17).

Theorem 6. ([5]) *Suppose that there exists a smooth solution of system* (17). *Then the two-dimensional vector* $\omega(I,t) = (\omega_1, \omega_2)$ *satisfies the following equations:*

$$\frac{\partial\omega}{\partial t} + Q\frac{\partial\omega}{\partial I} + R\omega = \nu\Big(D^2\frac{\partial^2\omega}{\partial I^2} + M\frac{\partial\omega}{\partial I} + Z\omega\Big). \tag{19}$$

Here the scalar function D^2 *and the entries of the* 2×2 *matrices* Q, R, M *and* Z *can be expressed via* v *and the coefficients of the Euclidean metric in the three-dimensional space of the fast variables* z *and their derivatives at points of an invariant torus* T *(which corresponds to the value* I *of the action variable).*

The Kirchhoff conditions at the vertices of Γ. Each edge of the graph Γ corresponds to a domain of the three-dimensional space smoothly foliated into two-dimensional tori. Consider an arbitrary vertex of this graph; it corresponds to a singular level set of the Bernoulli integral B. Consider a neighborhood of this singular set. Suppose that this neighborhood is a 3-atom in the sense of [7], that is, it is a compact connected orientable manifold with boundary equipped with the structure of a Seifert fibration. The boundary of the manifold consists of several tori. Since they all lie in R^3, it follows that exactly one of these tori is exterior (that is, the entire manifold lies in the interior of this torus), and all other tori are interior (the atom lies in the exterior of each of these tori).

Remark. In the topological theory of integrable Hamiltonian systems with two degrees of freedom [7], neighborhoods of singular leaves of the Liouville foliation are 3-atoms by virtue of the Morse–Bott condition for the additional integral and the topological stability requirement. Needless to say, we can also impose similar conditions on the Euler field v. However, it is not essential to us what conditions guarantee the "atomic" structure of singular sets, and so we assume that this structure is known *a priori.*

Now let us consider the corresponding interior vertex of the graph. It has several incident edges, of which exactly one is outgoing, and the others are incoming (the graph is oriented in the increasing direction of the parameter I that is equal to the volume of the corresponding solid torus). Namely, the outgoing edge corresponds to the exterior boundary torus of the atom, and the incoming edges correspond to interior boundary tori. Let (η, ζ) be admissible bases of cycles on the nonsingular tori of the atom (see [7]). We recall the definition. The first base cycle η on a Liouville torus is a fiber of the Seifert fibration; as the torus tends to the critical circle, the fiber also tends to this circle and goes around it once or twice depending on whether the Seifert fibration on the atom is trivial or not. The second cycle ζ supplements η to a base; if the Seifert fibration is trivial, then it is constructed as the intersection of the Liouville torus with a two-dimensional surface that is a global section of the Seifert fibration. If the fibration is not trivial, then the second cycle is defined as the intersection of the torus with a global section of the "trivialized" Seifert fibration obtained from the atom by deleting small neighborhoods of critical circles. Moreover, the section must satisfy the following condition: the intersection ζ' of this section with the boundary of a tubular neighborhood of the critical circle is related to the fiber η of the Seifert fibration and the vanishing cycle κ by the formula $\eta = \kappa + 2\zeta$.

The choice of a base of cycles determines the frequencies of the field v on the nonsingular tori of the Liouville foliation that belong to a given atom. Let ω_0 be the frequency corresponding to the cycle η (the fiber of the Seifert fibration) and ω' the frequency corresponding to ζ. Clearly, ω_0 is a smooth function on the atom and is constant on each Liouville torus. (Note that ω' is not smooth; the asymptotics of this function at the passage through the critical circle can be found, e.g., in [7].)

Theorem 7. *([5]) The functions $D^2\frac{\partial B}{\partial I}$, and $D^2\frac{\partial \omega_0}{\partial I}$ satisfy the Kirchhoff conditions*

$$\left(D^2\frac{\partial \omega_0}{\partial I}\right)_{out} = \left(D^2\frac{\partial \omega_0}{\partial I}\right)_{in}, \quad \left(D^2\frac{\partial B}{\partial I}\right)_{out} = \left(D^2\frac{\partial B}{\partial I}\right)_{in}$$

at each interior vertex of Γ (that is, a vertex of degree > 1 corresponding to a 3-atom). Here the subscript "out" indicates the limit of the corresponding function along the outgoing edge, and the subscript "in" indicates the sum of limits along incoming edges.

Corollary 1. *Suppose that all vertices of Γ represent 3-atoms. For each edge of Γ, let us choose a base of cycles on the Liouville tori lying in the corresponding domain of the three-dimensional space, thus defining the frequencies ω_1 and ω_2 of the field v. Next, we choose an admissible base (η, ζ) of cycles on the Liouville tori of each atom. To each vertex of Γ of degree $m+1$, we assign the sequence $A_0, A_1, \dots, A_m$ of integral two-dimensional vectors defined as follows. Consider the jth edge incident to that vertex (the index j runs from 0 to m, where the value $j = 0$ corresponds to the outgoing edge). On the Liouville tori corresponding to this edge, there is a base (γ_1, γ_2) of cycles generating the frequencies (ω_1, ω_2) and* *the base* (η, ζ). *These bases can be expressed via each other. Let*

$$\gamma_1 = A_j^1 \eta + A_j^2 \zeta$$

(note that the same coefficients express the frequency ω_0 *via* ω_1, ω_2*). Now the Kirchhoff conditions for* ω_0 *at the vertices can be rewritten in terms of* ω_1, ω_2*; namely, the preceding theorem gives at each vertex of* Γ *relations*

$$D^2 \frac{\partial}{\partial I}(A_0, \omega^0) = \sum_{j=1}^{m} D^2 \frac{\partial}{\partial I}(A_j, \omega^j),$$

where ω^j *is the frequency vector* $\omega = \{\omega_1, \omega_2\}$ *on the* j*th edge.*

References

[1] V.I. Arnold, B.A. Khesin: Topological methods in hydrodynamics. *Springer-Verlag* (1996).

[2] A.I. Shafarevich: Differential Equations on Graphs, Describing Asymptotic Solutions to the Navier–Stokes Equations, Localized in a Small Vicinity of a Curve. *Dif. Uravneniya* **34** (1998), No 8, 1119–1130.

[3] A.I. Shafarevich: Generalized Prandtl–Maslov Equations on Graphs, Describing Stretched Vortices in an Incompressible Fluid. *Dokl. RAN* **358** (1998), No 6, 752–756.

[4] A.I. Shafarevich: Asymptotic Description of Vortex Filaments in an Incompressible Fluid. *J. Appl. Math., Mech.* **64** (2000), No 2, 255–265.

[5] A.I. Shafarevich: Asymptotic Solutions of the Navier–Stokes Equations and Topological Invariants of Vector Fields. *Russian J. of Math. Phys.* **7** (2000), No 4, 401–449.

[6] V.P. Maslov: Asymptotic Methods for Solving Pseudo-Differential Equations. *Nauka, Moscow,* (1987).

[7] A.V. Bolsinov, A.T. Fomenko: Integrable Hamiltonian Sysytems. Topology. Classification, *Faktorial, Izhevsk* (1999).

[8] Moffatt H.K.: Magnetostatic equilibria and analogous Euler flows of arbitrary complex topology. *J.Fluid Mech.*, **159** (1985), 359–378.

[9] O.M. Penkin, Yu.V. Pokornyi: Some Qualitative Properties of Equations on a One-Dimensional CW-Complex. *Matem. Zametki*, **59**, (1996), No 5, 777–780.

[10] V.I. Zhuk, O.S. Ryzhov: On a Boundary Layer with Self-Induced Pressure on a Moving Surface. *Dokl. AN SSSR*, **248**, (1979), No 2, 314–318.

[11] L.D. Landau, E.M. Lifshitz: Hydrodynamics. *Nauka, Moscow* (1984)

Andrei I. Shafarevich
Dept. of Mech. and Math.
Moscow State University
Moscow, 119899, RUSSIA
e-mail: shafar@@mech.math.msu.su

Operator Theory:
Advances and Applications, Vol. 132, 361–385

Spectral Aspects of a Class of Differential Operators

Harold S. Shapiro

Abstract. Recently Gisli Masson and Boris Shapiro initiated the study of a class of differential operators defined as follows: for each monic polynomial Q, let T_Q be the operator mapping f to $(d/dz)^k f$, where $k = \deg Q$ and f belongs, say, to the set E of all entire functions of a complex variable. They showed (Q being now fixed) that for each non-negative integer m there is a unique monic polynomial f_m of degree m which is an eigenfunction of T_Q. Moreover the corresponding eigenvalue is positive, and depends only on m and k, but not on the specific choice of Q with degree k. They studied the location of the zeros of f_m. The goal of this paper is to study natural spectral questions arising from their findings. Our main results are: 1) In the space E there are no eigenfunctions of T_Q other than the $\{f_m\}$. 2) In a certain (explicitly given) Hilbert space H of entire functions all T_Q with $\deg Q = k$ are similar to one and the same self-adjoint positive operator. (This "explains" both why the eigenvalues are the same for all Q of given degree, and why they are real and positive). A closely related result is that $\{f_n\}$ do not merely span H, but are a Riesz basis for this space. These results are proved using standard tools of perturbation theory. In a concluding section attention is drawn to an operator in a sense dual to T_Q, whose eigenfunctions (which now are entire, but not in general polynomials) relate in interesting ways to $\{f_n\}$.

1. Introduction

Let

$$Q(z) = z^k + a_{k-1}z^{k-1} + \cdots + a_0 \tag{1.1}$$

be a polynomial with complex coefficients and E the set of entire functions of a complex variable z. Then by T_Q we denote the linear differential operator on E defined by

$$T_Q : f \mapsto D^k(Qf); (D = d/dz). \tag{1.2}$$

If E has its usual topology (whereby "convergence" is understood as uniform convergence on compact sets), T_Q is a continuous map of E to E. Of course, (1.2) is meaningful also for f in wider classes, e.g. space of all functions holomorphic in a given domain, but in the present study we shall focus on the action of T_Q restricted to E, as well as certain Hilbert spaces properly contained in E.

The study of the general class of operators T_Q was proposed by Boris Shapiro, and an account of some results, especially concerning location of the zeroes of its eigenfunctions, is in [MS]. We shall first develop a general framework, after which we more easily can formulate our main goals and results.

(1.3) **Theorem.** *T_Q is a continuous linear bijection of E.*

Proof. Clearly T_Q is a continuous linear map of E to E. It is also injective, because for f in E $T_Qf = 0$ implies Qf is in $\mathcal{P}_{k-1}$ where

(1.4) $\mathcal{P}_n$ denotes the set of polynomials of degree at most n,

thus $f = Qf/Q$ is an entire function which is $O(1/|z|)$ at infinity, and hence zero.

To show the surjectivity, we shall actually construct the inverse operator, that is, given $g \in E$ find f such that

$$D^k(Qf) = g. \tag{1.5}$$

Let G in E denote any k-fold primitive of g, i.e. $D^kG = g$. (Thus, G is determined modulo an additive polynomial of degree at most $k-1$; we shall see that the particular choice of G plays no role.) Then, (1.5) can be written $D^k(G - Qf) = 0$ so that

$$L := G - Qf \text{ is in } \mathcal{P}_{k-1}. \tag{1.6}$$

Thus, L is the (unique) Lagrange interpolating polynomial, which interpolates G at the zeroes (counting multiplicities) of Q. For greater specificity, let us denote L as $L(G;Q)$. Then

$$f = \frac{G - L(G;Q)}{Q} \tag{1.7}$$

is in E, and $T_Qf = g$. Since we know that T_Q is injective on E, f does not depend on the choice of the primitive G. This can of course be seen directly from (1.7): if G_1, G_2 are primitives of g, then $G_1 - G_2 \in \mathcal{P}_{k-1}$, so $G_1 - G_2 = L(G_1 - G_2; Q)$, hence $G_1 - L(G_1;Q) = G_2 - L(G_2;Q)$. We shall shortly give another interesting interpretation of the inversion formula (1.7).

To summarize: in order to compute $T_Q^{-1}g$ one performs the following steps

a) Compute k-fold primitive G of g;
b) Form the Lagrange interpolant $L(G;Q)$;
c) $f = T_Q^{-1}g$ is then given by (1.7).

Thus, we have proven Theorem 1.2. □

Another expression for $T_Q^{-1}g$, useful for some purposes, will now be presented. First, let us review the classical notion of divided difference operators (a good source for this material is [G]; we use somewhat different notations).

(1.8) **Definition.** *For f in E and a in $\mathbf{C}$, $[f;a]$ denotes the entire function $(f(z) - f(a)/(z-a)$.*

Clearly $f \mapsto [f;a]$ defines a continuous linear map of E, whose kernel is the constant functions.

Let Γ denote a positively oriented circular path in the complex plane, surrounding the point z. Then

$$f(z) = \frac{1}{2\pi i}\int_\Gamma \frac{f(\zeta)}{\zeta - z}dz,$$

so for a inside Γ we have

$$\begin{aligned} f(z) - f(a) &= \frac{1}{2\pi i}\int_\Gamma f(\zeta)\left[\frac{1}{\zeta - z} - \frac{1}{\zeta - a}\right]d\zeta \\ &= \frac{1}{2\pi i}\int_\Gamma \frac{z-a}{(\zeta - z)(\zeta - a)}f(\zeta)d\zeta, \end{aligned}$$

hence

(1.9) $$[f;a] = \frac{1}{2\pi i}\int_\Gamma \frac{f(\zeta)}{(\zeta - z)(\zeta - a)}d\zeta.$$

Now, $[f;a]$ is an entire function, and we may in turn form its divided difference with respect to a point b. We shall denote this by $[f;a,b]$ (here it is allowed that $b = a$). Using (1.9) and assuming that also b is surrounded by Γ, we get

$$(z-b)[f;a,b] = \frac{1}{2\pi i}\int_\Gamma \frac{f(\zeta)}{\zeta - a}\left[\frac{1}{\zeta - z} - \frac{1}{\zeta - b}\right]d\zeta$$

which after simplifying becomes

(1.10) $$[f;a,b] = \frac{1}{2\pi i}\int_\Gamma \frac{f(\zeta)}{(\zeta - a)(\zeta - b)(\zeta - z)}d\zeta.$$

From this we see (what is not tautological) that divided difference operators commute. Thus, if we have any k complex numbers z_1, z_2, ..., z_k we can successively form (in any order) the divided differences of f with respect to these numbers. Denoting the result by $[f;z_1,z_2,\ldots,z_k]$ we have

(1.11) $$[f;z_1,z_2,\ldots,z_k] = \frac{1}{2\pi i}\int_\Gamma \frac{f(\zeta)}{Q(\zeta)(\zeta - z)}d\zeta$$

where

(1.12) $$Q(\zeta) = (\zeta - z_1)(\zeta - z_2)\cdots(\zeta - z_1)$$

and Γ is a circle (or, of course, any oriented contour) surrounding ζ, $z_1,\ldots,z_k$ with winding number $+1$. For brevity, we shall also write

(1.13) $$[f;Q] := [f;z_1,z_2,\ldots,z_k]$$

where Q is given by (1.12).

Let us now denote $f - [f;Q]Q$ by h. Then, using (1.11)

$$h(z) = \frac{1}{2\pi i}\int_\Gamma \frac{f(\zeta)}{\zeta - z}\left[1 - \frac{Q(z)}{Q(\zeta)}\right] d\zeta,$$

Γ being as before. The integrand equals

$$\frac{f(\zeta)}{Q(\zeta)}\left[\frac{Q(\zeta) - Q(z)}{\zeta - z}\right]$$

which is a polynomial in z of degree $k-1$, for each ζ, and therefore $h \in \mathcal{P}_{k-1}$, also, $h - f$ is divisible by Q, so we conclude $f - [f,Q]Q = L(f,Q)$, i.e.

(1.14) **Proposition.** *The Lagrange interpolant $L(f,Q)$ equals $f - [f,Q]Q$, and we have (with Γ as above):*

$$L(f,Q) = \frac{1}{2\pi i}\int_\Gamma \left[\frac{Q(\zeta) - Q(z)}{\zeta - z}\right]\frac{f(\zeta)}{Q(\zeta)} d\zeta. \tag{1.15}$$

This is Hermite's classical formula for the Lagrange interpolant. Coming back now to (1.7) we have

(1.16) **Corollary.** *$T_Q^{-1}g$, for g in E, is equal to the divided difference $[G;Q]$ where G is any solution to $D^kG = g$.*

Again, since the kernel of the operator $f \mapsto [f;Q]$ is $\mathcal{P}_{k-1}$, it makes no difference which G is chosen.

2. Eigenfunctions of T_Q

For each $m = 0,1,2,\ldots$ T_Q is a bijection of $\mathcal{P}_m$. Indeed, it is clear from the definition that T_Q maps $\mathcal{P}_m$ to $\mathcal{P}_m$, and since it is injective, $T_Q\mathcal{P}_m = \mathcal{P}_m$.

(2.1) **Proposition.** (B. Shapiro) *For each m there is, apart from normalization, a unique eigenfunction of T_Q contained in $\mathcal{P}_m$. The associated eigenvalue $\lambda = \lambda_{m,k}$ depends only on m and k (and not on the coefficients a_j in (1.1)); moreover*

$$\lambda_{m,k} = (m+1)(m+2)\ldots(m+k). \tag{2.2}$$

For convenience, we include a proof (see also [MS]).

Proof. Introducing in $\mathcal{P}_m$ the basis $1,z,z^2,\ldots,z^m$ and noting that $T_Qz^n = \lambda_{n,k}z^n + \cdots$, where $\ldots$ denotes a polynomial of degree less than n, we see that $T_Q|_{\mathcal{P}_m}$ is represented by an upper triangular $(m+1)\times(m+1)$ matrix with diagonal elements $\lambda_{0,k},\ldots,\lambda_{m,k}$. These are then the eigenvalues of $T_Q|\mathcal{P}_m$ and since they are all distinct, this operator has $m+1$ linearly independent eigenvectors, so not all of these can lie in $\mathcal{P}_{m-1}$. Hence there is at least one eigenfunction which is a polynomial of precise degree m. By the same reasoning applied to $T_Q|_{\mathcal{P}_{m-1}}$, etc. there is at least one eigenvector of precise degree n for each $n = 0,1,2,\ldots,m$. Since there are $m+1$ of them in $\mathcal{P}_m$ altogether, there must be exactly one eigenfunction (apart from normalization) of degree precisely n, for each n. This concludes the proof. □

Let us henceforth denote by f_m the unique monic eigenfunction of degree m, for T_Q (we shall usually have a fixed Q throughout the discussion, so to keep notations simple we shall generally not indicate the dependence on Q in the notation).

Boris Shapiro found a striking property of the zeroes of f_m:

(2.3) **Theorem.** *The zeroes of f_m are contained in the convex hull of the set of zeroes of Q, for each m.*

We refer to [MS] for the proof, which is based on a clever application of the Gauss-Lucas theorem.

In particular, if Q has only real zeroes, contained in an interval $[a, b]$, the same is true for each f_m. As noted in [MS] many remarkable conjectures about the zeroes of f_m are supported by numerical calculations. We shall not enter into those matters here. One result, however, going slightly beyond [MS], is

(2.4) **Theorem.** *If Q has only real zeroes, say $\{x_1, \dots, x_k\}$, and $\{t_1, t_2, \dots, t_m\}$ are the zeroes of f_m, then for every convex function φ on* **R** we have

$$\frac{1}{m}\sum_{j=1}^{m}\varphi(t_j) \leq \frac{1}{k}\sum_{j=1}^{k}\varphi(x_k). \tag{2.5}$$

Proof. This is based on the following recent (and still unpublished) theorem of J. Aniansson:

(2.6) **Theorem.** *If f is a polynomial of degree n with all real zeroes $\alpha_1, \alpha_2, \dots, \alpha_n$ and f' has zeroes $\beta_1, \beta_2, \dots, \beta_{n-1}$ then for every convex function φ on* **R**,

$$\frac{1}{n-1}\sum_{j=1}^{n-1}\varphi(\beta_j) \leq \frac{1}{n}\sum_{j=1}^{n}\varphi(\alpha_j). \tag{2.7}$$

To deduce (2.4) from this, suppose f_m is an eigenfunction (of degree m), so

$$D^k(Qf_m) = \lambda_{m,k} f_m.$$

Thus, the k-th derivative of Qf_m (the roots of which are $\{x_1, x_2, \dots, x_k, t_1, t_2, \dots, t_m\}$) is f_m with roots $\{t_1, t_2, \dots, t_m\}$. By repeated application of (2.7) it follows that the mean value of φ over the latter set cannot exceed its mean value over the former set, so

$$\frac{1}{m}\sum_{i=1}^{m}\varphi(t_i) \leq \frac{1}{k+m}\left[\sum_{j=1}^{k}\varphi(x_j) + \sum_{i=1}^{m}\varphi(t_i)\right],$$

which implies (2.5). □

A similar, elementary argument, which we leave to the reader, also shows (for **all** polynomials Q):

(2.8) **Proposition.** *The arithmetic mean of the zeroes of f_m equals the arithmetic mean of the zeroes of Q.*

For another, new result on the zeroes of eigenfunctions, see Section 5.

It is not clear a priori whether T_Q has, besides its polynomial eigenfunctions, also entire transcendental ones. We shall now show:

(2.9) **Theorem.** *Every entire eigenfunction of T_Q is a polynomial.*

We will give two proofs of this.

First proof of (2.9). Suppose $f \in E$ and $\lambda \in \mathbf{C}$, and

$$D^k(Qf) = \lambda f.$$

Multiplying both sides by Q, and denoting Qf by g we get

$$QD^k g = \lambda g. \tag{2.10}$$

Now, the differential equation

$$u^{(k)} = \frac{\lambda u}{Q} \tag{2.11}$$

is, relative to the point $z = \infty$, of Fuchsian type since the coefficient of u vanishes to order k at ∞ (see [P, p.51]). Therefore no solution can have an essential singularity at ∞, so g must be a polynomial. □

Remark. If we consider T_Q as an operator on the larger class of functions holomorphic in a domain D, then there may well be plenty of other eigenfunctions: just choose any non-zero complex number λ, and let D be any simply connected domain not containing any zeroes of Q. Then (2.11) admits k linearly independent solutions holomorphic in D, which lead back to eigenfunctions of T_Q.

Second proof of (2.9). Again it is convenient to work with (2.10). We will show, **if g is entire and satisfies (2.10) it is a polynomial.** From (2.10) it follows that there are positive constants M, R such that

$$|z|^k |g^{(k)}(z)| \leq M|g(z)|, \quad \text{for } |z| \geq R. \tag{2.12}$$

That such an estimate implies that g is a polynomial follows from a general theorem of Clunie and Hayman (see [H, p.341, Theorem 12]; I am indebted to Walter Hayman for providing this reference). However, it is also possible to give a simple **ad hoc** proof, which we proceed to do.

Denote by $b_0 + b_1 z \cdots$ the Taylor expansion of g. Then, from (2.12), writing $z = re^{i\theta}$, we deduce that

$$\frac{1}{2\pi}\int_0^{2\pi} r^{2k}|g^{(k)}(re^{i\theta})|^2 d\theta \leq M^2 \left[\frac{1}{2\pi}\int_0^{2\pi} |g(re^{i\theta})|^2 d\theta\right] \tag{2.13}$$

holds for $r \geq R$, and performing the integration,

$$\sum_{m=0}^{\infty} \left[(m+1)(m+2)\cdots(m+k)\right]^2 |b_{m+k}|^2 r^{2m+2k} \leq M^2 \sum_{n=0}^{\infty} |b_n|^2 r^{2n} \tag{2.14}$$

for $r \geq R$. The left side can be written

$$\sum_{n=k}^{\infty} [n(n-1)\dots(n-k+1)]^2 |b_n|^2 r^{2n}.$$

Thus, if we introduce the notations

$$A_n = \begin{cases} 0, & n < k \\ (n(n-1)\dots(n-k+1))^2, & n \geq k, \end{cases}$$

$t = r^2$, and $p_n = |b_n|^2$ we can rewrite (2.14) as

$$\sum_{n=0}^{\infty} A_n p_n t^n \leq M^2 \sum_{n=0}^{\infty} p_n t^n, \ t \geq R^2. \tag{2.15}$$

For each m, the left-hand member is not less than

$$\sum_{n=m}^{\infty} A_n p_n t^n \geq A_m \sum_{n=m}^{\infty} p_n t^m$$

so (2.15) implies, for $t \geq R^2$

$$(A_m - M^2) \sum_{n=m}^{\infty} p_n t^n \leq M^2 \sum_{n=0}^{m-1} p_n t^n.$$

Retaining only the first term in the sum on the left, we get, for all m such that $A_m > M^2$

$$p_m \leq \frac{M^2 \sum_{n=0}^{m-1} p_n t^n}{(A_m - M^2) t^m}.$$

This holds for all $t \geq R^2$, so we can let $t \to \infty$ and deduce that $0 = p_m = |b_m|^2$ for all large enough m, hence g is a polynomial. □

A formula for the eigenfunctions, useful for computations

If f_n denotes the monic eigenfunction of degree n for T_Q, then $f_n = z^n + p$, where p is in $\mathcal{P}_{n-1}$ and so, for some complex numbers c_i

$$f_n = z^n + \sum_{j=0}^{n-1} c_j f_j.$$

Now, f_j is annihilated by $T_Q - \lambda_{j,k} I$, I denoting the identity operator, so the sum on the right is annihilated by V_{n-1}, where

$$V_{n-1} = V_{n-1,k} := \prod_{j=0}^{n-1} (T_Q - \lambda_{j,k} I).$$

Therefore

$$V_{n-1}z^n = V_{n-1}f_n = \prod_{j=0}^{n-1}(\lambda_{n,k} - \lambda_{j,k})f_n$$

which gives the formula

$$f_n = \frac{\prod_{j=0}^{n-1}(T_Q - \lambda_{j,k}I)z^n}{\prod_{j=0}^{n-1}(\lambda_{n,k} - \lambda_{j,k})}.$$

This is a useful formula for calculation of t_n: for given Q one first computes the action of T_Q on $1, z, \ldots, z^n$ and thereafter a simple iterative scheme enables us to compute in turn the polynomial $(T_Q - \lambda_{n-1,k})z^n$, then $T_Q - \lambda_{n-2,k}I$ applied to the preceding polynomial, and so on.

Having disposed of these preliminary matters, we can now outline our main goals.

Since all T_Q (operating on E) have only positive eigenvalues (which is remarkable, since T_Q in general does not have the appearance of self-adjointness) that suggests to look for a theoretical explanation. Now self-adjointness is a Hilbert space concept, so we must first of all introduce some Hilbert spaces of entire functions where such questions can be precisely formulated. To this end we shall define a scale of spaces depending on a positive parameter. The union of these spaces will encompass all entire functions of finite order. On each of these spaces T_Q will be a closed, unbounded operator (ignoring the trivial case $k = 0$), with a compact inverse.

Also, on the Hilbert spaces considered, the operator T_{Q_0}, where $Q_0(z) = z^k$ will trivially be self-adjoint, with $\{z^n,\ n \geq 0\}$ as eigenfunctions. Thus, it is natural to consider T_Q (when deg $Q = k$) as a perturbation of T_{Q_0}. By using rather advanced tools of perturbation theory we shall show along these lines that, for certain values of the scale parameter, T_Q and T_{Q_0} restricted to these spaces are similar, i.e. there is a bounded operator W with bounded inverse such that

$$T_Q = W^{-1}T_{Q_0}W \tag{2.16}$$

or, in terms of the compact inverse operators T_Q^{-1}, $T_{Q_0}^{-1}$

$$T_Q^{-1} = W^{-1}T_{Q_0}^{-1}W. \tag{2.17}$$

This will be our main result. First of all, it "explains" why all the T_Q, for fixed k, have the same spectrum (which is real and positive). Secondly, it implies that on the "good" Hilbert spaces where (2.16) holds, the eigenfunctions of T_Q not merely are linearly independent and span the space (which was obvious **a priori**), but they form a **Riesz basis**, i.e. are the image of an orthonormal basis under the map by a bounded, invertible operator, and this is an important information. (Actually, for technical reasons, we shall prove the "Riesz basis" version first, and deduce the other from it.)

Yet another equivalent form of our main result is that for each Q there are "good" Hilbert spaces of entire functions where T_Q is self-adjoint: it is a well-known elementary fact (details below) that if a, say bounded, operator T on a Hilbert space H is similar to a self-adjoint one, then one can introduce into H a new inner product, inducing the same topology, with respect to which T is self-adjoint (unfortunately though, our proof is highly non-constructive; it would be of great interest to give such spaces explicitly).

3. Some spaces of entire functions

Let $w = \{w_n\}_{n=0}^{\infty}$ be a sequence of positive numbers such that $r_n : w_{n+1}/w_n$ tend to infinity, and increase from some n onward. This implies easily that

$$\lim_{n\to\infty} w_n^{1/n} = \infty. \tag{3.1}$$

Consequently, a power series

$$f(z) = \sum_{n=0}^{\infty} b_n z^n \tag{3.2}$$

for which

$$\|f\|_w := \left(\sum_{n=0}^{\infty} w_n |b_n|^2\right)^{1/2} \tag{3.3}$$

is finite converges on the whole complex plane, and defines an entire function f. It is easily verified that the set of all entire functions with finite $\|\cdot\|_w$ norm is a (complete) Hilbert space with respect to this norm.

Let us give three (closely related) examples of such spaces:

(3.4) **Definition** *$\mathfrak{F}_p$, for $0 < p < \infty$, is the space H_w with $w_n = (n!)^p$.*

For $p = 1$, this is the classical Fock space (also called by many other names). Observe that the spaces $\mathfrak{F}_p$ become smaller as p increases. We leave it to the reader to verify that

(i) All f in $\mathfrak{F}_p$ are of order at most $2/p$.

(ii) All entire functions of order less than $2/p$ are in $\mathfrak{F}_p$.

Thus, the scale of spaces $\mathfrak{F}_p$ has as its union, all entire functions of finite order.

Although we shall henceforth work only in the spaces $\mathfrak{F}_p$, let us mention two alternative possibilities, each of which is useful in certain connections: these are

(a) H_w with $w_n = n^{np}$

(b) The Hilbert space of entire functions f such that

$$\|f\|_t^2 := \int_{\mathbf{C}} |f(z)|^2 \exp(-|z|^t) dA \ < \infty \tag{3.5}$$

where t is a positive parameter and dA denotes area measure on the complex plane. It is easily seen that this is a space of the H_w type with

$$w_n = C(t)\Gamma\left(\frac{2n+2}{t}\right) \tag{3.6}$$

where $C(t)$ is a constant depending only on t. For $t = 2$ this space is identical with the Fock space $\mathfrak{F}_1$.

Domains. Consider now any linear differential operator T whose coefficients are polynomials. Then, on the space E of all entire functions (with usual topology) T is an (everywhere defined) continuous linear operator. If H is any Hilbert space of entire functions, with topology stronger than that of E, there is a natural dense linear manifold $D_T \subset H$ such that the restriction of T to D_T is a closed (generally, unbounded) operator, namely the set of all f in H such that Tf (which is a well-defined element of E) is in H. **We shall always tacitly assume, in discussing unbounded differential operators on our Hilbert spaces, that their domains are specified in this way**. (Or, in other words: the graph of T in $H \times H$ is defined to be the intersection of its graph (as operator on E) in $E \times E$ with $H \times H$. Since the graph in the larger space is trivially closed in $E \times E$ (because T is continuous) its intersection with $H \times H$ is closed in the graph norm. Thus we may always suppose our operators closed.)

Besides differential operators, we shall also use the **backward shift operator**. In any topological vector space of holomorphic functions on a domain including 0, this is the operator

$$B : f \mapsto \frac{f(z) - f(0)}{z} \tag{3.7}$$

or, what is the same, if $f(z) = \sum_{n=0}^{\infty} b_n z^n$

$$\sum_{n=0}^{\infty} b_n z^n \mapsto \sum_{n=0}^{\infty} b_{n+1} z^n. \tag{3.8}$$

(3.9) **Proposition.** *On a Hilbert space H_w of entire functions (where $w_{n+1}/w_n \to \infty$) the backward shift is compact.*

Proof. An orthonormal basis for H_w is $\{e_n\}_{n=0}^{\infty}$ where

$$e_n(z) = z^n/\sqrt{w_n}. \tag{3.10}$$

Thus, the backward shift maps $e_0 \mapsto 0$ and $e_n \mapsto \frac{z^{n-1}}{\sqrt{w_n}} = \frac{\sqrt{w_{n-1}}}{\sqrt{w_n}} e_{n-1}$, for $n \geq 1$. Hence it is unitarity equivalent to an operator $\tilde{B}$ on the space l^2 of square summable complex sequences $\{c_n\}_{n=0}^{\infty}$ (which we may interpret as formal sums $\sum_{n=0}^{\infty} c_n e_n$), whose action is

$$\tilde{B} : \sum_{n=0}^{\infty} c_n e_n \mapsto \sum_{n=0}^{\infty} c_{n+1} \varepsilon_n e_n \quad (\varepsilon_n = \sqrt{w_n/w_{n+1}}). \tag{3.11}$$

$\tilde{B}$ is the composition of two operators: first a "pure" backward shift $c_n \to c_{n+1}$ which is a contraction of l^2, and following this a coordinatewise multiplication by $\{\varepsilon_n\}$, which is a compact operator on l^2 since $\varepsilon \to 0$. Thus $\tilde{B}$ is compact on l^2, so B is compact on H_w. □

We are now in the position to prove

(3.12) **Theorem.** *On every space H_w with $w_{n+1}/w_n \to \infty$, T_Q has a compact inverse (and so, in particular, on all the spaces $\mathfrak{F}_p$).*

Proof. We require first

(3.13) **Lemma.** *For $0 \le r \le k$, and f in E*

$$D^k(z^r f) = D^k(z^k B^{k-r} f) \tag{3.14}$$

where $D = d/dz$, and B denotes the backward shift (here B^0 is understood to be the identity map).

It is readily checked that

$$B^r f = (f(z) - s_{r-1}(z))/z^r, \ /; (r = 1, 2, \ldots) \tag{3.15}$$

where s_n denotes the partial sum of rank n of the Taylor expansion of f. Hence, the right member of (3.14) (since (3.14) is trivially true for $r = k$, we may suppose $r < k$) is equal to $D^k(z^k(f - s_{k-r-1})/z^{k-r}) = D^k(z^r f - z^r s_{k-r-1}) = D^k(z^r f)$. □

Proof of Theorem (3.12). It is usefull here and later to have a convenient designation for the operator T_Q in the special case where $Q(z) = z^k$, so let us denote by

$$S_k : f \mapsto D^k(z^k f). \tag{3.16}$$

this operator. As an operator on H_w it is self-adjoint admitting the e_n in (3.10) as eigenfunctions, with corresponding eigenvalues $\lambda_{n,k}$ given by (2.2). Then, for $Q = z^k + a_{k-1}z^{k-1} + \cdots + a_0$ we have

$$\begin{aligned} T_Q f &= D^k(z^k + a_{k-1}z^{k-1} + \cdots + a_0)f \\ &= (S_k + a_{k-1}S_k B + a_{k-2}S_k B^2 + \cdots + a_0 S_k B^k)f \end{aligned}$$

using (3.14), hence

$$T_Q = S_k(I + a_{k-1}B + \cdots + a_0 B^k). \tag{3.17}$$

Now, B is compact on H_w by (3.9). Also, it has no non-zero eigenvalues, since if $Bf = \lambda f$, f must be a constant multiple of $1/(1 - \lambda z)$ which is not entire if $\lambda \ne 0$. Hence the spectrum of B is $\{0\}$, and by the spectral mapping theorem the spectrum of the second factor in (3.17) is $\{1\}$. Denoting by V this operator, V is invertible in H_w, so $T_Q^{-1} = V^{-1}S_k^{-1}$. But, S_k^{-1} is a compact (self-adjoint) operator on H_w, indeed even in trace class with plenty to spare, hence T_Q^{-1} is compact (and trace class). □

4. Some perturbation theory, and our main results

For a while, H shall denote any separable Hilbert space, and $L(H)$ the set of bounded linear operators on H. We denote by I the identity operator, and by $\sigma(A), \rho(A)$ respectively the spectrum and the resolvent set of an operator A. (See [K] for a complete exposition of all this.) Thus, the complex number ζ is in $\rho(A)$ if and only if $\zeta I - A$ has a bounded inverse, and this inverse

$$R(\zeta, A) := (\zeta I - A)^{-1}, \quad \zeta \in \rho(A) \tag{4.1}$$

called the resolvent of A, is an H-valued holomorphic function on $\rho(A)$. We recall a few standard results from [K]:

(4.2) **Lemma.** *Let T be a (possibly unbounded) normal operator on H, A in $L(H)$ and $\zeta \in \mathbf{C}$. Then if*

$$\| A \| < d(\zeta) \tag{4.3}$$

where

$$d(\zeta) := \operatorname{dist}(\zeta, \sigma(T)), \tag{4.4}$$

$T + A - \zeta I$ is invertible, and

$$\| R(\zeta, T + A) \| \leq (d(\zeta) - \| A \|)^{-1}. \tag{4.5}$$

Proof. For ζ in $\rho(T)$, $(T - \zeta I)^{-1}$ is a bounded normal operator with spectrum equal to $\{1/(\lambda - \zeta) : \lambda \in \sigma(T)\}$ so

$$\| (T - \zeta I)^{-1} \| = 1/d(\zeta). \tag{4.6}$$

Thus, (4.3) implies that $AR(\zeta, T)$ has norm less than one, so $(I - AR(\zeta, T))^{-1} =: B$ is in $L(H)$ and invertible, and

$$B^{-1}(\zeta I - T) = \zeta I - T - A.$$

Since each factor on the left is invertible, so is $\zeta I - T - A$. Its inverse equals $(\zeta I - T)^{-1} B$, the norm of which does not exceed $\| B \| / d(\zeta)$ (from (4.6)). Also, using (4.6) again,

$$\| B \| \leq (1 - \| A \| \| R(\zeta, T) \|)^{-1} \leq d(\zeta)/(d(\zeta) - \| A \|),$$

so

$$\| R(\zeta, T + A) \| \leq \| B \| / d(\zeta) \leq (d(\zeta) - \| A \|)^{-1},$$

as asserted. □

Using the key estimate (4.5) we can now, following a technique from [DS, Theorem 19.7] deduce resolvent estimates also for unbounded perturbations of T.

(4.7) **Lemma** *Let T be an unbounded, self-adjoint and positive operator, having a compact inverse. Suppose moreover C is a (possibly unbounded) closed operator, $0 \leq \alpha < 1$, and*

$$A := CT^{-\alpha} \tag{4.8}$$

is bounded. Define, for ζ in $\rho(T)$

$$\delta(\zeta) := \sup_{\lambda \in \sigma(T)} \frac{\lambda^\alpha}{|\zeta - \lambda|}. \tag{4.9}$$

Then, if

$$\delta(\zeta) \parallel A \parallel < 1, \tag{4.10}$$

ζ is in the resolvent set of $T + C$ and

$$\parallel R(\zeta, T + C) \parallel < (d(\zeta)(1 - \delta(\zeta) \parallel A \parallel))^{-1}. \tag{4.11}$$

Proof. The spectrum of the normal operator $T^\alpha R(\zeta, T)$ is the set $\{\lambda^\alpha/(\zeta - \lambda) : \lambda \in \sigma(T)\}$, so

$$\parallel T^\alpha R(\zeta, T) \parallel = \delta(\zeta). \tag{4.12}$$

Thus, in view of assumption (4.10), $AT^\alpha R(\zeta, T)$ has norm less than one, so the series

$$\sum_{n=0}^{\infty} (AT^\alpha R(\zeta, T))^n = \sum_{n=0}^{\infty} (CR(\zeta, T))^n \tag{4.13}$$

converges in norm to a bounded operator V with

$$\begin{aligned} \parallel V \parallel &\le (1 - \parallel AT^\alpha R(\zeta, T) \parallel)^{-1} \\ &\le (1 - \parallel A \parallel \delta(\zeta))^{-1} \end{aligned} \tag{4.14}$$

from (4.12). Now, from (4.13)

$$\begin{aligned} V = I + VCR(\zeta, T), &\quad \text{so} \\ V(\zeta I - T) = \zeta I - T + VC& \\ V(\zeta I - T - C) = \zeta I - T, &\quad \text{or} \\ R(\zeta, T)V(\zeta I - T - C) = I.& \end{aligned}$$

Since $R(\zeta, T)$ and V are bounded, this shows that ζ is in the resolvent set of $T + C$, and

$$\parallel R(\zeta, T + C) \parallel \le \parallel V \parallel \parallel R(\zeta, T) \parallel$$

so finally, using (4.14) and (4.6) we obtain the desired estimate (4.11). □

With these estimates in hand, we now return to the study of T_Q. **Henceforth the underlying Hilbert space shall be $\mathfrak{F}_p$**, where conditions shall be placed on p as the need arises. Our "unperturbed" operator shall be S_k (defined in (3.16)) and the perturbations will be

$$C : f \to D^k(a_{k-1}z^{k-1} + \cdots + a_0)f. \tag{4.15}$$

To use the preceding estimates, we have first to determine for what values of α and p, the operator $CS_k^{-\alpha}$ is bounded in $\mathfrak{F}_p$. (This corresponds to the hypothesis that A in (4.8) is bounded.)

It is easy to see that it is enough to check the case

(4.16) $$C = C_0 : f \to D^k(z^{k-1}f).$$

Now, in terms of the orthonormal basis for $\mathfrak{F}_p$ given by

(4.17) $$e_n(z) = z^n/(n!)^{p/2}$$

the action of C_0 is to send z^n to $(n+k-1)(n+k-2)\cdots(n-1)z^{n-1}$, which also can be expressed by

(4.18) $$C_0 : e_n \to (n!)^{-p/2}(n+k-1)\cdots(n-1)\left((n-1)!\right)^{p/2} e_{n-1} =: \mu_n e_{n-1}$$

for $n \geq 1$. The action of $C_0 S_k^{-\alpha}$ is thus to send e_n to $\lambda_{n,k}^{-\alpha}\mu_n e_{n-1}$ for $n \geq 1$ so this operator is bounded if and only if $\sup_n \lambda_{n,k}^{-\alpha}\mu_n$ is finite. Asymptotically, $\lambda_{n,k}^{-\alpha}\mu_n$ is $n^{-k\alpha}n^{-p/2}n^k = n^{k(1-\alpha)-(p/2)}$ so we have

(4.19) **Lemma.** *$C_0 S_k^{-\alpha}$ is bounded in $\mathfrak{F}_p$ if and only if*

(4.20) $$k(1-\alpha) \leq p/2.$$

If we are willing to introduce an inessential normalizing assumption about Q, we can do better. Suppose, namely that

(4.21) $$a_{k-1} = 0,$$

i.e. the coefficient of z^{k-1} in Q (and hence, the sum of the roots of Q) vanishes. Then the boundedness of $CS_k^{-\alpha}$ is controlled by that of $C_1 S_k^{-\alpha}$ where C_1 denotes the operator

$$C_1 : f \to D^k(z^{k-2}f).$$

Now, in the boundedness condition the factor

$$((n-1)!)^{p/2} \qquad \text{is replaced by} \qquad ((n-2)!)^{p/2},$$

so reasoning as before we get

(4.22) **Lemma.** *$C_1 S_k^{-\alpha}$ is bounded in $\mathfrak{F}_p$ if and only if*

(4.23) $$k(1-\alpha) \leq p.$$

Let us summarize the essence of these last estimates as

(4.24) **Lemma.** *If (4.20) holds, then the "perturbation" $C := T_Q - S_k$ satisfies the condition*

(4.25) $$CS_k^{-\alpha} \text{ is bounded in } \mathfrak{F}_p.$$

If moreover the coefficient of z^{k-1} in Q vanishes, this conclusion follows from the weaker assumption (4.23). (In like manner, if $a_{k-1} = a_{k-2} = \cdots = a_{k-r} = 0$, the conclusion follows provided $k(1-\alpha) \leq (r+1)p/2$. This in turn implies that our main theorem is valid in $\mathfrak{F}_p$ if $p > 3/r$.)

(4.26) **Remark.** The normalization (4.21) is really no essential loss of generality, because if

$$D^k(Qf) = \lambda f$$

the same relation will hold if Q, f are replaced by $Q(z+w), f(z+w)$ for any complex number w. Choosing $w = -a_{k-1}/k$ achieves the normalization (4.21). The only effect on the eigenfunctions is that they, too, are shifted by w. Taking into account (2.8), we then see that **for every n the eigenfunction f_n of degree n has a vanishing coefficient of z^{n-1}** (or, what is equivalent, the sum of its roots, too, is zero).

In view of this, **let us tacitly assume henceforth the normalization (4.21) for Q.**

We are now in the position to prove our main result:

(4.27) **Theorem.** *If $p > 3/2$, T_Q as operator on $\mathfrak{F}_p$, is similar to the self-adjoint operator T_{Q_0}, where $Q_0(z) = z^k$.*

We shall unravel further consequences of this later, but now let us prove the theorem. For convenience, we again denote T_{Q_0} by S_k.

We shall make extensive use of the Riesz-Dunford functional calculus, and resolvents ([K], [DS]). Recall first that for any closed operator Y on a Hilbert space, to each compact K of $\sigma(Y)$ there corresponds a corresponding idempotent operator J_K, which is expressible as a "Cauchy integral"

(4.28) $$J_K = (1/2\pi i)\int_\Gamma R(\zeta, Y)d\zeta,$$

over the contour Γ lying in $\rho(Y)$ and surrounding K, with winding number $+1$, but no other points of $\sigma(Y)$. We will call J_K **the canonical idempotent associated to** K (relative to the operator Y).

By Γ_n we denote the positively oriented boundary of the square centered at the origin, one side of which is the segment joining the points $A_n + iA_n/2$ and $A_n - iA_n/2$, where

(4.29) $$A_n = (\lambda_{n,k} + \lambda_{n+1,k})/2.$$

By γ_n we denote the positively oriented boundary of the square centered at $\lambda_{n,k}$ and with side length B_n, where

(4.30) $$B_n = (\lambda_{n+1,k} - \lambda_{n,k})/2.$$

Observe that Γ_n encloses all the points $\lambda_{j,k}$ with $0 \leq j \leq n$, but none with $j \geq n+1$, whereas γ_n encloses only that $\lambda_{j,k}$ with $j = n$.

(4.30) **Lemma.** *Suppose $k(1-\alpha) \leq p$, and that $\zeta \in \rho(S_k) \cap \rho(T_Q)$. Then, for some constant K independent of ζ we have*

(4.31) $$\| R(\zeta, S_k) - R(\zeta, T_Q) \| \leq \frac{K\delta(\zeta)}{d(\zeta)(1 - K\delta(\zeta))}$$

where

(4.32) $$d(\zeta) - \operatorname{dist}(\zeta, \sigma(S_k))$$

and

(4.33) $$\delta(\zeta) = \sup\left(\frac{\lambda^\alpha}{|\lambda - \zeta|}\right) \quad \text{over } \lambda \in \sigma(S_k).$$

Proof. We have

$$R(\zeta, S_k) - R(\zeta, T_Q) = (S_k - T_Q)R(\zeta, S_k)R(\zeta, T_Q). \tag{4.34}$$

Now write, as usual, $C = T_Q - S_k = AS_k^\alpha$ where A is bounded (we know that this is possible by assumption (4.23)). Then, the right-hand member of (4.34) can be written $-AR(\zeta, T_Q)\,[S_k^\alpha R(\zeta, S_k)]$. The norm of the bracketed operator is $\delta(\zeta)$ as given by (4.33) and $\| R(\zeta, T_Q) \| = \| R(\zeta, S_k + C) \|$ does not exceed $[d(\zeta)\,(1 - \delta(\zeta) \| A \|)]^{-1}$ by (4.11). Hence, the norm of the right-hand member of (4.34) does not exceed

$$\frac{\| A \| \, \delta(\zeta)}{d(\zeta)(1 - \delta(\zeta) \| A \|)}$$

which implies (4.31) with $K = \| A \|$. □

Now, let $\{p_n\}_{n=0}^\infty$ denote the canonical idempotents associated with the $\lambda_{n,k}$ for the operators S_k. Thus, P_n is just the orthogonal projector onto the one-dimensional space spanned by z^n. Since $\lambda_{n,k}$ is also an isolated point of the spectrum of T_Q (because T_Q has a compact inverse, whose spectrum consists precisely of $\{\lambda_{n,k}^{-1}\}_0^\infty$ and 0) there is associated to the point $\lambda_{n,k}$ a canonical idempotent Π_n relative to T_Q. This is not in general self-adjoint, but some "oblique" projector onto the one-dimensional space spanned by the eigenvector f_n. It follows from what already has been done that T_Q has no other spectrum. This can also be confirmed in an instructive way, as follows. It is easily checked that, for $\zeta \in \Gamma_n$, we have, at least for the narrower range $p > 2$, $k \geq 3$

$$\delta(\zeta) \leq c_1 n^{k\alpha - (k-1)} \quad (\zeta \in \Gamma_n) \tag{4.35}$$

and

$$c_2 n^{k-1} \leq d(\zeta) \leq c_3 n^k \quad (\zeta \in \Gamma_n) \tag{4.36}$$

where c_i are positive constants independent of ζ. Suppose now that α and p satisfy, besides (4.23), also

$$k(1 - \alpha) > 3/2. \tag{4.37}$$

(Indeed, if we **assume** $p > 3/2$, **as we shall henceforth,** then we can always find $\alpha, 0 < \alpha < 1$ satisfying both (4.23) and (4.37), at least if we assume also that $k \geq 2$ (which, to avoid trivialities, we do): for, it is easily checked that these requirements are equivalent to being able to choose α satisfying

$$1 - \frac{p}{k} \leq \alpha < 1 - \frac{3}{2k}. \tag{4.38}$$

The right-hand member of this inequality is strictly between 0 and 1, and because $p > 3/2$ the left-hand member is smaller than this, which gives us a subinterval of $(0, 1)$ in which we are free to choose α, and (4.23) and (4.37) hold.)

Then from (4.35), $\delta(\zeta) = O(n^{1-k(1-\alpha)}) = O(n^{-\beta})$ for some $\beta > 1/2$, because of (4.37), when $\zeta \in \Gamma_n$. Hence from the estimate (4.31) we easily see that

$$(4.39) \qquad \| \frac{1}{2\pi i}\int_{\Gamma_n} R(\zeta, S_k)d\zeta - \frac{1}{2\pi i}\int_{\Gamma_n} R(\zeta, T_Q) \| = O(n^{2-k(1-\alpha)})$$

as $n \to \infty$. But, the respective integrals are the canonical idempotents for S_k and T_Q associated with the subsets of their spectra enclosed by Γ_n (see [K] for all necessary background). The difference of these idempotents has norm less than one, for all n large enough, if $\alpha > 2/k$ because of (4.39). Hence, in consequence of Theorem 6.35 and 6.32 of [K], the ranges of these idempotents have the same dimension, which we know in the case of S_k is $n+1$. This shows the range of the canonical idempotents for T_Q, associated to the part of $\sigma(T_Q)$ interior to Γ_n, has dimension $n+1$. Since this dimension is already accounted for by the eigenfunctions corresponding to $\lambda_{j,m} (j = 0, 1, \ldots, n)$, there cannot be any other spectrum of T_Q inside Γ_n (and hence, varying n, anywhere in **C** but at points $\lambda_{j,m}$) and moreover all these eigenvalues must be simple (which we already knew, but it is instructive to see that this follows from resolvent estimates, which we in any case need to prove our main theorem).

So, we now know that all canonical projectors for T_Q associated to compact portions of $\sigma(T_Q)$ are finite sums of the Π_n. Now, a similar estimate to (4.39) is valid if we replace Γ_n by γ_n, so we have

$$(4.40) \qquad \| P_n - \Pi_n \| \le cn^{-\beta}$$

for some constant c, and some $\beta > 1/2$. Hence

$$(4.41) \qquad \sum_{n=0}^{\infty} \| P_n - \Pi_n \|^2 < \infty.$$

To continue the proof of Theorem (4.27) we must first quote a special case of Lemma 4.17a of [K]:

(4.42) **Lemma.** *Let $\{J_m\}_{m=0}^{\infty}$ be self-adjoint projectors onto mutually orthogonal subspaces of a Hilbert space H, which span H. Let $\{V_m\}_{m=0}^{\infty}$ be idempotents (not necessarily self-adjoint) with mutually disjoint ranges. Suppose moreover that*

$$(4.43) \qquad \dim \text{range}\,(J_0) = \dim\, \text{range}\,(V_0) < \infty$$

and

$$(4.44) \qquad \sum_{m=1}^{\infty} \| J_m - V_m \|^2 < 1.$$

Then, there exists $W \in L(H)$ with bounded inverse such that

$$(4.45) \qquad V_m = W^{-1} J_m W \;\; (m = 0, 1, \ldots).$$

To apply this, choose N so large that

$$(4.46) \qquad \sum_{n=N+1}^{\infty} \| P_n - \Pi_n \|^2 < 1$$

(this is possible by (4.41)). Then, defining $J_0 = P_0 + P_1 + \cdots + P_N$, $J_m = P_{N+m}$ for $m \geq 1$ and $V_0 = \Pi_0 + \Pi_1 + \cdots + \Pi_N$, $V_m = \Pi_{N+m}$ for $m \geq 1$ the hypotheses of Lemma (4.42) are fulfilled. Therefore, we get an invertible W such that

$$\Pi_0 + \Pi_1 + \cdots + \Pi_N = W^{-1}(P_0 + P_1 + \cdots + P_N)W \tag{4.47}$$

$$\Pi_n = W^{-1}P_n W, \quad n \geq N+1. \tag{4.48}$$

Now, if we knew (4.48) for **all** n, we could easily conclude that T_Q and S_k are similar — but (4.48) only implies the similarity modulo operators of rank $N+1$ so there still is work to be done! The last phase of the proof uses heavily results about **Riesz bases**, which may be found in [GK, Chapter 6]. Recall that **a Riesz basis for a Hilbert space is the image of an orthonormal basis under a bounded invertible linear transformation**. (There are many other equivalent characterizations, see [GK].) The following is a Theorem of Nina Bari (see [GK,p.317]):

(4.49) **Theorem (N. Bari).** *Suppose $\{g_n\}_{n=0}^{\infty}$ are vectors in a Hilbert space H satisfying*

$$a \leq \| g_n \| \leq b \text{ for all } n \tag{4.50}$$

where a, b are positive numbers, as well as the condition

(4.51) *If $\{c_n\}$ is any square summable complex sequence such that $\| \sum_{n=0}^{M} c_n g_n \| \to 0$ as $M \to \infty$, then all c_n are zero.*

Then, if there exists an orthonormal basis $\{e_n\}$ for H such that

$$\sum \| g_n - e_n \|^2 < \infty, \tag{4.52}$$

$\{g_n\}$ is a Riesz basis for H.

To apply this, let $\{f_n\}_0^{\infty}$ be eigenfunctions of T_Q (f_n is a monic polynomial of degree n) and $\{e_n\}_0^{\infty}$ normalized eigenfunctions of $S_k : e_n(z) = z^n/(n!)^{p/2}$. We now introduce the normalized eigenfunctions $g_n = w_n f_n / \| f_n \|$ where w_n are complex numbers of modulus one so chosen that

$$\langle g_n, e_n \rangle \geq 0, \quad n = 0, 1, 2, \ldots \tag{4.53}$$

(The reasons for imposing this condition will become clear later.) Obviously

$$\| g_n \| = 1, \quad n = 0, 1, 2, \ldots \tag{4.54}$$

Our next goal is to prove, using (4.49) that $\{g_n\}$ **is a Riesz basis for** $\mathcal{F}_p$.

(4.55) **Lemma.** *The $\{g_n\}$ just defined satisfy (4.51).*

Proof. Suppose $\{c_n\}$ is a square-summable sequence such that $\sum_0^{\infty} c_n g_n = 0$. We must show all c_n are zero. From (4.48) we deduce

$$W\Pi_n g_n = P_n W g_n, \quad n \geq N+1. \tag{4.56}$$

But, $\Pi_n g_n = g_n$, so (4.56) implies that $W g_n$ is left fixed by P_n, and consequently

$$W g_n = b_n e_n, \quad n \geq N+1 \tag{4.57}$$

where $\{b_n\}$ are **non-zero** complex numbers (since W is invertible). Likewise, from (4.47), we see that $W\left(\sum_0^N c_n g_n\right)$ is left fixed by $P_0 + \cdots + P_N$, and so

(4.58) $$W\left(\sum_0^N c_n g_n\right) \text{ is a polynomial of degree at most } N.$$

Now, by assumption $\sum_0^M c_n g_n \to 0$ as $M \to \infty$, and hence also $W\left(\sum_0^M c_n g_n\right) \to 0$, so (for $M > N$)

(4.59) $$W\left(\sum_0^N c_n g_n\right) + W\left(\sum_{N+1}^M c_n g_n\right) \to 0 \text{ as } M \to \infty.$$

But, by (4.57) and (4.58) the two summands in (4.59) are orthogonal! Hence $W\left(\sum_0^N c_n g_n\right)$ must be 0, whence $\sum_0^N c_n g_n = 0$, and $c_n = 0$ for $0 \le n \le N$, since the g_n are linearly independent. Moreover, $W\left(\sum_{N+1}^M c_n g_n\right) \to 0$ as $M \to \infty$, whence (from (4.57)) $\sum_{N+1}^M c_n b_n e_n \to 0$ as $M \to \infty$. Since the summands are orthogonal, all $b_n c_n$ must vanish, and since $b_n \neq 0, c_n = 0$ for all $n \geq N+1$. Thus, all c_n are 0. □

(4.60) **Lemma.** *The g_n satisfy (4.52).*

Proof. As we showed earlier

(4.61) $$\sum_0^\infty \| P_n - \Pi_n \|^2 < \infty.$$

Let now Q_n denote the **orthogonal** projection onto the range of Π_n, then $\| P_n - Q_n \| \le \| P_n - \Pi_n \|$, by virtue of [K, Theorem 6.35], so

(4.62) $$\sum_0^\infty \| P_n - Q_n \|^2 < \infty.$$

Now, we claim

(4.63) $$\| e_n - g_n \|^2 \le 2 \| P_n - Q_n \|^2 .$$

Assuming this, $\sum \| e_n - g_n \|^2 < \infty$ and we are done. So it remains to prove (4.63). Again, we require a lemma (see [GK, pp. 336–337]).

(4.64) **Lemma.** *Let P, Q be orthogonal projectors onto one-dimensional subspaces of a Hilbert space H, spanned by the unit vectors e, f respectively. Then*

(4.65) $$\| P - Q \|^2 = 1 - |\langle e, f\rangle|^2.$$

Since [GK] glosses over this, let us sketch a proof. Since the result is trivial if e and f are linearly dependent, we assume this is not the case. For any x in H,

$$Px = \langle x, e\rangle e, \quad Qx = \langle x, f\rangle f \quad \text{so} \quad \| Px - Qx \|^2 = \| \langle x, e\rangle e - \langle x, f\rangle f \|^2$$

and we must prove that the supremum of the ratio

(4.66) $$\| \langle x,e\rangle e - \langle x,f\rangle f \|^2 / \| x \|^2$$

as x ranges over all non-zero vectors in H, equals $1 - |\langle e, f\rangle|^2$. Observe that, if x is orthogonal to e and f, the ratio in (4.66) is zero, so in determining the supremum (which certainly is positive, since $P - Q$ is not the zero operator) we may assume that x is not orthogonal to e and f. If we now replace x by its orthogonal projection on the space M spanned by e and f, the ratio (4.66) clearly cannot decrease, indeed unless x is in M it must **increase**. Hence, the ratio (4.66) attains a maximum on H for some x in M. Thus, writing $x = ae + bf$, we have $Px = (a + b\bar{\alpha})e$, and $Qx = (a\alpha + b)f$ where $\alpha = \langle e, f\rangle$, while $\| x \|^2 = |a|^2 + |b|^2 + a\bar{b}\alpha + ab\bar{\alpha}$ so (4.65) will follow if

(4.67) $$\| (a + b\bar{\alpha})e - (a\alpha + b)f \|^2 \leq (1 - |\alpha|^2)\left(|a|^2 + |b|^2 + a\bar{b}\alpha + ab\bar{\alpha}\right)$$

for all complex numbers a, b, α. But a calculation, which we omit, shows that the left and right members of (4.67) are in fact **equal**! (In other words, the ratio (4.66) is **constant** for x in M. In $\mathbf{R}^2$ this is equivalent to the elementary geometric theorem: the distance between the feet of the perpendiculars from a point P on a circle to two fixed diameters is independent of P.) This proves the lemma. □

We now deduce (4.63): Recall from (4.53) that $\langle e_n, g_n\rangle \geq 0$, so

$$\| e_n - g_n \|^2 = 2 - 2\langle e_n, g_n\rangle \leq 2 - 2\langle e_n, g_n\rangle^2 = 2 \| P_n - Q_n \|^2,$$

where we used (4.65). Thus, we have proved the principal result:

(4.68) **Theorem.** *If $p > 3/2$, the eigenfunctions of T_Q (where Q is any monic polynomial of degree $k \geq 2$, the coefficient in which of z^{k-1} vanishes), if normalized to have norm 1, are a Riesz basis for $\mathfrak{F}_p$.*

It remains to deduce (4.27) from this. This deduction is based on the general (trivial) lemma which follows.

(4.69) **Lemma.** *Suppose $\{e_n\}_0^\infty$ is an orthonormal basis of a Hilbert space, W is a bounded invertible linear operator, and $g_n = We_n$. Suppose A and B are bounded linear operators, and for some sequence $\{\lambda_n\}_0^\infty$ we have*

(4.70) $$Ae_n = \lambda_n e_n, \quad Bg_n = \lambda g_n, \quad n = 1, 2, \ldots$$

then $A = W^{-1}BW$.

Proof. From (4.70), and our assumptions

$$\lambda_n We_n = \lambda_n g_n = Bg_n = BWe_n, \quad \text{so}$$

$$W^{-1}BWe_n = \lambda_n e_n = Ae_n, \quad n = 0, 1, \ldots$$

Thus, the bounded operator $A - W^{-1}BW$ is zero on all the e_n, hence it is the zero operator.

To deduce (4.27), we apply (4.69) to the Hilbert space $\mathfrak{F}_p$ with $A = S_k^{-1}$, $B = T_Q^{-1}$, where the e_n and g_n in (4.69) are the elements denoted by those symbols in the preceding discussion and $\lambda_n = \lambda_{n,k}^{-1}$. (Note that we had to pass to the operators **inverse** to S_k and T_Q to obtain **bounded** ones.)

We conclude that for the operator W mapping e_n on g_n, $S_k^{-1} = W^{-1}T_Q^{-1}W$, which is equivalent to $S_k = W^{-1}T_Q W$. (Note that the W here need not be identical with the W in (4.47) and (4.48).)

Thus, Theorem 4.27 is completely proved. Theorems (4.27) and (4.68) are actually equivalent. □

Still another face of our main result is

(4.71) **Corollary.** *If $p > 3/2$, $\mathfrak{F}_p$ can be given a new inner product, inducing an equivalent norm, with respect to which T_Q is self-adjoint.*

Proof. The deduction of this from (4.27) is by means of the following simple general observation (it is related to so-called **symmetrizable operators**, see [Z] for very detailed discussion). Suppose T and S are bounded linear operators on a Hilbert space H, and S is self-adjoint, and $T = A^{-1}SA$ for some invertible operator A. If $\langle \cdot, \cdot \rangle$ denotes the inner product in H, we introduce a new inner product, denoted $\ll \cdot, \cdot \gg$ by

$$\ll f, g \gg = \langle Af, Ag \rangle = \langle A^*Af, g \rangle$$

for all f and g in H. The corresponding new norm of f is then the original norm of Af, which is equivalent to the original norm since

$$\left(\| A^{-1} \|\right)^{-1} \| f \| \leq \| Af \| \leq \| A \| \| f \| .$$

Moreover, T is self-adjoint with respect to $\ll \cdot, \cdot \gg$. We have only to verify $\ll Tf, g \gg = \ll f, Tg \gg$ for all f, g, that is

$$\langle ATf, Ag \rangle = \langle Af, ATg \rangle$$

which is equivalent to $A^*AT = T^*A^*A$, i.e. that A^*AT is self-adjoint (with respect to $\langle \cdot, \cdot \rangle$). But

$$A^*AT = A^*AA^{-1}SA = A^*SA \tag{4.72}$$

which is self-adjoint. Applying this to the space $\mathfrak{F}_p$ with T_Q^{-1}, S_k^{-1} in the roles of T, S we deduce (4.71) for T_Q^{-1}, and hence for T_Q. □

In passing, observe that the similarity of T to a self-adjoint operator implies also that T can be represented as $P\Sigma$ where P is an invertible **positive** operator, and Σ is self-adjoint. This follows from (4.72), with $P = (A^*A)^{-1}$ and $\Sigma = A^*SA$. Conversely, from the representation $T = P\Sigma$ follows $P^{-1/2}TP^{1/2} = P^{1/2}\Sigma P^{1/2}$ and the last operator is self-adjoint, so the $P\Sigma$ representation implies that T is similar to a self-adjoint operator.

5. Concluding remarks

Our main result says that for $p > 3/2$, T_Q and S_k are similar as operators on $\mathfrak{F}_p$. What happens for smaller p? I have no counterexamples showing that there is **any** p for which the assertion is false.

It is known, in general, that even if a compact operator (like our T_Q^{-1}) on a Hilbert space has a system of eigenvectors that span the space, these eigenvectors can be very "pathological" e.g. they need not span all invariant subspaces of the operator. See [N] for a discussion of this phenomenon, and further references.

Multivariable generalizations. The operator T_Q is a very special case of operators $f \mapsto P(D)(Qf)$ where P and Q are polynomials in n variables, $P(D)$ denoting the partial differential operator $P(\frac{\partial}{\partial z_1}, \ldots, \frac{\partial}{\partial z_n})$ and f is in some space of holomorphic functions. I discussed such operators in [S]. Further studies of them may be found e.g. in [KS], [ES1], [ES2] and [A], and I am told of current work by mathematicians in Ufa, Russia (Krasichkov-Ternovski, Napalkov, Popenov). But, although a great deal is known about questions like injectivity, surjectivity, closed range etc. of such operators in various spaces, **nearly nothing** appears to be known about their **eigenfunctions and spectral properties**, apart from some rudimentary results in [S]. The fact that, already in the simple one-dimensional framework of the present paper (where injectivity and surjectivity etc. are easy to settle) the **eigenvalue problem** leads into such deep matters, suggest that the eigenvalue problem, even for very special classes of partial differential operators of type $f \mapsto P(D)(Qf)$ deserves further study. One can see from simple examples that new phenomena of great complexity can arise. For example, in the case $P(z) = Q(z) = z_1^2 + \cdots + z_n^2$ (which even gives a self-adjoint operator in the n-variable Fock space) the spectrum and eigenfunctions were given in S. Although the inverse is compact, there are when $n > 2$ spectral points of arbitrary high multiplicity!

In connection with the study of linear differential operators (in one or several variables) on spaces of holomorphic functions, a very important role is played by a notion of **duality** between two spaces based on the bilinear form which (in one variable) is $\sum_{n=0}^{\infty} n! a_n b_n$, where a_n, b_n are the Taylor coefficients of functions f, g in the respective spaces. We cannot here go into details, there is an enormous literature on such matters, going back to classical work on infinite order differential equations (for an orientation and references, see e.g. [M], and [ES1] and [A] have discussion and references for the multivariable case). The key feature of this duality is that it associates to an operator $f \mapsto P(D)(Qf)$ on one space a dual "transposed", or "adjoint" operator $g \mapsto Q(D)(Pg)$ on the dual space, and properties of the one imply corresponding properties of the other, according to a certain dictionary.

For example, the dual space (in the aforementioned sense) to E is the space $\mathbf{X}$ of entire functions of exponential type, with a suitable topology. The operator dual to Boris Shapiro's operator T_Q (acting on E) is then

$$g \mapsto Q(D)(z^k g), \quad g \in \mathbf{X}. \tag{5.1}$$

Thus, the bijectivity of T_Q on E is equivalent to the bijectivity of (5.1) on **X**, which is precisely the assertion that the classical Cauchy initial value problem

$$Q(D)u = h,\ u(0) = u'(0) = \cdots u^{(k-1)}(0) = 0 \tag{5.2}$$

has a unique solution u in **X**, for each h in **X**. This can of course easily be established directly.

But, what about the eigenvalue problem for (5.1)? In contrast to T_Q, these operators admit transcendental entire eigenfunctions. Their study should be rewarding. Let us, in conclusion, briefly derive some properties of the "dual eigenfunctions" and show how, with their help, one can deduce a new, nontrivial property of the zeroes of f_n.

For two power series $f = \sum_0^\infty a_n z^n$, $g = \sum_0^\infty b_n z^n$ we shall write

$$\{f, g\} = \sum_0^\infty n! a_n b_n \tag{5.3}$$

assuming this series is absolutely convergent. It is then easy to check that the set of g (initially thought of just as formal power series) for which we have absolute convergence for every f in $\mathfrak{F}_p$ is precisely $\mathfrak{F}_{2-p}$. Let us, in the present discussion, assume that $0 < p < 2$, so that $\mathfrak{F}_{2-p}$ is again a space of entire functions. This gives an identification of the dual space to the Hilbert space $\mathfrak{F}_p$, which is different from the usual one based on the inner product in $\mathfrak{F}_p$ as the basic bilinear (more precisely, sesquilinear) form whereby $\mathfrak{F}_p$ is its own dual. The main reason why this new dual (there does not seem to be a generally accepted name for it, the formalism can be traced back to E. Borel and G. Pólya, so let's provisionally call it the Borel-Pólya dual) is important is that the operator "multiplication by z" in one of these spaces is adjoint to d/dz in the other, and vice-versa.

Suppose now $3/2 < p < 2$, so $\{f_n\}$ are a Riesz basis for T_Q in $\mathfrak{F}_p$. It is then easy to show that there are unique elements $\{g_n\}_0^\infty$ in $\mathfrak{F}_{2-p}$ satisfying

$$\{f_m, g_n\} = \delta_{mn}. \tag{5.4}$$

Moreover g_n is an eigenfunction for the operator (5.1) with eigenvalue $\lambda_{n,k} = (n+1)\cdots(n+k)$, and $\{g_n\}$ form a Riesz basis for the space $\mathfrak{F}_{2-p}$. In general g_n are not polynomials. For h in $\mathfrak{F}_{2-p}$, the solution of the Cauchy problem (5.2) is then

$$u(z) = z^k \sum_{n=0}^\infty (c_n/\lambda_{n,k}) g_n(z) \tag{5.5}$$

where $\sum_0^\infty c_n g_n$ is the expansion for h in the basis $\{g_n\}$.

We illustrate the use of the Borel-Pólya dual by proving

(5.6) **Theorem.** *For any complex a, there are infinitely many eigenfunctions f_n for T_Q, which do not vanish at a, except for the case $Q(z) = (z-a)^k$, when $f_n(z) = (z-a)^n$.*

Proof. Suppose that $f_n(a) = 0$ for all $n, n > N$. Let g_n denote the dual eigenfunctions defined above (for definiteness, we can assume, say $p = 7/4$ throughout) and let

$$e^{az} = \sum_{n=0}^{\infty} c_n g_n(z) \tag{5.7}$$

be the expansion of e^{az} in this basis. From (5.4), $c_n = \{f_n, e^{az}\} = f_n(a)$ so the summands in (5.7) are 0 for $n > N$, i.e. e^{az} **is a linear combination of** $g_0, g_1, \ldots, g_N$. Let now $V = V_Q$ denote the operator $g \mapsto Q(D)(z^k g)$. The key to the proof is

(5.8) **Lemma.** *For* $m = 0$, *and assuming* $Q(z) \neq (z-a)^k$,

$$V^m e^{az} = p_m(z) e^{az} \tag{5.9}$$

where p_m *is a polynomial, and* $\deg p_m < \deg p_{m+1}$.

The proof is by induction. We note first the general identity (corollary of the Leibniz rule for differentiating products)

$$Q(D)(e^{az}u) = [Q(D+a)u]e^{az}. \tag{5.10}$$

Now, for $m = 0$ we have (5.9) with $p_0(z) = 1$. Suppose for some m (5.9) holds. Then

$$V^{m+1}e^{az} = Q(D)\left(z^k p_m(z) e^{az}\right) = \left[Q(D+a)(z^k p_m(z))\right] e^{az}$$

using (5.10). Let now d denote $\deg p_m$. Then unless $Q(D+a)$ is D^k (that is, $Q(z) = (z-a)^k$), $Q(D+a)$ can reduce the degree of $z^k p_m$ (which is $k + d$) by at most $k-1$ so $V^{m+1}e^{az} = p_{m+1}e^{az}$ with $\deg p_{m+1} \geq d+1$. This completes the inductive step, proving (5.8).

Proof of Theorem concluded. If the sum in (5.7) terminates after $n = N$, we must have $R(V)e^{az} = 0$, where R is the polynomial

$$R(w) := \prod_{n=0}^{N} (w - \lambda_{n,k}) = w^N + \ldots$$

where $\ldots$ denotes terms of degree less than N. But, assuming now that $Q(z)$ is not $(z-a)^k$, the lemma tells us that $R(V)e^{az} = V^N e^{az} + \ldots$ equals $e^{az}(p_N(z) + \ldots)$ where $\ldots$ denotes a finite sum of polynomials, all of degree less that $\deg p_N$. Such a sum cannot be 0, and we have reached a contradiction, establishing Theorem (5.6). □

Acknowledgements. I wish to thank Boris Shapiro and Gisli Masson for inspiring my interest in the polynomials which are the main subject of this paper. Pavel Kurasov and Sergei Naboko provided me with useful literature and information about perturbation theory. To Mihai Putinar I am also indebted for valuable discussions and references.

References

[A] Aniansson, J., Some Integral Representations in Real and Complex Analysis, Dissertation, Royal Institute of Technology, Stockholm, 1999.

[DS] Dunford, N. and J. Schwartz, Linear Operators, Wiley Classics ed., 1988.

[ES1] Ebenfelt, P. and H.S. Shapiro, The mixed Cauchy problem for holomorphic partial differential operators, *J.d'Analyse Math.* **65** (1996) 237–295.

[ES2] Ebenfelt, P. and H.S. Shapiro, The Cauchy-Kovalevskaya Theorem and generalizations, *Commun. in PDE* **20** (1995) 939–960.

[G] Gelfond, A.O. The calculus of Finite Differences, FIZMATGIZ; Moscow, 1959 (Russian).

[GK] Gokhberg, I. and M.G. Krein, Introduction to the Theory of Linear Nonselfadjoint operators, Transl. of Math. Monographs, vol. 18, AMS 1969.

[H] Hayman, W., The local growth of power series: a survey of the Wiman-Valiron method, *Canad. Math. Bull.* **17** (1974) 317–358.

[K] Kato, T., Perturbation Theory for Linear Operators, second ed. corrected, Springer-Verlag, 1980.

[KS] Khavinson, D. and H.S. Shapiro, The Dirichlet problem when the data is an entire function, *Bull.London Math. Soc.* **24** (1992) 456–468.

[M] Muggli, H., Differentialgleichungen unendlich hoher Ordnung mit konstanten Koeffizienten, *Commentarii Math. Helvetici* **11** (1938) 151–179.

[MS] Masson, G. and B. Shapiro, A note on polynomial eigenfunctions of $(d/dx)^k Q_k(x)$, preprint, Stockholm, 1999.

[N] Nikolski, N.K., Complete extensions of Volterra operators, *Izv. Akad. Nauk SSSR* **33** (1966), English translation in *Math. USSR - Izvestija*, **3** (1969) 1271–1276.

[P] Plemelj, J., Problems in the sense of Riemann and Klein, Wiley-Interscience, 1964.

[S] Shapiro, H.S., An algebraic theorem of E. Fischer and the holomorphic Goursat problem, *Bull. London Math. Soc.* **21** (1989) 513–537.

[Z] Zaanen, A.C., Linear Analysis, P. Nordhoff, Groningen, 1953.

Harold S. Shapiro
Dept. of Mathematics
KTH
S-10044 Stockholm, SWEDEN
e-mail: shapiro@math.kth.se

References

[illegible]

Operator Theory:
Advances and Applications, Vol. 132, 387–394

Spectral Properties of Jacobi Matrices with Rapidly Growing Power-like Weights

Luis O. Silva

Abstract. The semi-classical method and a grouping in blocks technique are jointly applied, within the framework of the Subordinacy Theory, to the spectral analysis of Jacobi matrices with rapidly growing weights. In this paper it is proved that Discrete String Operators with weights given by $\lambda_n = n^\alpha\left(1+\frac{c_n}{n}\right)$, where $\alpha > 1$ and $\forall n \in \mathbb{N}\ c_{n+2} = c_n$ have pure point spectrum even in the limit point case. The technique developed can be used for studying such Jacobi matrices with more complicated spectrum.

1. Introduction

In the Hilbert space $l^2(\mathbb{N})$ let us single out the dense subset $l_{fin}(\mathbb{N})$ of sequences which have a finite number of non-zero elements. Consider the operator

$$J : l_{fin}(\mathbb{N}) \to l^2(\mathbb{N})$$

defined by means of the recurrence relation

$$(Ju)_n = \lambda_{n-1}u_{n-1} + \lambda_n u_{n+1} \quad n \in \mathbb{N}, \tag{1}$$

where u stands for the sequence $\{u_n\}_{n=1}^\infty$, λ_n are positive real numbers for all $n \in \mathbb{N}$, and $\lambda_0 = u_0 = 0$. This operator has the following matrix form with respect to the canonical basis in $l^2(\mathbb{N})$:

$$\begin{pmatrix} 0 & \lambda_1 & 0 & 0 & \cdots \\ \lambda_1 & 0 & \lambda_2 & 0 & \cdots \\ 0 & \lambda_2 & 0 & \lambda_3 & \cdots \\ \vdots & \vdots & \vdots & & \end{pmatrix} \tag{2}$$

Being represented by a Hermitian Jacobi matrix, this operator is symmetric and the only possible values of the deficiency indices are $(1,1)$ and $(0,0)$ (these cases correspond to the limit circle case and the limit point case respectively). Since the operator J is symmetric, it is closable and in the sequel whenever we refer to the spectrum of J we shall have in mind its closure's spectrum. Operator

The author appreciates the support of the Consejo Nacional de Ciencia y Tecnología. SEP, México.

J belongs to a subclass of Jacobi matrices which are the discrete analogue of string operators and, therefore, they are often referred to as Discrete String Operators.

Jacobi matrices with slowly growing weights λ_n have been studied in a series of papers ([3], [4], [6], [7], [8]). However, as it seems to the author, the spectral properties of Discrete String Operators with rapidly growing weights have not been thoroughly studied. In this first approach to the matter, the following structure for the weights proved to arise in a quite natural way:

$$\lambda_n = n^\alpha\left(1 + \frac{c_n}{n}\right), \tag{3}$$

where $\alpha > 1$ and $c_{n+2} = c_n$ for any $n \in \mathbb{N}$. Indeed if we considered $\lambda_n = n^\alpha$, $\alpha > 1$ we would be in the limit circle case and therefore J would have discrete spectrum. We then add a small perturbation with periodic character so that J is in the limit point case. Below we show that when the coefficients c_n in (3) satisfy $|c_1 - c_2| \geq \alpha - 1$ we have the limit point case. However, when the coefficients have odd periodicity (i.e. $c_{n+2j-1} = c_n$, $j \in \mathbb{N}$), J is always in the limit circle case.

In the present work we shall be restricted to the limit point case, in which Discrete String Operators may have not only discrete spectrum, as in the limit circle case, but absolutely continuous or singular continuous spectra. In the limit point case we also are able to apply useful techniques developed on the basis of the so-called theory of subordinacy for the spectral analysis of Jacobi matrices. We use a combination of these techniques to disclose the spectral properties of J defined by (1) and (3), moreover, we elaborate the proper method for studying more complicated examples of Jacobi matrices with rapidly growing weights, which is one of the goals of this paper.

Note that all our results will be obtained for the case when $c_{n+2} = c_n$, however it is not difficult to extend them to the general case of even periodicity.

2. Preliminaries

Consider the recurrence equations with the real parameter λ:

$$\lambda_{n-1}u_{n-1} + \lambda_n u_{n+1} = \lambda u_n \quad n > 1. \tag{4}$$

The solutions of this system are called the generalized eigenvectors of the operator J defined by (1) corresponding to λ. Notice that, although this recurrence relation resembles the operator spectral equation $Ju = \lambda u$, it does not contain the initial condition $\lambda_1 u_2 = \lambda u_1$.

The theory of subordinacy, the cornerstone of the present work, was first applied to the study of Jacobi matrices in the paper [9], whose fundamentals were previously established in [5], a work on differential equations. A key notion in the theory is the following definition:

Definition 1. *For some $\lambda \in \mathbb{R}$, a non trivial solution u of (4) is called subordinate if*

$$\lim_{N\to\infty} \frac{\sum_{n=1}^{N} |u_n|^2}{\sum_{n=1}^{N} |v_n|^2} = 0 \tag{5}$$

for every solution v of (4) which is linearly independent of u.

Provided that J is in the limit point case, the Gilbert-Pearson theory of subordinacy maintains that if for almost all λ in a certain interval (a, b) there is no subordinate solution of (4), then the spectrum of J has an absolutely continuous component filling the interval (a, b). On the other hand if a subordinate solution exists for a λ and satisfies the initial condition $\lambda_1 u_2 = \lambda u_1$ then this λ is either in the pure point component of the spectrum provided the subordinate solution is in l^2 or in the singular continuous component if the subordinate solution is not in l^2. A consequence of these results is that the asymptotic behavior of the generalized eigenvectors determines the spectral properties of the corresponding operators.

Equation (4) is analysed here by means of its equivalent matrix equation

$$\vec{u}_{n+1} = B_n(\lambda)\vec{u}_n, \tag{6}$$

where $\vec{u}_n = \binom{u_{n-1}}{u_n}$ and $B_n(\lambda) = \begin{pmatrix} 0 & 1 \\ -\frac{\lambda_{n-1}}{\lambda_n} & \frac{\lambda}{\lambda_n} \end{pmatrix}$ is the so-called transfer matrix. From (6) we deduce that

$$\vec{u}_{n+1} = \prod_{k=2}^{n} B_k(\lambda)\vec{u}_2, \tag{7}$$

where the product, as in the sequel, has been taken in chronological order.

3. Essentially self-adjointness

Let us find out the conditions that must be imposed on the coefficients c_n in order that the closure of operator J be self-adjoint. The criterion for establishing the essentially self-adjointness of J is straightforwardly deduced from [10, Theorem 1.1] and it affirms that a Jacobi matrix of the form (2) is in the limit circle case if and only if its weights satisfy

$$\sum_{n=1}^{\infty}\Big[\prod_{k=1}^{n}\frac{\lambda_{2k-1}}{\lambda_{2k}}\Big]^2 < \infty \text{ and } \sum_{n=1}^{\infty}\Big[\prod_{k=1}^{n}\frac{\lambda_{2k}}{\lambda_{2k+1}}\Big]^2 < \infty. \tag{8}$$

Using this criterion it is not difficult to obtain the following results

Theorem 2. *The operator J defined by (1) and (3) is essentially self-adjoint if and only if $|c_1 - c_2| \geq \alpha - 1$*

Theorem 3. *Let J be defined as in Theorem 2, but suppose that the periodicity of the sequence $\{c_n\}_{n=1}^{\infty}$ is odd ($c_{n+2j-1} = c_n$, $j \in \mathbb{N}$), then operator J is in the limit circle case.*

4. Asymptotic behavior of the generalized eigenvectors

The asymptotic analysis of the solutions of the recurrence equation (6) is carried out here in the same way as we usually investigate the solutions' asymptotic behavior in the case of linear systems of differential equations. Actually we base our analysis on the discrete analogue of the Levinson theorem for differential equations [2]. A particular case of this theorem is the following assertion.

Theorem 4. *(N.Levinson) Let A be a 2×2 invertible matrix with two different eigenvalues $\mu_+ \neq \mu_-$ and let R_n be a sequence of real 2×2 matrices such that $\{\|R_n\|\} \in l^1$.*

Then there exist two non-zero solutions $\vec{x}_n^+$, $\vec{x}_n^-$ of the system of equations

$$\vec{x}_{n+1} = (A + R_n)\vec{x}_n \tag{9}$$

such that

$$\vec{x}_n^{\pm} = \mu_{\pm}^n(\vec{e}_{\pm} + o(1)), \quad \text{as} \quad n \to \infty, \tag{10}$$

where $\vec{e}_{\pm}$ are the eigenvectors of A corresponding to the eigenvalues $\mu_{\pm}$.

It turns out that it is possible to find an equation of the form (9) equivalent to (6). Nevertheless Theorem 4 cannot be directly applied to this equation, since the constant matrix corresponding to A would have equal eigenvalues. This situation is surmountable if we use equation (7) instead and take the product there grouping in blocks as will be explained in the proof of this section's main result. Here it is worth mentioning that grouping in blocks procedures have been applied earlier to the spectral analysis of Jacobi matrices in [6] and [8].

The main result of this section is the following

Theorem 5. *Let J be defined by (1) and (3), then the generalized eigenvectors corresponding to the operator J have the following asymptotic behavior:*

$$u_n^j = n^{\beta_j}(1 + o(1)) \qquad j = 1, 2, \tag{11}$$

where $\beta_1 = \frac{c_1 - c_2 - \alpha}{2}$ and $\beta_2 = \frac{c_2 - c_1 - \alpha}{2}$

Proof. First of all let us take advantage of the periodicity of the sequence $\{c_n\}_{n=1}^{\infty}$ introducing the matrices $M_k = B_{2k+1}B_{2k}$. By means of a straightforward algebraic calculation one obtains the following asymptotics for the sequence of matrices $\{M_k\}_{k \in \mathbb{N}}$ as $k \to \infty$:

$$M_k = \begin{pmatrix} -1 - \frac{\beta_1}{k} + O(k^{-2}) & O(k^{-\alpha}) \\ O(k^{-\alpha}) & -1 - \frac{\beta_2}{k} + O(k^{-2}) \end{pmatrix} \tag{12}$$

Now instead of (7) we have the following equation

$$\vec{u}_{2n+2} = \prod_{k=2}^{n} M_k \vec{u}_2. \tag{13}$$

The product in this equation is analysed grouping in blocks as follows

$$\prod_{k=1}^{n} M_k = M_n \dots M_1 = \prod_{l} \prod_{k=n_l}^{n_{l+1}-1} M_k,$$

where the length of each block is defined by the conditions

$$\sum_{k=n_l}^{n_{l+1}-1} \frac{1}{k} \leq 1 \quad \text{and} \quad \sum_{k=n_l}^{n_{l+1}} \frac{1}{k} > 1, \tag{14}$$

so we have a well-defined partition of the natural numbers $\cup_l [n_l, n_{l+1}) = \mathbb{N}$. From (14) it is easy to deduce that, as $n_l \to \infty$,

$$\ln\left(\frac{n_{l+1}-1}{n_l}\right) = 1 + O(n_{l+1}^{-1})$$

and

$$n_l = e^l + O(1). \tag{15}$$

Introducing the matrices D_k and R_k as follows

$$D_k = diag(-\frac{\beta_1}{k} + O(k^{-2}), -\frac{\beta_2}{k} + O(k^{-2})),$$

$$R_k = \begin{pmatrix} 0 & O(k^{-\alpha}) \\ O(k^{-\alpha}) & 0 \end{pmatrix}$$

we are able to write (12), in the form $M_k = -I + D_k + R_k$, where I is the identity 2×2 matrix. Now, we compute the product of matrices M_k within each block $[n_l, n_{l+1})$, i.e., $\prod_{k=n_l}^{n_{l+1}-1} M_k$. First of all we rewrite this product in such a way that the diagonal and anti-diagonal terms are separated

$$\prod_{k=n_l}^{n_{l+1}-1} M_k = \prod_{k=n_l}^{n_{l+1}-1} (-I + D_k + R_k) = \prod_{k=n_l}^{n_{l+1}-1} (-I + D_k)$$

$$\times \prod_{k=n_l}^{n_{l+1}-1} \left\{ I + (\prod_{j=n_l}^{k} (-I + D_j))^{-1} R_k \prod_{j=n_l}^{k-1} (-I + D_j) \right\}.$$

The first multiplier in the last equality is a diagonal matrix:

$$\prod_{k=n_l}^{n_{l+1}-1} (-I + D_k) = \prod_{k=n_l}^{n_{l+1}-1} \begin{pmatrix} -1 - \frac{\beta_1}{k} + O(k^{-2}) & 0 \\ 0 & -1 - \frac{\beta_2}{k} + O(k^{-2}) \end{pmatrix}.$$

We evaluate how the first element of this matrix behaves asymptotically.

$$\prod_{k=n_l}^{n_{l+1}-1} (1 + \frac{\beta_1}{k} + O(k^{-2})) = (-1)^{n_{l+1}-n_l} \exp\left\{ \sum_{k=n_l}^{n_{l+1}-1} \ln(1 + \frac{\beta_1}{k} + O(k^{-2})) \right\}$$

$$= (-1)^{n_{l+1}-n_l} \exp\left\{ \beta_1 \ln\left(\frac{n_{l+1}-1}{n_l}\right) + O(n_{l+1}^{-1}) \right\}.$$

Treating the second element of the diagonal matrix in the same way we obtain

$$\prod_{k=n_l}^{n_{l+1}-1}(-I+D_k)=\pm(1+O(n_{l+1}^{-1}))\begin{pmatrix} e^{\beta_1} & 0 \\ 0 & e^{\beta_2}\end{pmatrix}, \tag{16}$$

where the sign obviously depends on the length of the l-the block and it is not important in our present discussion. Now, using (16), it is easy to show that

$$(\prod_{j=n_l}^{k}(-I+D_j))^{-1}R_k\prod_{j=n_l}^{k-1}(-I+D_j)=\begin{pmatrix} 0 & O(k^{-\alpha}) \\ O(k^{-\alpha}) & 0\end{pmatrix}.$$

Let the sequence of matrices $\{S_k\}_{k=1}^{\infty}$ be defined by $S_k=\begin{pmatrix} 0 & O(k^{-\alpha}) \\ O(k^{-\alpha}) & 0\end{pmatrix}$. Clearly $\{\|S_k\|\}_{k=1}^{\infty}\in l^1$. Thus we have obtained the following expression for the product of the matrices M_k within each block:

$$\prod_{k=n_l}^{n_{l+1}-1}M_k=(1+O(n_{l+1}^{-1}))\begin{pmatrix} e^{\beta_1} & 0 \\ 0 & e^{\beta_2}\end{pmatrix}\prod_{k=n_l}^{n_{l+1}-1}(I-S_k)$$

Let now the matrix Γ_l be defined by $\Gamma_l=\prod_{k=n_l}^{n_{l+1}-1}(I+S_k)-I$, then we could write

$$\prod_{k=n_l}^{n_{l+1}-1}M_k=A+R_l,$$

where $A=\begin{pmatrix} e^{\beta_1} & 0 \\ 0 & e^{\beta_2}\end{pmatrix}$ and $R_l=\begin{pmatrix} O(e^{-l}) & 0 \\ 0 & O(e^{-l})\end{pmatrix}+\Gamma_l\begin{pmatrix} e^{\beta_1}+O(e^{-l}) & 0 \\ 0 & e^{\beta_2}+O(e^{-l})\end{pmatrix}$.

It is not difficult to prove, using the following known inequality

$$\|\prod_{k=1}^{N}(I+S_k)-I\|\le\prod_{k=1}^{N}(1+\|S_k\|)-1,$$

that if $\{\|S_k\|\}_{k=1}^{\infty}\in l^1$, then $\{\|\Gamma_l\|\}_{l=1}^{\infty}\in l^1$ and, therefore, $\{\|R_l\|\}_{l=1}^{\infty}$ is also in l^1.

We can now apply Theorem 4 to the equation

$$\begin{pmatrix} u_{2n_{l+1}-1} \\ u_{2n_{l+1}}\end{pmatrix}=\prod_{k=n_l}^{n_{l+1}-1}M_k\begin{pmatrix} u_{2n_l-1} \\ u_{2n_l}\end{pmatrix},$$

which is of course equivalent to (13) and has the form (9) assuming that $\vec{x}_l=\begin{pmatrix} u_{2n_l-1} \\ u_{2n_l}\end{pmatrix}$. According to (10) we obtain the following asymptotic behavior as $l\to\infty$ for the solutions of equation (6):

$$\vec{x}_l^{\,k}=e^{l\beta_k}(\vec{e}_k+o(1)),\qquad (k=1,2). \tag{17}$$

Taking into account (15) and the definition of $\vec{x}_l$ it is not difficult to show that the solutions of (4) have the following asymptotic behavior

$$u_n^{(k)}=n^{\beta_k}(1+o(1)),\qquad (k=1,2).$$

This completes the proof. □

5. The spectrum

Our first result in the spectral analysis of J establishes the existence of subordinate solutions.

Theorem 6. *Let J be in the limit point case, i.e., $|c_1 - c_2| \geq \alpha - 1$. Then the corresponding recurrence equation (4) has a subordinate solution for any $\lambda \in \mathbb{R}$. This solution is always an element of the space $l^2(\mathbb{N})$.*

The proof of this result readily follows from Theorem 5 and the fact that $\alpha > 1$. Indeed, the existence of a subordinate solution is a consequence of the different growth rate of the generalized eigenvectors' asymptotics and $\alpha > 1$ implies that this subordinate solution is always in l^2.

Using results from the theory of the subordinacy together with Theorem 6 we obtain the following

Corollary 7. *The operator J defined by (1) and (3) has pure point spectrum in the limit point case, i.e. when $|c_1 - c_2| \geq \alpha - 1$.*

As we mentioned before in the limit circle case J has discrete spectrum. This fact is easily deduced from the general theory.

Acknowledgments. The author expresses his gratitude to Prof. S. Naboko for his constant and invaluable support.

References

[1] Yu.M. Berezansky, *Eigenfunction Expansions of Self-Adjoint Operators.* Naukova dumka, Kiev, 1965, (in Russian).

[2] E. Coddington and N. Levinson, *Theory of Ordinary Differential Equations.* McGraw-Hill, New York, 1955.

[3] J. Dombrowski, Absolutely continuous measures for systems of orthogonal polynomials with unbounded recurrence coefficients. *Constr. Aprox.* **8**(1992) 161–167.

[4] J. Dombrowski and S. Pedersen, Orthogonal polynomials, spectral measures, and absolute continuity. *J. Comp. Appl. Math* **65**(1995) 115–124.

[5] D. Gilbert and D. Pearson, On subordinacy and analysis of the spectrum of one dimensional Schrödinger operators. *J. Math. Anal. Appl.* **128**(1987) 30–56.

[6] J. Janas and S.N. Naboko, Jacobi matrices with power-like weights, grouping in blocks approach. *J. Functional Analysis* **166**(1999) 218–243.

[7] J. Janas and S.N. Naboko, Multithreshold spectral phase transition examples in a class of unbounded Jacobi matrices. *Res. Reports in Math. Stockholm Univ.* (1999), 7.

[8] J. Janas and S.N. Naboko, Asymptotics of generalized eigenvectors for unbounded Jacobi matrices with power-like weights, Pauli matrices commutation relations and Cesaro averaging. *Operator Theory: Adv. and Appl.* **117**(2000), volume dedicated to M.G. Krein.

[9] S. Khan and D. Pearson, Subordinacy and spectral theory for infinite matrices. *Helv. Phys. Acta* **65**(1992) 505–527.

[10] A.G. Kostuchenko and K. Mirzoev, Generalized Jacobi matrices and deficiency indices of differential operators with polynomial coefficients. *Funct. Anal. Appl.* **33**(1999) 30–45.

[11] E.C. Titchmarsh, *Eigenfunction Expansions Associated with Second Order Ordinary Differential Equations, Part 1.* Oxford University, Oxford, 1962.

Luis O. Silva
66/3 Botanicheskaya, k.119/2
Stary Peterhof
St. Petersburg, Russia.
e-mail: silva@paloma.spbu.ru

Operator Theory:
Advances and Applications, Vol. 132, 395–402

Kovalevskaya's Dynamics and Schrödinger Equations of Heun Class

S.Yu. Slavyanov

In previous publications of the author [6], [7], [8] the Painlevé equations have been derived from linear second-order equations related to the Heun class. This has been done twofold. One approach is based on a phenomenological transformation from a quantum hamiltonian, related to Heun's class equation, to a classical hamiltonian related to Painlevé equations. The other approach is based on isomonodromie conditions formulated for auxiliary linear equations where an additional apparent singularity is added. The latter is an expansion of the idea first proposed by R. Fuchs [1] [1]. However, these studies are based on specific forms of linear equations and therefore are not invariant. Here, firstly, the links of the proposed approach to classical dynamics and studies of S. Kovalevskaya are traced; hence, the isomonodromic conditions play more the role of a conservation law. Secondly, the derivations are presented in invariant terms.

1. Kovalevskaya's dynamics

Suppose we study a one-dimensional movement of a particle determined by a hamiltonian $H(p,q,t)$ where q, p are the space and the momentum coordinates respectively, and t is time with the restriction that $H(p,q,t)$ is a quadratic function of p and a polynomial function of q and t. The functions $q(t)$, $p(t)$ are solutions of the Hamilton system of equations

$$\frac{dq}{dt} = \frac{\partial H(q,p,t)}{\partial p}, \quad \frac{dp}{dt} = -\frac{\partial H(q,p,t)}{\partial q} \tag{1}$$

with initial conditions

$$q(t_0) = q_0, \quad p(t_0) = p_0 \tag{2}$$

It is clear that the functions $p(t)$, $q(t)$ allow analytic continuation onto the complex plane of independent variable t. The singularities of solutions $p(t)$ and $q(t)$ of (1–2) which change their position if initial conditions (2) change are called moving singularities. Among these are distinguished moving critical points which include branching points and essential singularities. The problem posed by S. Kovalevskaya was: under what conditions are there no moving critical points [2] for $p(t)$ and $q(t)$

[1] While this publication was already prepared for the press the author became familiar with the paper by K. Okamoto [5] containing similar considerations.

so that moving singularities reduce to poles. She proposed a test how to solve this problem for a given equation. Her theory of motion of the top based on this method was awarded the prize of the Paris Academy of Science in 1888. The historical and mathematical aspects of this research are deeply discussed in the publiation by P. Clarkson and M. Kruscal [3].

Later the absence of moving critical points was called the Painlevé property; the Euler-Lagrange equations related to (1) are known as the Painlevé equations in this case.

The Riemann surfaces (namely, two of them related to the space and to the momentum coordinates) corresponding to the considered analytic continuation may be studied. Painlevé property leads to the consequence that this Riemann surface is fixed by equations (1) only (by its singularities!) and not depending on initial conditions (2). We call movement on this fixed Riemann surface as *Kovalevskaya's holomorphic dynamics.*

Our goal is to introduce a one-dimensional Schrödinger equation which in some sense generates Kovalevskaya's dynamics.

2. Heun's and deformed Heun's equations

Heun's equation is a Fuchsian second-order equation with four regular singularities. In general, it can be denoted by the following Riemann symbol

$$\begin{pmatrix} z_1 & z_2 & z_3 & z_4 & ; z \\ \rho_{11} & \rho_{12} & \rho_{13} & \rho_{14} & ; \lambda \\ \rho_{21} & \rho_{22} & \rho_{23} & \rho_{24} & \end{pmatrix} \tag{3}$$

Here z_j, $j = 1, 2, 3, 4$ denote the location of singularities of the equation and ρ_{mj}, $m = 1.2$, $j = 1, 2, 3, 4$ denote the characteristic exponents of Frobenius solutions at these points; solutions then have the form

$$y_m(z, z_j) = (z - z_j)^{\rho_{mj}} \sum_{k=0}^{\infty} c_{mjk}(z - z_j)^k \tag{4}$$

These characteristic exponents are not independent; they have to satisfy Fuchs' identity

$$\sum_{j=1}^{4} (\rho_{1j} + \rho_{2j}) = 2 \tag{5}$$

The parameter λ is called an accessory parameter and is a "global" parameter which is not copncerned with the principal local behavior of solutions at the singularities.

In this paper also the Fuchsian equation with five regular singularities is studied which can be denoted by the following Riemann symbol

$$\begin{pmatrix} z_1 & z_2 & z_3 & z_4 & z_5 & ; z \\ \rho_{11} & \rho_{12} & \rho_{13} & \rho_{14} & \rho_{15} & ; \lambda \\ \rho_{21} & \rho_{22} & \rho_{23} & \rho_{24} & \rho_{25} & ; \mu \end{pmatrix} \tag{6}$$

with two accessory parameters λ and μ. Fuchs' identity for the characteristic exponents changes to

$$\sum_{j=1}^{5}(\rho_{1j}+\rho_{2j})=3 \tag{7}$$

There is no unique correspondence between the Riemann symbols (3) and (5), and the Fuchsian equations under consideration. The reason is that the accessory parameter λ may be replaced by any linear combination $r_1\lambda+r_2$ where r_1 and r_2 are arbitrary functions of other parameters of the equation. We choose the appropriate equations according to the needs of the specific problem under consideration. Without loss of generality, the singularities of the Heun equation are assumed to be located as

$$z_1=0, \quad z_2=1, \quad z_3=t, \quad z_4=\infty. \tag{8}$$

However the notations z_j, $j=1,2,3$ are kept until the end of the paper for unification. The characteristic exponents at $z=\infty$ are replaced by κ_m, $m=1,2$. According to Fuchs' identity only one of these two quantities is arbitrary. Heun's equation is then written as

$$\sigma(z)y''(z)+\sum_{j=1}^{3}(b_j+1)\sigma_j(z)y'(z)+\left[\sum_{j=1}^{3}\frac{a_j\sigma_j(z)}{z-z_j}+\right.$$
$$\left.d_\infty(z-z_3)+\sigma_3(z_3)\left(\frac{-\lambda}{z_2-z_1}+\varphi(z_3)\right)\right]y(z)=0 \tag{9}$$

Here

$$b_j=-(\rho_{1j}+\rho_{2j}), \quad a_j=\rho_{1j}\rho_{2j}, \quad j=1,2,3, \quad a_\infty=\kappa_1\kappa_2$$
$$d_\infty=a_\infty-\sum_{j=1}^{3}a_j$$

and

$$\sigma(z)=\prod_{j=1}^{3}(z-z_j), \quad \sigma_j(z)=\frac{\sigma(z)}{z-z_j}$$

with

$$\varphi(z_3)=\frac{1}{2}\sum_{j=1}^{2}\frac{(b_3+1)(b_j+1)}{z_3-z_j}$$

If $\rho_{1j}=0$ hold for finite singularities, which implies $a_j=0$, then equation (9) is called to be in the canonical form; actually, this canonical form is often considered as Heun's equation [4].

The form of proposed equation (9) may seem to be somewhat unusual. Firstly, the symmetry between the singularities is broken; secondly, the accessory parameter λ is introduced in a rather artificial manner. However, this presentation of the equation is underlining the fact that the points z_1, z_2 are considered as fixed and the point z_3 is considered as a varying parameter of the equation.

Lemma 1. *The quantity* λ *in equation* (9) *remains invariant under the linear transformations of independent variable* z

$$z := a\zeta + b \tag{10}$$

and the s-homotopic transformations of dependent variable y

$$y := (z - z_k)^{\gamma} w \tag{11}$$

Proof: The lemma is proved by direct calculations. Testing linear teansformations is trivial. Now test (11). Let the values a_k, b_k d_∞ and $\varphi(z_3)$ be $\tilde{a}_k$, $\tilde{b}_k$, $\tilde{d_\infty}$ and $\tilde{\varphi}(z_3)$ after transformation (11) with fixed k. Then

$$\tilde{a}_k = a_k + \gamma b_k + \gamma^2, \quad \tilde{b}_k = b_k + 2\gamma, \quad \tilde{d_\infty} = d_\infty + \gamma \sum_{j=1, j\neq k}^{3} (b_j + 1) \tag{12}$$

For transformations of φ it is necessary to distinguish values of k. If $k = 3$ then

$$\tilde{\varphi}(z_3) = \varphi(z_3) + \gamma \sum_{j=1}^{2} \frac{b_j + 1}{z_3 - z_j} \tag{13}$$

If $k \neq 3$ then

$$\tilde{\varphi}(z_3) = \varphi(z_3) + \gamma \frac{b_3 + 1}{z_3 - z_k} \tag{14}$$

On the other hand equation (9) transforms to

$$\sigma(z) w''(z) + \left(\sum_{j=1, j\neq k}^{3} (b_j + 1)\sigma_j(z) + (b_k + 2\gamma + 1)\sigma_k(z) \right) w'(z)$$
$$+ \left[\sum_{j=1}^{3} \frac{a_j \sigma_j(z)}{z - z_j} + \frac{(\gamma^2 + \gamma b_k)\sigma_k}{z - z_k} + \gamma \sum_{j=1, j\neq k}^{3} \frac{(b_j + 1)\sigma_j(z)}{z - z_k} \right.$$
$$\left. + d_\infty (z - z_3) + \sigma_3(z_3) \left(\frac{-\lambda}{z_2 - z_1} + \varphi(z_3) \right) \right] y(z) = 0 \tag{15}$$

Substituting (12–14) into (15) we arrive to the required result. □

In a somewhat weaker formulation this result can be found in the appendix to the book [4]. The definition and the properties of the s-homotopic transformation are also discussed there. The other quantities which are conserved under transformations (10–11) are the differences between characteristic exponents. In applications it is preferable to take the squares of these differences

$$\Delta_j = (\rho_{1j} - \rho_{2j})^2, \; j = 1, 2, 3 \quad \Delta_\infty = (\kappa_1 - \kappa_2)^2. \tag{16}$$

Consider an equation with five regular singularities of the form

$$\sigma(z)y''(z) + \left(\sum_{j=1}^{3}(b_j+1)\sigma_j(z) - \frac{\sigma(z)}{z-q}\right)y'(z) + \left[\sum_{j=1}^{3}\frac{a_j\sigma_j(z)}{z-z_j}\right.$$

$$\left. + d_\infty(z-z_3) + \sigma_3(z_3)\left(\frac{-\lambda}{z_2-z_1} + \varphi(z_3)\right) + \sigma_3(q)\frac{q-z_3}{z_2-z_1}\frac{p}{z-q}\right]y(z) = 0 \quad (17)$$

Here the characteristic exponents at the singularity $z = q$ are chosen as

$$\rho_{14} = 0, \quad \rho_{24} = 2$$

which yelds

$$b_4 = -(\rho_{14} + \rho_{24}) = -2, \quad a_4 = \rho_{14}\rho_{24} = 0,$$

Other notations are the same as in (9).

Consider a particular case of equation (17) determined by the condition that the general solution of it is a holomorphic function at the point $z = q$ whereas it remains a singularity. In this case the point $z = q$ is an apparent singularity of (17).

Lemma 2. *The condition*

$$\lambda = \frac{z_2-z_1}{\sigma_3(z_3)}\left[\frac{\sigma(q)}{(z_2-z_1)^2}p^2 + \frac{p}{z_2-z_1}\sum_{j=1}^{3}(b_j+1)\sigma_j(q)\right.$$

$$\left. + \sum_{j=1}^{3}\frac{a_j\sigma_j(q)}{q-z_j} + d_\infty(q-z_3)\right] + (z_2-z_1)\varphi(z_3) \quad (18)$$

is sufficient for the point $z = q$ to be an apparent singularity.

Proof: In a neighbourhood of the apparent singularity $z = q$ the general solution of equation (17) is expanded in the Taylor series

$$y(z,q) = \sum_{k=0}^{\infty} r_k(z-q)^k \quad (19)$$

By substituting (19) into (17) and equating coefficients for degrees of $z - q$ we arrive to the following first two equations in the recurrent system

$$r_1 = \frac{p}{z_2-z_1}r_0, \quad (20)$$

$$\left(\frac{\sigma(q)}{z_2-z_1}p + \sum_{j=1}^{3}(b_j+1)\sigma_j(q)\right)r_1$$

$$+ \left(\sum_{j=1}^{3}\frac{a_j\sigma_j(q)}{q-z_j} + d_\infty(q-z_3) + \sigma_3(z_3)\varphi(z_3) - \frac{\sigma_3(z_3)\lambda}{z_2-z_1}\right)r_0 = 0 \quad (21)$$

The compatibility of (20–21) leads to condition (18) as required. □

We call equation (17) under condition (18) the *deformed Heun's equation.* Condition (18) is equivalent to the following process. Instead D and z in equation (9) we formally take p and q with an additional modification of the value of the parameter d_∞ due to the modified Fuchs condition (7).

3. Heun's equation and Painlevé sixth equation

Assume that Heun equation is written in a Schrödinger form

$$H(\hat{p},\hat{q},t)\psi = \lambda\psi$$

where t is a parametric time (adiabatic approach) and $\hat{p},\hat{q}$ are the operators of the quantum momentum and space coordinates

$$\hat{p} \equiv \frac{d}{dz}, \quad \hat{q} \equiv z.$$

The function $H(\hat{p},\hat{q},t)$ is a quantum hamiltonian which in terms of (9) may be expressed as

$$H(\hat{p},\hat{q},t) = \frac{z_2 - z_1}{\sigma_3(z_3)}\left[\sigma(z)\hat{p}^2 + \sum_{j=1}^{3}(b_j+1)\sigma_j(z)\hat{p} + \sum_{j=1}^{3}\frac{a_j\sigma_j(z)}{z - z_j} + d_\infty(z-t)\right] + \varphi(z_3) \tag{22}$$

Corresponding to (22) the hamiltonian of the classical motion takes the form

$$H(p,q,t) = \frac{1}{\sigma_3(t)}\left[\sigma(q)p^2 + \sum_{j=1}^{3}(b_j+1)\sigma_j(q)p + \sum_{j=1}^{3}\frac{a_j\sigma_j(q)}{q - z_j} + d_\infty(q-t)\right] + \varphi(t) \tag{23}$$

According to Lemma 2 this form of the hamiltonian is induced by study of the deformed Heun's equation (17). The Legendre transformation generates the relationship between momentum and velocity as follows

$$p = \frac{1}{2\sigma(q)}(\dot{q}\sigma_3(t) - \sum_{j=1}^{3}(b_j+1)\sigma_j(q)) \tag{24}$$

The corresponding Lagrangian reads

$$\mathcal{L}(\dot{q},q,t) = \dot{q}p - H(p,q,t) = \frac{1}{2\sigma(q)\sigma_3(t)}(\dot{q}\sigma_3(t) - \sum_{j=1}^{3}(b_j+1)\sigma_j(q))^2$$
$$-\frac{1}{\sigma_3(t)}\sum_{j=1}^{3}\frac{a_j\sigma_j(q)}{q-z_j} - \frac{d_\infty(q-t)}{\sigma_3(t)} - \varphi(t) \qquad (25)$$

The equation of the corresponding classical motion (Euler-Lagrange) is

$$\frac{d}{dt}\frac{\partial\mathcal{L}}{\partial\dot{q}} - \frac{\partial\mathcal{L}}{\partial q} = 0 \qquad (26)$$

After differentiation and appropriate simplification we arrive to the following equation

$$\frac{2\sigma_3(t)}{\sqrt{\sigma(q)}}\frac{d}{dt}\frac{\dot{q}\sigma_3(t)}{\sqrt{\sigma(q)}} + \frac{\dot{q}\sigma_3^2(t)}{\sigma(q)(q-t)} + \left[A_\infty + \sum_{j=1}^{3}\frac{A_j\sigma_j(z_j)}{(q-z_j)^2}\right] = 0 \qquad (27)$$

with

$$\begin{aligned} A_\infty &= -(\kappa_1-\kappa_2)^2 \\ A_j &= \Delta_j + 1 + 2b_j,\ j=1,2, \\ A_3 &= \Delta_3 - 1. \end{aligned} \qquad (28)$$

This is the general form of the Painlevé equation P^6 generated by Heun's equation and resulting in Kovalevskaya's dynamics. It is clear that only the parameters A_j in equation (27) are varying. Of course the location of the fixed singularities may also be changed. The particular forms of Heun's equation are:

i) the canonical form with $\rho_{2j}=0$, $b_j=-\rho_{1j}$,
ii) the normal form (without the first derivative) with $b_j=-1$,
iii) the self-adjoint form with $b_j=0$.

These forms lead to the particular values of parameters in the Painlevé equation. Moreover, explicit solutions of the Heun's equation at half-integer values of $\rho_{1j}-\rho_{2j}$ may lead to explicit solutions of the corresponding Painlevé equations.

In a similar manner the confluent cases of Heun's equations can be considered.

Acknowledgements. The author is pleased to express his gratitude to Prof. W.N. Everitt and Dr. P. Kurasov for timely support while this text was in preparation and thank Dr. A. Kitaev for drawing the author's attention to paper [5]. This research was partly sponsored by the Swedish Royal Academy.

References

[1] R. Fuchs, Math. Ann., 63, 301–321, (1907)

[2] S. Kovalevski, Acta Math. 12, 177–232 (1889), S. Kovalevski, Acta Math. 14, 81–93 (1889).

[3] M. Kruscal, P. Clarkson, Studies in Appl. Math. 86, 87–165, (1992).

[4] Heun's Differential Equations, ed. A. Ronveaux, Oxford University Press, 1996.

[5] K. Okamoto, in *Painlevé transcendents*, ed. B. Levi and P. Winternitz, Plenum Press, New York, (1992).

[6] S. Slavyanov, J. Phys. A., 29, 7329–7335, (1996).

[7] S. Slavyanov, Theor. Math. Phys., 123, 744–753, (2000).

[8] S.Yu. Slavyanov, W. Lay. Special Functions: Unified Theory based on Singularities, Oxford University Press, 2000.

S.Yu. Slavyanov
Institute of Physics
St. Petersburg State University
Peterhoff
Ulianovskaya st., 1
198904, Russia, St. Petersburg

Operator Theory:
Advances and Applications, Vol. 132, 403–410

Effective Quantum Number for Centrally Symmetric Potentials

Yu.V. Tarbeyev, N.N. Trunov, A.A. Lobashev and V.V. Kukhar

Abstract. An effective quantum number as some linear combination of the radial and orbital quantum numbers with the coefficient depending on the specific form of centrally symmetric potentials is introduced. This quantum number defines the order of the level succession and the values of their energies by means of a modified semiclassical quantization condition. The developing method presents some perturbation theory treating given potentials in comparison with the oscillator and Coulomb ones.

1. Introduction

A number of problems appearing in atomic physics, theoretical models of nuclei [1], quarkonium models [2, 3], models of the metallic clusters [4] using a selfconsistent potential etc. include determining of the bound state spectra. The lack of analytic solutions for the most potentials calls for developing methods approximating the spectra. Such method should be simple enough, universal and physically transparent.

At least two different variants of questions arise in such problems. The first one is obtaining of the entire spectrum for a fixed potential or, in a simplified version, determining of the order of the level succession with increasing value of E. Another one is studying of the appearence of new states (with zero energy) for the increasing amplitude of the potential. Such special case of the Sturm problem takes place for the periodic system of the elements: new levels (n_r, l) where n_r and l are the radial and orbital quantum numbers correspondingly, appear for the increasing nuclear charge Z in the selfconsistent atomic potential $V(r) = -Z\zeta(r, Z)/r$, with $\zeta(r, Z)$ being some screening function on which the levels ordering depends.

In semiempiric determining of the spectra one often expect that the energy values depend on some linear combinations n_r and l, i.e. $E(n_r, l) = E(\alpha n_r + \beta l)$. E.g. the Madelung-Kletchkovsky rule predicts appearing of new shells (n_r, l) in the periodic system with increasing $n_r + 2l$ [5, 6, 7]; for the metallic clusters it is expected $E = E(3n_r + l)$ [8], a similar quantum number for nuclei was proposed in [1].

Supported by The Royal Swedish Academy of Sciences.

However, such dependence of the exact spectra on some linear combination is known only for the Coulomb (E_c) and oscillator (E_{osc}) potentials:

$$E_c = \frac{\mathrm{const}}{(n_r + l + 1)^2} \equiv \frac{\mathrm{const}}{n^2}; \qquad E_{\mathrm{osc}} = \mathrm{const}(n_r + \frac{1}{2}l + \frac{3}{4}) \equiv \mathrm{const} \cdot n_{\mathrm{osc}} \quad (1)$$

Besides, it is worth mentioning the condition of appearing of the new level (n_r, l) with zero energy in the Lentz potentials $V_\sigma(r)$ [5] with the increasing amplitude b:

$$V_\sigma(r) = -\frac{b}{r^2\left(r^\sigma + r^{-\sigma}\right)^2}, \quad b = 2\sigma^2\left(n_\sigma^2 - \frac{1}{4}\right); \quad n_\sigma = n_r + \frac{1}{2} + \frac{1}{\sigma}\left(l + \frac{1}{2}\right). \quad (2)$$

Thus the closeness of the actual argument in $E(n_r, l)$ to some linear combination $\alpha n_r + \beta l$ in many cases calls for searching a method for explaining this fact, determining the optimum value α/β and estimating accuracy of the method.

In the present work we develop such method based on the usual semiclassical quantization condition for centrally symmetric problems [9, 10]. It should be mentioned here that this condition leads to the exact spectra for our two reference cases, i.e. the oscillator and Coulomb potentials.

2. The linear quantization condition

Our analysis in what follows is based on the usual WKB quantization condition for centrally symmetric potentials [9, 10]

$$\Phi(E, \lambda) = \frac{1}{\pi\hbar}\int_0^\infty \left[2\left(E - V(r)\right) - \frac{\lambda^2}{r^2}\right]_+^{1/2} dr = n_r + \frac{1}{2} \equiv \nu, \quad (3)$$

where $\lambda = l + 1/2$ and $f_+ \equiv (f + |f|)/2$ (i.e. the coordinate integrals are taken between the turning points). It is easy to find that for the oscillator and Coulomb potentials

$$\Phi(E, 0) - \Phi(E, \lambda) = \phi\lambda \quad (4)$$

with the constants $\phi = \phi_{\mathrm{osc}} = 1/2$ for the case of the oscillator potential and $\phi = \phi_c = 1$ for the case of the Coulomb potential (independently on E). We treat λ as an continuous parameter in intermediate calculations.

Thus the condition (3) is equivalent for the above mentioned potentials to a modified condition

$$\Phi(E, 0) = \frac{1}{\pi\hbar}\int \left[2\left(E - V(r)\right)\right]_+^{1/2} = \nu + \phi\lambda \equiv T(\nu, \lambda, \phi) \quad (5)$$

involved the elimination of the centrifugal potential λ^2/r^2 from the phase integral.

Using (5) we obtain the exact spectra (1) with $T(\nu, \lambda, \phi_{\mathrm{osc}}) = n_{\mathrm{osc}}$ and $T(\nu, \lambda, \phi_c) = n$ (as well as for (2) $T(\nu, \lambda, \phi_\sigma) = n_\sigma$ with $\phi_\sigma = 1/\sigma$ for $E = 0$). We interpret $T(\nu, \lambda, \phi)$ as an effective quantum number also for arbitrary potentials $V(r)$ with the value of ϕ depending on $V(r)$ functionally.

In what follows, it is useful to introduce the new variables $x = \ln r$ and

$$W(x,E) = 2\left[(E - V(e^x))\, e^{2x}\right]_+ \tag{6}$$

and to work with the reduced variables $U(x)$ and Λ

$$\begin{aligned} 0 \le U(x,E) &= \frac{W(x,E)}{A^2} \le 1, \quad 0 \le \Lambda = \frac{\lambda}{A} \le 1; \\ A^2(E) &= \max_x W(x,E). \end{aligned} \tag{7}$$

We assume that $U(x)$ has the sole maximum and often omit the argument E where it is possible.

The original condition (4) and its linear in Λ approximation (5) look now correspondingly as

$$\frac{1}{\pi\hbar}\int \left[U(x) - \Lambda^2\right]_+^{1/2} dx \equiv \frac{1}{\pi\hbar}\int U^{1/2}(x)\, dx - \int_0^{\Lambda} \psi(\Lambda) d\Lambda = \nu / A \tag{8}$$

$$\frac{1}{\pi\hbar}\int U^{1/2}(x)\, dx - \phi\Lambda = \nu/A. \tag{9}$$

The equation (8) defines a new function $\psi(\Lambda)$ coinciding with ϕ if $\psi(\Lambda)$ is independent on Λ.

Comparing (8) and (9) we define the best ϕ as the mean square value:

$$I(\phi) = \int_0^1 d\Lambda \left[\int_0^{\Lambda} \psi(\Lambda) d\Lambda - \phi\Lambda\right]^2 = \min. \tag{10}$$

After simple calculations it leads to

$$\phi = \frac{3\chi_1 - \chi_3}{2}, \qquad \chi_p \equiv p\int_0^1 \psi(\Lambda)\Lambda^{p-1} d\Lambda. \tag{11}$$

The function χ_p (11) is defined so that the integral operator becomes the identity if $\psi(\Lambda) = \text{const}$ and χ_p/p is the Mellin transform of $\psi(\Lambda)$ (we extend $\phi(\Lambda)$ by zero outside the interval $(0,1)$). It follows from the general properties of the Mellin transform that

$$\psi(\Lambda = 0) = \chi_0, \qquad \psi(\Lambda = 1) = \chi_\infty.$$

For the oscillator and Coulomb potentials $\phi = \psi = \chi_p = \text{const}$ independently on p and E. For arbitrary potentials these function may depend on E and χ_p also on p. In any case $\phi - \chi_3 = 3(\phi - \chi_1)$ so that we have $\phi < \chi_1 < \chi_3$ or $\phi > \chi_1 > \chi_3$; of course our approximation requires what $\phi > 0$, i.e. $3\chi_1 > \chi_3$ what is really always fulfilled. Note that χ_1 is the mean value of $\phi(\Lambda)$, so that the dependence of χ_p on p may be evaluated by the dimensionless parameter

$$\eta = \frac{|\phi - \chi_1|}{\chi_1}.$$

Deviation of the WKB condition (8) from the linear in Λ one (9) increases together with the increasing value of η. For our reference potentials $\eta = 0$. It should be noted that a possible dependence ϕ on E does not prevent the equivalency of (8) and (9) but really it is accompanied by the dependence χ_p on p. The inverse statement is not valid, see (15), (16).

Differentiating Eq. (9) with respect to Λ, we obtain

$$\frac{\psi(\Lambda)}{\Lambda} = \int \frac{dx}{[U(x) - \Lambda^2]_+^{1/2}}. \tag{12}$$

To find χ_p (11) as a functional of $U(x)$, we multiply both sides of (12) by Λ^{p-1} and integrate with respect to Λ from zero to one. Changing the order of integrations in the right-hand side, we obtain

$$\chi_p = \frac{M_p}{m_p}, \quad M_p = \int U^{p/2}(x)dx; \quad m_p = B(p/2, 1/2), \tag{13}$$

where $B(u, v)$ is the beta function.

For a broad class of power-low potentials

$$V(r) = br^{\mu}, \qquad b\mu > 0 \tag{14}$$

with $-2 < \mu < \infty$ the function χ_p (and $\psi(\Lambda)$) are monotonic and not depending on E

$$\chi_p(\mu) = \left(\frac{2+\mu}{2}\right)^{\frac{2+\mu}{2}\frac{p}{\mu}} \frac{2^{p/2}}{\mu^{p/2+1}} \frac{B(p/\mu, p/2+1)}{B(p/2, 1/2)} \qquad (\mu > 0), \tag{15}$$

$$\chi_p(\mu) = \left(\frac{2}{2-|\mu|}\right)^{\frac{2-|\mu|}{2}\frac{p}{|\mu|}} \frac{2^{p/2}}{|\mu|^{p/2+1}} \frac{B(p\frac{2-|\mu|}{2|\mu|}, \frac{p}{2}+1)}{B(p/2, 1/2)} \qquad (-2 < \mu < 0). \tag{16}$$

In the limiting case $p \to \infty$ we obtain from (15), (16) for all $\mu > -2$

$$\chi_\infty(\mu) = \frac{1}{\sqrt{\mu+2}}. \tag{17}$$

Naturally χ_p (15), (16) do not depend on p if $\mu = 2$, $\mu = -1$.

The Schrödinger equation for the power-law potentials (14) is invariant under the transformations

$$\begin{pmatrix} \mu \\ \Psi(r) \\ E \\ b \\ \lambda \end{pmatrix} \to \begin{pmatrix} \mu' \\ \tilde{\Psi}(r) = r^q \Psi(r^\kappa) \\ E' = -\kappa^2 b \\ b' = -\kappa^2 E \\ \lambda' = \kappa\lambda \end{pmatrix}; \tag{18}$$

$$(2+\mu)(2+\mu') = 4; \quad \kappa \equiv |\mu'/\mu|, \quad q = (1-\kappa)/2$$

with $\Psi(r)$ and $\tilde{\Psi}(r)$ being the old and new wave function respectively (see, for example, [3]). This, in particular, reproduces the known duality between the oscillator ($\mu = 2$) and Coulomb ($\mu' = -1$) potentials.

If our linear in λ condition (5) satisfies to (18), it means in particular that $\phi(\mu')\lambda' = \phi(\mu)\lambda$ with $\lambda' = \kappa\lambda$. This fact really takes place as it follows from (15), (16) that $\chi_p(\mu')\lambda' = \chi_p(\mu)\lambda$ for each value of p. Moreover, it is easily verified that the energy levels $E(\nu, \lambda)$ calculated for μ and μ' by means of (9)

$$E(\nu,\lambda) = b^{\frac{2}{2+\mu}} \left[(\nu + \phi\lambda)\frac{\mu\pi}{\sqrt{2}B(3/2, 1/\mu)}\right]^{\frac{2\mu}{2+\mu}} \qquad (\mu > 0),$$

$$|E(\nu,\lambda)| = |b|^{\frac{2}{2-|\mu'|}} \left[\frac{1}{\nu + \phi\lambda} \frac{\sqrt{2}B\left(\frac{3}{2}, \frac{2-|\mu'|}{2|\mu'|}\right)}{|\mu'|\pi}\right]^{\frac{2|\mu'|}{2-|\mu'|}} \qquad (-2 < \mu' < 0)$$

also satisfy to (18). Thus, for all power-law potentials our linear condition (9) reproduces the exact relations (18).

3. Universal asymptotical behavior of χ_p and $\psi(\Lambda)$

By the substituting $U = \exp S$, integral (13) is brought to the standard form for the asymptotic Laplace expansion:

$$M_p = \int dx U^{p/2}(x) \simeq \sqrt{\frac{4\pi}{|U^{(2)}|p}} \left(1 + \frac{a_1}{p} + \frac{a_2}{p^2} + \dots\right), \tag{19}$$

all the derivatives $U^{(n)}$ are taken at the maximum point. Returning to U we have

$$a_1 = \frac{i_1(U) - 3}{4}, \qquad i_1 = \frac{U^{(4)}}{[U^{(2)}]^2} - \frac{5}{3}\frac{[U^{(3)}]^2}{[U^{(2)}]^3} \tag{20}$$

and so on. Expanding m_p in the denominator in (15) similarly to (20) we obtain an asymptotic expansion of χ_p, which can be conveniently written in the exponential form

$$\frac{\chi_p}{\chi_\infty} = \exp\left(\sum_{k=1}^{\infty} \frac{b_k}{p^k}\right); \quad \chi_\infty = \sqrt{\frac{2}{|U^{(2)}|}}; \quad b_1 = \frac{i_1}{4} - 1. \tag{21}$$

The asymptotic regime is rapidly reached with increasing p if $b_1 < 1$, $b_2 \ll b_1$ etc., especially if $b_1 b_2 < 0$. This is really the case with the Γ function through which M_n are expressed not only for power-law but also for many other potentials. The ratio of $\Gamma(p)$ to its asymptotic Stirling expression $\Gamma_0(p)$ has form (21) with $b_1 = 1/12$ and $b_2 = -1/360$.

The smallness of the characteristic parameter $p^* < 1$ with which the asymptotic behavior starts is as follows: $\Gamma(p)$ can be represented as [11]

$$\frac{d^2}{dp^2} \log \frac{\Gamma(p)}{\Gamma_0(p)} = J(p) = 2p \int_0^\infty \frac{\arctan t \, dt}{e^{2\pi pt} - 1},$$

where J is some typical integral of the Abel-Planá method, so that the fall-off parameter $p^* \simeq (2\pi)^{-1} < 0.2$ (see examples in [12]). Thus the individual features of $U(x)$ are lost in χ_p for $p > p^*$ so that $\chi_p, \phi, \psi(\Lambda)$ have an universal behaviour. It means that the whole conception of the present method is also valid for arbitrary potentials.

For power-law potentials we obtain the expansion (21) with χ_∞ (17) and

$$b_1(\mu) = \frac{(\mu+4)^2}{12(\mu+2)} - \frac{3}{4}, \; b_3(\mu) = \frac{1}{360}\left[7 + \frac{8\mu^3}{(\mu+2)^2} - \mu^3\right], \; b_2 = b_4 \equiv 0.$$

The values of b_1 and b_3 are equal to zero for the oscillator and Coulomb potentials and are small in their vicinity with $|b_3| \ll |b_1|$, $b_1 b_3 < 0$. For instance, $b_1 = -1/12$ and $b_3 = 17/972 < 0.02$ for $\mu = 1$.

4. Discussion

Actually we have developed above a specific form of the perturbation theory with the smallness parameter $\delta(\Lambda) = (d\psi/d\Lambda)\psi^{-1}$. Since $\psi(\Lambda) = \text{const} = \phi$ and thus $\delta \equiv 0$ only for the oscillator and Coulomb potentials [13], at the same time we take into account the difference between the studied potential and one of these reference cases (the nearest to the studied one). The function $\delta(\Lambda)$ as well as $\psi(\Lambda)$ is in 1:1 correspondence with the function χ_p (15), (16) or even with the χ_k for $k = 1, 2, 3, \ldots$ (holomorphic and bounded in the right half-plane of p function χ_p). The pair (χ_1, χ_3) produces the linear quantization condition (9) with the effective quantum number $T(\nu, \lambda, \phi)$. Taking into account three (or more) parameters χ_k we can get non-linear corrections to the quantization condition and the spectra. Actually a very good approximation for χ_p is

$$\tilde{\chi}_p = \chi_\infty \frac{p+A}{p+B} \tag{22}$$

with A and B determining from $\tilde{\chi}_p = \chi_k$ for two chosen values of k (often we use $k = 1$ and $k = 3$) [13, 14]. It should be noted that functions $\tilde{\chi}_p$ received with different choices of k are close enough. According to (22), the universal form of the approximate $\psi(\Lambda)$ is

$$\tilde{\psi}(\Lambda) = C + D\Lambda^F. \tag{23}$$

The simplest linear condition is actually exact enough to predict the correct order of the level succession in all potentials (14) and in particular the correct splitting of levels degenerated for the oscillator by any small difference $|\mu - 2|$ in (14). The accuracy of spectra calculated according to (5), (11), (15) for $V(r) = r$ is $0.3 \div 0.5\%$ and even in the worst case of the non-analytic regtangular potential well ($\mu \to \infty$) errors do not exceed $3 \div 5\%$ [14].

The coefficients of the asymptotic expansion (21) also are in 1:1 correspondence with χ_p so that we can use the set $\{\chi_\infty, b_k\}$ with $k = 1, 2, 3, \ldots$ instead of χ_p or some subset χ_n as its representation. Since values of b_k fall down with increasing k, we obtain a set of the subordinate parameters. In accordance with our conception of the reduced description [13] we can define the first correction to $\phi(\Lambda) = \phi = \text{const}$ neglecting all b_k with $k \geq 2$, the second one — with $k \geq 3$

and so on. Restricting to the terms (χ_∞, b_1) in (21) and inserting them in (11) we obtain linear in b_1 expressions

$$\phi = \chi_\infty \left(1 + \frac{4b_1}{3}\right) \simeq \chi_{p=3/4}; \qquad \eta = \frac{|b_1|}{3}. \tag{24}$$

It is interesting to find ϕ not as a linear combination χ_1 and χ_3, but as some product $\hat{\phi} = \chi_1^\sigma \chi_3^\tau$, restricting to the coefficient b_1 in (21). As ϕ is proportional to χ_∞ we obtain $\sigma + \tau = 1$. Comparing this product with (24) and using the linear in b_1/p approximation we have $\alpha + \beta/3 = 4/3$ and

$$\hat{\phi} \equiv \frac{\chi_1^{3/2}}{\chi_3^{1/2}}. \tag{25}$$

This combination $\hat{\phi}$ indeed appears as the best estimate for ϕ in another variant of the presented method [13] and is very close to ϕ.

It is worth mentioning that the moments M_n (13) are proportional to the Weyl estimates $N_n[V;E]$ [15] for the bound states numbers in n-dimensional spherically symmetric problem with the given potential $V(r)$ and energies not exceeding E. Thus according to (25)

$$\hat{\phi} = \text{const}\, \frac{N_1^{3/2}}{N_3^{1/2}}. \tag{26}$$

For the power-law potentials discussed above all parameters ϕ, $\hat{\phi}$, $\psi(\Lambda), \chi_p$ and b_k do not depend on E, but such dependence takes place for other potentials. If $|d\phi/dE| \ll 1$, the energy levels may be determined from the linear condition (5) with $T = T(\nu, \lambda, \phi(E))$ by the means of an iteration procedure.

The above method is based on the quantization condition (3) which is the leading approximation of the exact condition [10]. Taking into account the correction proportional to $\hbar^2$ [13] and using (8) and (13), we obtain

$$\frac{A\chi_1}{\hbar}\left(1 - \frac{\hbar^2}{24\pi}\frac{R(\Lambda^2)}{\chi_1 A^2}\right) = \nu + \int_0^\Lambda \psi(\Lambda) d\Lambda \tag{27}$$

$$R(\tau) = \frac{d^2 I(\tau)}{d\tau^2}, \quad I(\tau) = \int \frac{dx}{[U(x) - \tau]_+^{1/2}} \left(\frac{dU(x)}{dx}\right)^2.$$

where we set $\Lambda^2 = \tau$ for brevity. As it is seen from (28), if $\Lambda \to 0$ we have $\nu_{\max} \simeq A\chi_1/\hbar$, so that the relative smallness of the correction is

$$\rho = \frac{\hbar^2 R}{24\pi\chi_1 A^2} \simeq \frac{R\chi_1}{24\pi\nu_{\max}^2}. \tag{28}$$

It depends not separately on R, i.e. on the form of studied potentials or on the number of states $\nu_{\max}$, but only on their combination (28). Thus for our reference cases $\mu = 2, -1$ when $R \equiv 0$ the usual condition (3) or (8) is always exact. For other potentials with $R \neq 0$ its accuracy increases with increasing $\nu_{\max}$ and in physically actual cases this correction is small enough if $\nu_{\max} \geq 2$.

References

[1] A. Bohr, B.R. Mottelson, Nuclear Structure, vol. II: Nuclear Deformations, Ch. 6, W.A. Benjamin Inc., New York, Amsterdam, 1974.

[2] A. Martin, Quarkonium, *Comments Nucl. Part. Phys.*, **16**, 249 (1986).

[3] C. Quigg, J.L. Rosner, Quantum mechanics with application to quarkonium, *Phys. Rep.*, **56**, 167 (1979).

[4] B.A. Kotos and M.E. Grypeos, Parametrization of the effective potential in sodium clusters, in Atomic and Nuclear Clusters, G.S. Anagnostatos, W. von Oertzen (Eds.), Proc. of the Second Intern. Conf. at Santorini, Greece, Springer-Verlag, Berlin, Heidelberg, 1995.

[5] Yu.N. Demkov and V.N. Ostrovsky, Law of $n + l$ filling in the periodic system and focussing potentials, *Zh. Eksp. Teor. Fiz.*, **62**, 125 (1972) [*Sov. Phys. JETP*, **35**, 66 (1972)].

[6] V.M. Klechkovskii, The Distribution of Atomic Electrons and the Filling Rule for $(n + l)$-groups [in Russian], Atomizdat, Moscow (1968).

[7] Yu.V. Tarbeyev, N.N. Trunov, A.A. Lobashev, V.V. Kukhar, Extraction of information from the periodic system of the elements as an inverse problem, *Zh. Eksp. Teor. Fiz.*, **112**, 1226 (1997) [*Sov. Phys. JETP*, **85**, 667 (1997)].

[8] V.N. Ostrovsky, $3n_r + l$ grouping of energy levels and clusters, *Phys. Rev. A*, **65**, 626 (1997).

[9] *L.D. Landau, E.M. Lifshitz.* Quantum Mechanics. Perganom, Oxford, 1977.

[10] S.Yu. Slavyanov, Asymptotic Solutions of the One-Dimensional Schrödinger Equation, Translations of Mathematical Monographs, v. **151**, 1996.

[11] E.T. Whittaker and G.N. Watson, A Course of Morden Analysis, Part II, Sect. 12.33, Cambridge University Press, 1927.

[12] V.M. Mostepanenko and N.N. Trunov, The Casimir Effect and its Applications, Clarendon, Oxford, 1997.

[13] A.A. Lobashev, N.N. Trunov, The reduced semiclassical description method, *Teor. Mat. Fiz.*, **120**, 99 (1999) [*Theor. Math. Phys.*, **120**, 869, (1999)].

[14] A.A. Lobashev, N.N. Trunov, An integral semiclassical method for calculating the spectra for centrally symmetric potentials, *Teor. Mat. Fiz.*, **124**, N 3, 463 (2000) [*Theor. Math. Phys.*, **124**, N 3, 1250 (2000)].

[15] M. Reed, B. Simon, Methods of Modern Mathematical Physics, vol. IV: Analysis of Operators, (§XIII.15), Acad. Press, New York (1978).

Yu.V. Tarbeyev, N.N. Trunov, A.A. Lobashev and V.V. Kukhar
D.I. Mendeleev Research Institute of Metrology
Moskowsky pr. 19
198005, St. Petersburg, Russia
e-mail: lobashev@vniim.ru

Operator Theory:
Advances and Applications, Vol. 132, 411–417

List of Participants

- **Mikhail Agranovich**
  Moscow State Institute of Electronics and Mathematics
  (MGIEM) Moscow 109028, Russia
  msa.funcan@mtu-net.ru

- **Sergio Albeverio**
  Institute of Applied Mathematics, University of Bonn
  53115 Bonn, Germany
  albeverio@uni-bonn.de

- **Tatyana Alferova**
  Fysikum, Stockholm University
  106 91 Stockholm, Sweden
  alferova@physto.se

- **Samir Al-Mulla**
  Högskolan i Borås, Allegatan 1,
  501 90 Borås, Sweden
  Samir.Almulla@hb.se

- **Ivan Andronov**
  Dept. of Math. Physics, St. Petersburg University
  198904 St. Petersburg, Russia
  iva@aa2628.spb.edu

- **Jockum Aniansson**
  Institutionen för Matematik, KTH,
  100 44 Stockholm, Sweden
  jockum@math.kth.se

- **Erik Aurell**
  Dept. of Mathematics, Stockholm University
  106 91 Stockholm, Sweden
  eaurell@matematik.su.se

- **Massimo Bertini**
  Dipartimento di Matematica, Università di Milano
  Via Saldini 51,
  20100 Milano, Italy
  bertini@mat.unimi.it

- **Jan-Erik Björk**
  Dept. of Mathematics, Stockholm University
  106 91 Stockholm, Sweden
  jeb@matematik.su.se

- **Jan Boman**
  Dept. of Mathematics, Stockholm University
  106 91 Stockholm, Sweden
  jabo@matematik.su.se

- **Anne Boutet de Monvel**
  Mathématiques, case 7012, Université Paris 7,
  2 place Jussieu 75251, Paris Cedex 05 France
  aboutet@math.jussieu.fr

- **Johannes Brasche**
  Matematiska Intitutionen, CTH & GU
  41296 Göteborg, Sweden
  brasche@math.chalmers.se

- **Jaime Cruz Sampedro**
  Departamento de Matematicas, Universidad de las Americas
  Cholula, Pue, 72820 Mexico
  sampedro@mail.udlap.mx

- **Anders Dahlner**
  Lunds Universitet, Matematiska Institutionen, Box 118
  221 00 Lund, Sweden
  Anders.Dahlner@math.lu.se

- **Yuri Demkov**
  Dept. of Physics, St. Petersburg University,
  198904 St. Petersburg, Russia

- **Vladimir Derkach**
  Dept. of Mathematics, Donetsk State University,
  Universitetskaja str. 24, 83055, Donetsk, Ukraine
  derkach@cc.helsinki.fi

- **Aad Dijksma**
  University of Groningen, Dept. of Math., P.O.Box 800,
  9700 AV Groningen, The Netherlands
  A.Dijksma@math.rug.nl

- **Peter Ebenfelt**
  Institutionen för Matematik, KTH,
  100 44 Stockholm, Sweden
  ebenfelt@math.kth.se

- **Nils Elander**
  Fysikum, Stockholm University
  106 91 Stockholm, Sweden
  elander@physto.se

- **W.Norrie Everitt**
  School of Mathematics, University of Birmingham,
  Edgbaston, Birmingham B15 2TT, England, UK
  w.n.everitt@bham.ac.uk

- **Shao-Ming Fei**
  Institute of Applied Mathematics, University of Bonn
  53115 Bonn, Germany
  fei@wiener.iam.uni-bonn.de

- **Björn Gustafsson**
  Institutionen för Matematik, KTH,
  100 44 Stockholm, Sweden
  gbjorn@math.kth.se

- **Seppo Hassi**
  Dept. of Statistics/Mathematics
  P.O. Box 54, 00014 University of Helsinki,
  Helsinki, Finland
  hassi@cc.helsinki.fi

- **Astrid Hilbert**
  Luleå University of Technology,
  197 87 Luleå, Sweden
  Astrid.Hilbert@sm.luth.se

- **Svante Jansson**
  Dept. of Mathematics, Uppsala University,
  Box 480, 751 06 Uppsala, Sweden
  Svante.Janson@math.uu.se

- **Jan Janas**
  Institute of Mathematics PAN,
  Instytut Matematyczny PAN,31-027 Cracow,
  ul.Sw.Tomasza 30, Poland
  najanas@cyf-kr.edu.pl

- **Lennart Jönsson**
  Högskolan i Borås, Allegatan 1,
  501 90 Borås, Sweden
  Lennart.Jonsson@hb.se

- **Valery Kapshai**
  Department of Physics, Gomel State university
  Sovetskaia 102
  246699 Gomel, Belarus
  kvn@gsu.unibel.by

- **Yulia Karpeshina**
  Dept. of Mathematics, The University of Alabama at Birmingham
  Campbell Hall, 1330 University BLvd
  Birmingham AL 35294-1170, USA
  karpeshi@vorteb.math.uab.edu

- **Witold Karwowski**
  University of Wroclaw, Institute of Theoretical
  Physics, 50-204 Wroclaw, pl. Maxa Borna 9, Poland
  wkar@ift.uni.wroc.pl

- **Andrei Khrennikov**
  Institute of Mathematics, Växjö University,
  S-35195 Växjö, Sweden
  Andrei.Khrennikov@msi.vxu.se

- **Alexander Kiselev**
  Dept. of Math. Physics, St. Petersburg University
  198904 St. Petersburg, Russia
  kisa@AK2570.spb.edu, akiselev@mph.phys.spbu.ru

- **Pavel Kurasov**
  Dept. of Mathematics, Stockholm University
  106 91 Stockholm, Sweden
  kurasov@maths.lth.se

- **Sergei Levin**
  Fysikum, Stockholm University
  106 91 Stockholm, Sweden
  levin@physto.se

- **Alexei Lobashev**
  D.I.Mendeleev Research Institute of Metrology,
  Moskowsky pr. 19, 198005, St. Petersburg, Russia
  lobashev@friedman.usr.lgu.spb.su

- **Hans Lundmark**
  Matematiska Institutionen, Linköpings Universitet
  581 83 Linköping, Sweden
  halun@mai.liu.se

- **Gisli Masson**
  Dept. of Mathematics, Stockholm University
  106 91 Stockholm, Sweden
  gisli@matematik.su.se

- **Kirsti Mattila**
  Institutionen för Matematik, KTH,
  100 44 Stockholm, Sweden
  kirsti@math.kth.se

- **Yuri Melnikov**
  International Solvay Institutes for Physics and Chemistry,
  Campus Plaine ULB, C.P.231, Boulevard du Triomphe,
  Brussels 1050, Belgium
  Iouri.Melnikov@ulb.ac.be

- **Sergei Naboko**
  Depart.of Math.Physics, Institute of Physics, St. Petersburg University,
  St. Petergoff, St. Petersburg, 198904, Russia.
  naboko@snoopy.phys.spbu.ru

- **John M. Noble**
  Matematiska Institutionen, KTH
  S - 100 44 Stockholm, Sweden
  noble@math.kth.se

- **Anders Olofsson**
  Dept. of Mathematics, Stockholm University
  106 91 Stockholm, Sweden
  anderso@matematik.su.se

- **Karel Packalen**
  Lund University, Sweden
  Saarisvägen 14a, 4tr,
  212 19 Malmö, Sweden
  karel@telia.com, karel@lufysik.fysik.lul.se

- **Mikael Passare**
  Dept. of Mathematics, Stockholm University
  106 91 Stockholm, Sweden
  passare@matematik.su.se

- **Boris Pavlov**
  Dept. of Mathematics, The University of Auckland,
  Private Bag 92019 Auckland, New Zealand
  pavlov@math.auckland.ac.nz

- **Yuri Pavlov**
  Institute of Mechanical Engineering, Russian Academy of Sciences
  61 Bolshoy, V.O., St. Petersburg, 199178, Russia
  pavlov@ipme.ru

- **Alexei Pokrovski**
  Laboratory of Complex Systems Theory,
  Institute for Physics, Ulyanovskaya, 1
  198904 St. Petersburg, Russia
  Alexis.Pokrovski@pobox.spbu.ru

- **Alexei Poltoratski**
  Dept. of Math., Texas A and M University,
  College Station, TX 77845, USA
  alexeip@math.tamu.edu

- **Andrea Posilicano**
  Università dell'Insubria, Como,
  via Lucini 3, 22100, Como, Italy
  Posilicano@mat.unimi.it

- **Roman Romanov**
  School of Mathematical Sciences, University of Sussex,
  Brighton, Falmer BN1 9QH, UK
  R.Romanov@sussex.ac.uk

- **Alexei Rybkin**
  Department of Mathematical Sciences, University of
  Alaska Fairbanks, Fairbanks, AK 99775-6660, USA
  ffavr@uaf.edu

- **Harold Shapiro**
  Institutionen för Matematik, KTH,
  100 44 Stockholm, Sweden
  shapiro@math.kth.se

- **Steve Sontz**
  Universidad Autonoma Metropolitana-Iztapalapa
  Depto. de Matematicas,
  Col. Vicentina
  Mexico DF 09340, Mexico
  sontz@xanum.uam.mx

- **Sergei Slavyanov**
  Dept. of Math. Physics, St. Petersburg University
  198904 St. Petersburg, Russia
  slav@slav.usr.pu.ru

- **Gunter Stolz**
  Dept. of Mathematics, University of Alabama, Birmingham
  AL 35294-1170, USA
  stolz@vorteb.math.uab.edu
- **Alessandro Teta**
  Dipartimento di Matematica Pura e Applicata,
  Università di L'Aquila, Via Vetoio - Loc. Coppito, 67100 L'Aquila, Italy
  teta@univaq.it
- **Francoise Truc**
  Université de Grenoble 1. Institut Fourier,
  B.P 74 38402 St Martin d'Heres Cedex, France
  Francoise.Truc@ujf-grenoble.fr
- **Leonid Volevich**
  Keldysh Institute of Applied Mathematics,
  Russian Academy of Sciences,
  Miusskaya sqr 4, 125047 Moscow, Russia
  volev@orc.ru
- **Claes Waksjö**
  Dept. of Math., Linköping University,
  581 83 Linköping, Sweden
  claes.waksjo@math.liu.se

Operator Theory:
Advances and Applications, Vol. 132, 419–422

List of Lecture Titles

- **M. Agranovich**
  Spectral properties of some potential type operators on Lipschitz surfaces
- **S. Albeverio**
  Some New Developments Concerning Singular Operators (in Finite and Infinite Dimensions)
- **I. Andronov**
  Generalized point models in boundary contact value problems of hydroelasticity
- **E. Aurell**
  An application of large deviation theory to compute the stability of a genetic switch
- **J.-E. Björk**
  Sonja Kovalevsky
- **A. Boutet de Monvel**
  Surface Waves
- **J. Brasche**
  On the construction of self-adjoint extensions with preassigned spectral properties
- **J. Cruz Sampedro**
  Generalized Fourier Transform for Schrödinger Operators with Potentials of Order Zero
- **Yu. Demkov**
  Generalized Landau-Zener Problem and Factorization of the S-Matrix
- **V. Derkach**
  Operator models associated with singular perturbations
- **A. Dijksma**
  Singular Point-like Perturbations of the Bessel Operator in a Pontryagin Space
- **P. Ebenfelt**
  Complete systems of differential equations for CR mappings
- **W.N. Everitt**
  Linear quasi-differential operators in locally integrable spaces on the real line

- **Sh-M. Fei**
  Many Body Problems with Contact Interactions
- **B. Gustafsson**
  A converse of the Cauchy-Kovalevskaya theorem: analyticity of some free boundaries
- **S. Hassi**
  Regular perturbations of selfadjoint operators in a Pontryagin space
- **A. Hilbert**
  Nagel Stein Wainger Estimates for Balls Associated with the Bismut Condition
- **J. Janas**
  Spectral properties of some unbounded Jacobi operators
- **S. Jansson**
  On complex hypercontractivity
- **L. Jönsson**
  Local-density approximations to the elastic scattering of electrons from Mg and Ba atoms
- **Yu. Karpeshina**
  Spectral properties of the Schrödinger operator with a Periodic Potential
- **W. Karwowski**
  Null set perturbation of the Schrödinger operators as the boundary conditions
- **A. Khrennikov**
  Theory of ultradistributions and the Cauchy problem for PDE on the superspace
- **A. Kiselev**
  Similarity problem for the nonselfadjoint operators with absolutely continuous spectrum: functional model approach
- **T. Kolsrud**
  The heat Lie algebra and conserved quantities for heat equations
- **S. Levin**
  Dependence of the $NeICl$ van der Waals complex resonance width on the total angular momentum value
- **A. Lobashev**
  Effective Quantum Number for Centrally Symmetric Potentials
- **H. Lundmark**
  A new class of integrable mechanical systems and nonstandard separation of variables

- **G. Masson**
  Polynomial eigenfunctions of $(dk/dxk)Qk(x)$
- **K. Mattila**
  Relatively tauberian operators and boundary points of the spectrum
- **Yu. Melnikov**
  Scattering on graphs and one-dimensional approximation of N-dimensional Schrödinger dynamics
- **S. Naboko**
  Essential Spectrum of a System of Singular Differential Operators and the Asymptotic Hain-Lust Operators
- **J. Noble**
  On the action functional of Brownian motion and the Inviscid Burgers' equation
- **K. Packalen**
  Inverse Scattering Problem for Positons
- **M. Passare**
  Amoebas and Cauchy-Kovalevsky theorem
- **Yu. Pavlov**
  The Hamilton Operator and Quantum Vacuum for Nonconformal Scalar Fields in the Homogeneous and Isotropic Space
- **A. Pokrovski**
  Lax-Phillips resonance theory for a model system with infinite scatterer
- **A. Poltoratski**
  Finite-rank perturbations of spectra
- **A. Posilicano**
  Singular Perturbations of Self-adjoint Operators
- **R. Romanov**
  Spectral singularities and asymptotic behavior of contractive semigroups
- **A. Rybkin**
  High energy Weyl m-function asymptotics and a unified approach to trace formulas for Schrödinger operators on the line
- **H. Shapiro**
  Analytic Continuation of Solutions to Partial Differential Equations
- **S. Slavyanov**
  Correlation between Painlevé equations and Heun class equations
- **S. Sontz**
  On reverse hypercontractivity and reverse coercivity inequalities

- **G. Stolz**
  Some Recent Results and Open Problems in Random Schrödinger Operators
- **A. Teta**
  Blow-up Solutions for the Schrödinger Equation with Concentrated Nonlinearity
- **F. Truc**
  Semi-classical Eigenvalue Asymptotics for a Schrödinger Operator with a Degenerate Potential
- **L. Volevich**
  The Newton Polyhedron in the Cauchy and Mixed Problem
- **C. Waksjö**
  A Criterion of Separability for Natural Hamiltonian Systems

Operator Theory - Advances and Applications

Previously published titles

OT 114: **Ramírez de Arellano, E. / Shapiro, M.V. / Tovar, L.M. / Vasilevski, N.L.** (Eds.) Complex Analysis and Related Topics (1999) ISBN 3-7643-6228-6

OT 113: **Havin, V.P. / Nikolski, N.K.** (Eds.) Complex Analysis, Operators, and Related Topics (1999) ISBN 3-7643-6214-6

OT 112: **Maz'ya, V. / Nazarov, S. / Plamenevskij, B.** Asymptotic Theory of Elliptic Boundary Value Problems in Singularly Pertubed Domains,Volume II (2000) ISBN 3-7643-6398-3

OT 111: **Maz'ya, V. / Nazarov, S. / Plamenevskij, B.** Asymptotic Theory of Elliptic Boundary Value Problems in Singularly Pertubed Domains, Volume I (2000)ISBN 3-7643-6397-5

OT 110: **Rossmann, J. / Takác, P. / Wildenhain, G.** (Eds.) The Maz'ya Anniversary Collection,Volume 2 (1999) ISBN 3-7643-6202-2

OT 109: **Rossmann, J. / Takác, P. / Wildenhain, G.** (Eds.) The Maz'ya Anniversary Collection, Volume 1 (1999) ISBN 3-7643-6201-4

OT 108: **Dittrich, J. / Exner, P. / Tater, M.** (Eds.) Mathematical Results in Quantum Mechanics (1999) ISBN 3-7643-6097-6

OT 107: **Sakhnovich, L.A.** Spectral Theory of Canonical Differential Systems (1999) ISBN 3-7643-6057-7

OT 106: **Dijksma, A. / Gohberg, I. / Kaashoek, M.A. / Mennicken, R.** (Eds.), Contributions to Operator Theory in Spaces with an Indefinite Metric. The Heinz Langer Anniversary Volume (1998) ISBN 3-7643-6003-8

OT 105: **Bagdasarov, S.**, Chebyshev Splines and Kolmogorov Inequalities (1998) ISBN 3-7643-5984-6

OT 104: **Bercovici, H. / Foias, C.** (Eds.), Nonselfadjoint Operator Algebras, Operator Theory, and Related Topics. The Carl M. Pearcy Anniversary Volume (1998) ISBN 3-7643-5954-4

OT 103: **Gohberg, I. / Mennicken, R. / Tretter, C.** (Eds.), Recent Progress in Operator Theory. International Workshop on Operator Theory and Applications, IWOTA 95, in Regensburg, July 31–August 4, 1995 (1998) ISBN 3-7643-5891-2

OT 102: **Gohberg, I. / Mennicken, R. / Tretter, C.** (Eds.), Differential and Integral Operators. International Workshop on Operator Theory and Applications, IWOTA 95, in Regensburg, July 31–August 4, 1995 (1998) ISBN 3-7643-5890-4

OT 101: **Eidelman, S.D. / Zhitarashu, N.V.**, Parabolic Boundary Value Problems (1998) ISBN 3-7643-2972-6

OT 100: **Foias, C.I. / Frazho, A.E. / Gohberg, I. / Kaashoek, M.A.** (Eds.), Metric Constrained Interpolation, Commutant Lifting and Systems (1998) ISBN 3-7643-5889-0

OT 99: **Ayerbe Toledano, J.M. / Dominguez Benavides, T. / López Acedo, G.** Measures of Noncompactness in Metric Fixed Point Theory (1997) ISBN 3-7643-5794-0

OT 98: **Gohberg, I. / Lyubich, Y.** (Eds.), New Results in Operator Theory and Its Applications. The Israel M. Glazman Memorial Volume (1997) ISBN 3-7643-5775-4

OT 97: **Gorbachuk, M.L. / Gorbachuk, V.I.**, M.G. Krein's Lectures on Entire Operators (1997) ISBN 3-7643-5704-5

OT 96: **Alpay, D. / Dijksma, A. / Rovnyak, J. / Snoo, H. de** (Eds.), Schur Functions, Operator Colligations, and Reproducing Kernel Pontryagin Spaces (1997) ISBN 3-7643-5763-0

OT 95: **Dym, H. / Fritzsche, B. / Katsnelson, V. / Kirstein, B.** (Eds.), Topics in Interpolation Theory (1997) ISBN 3-7643-5723-1

OT 94: **Kadets, M.I. / Kadets, V.M.**, Series in Banach Spaces. Conditional and Unconditional Convergence (1997) ISBN 3-7643-5401-1

OT 93: **Egorov, Y. / Schulze, B.W.**, Pseudo-Differential Operators, Singularities, Applications (1997) ISBN 3-7643-5484-4

OT 92: **Shklyar, A.Y.**, Complete Second Order Linear Differential Equations in Hilbert Spaces (1997) ISBN 3-7643-5377-5

OT 91: **Skubachevskii, A.L.**, Elliptic Functional Differential Equations and Applications (1997) ISBN 3-7643-5404-6

OT 90: **Böttcher, A. / Gohberg, I.** (Eds.), Singular Integral Operators and Related Topics. Joint German-Israeli Workshop, Tel Aviv, March 1-10, 1995 (1996) ISBN 3-7643-5466-6

OT 89: **Egorov, V. / Kondratiev, V.**, On Spectral Theory of Elliptic Operators (1996) ISBN 3-7643-5390-2

OT 88: **van Neerven, J.**, The Asymptotic Behaviour of Semigroups of Linear Operators (1996) ISBN 3-7643-5455-0

OT 87: **Gohberg, I. / Lancaster, P. / Shivakumar, P.N.** (Eds.), Recent Developments in Operator Theory and Its Applications. International Conference in Winnipeg, October 2–6, 1994 (1996) ISBN 3-7643-5413-5

OT 86: **Berezansky, Y.M. / Sheftel, Z.G. / Us, G.F.**, Functional Analysis. Vol. II (1996) ISBN 3-7643-5345-7

OT 85: **Berezansky, Y.M. / Sheftel, Z.G. / Us, G.F.**, Functional Analysis. Vol. I (1996) ISBN 3-7643-5344-9

OT 83: **Antonevich, A.**, Linear Functional Equations. Operator Approach (1996) ISBN 3-7643-2931-9

OT 82: **Constantinescu, T.**, Schur Parameters, Factorization and Dilation Problems (1996) ISBN 3-7643-5285-X

OT 81: **Upmeier, H.**, Toeplitz Operators and Index Theory in Several Complex Variables (1996) ISBN 3-7643-5282-5

OT 80: **Gohberg, I. / Langer, H.** (Eds.), Operator Theory and Boundary Eigenvalue Problems. International Workshop in Vienna, July 27-30, 1993 (1995) ISBN 3-7643-5275-2

OT 79: **Gohberg, I. / Kaashoek, M.A. / van Schagen, F.**, Partially Specified Matrices and Operators. Classification, Completion, Applications (1995) ISBN 3-7643-5259-0

OT 78: **Demuth, M. / Schulze, B.W.** (Eds.), Partial Differential Operators and Mathematical Physics. International Conference in Holzhau (Germany), July 3 – 9, 1994 (1995) ISBN 3-7643-5208-6

OT 77: **Lindenstrauss, J. / Milman, V.D.** (Eds.), Geometric Aspects of Functional Analysis. Israel Seminar (GAFA) 1992–94 (1995) ISBN 3-7643-5207-8

OT 76: **Krasnosel'skij, A.M.**, Asymptotics of Nonlinearities and Operator Equations (1995) ISBN 3-7643-5175-6

OT 75: **Huijsmans, C.B. / Kaashoek, M.A. / Luxemburg, W.A. / Pagter, B. de** (Eds.), Operator Theory in Function Spaces and Banach Lattices. The A.C. Zaanen Anniversary Volume (1995) ISBN 3-7643-5146-2

OT 74: **Hagen, R. / Roch, S. / Silbermann, B.**, Spectral Theory of Approximation Methods for Convolution Equations (1994) ISBN 3-7643-5112-8

OT 73: **Feintuch, A. / Gohberg, I.** (Eds.), Nonselfadjoint Operators and Related Topics. Workshop on Operator Theory and Its Applications, Beersheva, February 24–28, 1992 (1994) ISBN 3-7643-5097-0

OT 72: **Gohberg, I. / Sakhnovich, L.A.** (Eds.), Matrix and Operator Valued Functions. The Vladimir Petrovich Potapov Memorial Volume (1994) ISBN 3-7643-5091-1

OT 71: **Basor, E.L. / Gohberg, I.** (Eds.), Toeplitz Operators and Related Topics. The Harold Widom Anniversary Volume. Workshop on Toeplitz and Wiener-Hopf Operators, Santa Cruz, California, September 20–22, 1992 (1994) ISBN 3-7643-5068-7

OT 70: **Demuth, M. / Exner, P. / Neidhardt, H. / Zagrebnov, V.** (Eds.), Mathematical Results in Quantum Mechanics. International Conference in Blossin (Germany), May 17-21, 1993 (1994) ISBN 3-7643-5025-3

OT 69: **Ashyralyev, A. / Sobolevskii, P.E.**, Well-Posedness of Parabolic Difference Equations (1994) ISBN 3-7643-5024-5

OT 68: **Halanay, A. / Ionescu, V.**, Time-Varying Discrete Linear Systems. Input-Output Operators. Riccati Equations. Disturbance Attenuation (1994) ISBN 3-7643-5012-1

OT 67: **Govorov, N.V.**, Riemann's Boundary Problem with Infinite Index (1994) ISBN 3-7643-2999-8

OT 66: **Khatskevich, V. / Shoiykhet, D.**, Differentiable Operators and Nonlinear Equations (1993) ISBN 3-7643-2929-7

OT 65: **Djrbashian, M.M.**, Harmonic Analysis and Boundary Value Problems in the Complex Domain (1993) ISBN 3-7643-2855-X

OT 64: **Gohberg, I.** (Ed.), New Aspects in Interpolation and Completion Theories (1993) ISBN 3-7643-2948-3

OT 63: **Gohberg, I. / Goldberg, S. / Kaashoek, M.A.**, Classes of Linear Operators. Vol. 2 (1993) ISBN 3-7643-2944-0

OT 62: **Furuta, T. / Gohberg, I. / Nakazi, T.** (Eds.), Contributions to Operator Theory and its Applications. The Tsuyoshi Ando Anniversary Volume (1993) ISBN 3-7643-2928-9

OT 61: **Gheondea, A. / Timotin, D. / Vasilescu, F.H.** (Eds.), Operator Extensions, Interpolation of Functions and Related Topics. 14th International Conference on Operator Theory Timisoara, June 1-5, 1992 (1993) ISBN 3-7643-2902-5

OT 60: **Kuchment, P.A.**, Floquet Theory for Partial Differential Equations (1993) ISBN 3-7643-2901-7

OT 59: **Ando, T. / Gohberg, I.** (Eds.), Operator Theory and Complex Analysis. Workshop on Operator Theory and Complex Analysis, Sapporo (Japan), June 1991 (1992) ISBN 3-7643-2824-X

OT 58: **Gohberg, I.** (Ed.), Continuous and Discrete Fourier Transforms, Extension Problems and Wiener-Hopf Equations (1992) ISBN 3-7643-2809-6

OT 57: **Demuth, M. / Gramsch, B. / Schulze, B.W.**, Operator Calculus and Spectral Theory. Symposium on Operator Calculus and Spectral Theory, Lambrecht (Germany), December 1991 (1992) ISBN 3-7643-2792-8

MIX
Papier aus verantwortungsvollen Quellen
Paper from responsible sources
FSC® C105338

If you have any concerns about our products,
you can contact us on
ProductSafety@springernature.com

In case Publisher is established outside the EU,
the EU authorized representative is:
Springer Nature Customer Service Center GmbH
Europaplatz 3, 69115 Heidelberg, Germany

Printed by Libri Plureos GmbH
in Hamburg, Germany